Natural Hazards - New Insights

Edited by Mohammad Mokhtari

Published in London, United Kingdom

Natural Hazards - New Insights
http://dx.doi.org/10.5772/intechopen.104005
Edited by Mohammad Mokhtari

Contributors
Mustafa Yükseler, Jale Yazgan, Muhammad Arshad K. Khalafzai, Lingampally Sai Vinay, Ram Madhab Bhattacharjee, Nilabjendu Ghosh, Nima Khakzad, Mohammad Mahbubur Rahman, Mohammad Harun-Or-Rashid, Oğuzhan Çetindemir, Abdullah Can Zülfikar, Rodrigo Martins Moreira, Maria Paula Cardoso Yoshii, Hadis Amiri, Azra Jahanitabesh, Yonas Oyda, Hailu Regasa, Emilio Ramírez Juidías, Emilio José Cabello-Franco, Tse-Shan Hsu, Raju Attada, Nischal Sharma, Akash Pathaikara, Pravin Punde, Rohtash Saini, Sreehari K, Judith G. Ramos, Jesús Gracia-Sánchez, Liliana Marrufo Vázquez, Javier Osnaya Romero, Víctor M. Ortíz-Martínez, Kazuki Karashima, Akira Ohgai, Jorge Quezada, Arturo Belmonte, Cheikh Modou Noreyni Fall, Adama Faye, Mbaye Diop, Babacar Faye, Amadou Thierno Gaye, Roberto Richardson Varas, Mohsen Kazemnia Kakhki, Webe João Mansur, Shengfeng Zhang, Yongxian Zhang, Xue Yuan, Hu Dan, Ye Qiuyin, Zeng Wenjun, Yang Jing, Rao Min, Mahnaz Nedaei, Alonso Gómez-Bernal, Antonio Romero Peña, Jonathan de Anda Gil, Hnin Wuit Yee Kyaw, Phone Pyai Tun, Ashish Kumar Parashar, Mohammad Mokhtari, Parvaneh Faridi, Mehdi Masoodi, Seyed Mehran Ahmadi

First published in London, United Kingdom, 2023 by IntechOpen
IntechOpen is the global imprint of INTECHOPEN LIMITED, registered in England and Wales, registration number: 11086078, 5 Princes Gate Court, London, SW7 2QJ, United Kingdom

British Library Cataloguing-in-Publication Data
A catalogue record for this book is available from the British Library

Additional hard and PDF copies can be obtained from orders@intechopen.com

Natural Hazards - New Insights
Edited by Mohammad Mokhtari
p. cm.
Print ISBN 978-1-80355-849-3
Online ISBN 978-1-80355-850-9
eBook (PDF) ISBN 978-1-80355-851-6

For EU product safety concerns:
IN TECH d.o.o., Prolaz Marije Krucifikse Kozulić 3, 51000 Rijeka, Croatia,
info@intechopen.com or visit our website at intechopen.com.

Meet the editor

Mohammad Mokhtari, a highly accomplished geophysicist, holds a BSc from Azarabadegan University, Iran, an MSc from Southampton University, United Kingdom and a Ph.D. from the University of Bergen, Norway. His expertise extends to Utrecht University, Netherlands, where he conducted research, and to his role as a principal geophysicist at Norsk Hydro (Norway) and National Iranian Oil Company (Iran). With an impressive portfolio, he has made significant contributions as the director of the Seismological Research Center at the International Institute of Earthquake Engineering and Seismology and a member of the board of directors. Notably, he established the Iranian National Broadband Seismic Network, served as a co-founder and director of the National Center for Earthquake Prediction Center and the Risk Management Excellence Center, and was part of the Passive Seismic Equipment Expert Panel for The Comprehensive Nuclear-Test-Ban Treaty Organization. Moreover, he has contributed to GSA as a visiting researcher and the Indian Ocean Tsunami Hazard Assessment. He has also served as a member of the advisory board at IntechOpen publishing. Currently, Dr. Mokhtari is chairman of ICG/NWIO-WG, a founding member of TERC, director of the Maga Makran project at the University of Hormozgan Iran, and vice president of the SPIRM Institute. Throughout his career, he has supervised 32 MSc and 14 Ph.D. students, published more than 85 papers, delivered 120 conference presentations, and authored 9 books.

Contents

Preface

In a world constantly shaped by the forces of nature, humankind has always sought to understand and mitigate the impacts of natural hazards. These formidable events have the power to reshape landscapes, alter ecosystems, and dramatically affect human lives. From devastating floods that engulf entire communities to the relentless tremors of earthquakes that shake the very foundations of our existence, natural hazards have remained an enduring challenge throughout history.

Natural Hazards – New Insights explores the intricate realm of natural hazards, exploring the multifaceted nature of flooding, earthquakes, meteoritical hazards, mining-related disasters, and other major catastrophes that afflict our planet. In this book, we embark on a journey to gain a deeper understanding of these phenomena, unearthing novel perspectives and fresh insights into their causes, impacts, and the strategies employed to mitigate their effects. The book consists of six sections.

In Section 1, "The Hazards We Face: An In-Depth Exploration", we begin with a perspective chapter that offers readers a global view of natural hazards. This chapter serves as a foundation for understanding the diverse range of hazards that exist worldwide and the challenges they pose. Moving forward, we delve into the realm of seismic hazards. Chapter 2 focuses on probabilistic seismic vulnerability and loss assessment. By employing advanced statistical techniques and models, this chapter explores the estimation of potential damage and losses caused by earthquakes. Chapter 3 takes a more localized approach by presenting a seismic hazard assessment of North Chhattisgarh.

In Section 2, "Ground-Shaking Threats: A Closer Look at Earthquake Hazards", Chapter 4 analyzes and predicts earthquakes using the Seasonal Autoregressive Integrated Moving Average (SARIMA) model. This model aids in understanding earthquake patterns and characteristics. By forecasting earthquake occurrences, it provides valuable insights for preparedness and response efforts. Chapter 5 focuses on urban damage assessment following the Mw 5.8 Silivri earthquake in Turkey. Chapter 6 explores crustal faults that were reactivated during the 2010 Mw 8.8 Maule earthquake in Chile. It investigates the mechanisms and processes involved in fault reactivation and the resulting seismic hazards. Chapter 7 introduces the "natural time" method, a novel approach used to assess the potential for earthquake occurrence. Chapter 8 focuses on the seismic conditions required to cause structural failures. Chapter 9 presents new insights into major seismic events using Coulomb stress analysis. By studying the stress transfer and interactions between faults, this analysis provides valuable information on the occurrence and behavior of earthquakes.

In Section 3, "Water's Wrath: Examining Flooding as a Natural Hazard", we shift our focus to the devastating effects of flooding as a natural hazard. Chapter 10 delves into hydrological extremes in the Western Himalayas, analyzing trends and patterns in extreme precipitation and river discharge. Chapter 11 focuses on integrating Google

Earth and decametric sensors for flood mapping and analysis. Chapter 12 explores the interconnection among river flow levels, sediment transport, and flood hazards. Chapter 13 addresses risk reduction strategies for floods, emphasizing the importance of comprehensive approaches that combine structural and non-structural measures. Chapter 14 investigates the spatial-temporal relations of flood risk and its prediction.

Section 4, "From Climate Changes to Hurricanes: Exploring the Range of Climatological Hazards", focuses on climatological hazards and their diverse manifestations. Chapter 15 focuses on conducting a meticulous vulnerability assessment of process vessels in the face of hurricanes. Process vessels, including industrial infrastructure and storage facilities, are particularly susceptible to the destructive forces unleashed by these climatological hazards. Chapter 16 explores the intricate relationship between climate change and natural hazards. Chapter 17 highlights new remote sensing technologies and their applications in pre-disaster planning and post-disaster assessment. Chapter 18 focuses on the evolution of agroclimatic indicators in Senegal, shedding light on the impacts of climate change on agriculture and food security.

In Section 5, "Slopes in Motion: Exploring Landslides as Natural Hazards", we delve into the realm of landslides as significant natural hazards. Chapter 19 explores the role of polarization analysis in reducing natural hazards caused by landslides. This chapter highlights the potential of remote sensing technologies to enhance landslide risk management. Chapter 20 focuses on landslide assessment and hazard zonation in the Birbir Mariam district of the Gamo highlands. It presents a case study that combines field investigations, geotechnical analysis, and geospatial techniques to assess landslide susceptibility and delineate hazard zones. Chapter 21 delves into underground coal mining methods and their impact on slope stability.

Finally, Section 6, "When Disasters Strike: Navigating the Social Challenges of Natural Hazards", addresses the social challenges that arise when disasters strike. It focuses on the psychological reactions of individuals, the role of spontaneous volunteers, urban planning responses to disasters, and social vulnerability to earthquake hazards. Chapter 22 explores the psychological reactions that individuals experience after disasters. Chapter 23 examines the role of spontaneous volunteers in emergencies and disasters. Chapter 24 discusses trends in urban planning and responses to disasters. Chapter 25 analyzes social vulnerability to earthquake hazards. It examines the underlying factors that contribute to differential impacts and outcomes among different social groups during earthquakes.

It is my hope that this exploration of natural hazards will serve as a stepping stone towards a future where we can effectively navigate the challenges and minimize the risks of natural hazards and build a safer and more sustainable environment for generations to come.

Before we conclude, we would like to mention that the presented chapters have been peer-reviewed and accepted for publication. Hence, we would like to express our heartfelt gratitude to all the authors who have contributed their expertise and insights to this book. Their dedication and passion for understanding and addressing the hazards we face have been instrumental in shaping this comprehensive exploration.

Finally, we would also like to extend our appreciation to IntechOpen for initiating this book and for their invaluable support and guidance throughout the publication process. Their commitment to disseminating knowledge and fostering scholarly works has been pivotal in bringing this book to fruition.

Mohammad Mokhtari
Professor,
Tsunami and Earthquake Research Centre (TERC),
University of Hormozgan,
Bandar-Abbas, Iran

Section 1

The Hazards We Face: An In-Depth Exploration

Chapter 1

Perspective Chapter: A Global View of Natural Hazards Related Disasters

Mohammad Mokhtari, Parvaneh Faridi, Mehdi Masoodi and Seyed Mehran Ahmadi

Abstract

Natural hazards have become an increasingly prevalent threat to the world, with the frequency of recorded disasters rising in the recent years. Analyzing global data on natural hazards, identifying the most common and deadly events, and prioritizing decision-making on safe constructions, settlements, and factories are crucial to respond effectively to this trend. Using the EM-DAT database, this study examined primary natural hazards worldwide from 1900 to 2022, including earthquakes, volcanic eruptions, landslides, floods, droughts, wildfires, storms, and extreme temperatures. Statistical data show that storms have the highest frequency of occurrence (38%), droughts are the deadliest (53%), floods affect the most people (50%), and storms cause the most economic losses (41%). Asia has experienced the highest frequency, most deaths, and most total affected people due to natural hazards, followed by North America and Africa, respectively. North America has also seen the highest economic losses, followed by Asia and Europe. The study emphasizes the importance of databases like EM-DAT in advancing our knowledge of natural hazards worldwide, and supporting informed decision-making in risk assessment, disaster management, and mitigation efforts.

Keywords: affected people, death, direct economic losses, frequency, global view, natural hazards

1. Introduction

According to the emergency event database (EM-DAT) classification, there are two main types of disasters [1]: natural and human-made or technological. Natural hazards are defined as natural phenomena that have the potential to have a negative impact on humans, societies, and the environment. They are different from other types of hazards such as man-made hazards [2, 3]. Natural hazards can also lead to secondary events, creating additional hazards. For example, a volcanic eruption can trigger other hazards such as ash and lava flow. Although related, natural hazards and natural disasters are not identical. A natural hazard is a potential threat of an event that has the potential to cause a negative impact, while a natural disaster is the actual occurrence of that event, causing significant harm to a community. Natural hazards and disasters are caused by natural processes throughout the earth's history. However, the risk (the notion of risk is defined as the product of a hazard by a vulnerability) associated with environmental hazards is determined by physical conditions and events and human actions, conditions, decisions, and culture.

Human activities can directly contribute to physical events such as small- and medium-scale flooding, landslides, land subsidence, and drought in rural and urban areas due to environmental degradation, human intervention in ecosystems, and global climate change. These human-induced hazards occur at the intersection of natural and human processes and result in environmental degradation. Climate change introduces a new type of hazard modification, and it is essential to study these human-induced events and how they have affected the frequency of disasters in the past. This knowledge can be used to reduce risk in the future. Additionally, it is important to consider low probability but high-consequence hazards such as impact by near-Earth objects in risk analysis [4].

Natural hazards can be broadly categorized into six groups [1]: geohazards (such as earthquakes, tsunamis, and volcanic eruptions), meteorological hazards (such as storms and extreme temperatures), hydrological hazards (such as flooding and landslides), climatological hazards (such as drought and wildfires), biological hazards (such as epidemics and insect infestations), and extraterrestrial hazards (such as airbursts and space weather). The impacts of these hazards can be classified into three categories: primary, secondary, and tertiary. Primary impacts are the immediate effects of the hazard such as water damage from flooding or building collapse during an earthquake. Secondary impacts are consequences caused by the primary impact such as fires following an earthquake or disrupted electricity and water supply. Tertiary impacts are long-term outcomes resulting from the primary impact such as habitat loss due to flooding, changes in riverbed permeability, and crop failure due to volcanic eruptions.

Technological and human-made disasters can lead to the sudden and uncontrollable release of hazardous and explosive materials [1, 5] such as coal mine explosions, building fires, oil truck accidents in tunnels, train accidents, road accidents, bus collisions, fuel tank explosions, building collapses, shipwrecks involving migrants, plane crashes, bridge collapses, stampedes during concerts, and bomb accidents.

Technological disasters account for a significant proportion of disasters worldwide and demand close attention. However, this study exclusively focuses on natural hazards, analyzing the most significant events in the top ten countries worldwide over the past 123 years (1900–2022). To provide a better understanding of the location and types of hazards worldwide, this study proposes using databases such as EM-DAT as a source of information. The primary data from the EM-DAT database is used to analyze the most frequent, deadly, and impactful natural hazards, aiding earthquake occurrence, identifying high-risk areas, and supporting decisions on safe constructions, settlements, and factories. This paper provides an exemplary model of such a research type.

In this chapter, we will begin with a brief introduction followed by an explanation of the data and methodology used. Next, we will present the most common natural hazards in the world. Finally, we will provide a discussion of the findings and draw conclusions based on the analysis.

2. Data and method

The study aims to present an overview of the most frequent natural hazards that have occurred worldwide from 1900 to 2022. The information presented here was compiled from the EM-DAT emergency disaster database (public.emdat.be). The eight most common events are earthquakes, volcanic eruptions, landslides, floods, droughts, wildfires, storms, and extreme temperatures. Several other hazards have been omitted because they are too localized, and adequate data are unavailable. Additionally, it should be noted that tsunamis, a sub-group of earthquake events, can cause significant destruction to human life and economic losses in affected areas. A list of significant natural hazards in the following as comprehensively as possible is an attempt to present the frequency, deaths, total affected people, and economic losses of natural hazards events in the world.

3. The most common natural hazards in the world

In the twenty-first century, reporting natural hazards and their types has become more advanced and comprehensive on a global scale (**Figure 1**) [5, 6]. **Figure 2** and **Table 1** provide statistical analysis of natural hazards. The data from the top ten countries indicate that 6358 significant natural hazards (excluding biological and extraterrestrial disasters) occurred between 1900 and 2022.

Of all hazards, 2410 were caused by storms, representing the highest frequency at 38% of the total. 1840 were caused by floods, accounting for 29%. 894 were caused by earthquakes, accounting for 14%. 410 were caused by landslides, accounting for 7%. 270 were caused by wildfires, accounting for 4%. 255 were caused by extreme temperatures, accounting for 4%. 192 were caused by volcanic activities, accounting for 3%. And 87 were caused by drought, accounting for 1% (**Table 1**, **Figure 2a**).

4. Continent's natural hazards: 1900–2022

An estimated 45,000 people die yearly due to natural disasters, representing about 0.1% of global deaths. The number and proportion of fatalities caused

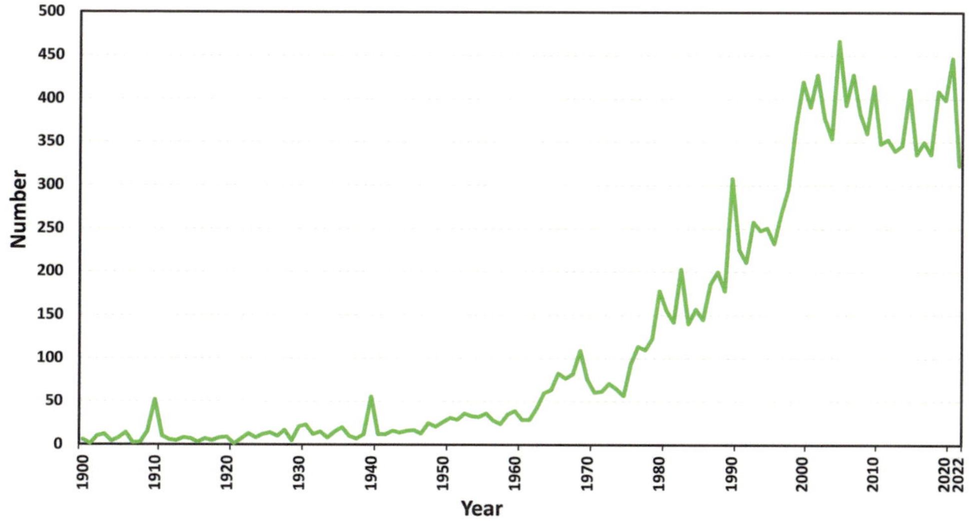

Figure 1.
The diagram shows the increased and completed of recorded natural hazards during 1900–2022.

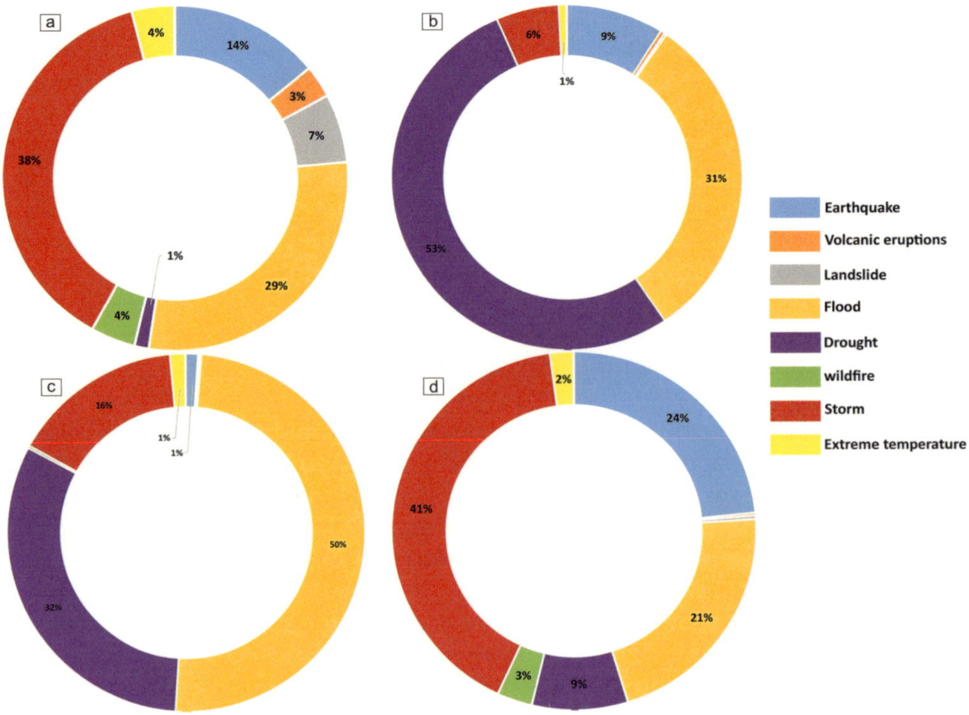

Figure 2.
(a) Frequency, (b) deaths, (c) total affected people, (d) direct economic losses (US dollars) per natural hazard types, based on the top ten countries in the world (1900–2022).

Disaster group	Disaster sub-group	Main disaster type	Frequency	Deaths	Total affected people	Economic losses (in million USdollars)
Natural	Geophysical	Earthquake	894	1,995,580	81,809,874	1,221,061
		Volcanic eruptions	192	85,319	9,230,986	12,411
	Hydrological	Landslide	410	48,532	13,505,420	17,476
		Flood	1840	6,884,491	3,676,715,666	1,109,311
	Climatological	Drought	87	11,715,980	2,365,801,967	444,784
		wildfire	270	3684	17,217,656	162,322
	Meteorological	Storm	2410	1,278,982	1,152,095,652	2,137,312
		Extreme temperature	255	165,880	101,216,729	109,314
		Total	6358	22,178,448	7,417,593,950	5,213,995

Table 1.
Statistical results on natural hazards, based on the top ten countries in the world (1900–2022).

by natural disasters have fluctuated in recent decades. While the number of deaths often falls below 10,000, the impact of major events, such as the famine and drought in Ethiopia in 1983–1985, the Indian Ocean 2004 earthquake and tsunami, the 2008 Cyclone Nargis, and the 2010 earthquake in Haiti, can be devastating, with over 200,000 deaths worldwide. While natural hazards such as earthquakes and tsunamis, which have low frequencies but high impacts, cannot be avoided, such human losses can be reduced through predictions, more resilient infrastructure, emergency preparedness, and response systems. Historical evidence shows that the world has significantly reduced disaster-related deaths due to these efforts [6].

Table 2 and **Figure 3** present the statistical results of natural hazards in various continents from 1900 to 2022 (based on data from the top ten countries). Of the 5734 natural events included in the statistics, Asia had the highest number of events with 4053,

Continent	Frequency	Deaths	Total affected people	Economic losses (in million USdollars)
Africa	0	820,337	113,350,602	0
Asia	4053	29,053,499	6,872,568,275	1,863,510
Europe	0	0	0	364,754
North America	1420	25,632	115,722,303	2,116,425
South America	261	0	11,736,425	86,415
Oceania	0	0	0	116,873
Total	5734	29,899,468	7,113,377,605	4,547,978

Table 2.
Statistical results on natural hazards in the continents (1900–2022).

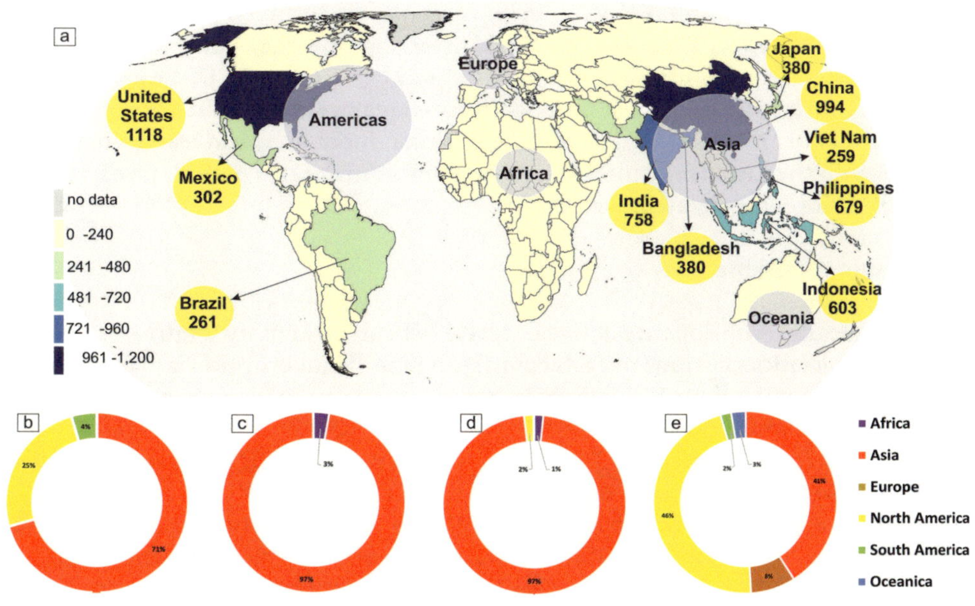

Figure 3.
(a) The number of naturel hazards in the continents and countries shown on the map, (b) frequency, (c) deaths, (d) total affected people, (e) direct economic losses (US dollars) per continents (1900–2022).

accounting for 71%, followed by North America and South America with 1420 (25%) and 261 (4%), respectively. Regarding deaths attributed to natural hazards, Asia had the highest number with 29,053,499, accounting for 97%, followed by Africa with 820,337 (3%) for the top ten countries in the world. In terms of affected people, Asia had the highest number with 6,872,568,275 (97%), followed by North America with 115,722,303 (2%) and Africa with 113,350,602 (1%). North America incurred the most economic losses due to natural hazards, reaching 2,116,425,136,000 US dollars, accounting for 46% of the total global economic losses. Asia came in second with 1,863,510,715,000 US dollars, accounting for 41%, followed by Europe with 364,754,712,000 US dollars, accounting for 8%, Oceania with 116,873,181,000 US dollars, accounting for 3%, and South America with 86,415,200,000 US dollars, accounting for 2%.

The frequency of the most common natural hazards around the world is explained as follows:

4.1 Earthquakes

Earthquakes have been a persistent global threat to human life, with occurrences reported daily by the United States Geological Survey (USGS), among other national and international seismological networks. While most earthquakes are not powerful enough to cause significant damage, major earthquakes with a magnitude of 7.5 or higher can result in significant loss of life and economic destruction. Historically, other natural disasters, such as floods, droughts, and epidemics, have resulted in higher death tolls. However, today, earthquakes and their resulting tsunamis often result in a high annual death toll. The deadliest earthquake in recorded history occurred in 1556 in Shaanxi, China, resulting in the deaths of 830,000 people. Recent devastating earthquakes such as the 2004 Sumatra earthquake and tsunami and the 2010 Haiti earthquake, resulted in 93% and 69% of deaths, respectively, with hundreds of thousands of deaths in total.

From 1900 to 2022, a total of 894 (14%) earthquakes occurred globally, resulting in 1,995,580 deaths (9% of the total) and affecting 81,809,874 people (1% of the total). The direct economic losses amounted to 1,221,061,032,000 US dollars (24% of the total). Asia had the highest number of earthquake events (74%), followed by Europe (10%) and North America (10%). China, Indonesia, and Iran were the countries with the most frequent earthquake events during this period. (see **Table 1** and **Figure 2** for more details).

4.2 Volcanic eruptions

Many volcanic eruptions are active but have little to no activity. **Figure 3** shows the significant volcanic eruptions in each country. A significant eruption is classified by a volcanic explosivity index (VEI) of six or larger, a tsunami or major earthquake, and over dollars1 million in damages.

From 1900 to 2022, a total of 3% of all volcanic events, or 192 eruptions, occurred globally. These eruptions resulted in 85,319 deaths, affected 9,230,986 people, and caused direct economic losses of 12,411,593,000 US dollars (as shown in **Table 1** and **Figure 2**). Asia, South America, and North America have the most volcanic eruption events in the world from 1900 to 2022. Notable eruptions from the past century include the 1902 Mount Pelée eruption in Martinique and the 1985 Nevado del Ruiz eruption in Colombia, which resulted in a high number of deaths.

4.3 Landslides

Landslides are a common geological phenomenon that can occur worldwide. They happen when a large amount of soil, rocks, and debris move down a slope as a result of natural events or human activities. Heavy rains, earthquakes, volcanoes, and droughts can all cause landslides. Climate change and warming have been linked to an increase in landslides, particularly in mountainous regions with snowy and icy conditions. Landslides can also have severe impacts on healthcare systems and essential services such as water supply and communication infrastructure.

The global distribution of landslide mortality risk highlights that densely populated mountainous areas such as the Andes in South America and the Himalayas in Asia are at the highest risk. From 1900 to 2022, a total of 7% of all landslide events, or 410 occurrences, took place globally. These landslides resulted in 48,532 deaths, affected 13,505,420 people, and caused direct economic losses of 17,476,352,000 US dollars (as shown in **Table 1** and **Figure 2**). Asia and South America had the highest number of landslide events in the world during this time period.

4.4 Tsunami

A tsunami is a type of wave that can strike any coast and can be generated in oceans, inland seas, and any large body of water. They are most common in Asia, the Pacific islands, and the United States West Coast, but experts say that seismic activity under the Atlantic Ocean could also trigger tsunamis on the East Coast. The last documented tsunami on the East Coast was in Canada in 1929. Tsunamis are most frequent in the Pacific Ocean's "Ring of Fire" (98%), the Atlantic Ocean and the Caribbean Sea (9%), the Mediterranean Sea (6%), the Indian Ocean (1%), and other seas (1%). Throughout history, tsunamis have occurred from approximately 1.4 million years ago (in Hawaii) to the present day (in Mexico on September 19, 2022). They can be caused by earthquakes, landslides, and volcanic activity. Tsunamis have caused damage and deaths in coastal areas worldwide. Some of the deadliest tsunamis in the world from 1900 to 2022 are the 2004 Indian Ocean earthquake and tsunami and the 2011 Tōhoku Japan Tsunami.

From 1900 to 2022, a total of 39 major tsunami events occurred worldwide. These tsunamis resulted in 266,629 deaths, affected 10,709,690 people, and caused direct economic losses of 334,985,586,000 US dollars (as shown in **Table 1** and **Figure 2**). The countries with the highest number of tsunami events in the world during this time period were Indonesia and Japan in the Asia continent.

4.5 Extreme temperature

Extreme weather or climate events include unusual or severe weather patterns that deviate from historical norms. These events can include heat waves, cold waves, and tropical cyclones, resulting in significant economic costs, loss of life, and ecosystem changes [7]. Extreme events are often based on a weather pattern recorded at a location and are defined as being in the most unusual 10%. [8]. The most important extreme weather types include heat waves, cold waves, and tropical cyclones. The effects of extreme weather events are reflected in increasing economic costs, loss of life, drought, flooding, landslides, and ecosystem changes.

From 1900 to 2022, 255 extreme temperature events occurred globally, causing 165,880 deaths, impacting 101,216,729 people, and resulting in direct economic losses of 109,314,953,000 US dollars. The continents with the highest frequency of these events include Asia, Europe, and North America, with India, the United States, and France having the highest frequency of events.

4.6 Wildfire

A wildfire, forest fire, bush fire, wildfire, or rural fire is an unplanned, uncontrolled, and unpredictable fire in an area of combustible vegetation [9, 10]. A wildfire can be more accurately identified as a bush fire (in Australia), desert fire, grass fire, hummock fire, peat fire, prairie fire, vegetation fire, or steppe fire, depending on the type of vegetation present [11]. Some natural forest ecosystems depend on wildfire [12]. Wildfires are the most common natural hazards in Siberia, California, and Australia [13–15]. Areas with a Mediterranean climate or in the taiga biome are particularly susceptible. On a global scale, human practices have exacerbated the effects of wildfires, with the area of land burned by wildfires doubling compared to natural levels. Wildfires have steadily decreased since records began in the early twentieth century. Humans have impacted wildfires through climate change, landslide change, and wildfire suppression [16].

A total of 270 (4%) wildfire events occurred from 1900 to 2022. However, the natural hazard caused 3684 deaths, accounting for $<0.5\%$ of the total; 17,217,656 people were affected, accounting for 1%; and direct economic losses were 162,322,664,000 US dollars, accounting for 3% (**Table 1**, **Figure 2**). Northern America (45%), Europe (23%), Oceania (14%), Asia (13%), and Southern America (5%) continents had the highest to the lowest number of wildfire events, respectively, in the world. The United States of America, Australia, and Spain, respectively, in the continents have had more frequency of events from 1900 to 2022.

4.7 Drought

Drought is a complex phenomenon characterized by a deficiency of water relative to average availability at a specific location and time of year. According to the IPCC's Sixth Assessment Report, drought is defined as drier than normal conditions [17]. The National Integrated Drought Information System (NIDIS) defines it as a prolonged lack of precipitation resulting in a water deficit. The NOAA (National Oceanic and Atmospheric Administration) defines it as a moisture deficiency that negatively impacts humans, animals, or vegetation over a large area [18]. However, due to its complex nature, drought is difficult to monitor and define [19]. In fact, by the early 1980s, over 150 different definitions of drought had been published, [20] reflecting different perspectives based on regions, needs, and disciplinary approaches.

A total of 87 (1%) drought events occurred from 1900 to 2022. However, the natural hazard caused 11,715,980 deaths, accounting for 53% of the total; 2,365,801,967 people were affected, accounting for 32%; and direct economic losses were 444,784,962,000 US dollars, accounting for 9% (**Table 1**). Africa (51%), Asia (29%), Northern America (20%), and Southern America (20%) continents had the highest to the lowest number of wildfire events, respectively, in the world from 1900 to 2022. Ethiopia, China, the United States of America, and Brazil countries, respectively, in the continents have had more frequency of the events.

4.8 Flood

A flood is an overflow of water or, less commonly, other liquids that typically inundate dry land [21]. The term can also refer to the inflow of tides. Floods are an important area of study in fields such as hydrology, agriculture, civil engineering, and public health. Human-caused changes to the environment, such as land use changes, changes to waterway courses or flood control measures, and larger environmental issues, can increase the intensity and frequency of floods. Additionally, climate change, increased rainfall, and extreme weather events can exacerbate flooding causes, resulting in more severe floods and an increased risk of flooding [22, 23].

There were 1840 major flood disasters during 1900–2022, accounting for 29% of the total number of significant natural hazards. 6,884,491 deaths were caused by flood disasters, accounting for 31% of the total deaths; 3,676,715,666 people were affected, accounting for 50%; and direct economic losses were 1,109,311,917,000 US dollars. India and other Asian countries have suffered severe floods (**Table 1**). The 1931 Yangtze-Huai River floods 1931 in China took place in China from June to August 1931 and were the deadliest floods, and this figure accounted for less than a quarter of all deaths in the first 100 days of the flood. The official report found 140,000 drowned and claimed 2 million people drowned or died of starvation during the flood [24].

4.9 Storm

A storm is any disturbance in the natural environment or atmosphere marked by strong winds, hail, thunder, tornadoes, lightning, heavy precipitation, ice storms, strong winds, and dust storms, among others. Storms are caused by the formation of a center of low pressure surrounded by high pressure. These events can have significant impacts on human life, agriculture, and navigation. One notable example is the 1970 Bhola cyclone, also known as the Great Cyclone of 1970, which struck Bangladesh (then East Pakistan) and resulted in an estimated death toll of between 150,000 and 550,000 people [25].

There were 2410 major storm disasters, accounting for 38% of the total number of major hazards; storm disasters caused 1,278,982 deaths (6%), 1,152,095,652 people were affected the direct (16%), and direct economic losses were 2,137,312,395,000 US dollars during 1900–2022.

5. Discussion

By considering the limitations of the data, such as potential underreporting or variations in reporting methods across different regions and time periods the following are important to be noted. The presented data shows a total of 22,178,448 deaths globally due to natural hazards between 1900 and 2022 (**Table 1**), with drought being the leading cause of death accounting for 53% (11,715,980 number of deaths), followed by floods at 31% (6,884,491 number of deaths), earthquakes at 9% (1,995,580 number of deaths), storms at 6% (1,278,982 number of deaths), extreme temperatures at 0.75% (165,880 number of deaths), volcanic activity at 0.38% (85,319 number of deaths), landslides at 0.22% (48,532 number of deaths), and wildfires at 0.02% (3648 number of deaths). The sources for the data presented in **Table 1** can be found in the references section.

A total of 7,417,593,950 people were affected by natural hazards globally between 1900 and 2022, with flooding being the leading cause of impact, accounting for 50% (3,676,715,666 affected), followed by drought at 32% (2,365,801,967 affected), storm at 16% (1,152,095,652 affected), the extreme temperature at 1% (101,216,729 affected), earthquake at 1% (81,809,874 affected), a landslide at 0.18% (13,505,420 affected), and volcanic eruptions at 0.12% (9,230,986 affected). The units used for each measurement are provided in parentheses.

A total of 5,213,995,868,000 US dollars in direct economic losses were caused by natural disasters globally between 1900 and 2022, with earthquakes being the leading cause of economic loss accounting for 41% (2,137,312,395,000 US dollars), followed by storms at 24% (1,221,061,032,000 US dollars), floods at 21% (1,109,311,917,000 US dollars), a drought at 9% (444,784,962,000 US dollars), wildfire at 3% (162,322,664,000 US dollars), the extreme temperature at 2% (109,314,953,000 US dollars), a landslide at 0.34% (17,476,352,000 US dollars), and volcanic activities at 0.24% (12,411,593,000 US dollars). The limitations of the data, such as potential underreporting or variations in reporting methods across different regions and time periods, should be considered when interpreting the results.

The study also found that climatological hazards were responsible for the most deaths caused by natural hazards (53%), followed by hydrological hazards (31%), geophysical hazards (9%), and meteorological hazards (7%). This indicates that despite having a lower frequency of occurrence, climatological hazards are the deadliest type of hazard. In terms of the number of people affected by natural hazards, hydrological hazards had the greatest impact, affecting 50% of the total (3,690,221,086 people), followed by meteorological hazards (17%), climatological hazards (32%), and geophysical hazards (1%).

It is important to note that the frequency and severity of natural hazards have changed over time, and the potential impacts of climate change on natural hazards should also be considered when interpreting these results.

6. Conclusions

This study utilized the EM-DAT database to investigate the primary natural hazards worldwide from 1900 to 2022. The results show that earthquakes, volcanic eruptions, landslides, floods, droughts, wildfires, storms, and extreme temperatures have been the primary natural hazards globally during the last 123 years. Regarding continents, Asia has experienced the highest frequency of natural hazards, with the most significant number of deaths and total affected people. North America follows in terms of natural hazard frequency, with Africa experiencing the highest number of fatalities and North America experiencing the most total affected people. North America has also suffered the highest economic losses due to natural hazards, followed by Asia and Europe. The highest frequency of natural hazards with the most significant economic losses is caused by meteorological events. Climatological events are the deadliest, while hydrological events affect the most people. The study's findings highlight databases such as EM-DAT are critical in enhancing our understanding of natural hazards worldwide. By providing comprehensive data on these events' type, location, and frequency, such databases can support informed decision-making in risk assessment, disaster management, and mitigation efforts. The results underscore the importance of continued monitoring of natural hazards and preparedness measures to mitigate their impact. Overall, the findings of this study contribute to advancing our knowledge of natural hazards and improving our ability to address the associated challenges.

Table 3 and **Figure 4** present the results of this study, which analyzed the top ten countries affected by natural hazards between 1900 and 2022. The data show that a total of 6358 natural hazard events occurred globally during this period. The majority of these events were caused by meteorological hazards (42%), followed by hydrological hazards (35%), climatological hazards (6%), and geophysical hazards (17%).

The economic impact of natural hazards was also significant, with a total of 5,213,995,868,000 US dollars in direct economic losses. Meteorological hazards accounted for the largest share of economic losses (43%), followed by geophysical hazards (23%), hydrological hazards (21%), and climatological hazards (12%). This suggests that while meteorological hazards may not cause the most deaths or affect the most people, they have the most significant economic impact.

Overall, these findings demonstrate the importance of understanding the distribution and impact of different types of natural hazards. By identifying which hazards are most frequent, deadly, and costly, policymakers and disaster management teams can better prepare and allocate resources to mitigate their effects.

Disaster group	Disaster sub- group	Frequency	Deaths	Total affected people	Economic losses (in million USdollars)
Natural	Geophysical	1086	2,080,899	91,040,860	1,233,472
	Hydrological	2250	6,933,023	3,690,221,086	1,126,788
	Climatological	357	11,719,664	2,383,019,623	607,107
	Meteorological	2665	1,444,862	1,253,312,381	2,246,627
	Total	6358	22,178,448	7,417,593,950	5,213,995

Table 3.
Statistical results on sub-group of natural hazards in the continents (1900–2022).

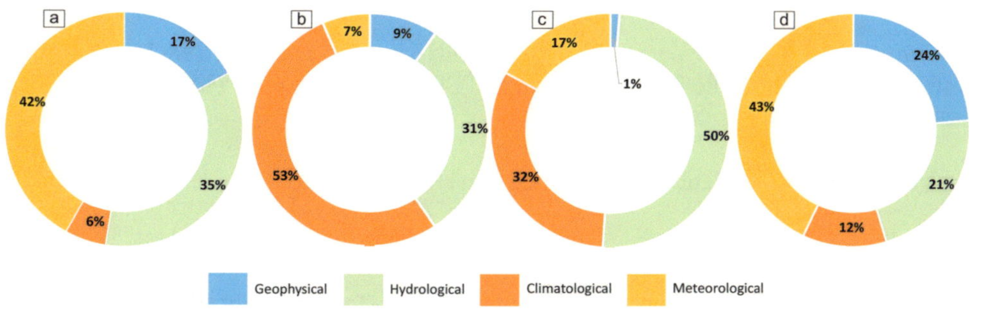

Figure 4.
(a) Frequency, (b) deaths, (c) total affected people, and (d) direct economic losses (US dollars) of the four main sub-group of natural hazards (1900–2022).

Conflict of interest

The author has no conflict of interest to declare. All co-authors have seen and agree with the manuscript's contents, and there is no financial interest to report.

Author details

Mohammad Mokhtari*, Parvaneh Faridi, Mehdi Masoodi and Seyed Mehran Ahmadi
Tsunami and Earthquake Research Centre (TERC), University of Hormozgan, Bandar-Abbas, Iran

*Address all correspondence to: m_7_mokhtari@yahoo.com

References

[1] Available from: http://public.emdat.be/

[2] Yigitcanlar T, Regona M, Kankanamge N, Mehmood R, D'Costa J, Lindsay S, et al. Detecting natural hazard-related disaster impacts with social media analytics: The case of Australian states and territories. Sustainability. 2022;**14**(2):810

[3] Rutgersson A, Kjellström E, Haapala J, Stendel M, Danilovich I, Drews M, et al. Natural hazards and extreme events in the Baltic Sea region. Earth System Dynamics. 2022;**13**(1):251-301

[4] A Science Plan for Integrated Research on Disaster Risk: Addressing the Challenge of Natural and human-induced environmental hazards. International Council for Science (ICSU). 2008. p. 66. ISBN 978-0-930357-66-5

[5] Available from: https://reliefweb.int/report/world/

[6] Available from: https://ourworldindata.org/

[7] Intergovernmental Panel on Climate Change (IPCC). Has Climate Variability, or Have Climate Extremes, Changed?. 2005. Available from: https://archive.ipcc.ch/ipccreports/tar/wg1/088.htm

[8] Hewer MJ. Determining the effect of extreme weather events on human participation in recreation and tourism: A case study of the Toronto zoo. Atmosphere. 2020;**11**:1-17

[9] Walter E, editor. Cambridge Advanced Learner's Dictionary. 3rd ed. Cambridge, United Kingdom: Cambridge University Press; 2008. pp. 1699. ISBN 978-0-521-71266-8

[10] Trucchia A, Meschi G, Fiorucci P, Provenzale A, Tonini M, Pernice U. Wildfire hazard mapping in the eastern Mediterranean landscape. International Journal of Wildland Fire; 2023. p. 417-434. DOI: 10.1071/WF22138

[11] Forest fire videos. BBC Earth. Available from: https://web.archive.org/web/20151016185535/http://www.bbc.co.uk/science/earth/natural_disasters/forest_fire

[12] Stephens SL, Collins BM, Fettig CJ, Finney MA, Hoffman CM, Knapp EE, et al. Drought, tree mortality, and wildfire in forests adapted to frequent fire. BioScience. 2018;**68**(2):77-88

[13] Main Types of Disasters and Associated Trends. Legislative Analyst's Office. Available from: https://lao.ca.gov/Publications/Report/3918

[14] Theresa M. The Far-Reaching Consequences of Siberia's Climate-Change-Driven Wildfires. Smithsonian Magazine. 2020. Available from: https://www.smithsonianmag.com/smart-news/siberian-wildfires-cause-record-pollution-send-smoke-us-180975275/

[15] Australia, Government Geoscience. Bushfire. 2017. www.ga.gov.au

[16] Yi-Lun HU, Guo-Xu JI, Ji-Hong LI, Ganjurjav HASBAGAN, Guo-Zheng HU, Qing-Zhu GAO. Interpretation of IPCC AR6: Terrestrial and freshwater ecosystems and their services. Advances in Climate Change Research. 2022;**18**(4):395

[17] Douville H, Raghavan K, Renwick J, Allan RP, Arias PA, Barlow M, et al, eidtors. Water cycle changes. In: Masson-Delmotte V, Zhai P, Pirani A,

Connors SL, Péan C, Berger S, editors. Climate Change: The Physical Science Basis. Contribution of Working Group I to 45 the Sixth Assessment Report of the Intergovernmental Panel on Climate Change. Cambridge, United Kingdom: Cambridge University Press; 2021. pp. 1055-1210. Available from: https://centaur.reading.ac.uk/101319/

[18] Drought Basics. NOAA National Integrated Drought Information System. Available from: https://www.drought.gov/what-is-drought/drought-basics

[19] Definition of Drought. www.ncei.noaa.gov. NOAA | National Centers for Environmental Information (NCEI). Retrieved September 16, 2022

[20] Types Drought. National Drought Mitigation Center. Available from: https://www.drought.unl.edu/Education/DroughtIn-depth/TypesofDrought.aspx

[21] Prinos P. Review of flood hazard mapping. Report Number: T03-07-01. 2009. pp. 62

[22] Hirabayashi Y, Mahendran R, Koirala S, Konoshima L, Yamazaki D, Watanabe S, et al. Global flood risk under climate change. Nature Climate Change. 2013;**9**(2013):816-821

[23] United Nations Environmental Program. How climate change is making record-breaking floods the new normal. 3 Mar 2020. Available from: https://reliefweb.int/report/world/how-climate-change-making-record-breaking-floods-new-normal

[24] Hsu ICY. The Rise of Modern China. 4th ed. New York: Oxford University Press; 1990. pp. 546-547

[25] Biswas S, Daly P. 'Cyclone not above politics': East Pakistan, disaster politics, and the 1970 Bhola cyclone. Modern Asian Studies. 2021;**55**(4):1382-1410

Chapter 2

Probabilistic Seismic Vulnerability and Loss Assessment of the Buildings in Mexico City

Alonso Gómez-Bernal, Antonio Romero Peña and Jonathan de Anda Gil

Abstract

This article presents a seismic vulnerability and risk assessment of buildings in Mexico City. A probabilistic seismic hazard analysis (PSHA) was carried out, which allowed the definition of seismic hazard curves as well as uniform hazard spectra (UHS) for several seismic zones. The seismic hazard includes the effects of all seismic sources located in an influence area with a radius of 500 km. Attenuation relationships were selected with basis in attenuation models of events affecting the areas of Central Mexico and were complemented by our own functions that include local soil effects. Already established the sources and attenuation functions, the seismic hazard is quantified throughout UHS, which calculated using a return period Tr = 100 years. For the vulnerability assessment, fragility curves were defined. Two groups of fragility curves were studied, the first for the first for buildings built before 1985, and the second for buildings built after 1985. In the first case, static nonlinear analyzes of selected buildings were performed to define the capacity spectra. In the second case, the capacity spectra were defined from design spectra of the Mexico City Building Code. The results showed a very good correlation with the seismic demands of the 2017 earthquake.

Keywords: Mexico earthquakes, vulnerability in Mexico City, seismic risk in Mexico City, seismic hazard in México

1. Introduction

1.1 The seismic vulnerability assessment

The ability to assess the vulnerability of residential infrastructure to earthquake damage is undoubtedly one of the most important challenges facing structural engineers. Two methods are typically used to predict earthquake damage: nonlinear finite element analysis (FE) and seismic vulnerability curves. Nonlinear FE analysis is particularly applicable when a detailed damage estimate is required only for a small

IntechOpen

number of structures. However, if an estimate is required for the many structures, the process becomes slow and inefficient. Seismic vulnerability curves provide a more efficient method for predicting damage to a class of similar structures. These curves often relate strong ground motion and structural properties to damage.

Vulnerability curves are generally constructed from statistical analyses of historical field data, as example Ref. [1], or analytically simulated data, as Ref. [2]. However, the number of parameters considered when creating vulnerability curves is usually very limited. Research on widespread methods of damage prediction has been scarce. Notable examples include Ref. [3, 4] who used discriminatory analysis to predict the damage to buildings in a classification of two or three damage degrees (DD). Vulnerability studies have recently been conducted using artificial neural networks [5], where the soil movement intensity parameter is correlated with the damage states of type structures, for large-scale risk analysis. In a vulnerability study [6], defined sets of single-degree freedom oscillators and a series of ground motion records using nonlinear time history analysis, and the resulting damage distributions were used to derive sets of fragility functions.

1.2 The two catastrophic earthquakes of September 19: 1985 and 2017

Many of the seriously damaged buildings during the September 19, 2017, Earthquake (M = 7.1) were medium-height buildings (4–10 stories), with natural periods ranging between 0.7 to 1.4 sec, located in the Lake-bed zone. **Figure 1a** shows the location of the buildings that suffered the greatest damage caused by the 2017

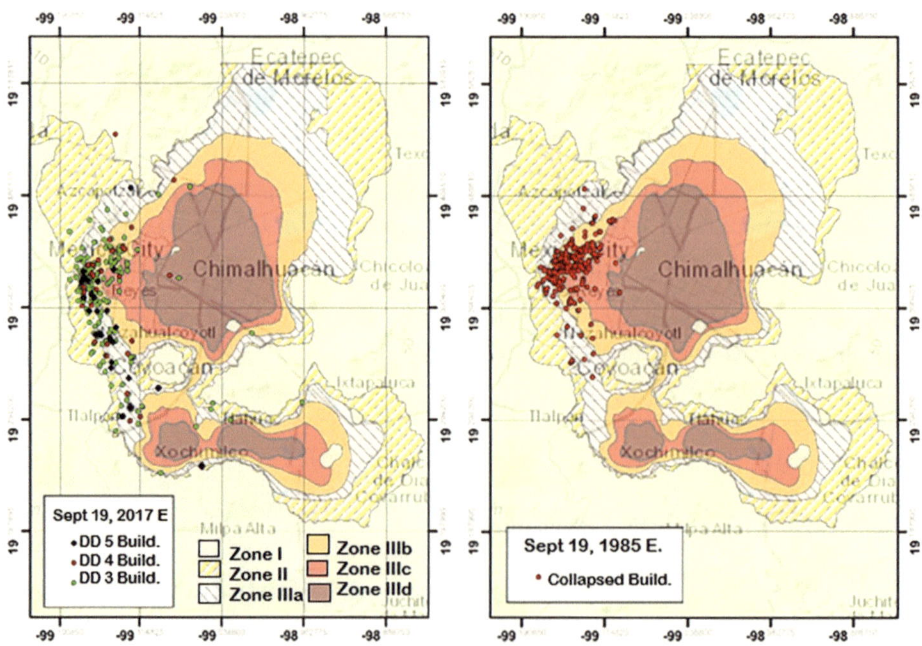

(a) (b)

Figure 1.
(a) Damaged buildings due to the September 19, 2017, Earthquake: Damage Degree 5 (DD5), Damage Degree 4 (DD4), Damage Degree 3(DD3). (b) Collapsed buildings during September 19, 1985, Earthquake.

earthquake (DD3, DD4, and DD5), while **Figure 1b** shows the location of the collapsed buildings (DD5) for the September 19, 1985, Michoacan Earthquake (M = 8.1); As can be seen, these damages occurred in the same area of the Valley of Mexico for both earthquakes. One of the most vulnerable typologies corresponds to buildings structured with columns of reinforced concrete without girders and supporting a waffle slab, which was built before the September 19, 1985, Earthquake. A complete information statistic of damaged buildings during September 19, 2017, Earthquake, could be founded for example in Ref. [7, 8]. Among the principal causes of buildings damaged were, a) lack of seismic strength; b) structural irregularities, in this case, most configuration problems were associated with the contribution of no-structural elements to the building response, especially in corner buildings; and c) tilting and foundation problems.

This study focuses on the vulnerability assessment of two groups of buildings, the first the structures built before 1985 when the 1976 Mexican Building Code [9] was applicable, and the second includes those built after that date.

2. Methodology

With the purpose of assigning structural vulnerability to the building population of México City, it is necessary to apply an analytical methodology for assessing the seismic response of a class of buildings, which is combined with a damage model in order to derive sets of fragility functions for different levels of seismic behavior. The seismic vulnerability assessment developed in this paper focuses mainly on the HAZUS methodology [10] and RISK-EU [11], which has been adapted for damaged buildings in Mexico City during the September 19, 2017, earthquake.

The methodology developed consists in general of the following steps: (1) seismic hazard assessment for Mexico City, using a probabilistic procedure, for the calculation of the uniform hazard spectra (UHS). (2) Definition of the capacity curves of structural systems, in this case, data from damaged buildings representative of the most vulnerable typology were used for the September 19, 1985 and 2017 earthquake, but analytical models representative of other common typologies of structures built after 1985 were also used; (3) Definition of the damage criteria adopted and development of capacity spectrums; (4) Estimation of the medium capacity curves, which are represented in their bilinear form to define the damage thresholds and thereby build the fragility curves and their damage probability matrices; (5) calculation of overall damage of the structure analytically. The fragility curves resulting from this study were calibrated with the event of September 19, 2017.

3. Seismic hazard analysis

3.1 Probabilistic seismic hazard analysis PSHA

The seismic hazard for Mexico City was developed with a probabilistic seismic hazard analysis (PSHA). The PSHA use of probabilistic concepts has allowed uncertainties in the location, size, and rate of recurrence of earthquakes and in the variation of ground motion characteristics with earthquake size and location to be explicitly considered in the evaluation of seismic hazards [12]. In this study, the seismic hazard settings were defined, using uniform hazard spectra (UHS) and seismic parameters. Three ground motion prediction equations (GMPEs) were revised and used, and a disaggregation study was performed. Site effects estimations were included in the prediction equations. UHS was established for both firm and soft soils for Mexico City. Finally, the computed seismic parameters were compared to those for the seismic design guidelines for Mexico City, and we found out that our UHS captured better the seismic settings.

The model characterization for the earthquake occurrence, the seismic sources, the magnitude-recurrence, and the attenuation laws are statistically evaluated (see for example [13, 14]). The probabilistic seismic hazard analysis is characterized by four stages [12]), these stages can be summarized as 1) identification and characterization of seismic sources, historical descriptions, seismic catalogs, isoseismal maps, and instrumental information; 2) Definition of recurrence relationships, it is necessary to ensure that data is homogeneous, independent, and complete (events with M > 5). The recurrence ratio, which specifies the average rate at which an earthquake of some size will be exceeded, is used to characterize the seismicity of each source zone. Gutenberg's law was used. 3) Definition of predictive relationships. 4) development of seismic hazard curves, for which, the uncertainties in earthquake location, earthquake size, and ground motion parameter prediction are combined to obtain the probability that the ground motion parameter will be exceeded during a particular time period.

The equation used to assess the seismic hazard according to the classical methodology is:

$$\lambda(y>Y) = \sum_{i=1}^{N} \lambda_i(y>Y) = \sum_{i=1}^{N} \upsilon_i \iiint P_i\left[y>Y|m,r,\epsilon\right]\, f_{Mi}(m)\, f_{Ri}(r)\, f_{\epsilon i}(\epsilon)\, dm\, dr\, d\epsilon \tag{1}$$

where $\lambda(y > Y)$ is the annual rate of exceedance of the level of ground motion Y, due to the occurrence of earthquakes in the N seismic sources, which is equal to the sum of the annual rates of exceedance in each source zone $\lambda i(y > Y)$, the same ones that present an annual rate of earthquakes vi; the term $Pi[y > Y \mid m, r, \epsilon]$ gives the probability of conditional exceedance to the trio of variables m, r, and ϵ representing magnitude, distance, and epsilon; $fMi\ (m) fRi\ (r)\ f\epsilon i(\epsilon)$ are the probability density functions of magnitude, distance, and epsilon [15].

3.2 Hazard curves and UHS for Mexico City

The identification of the seismotectonic regions is one of the most important stages in the seismic hazard procedure, in order to understand the seismicity of each region that is the characteristics of earthquake occurrence and to identify whether a source, close to the site, is active or not, as well as the definition failure mechanisms. We considered two types of seismic sources capable of producing earthquakes in Central Mexico: faults and areas. Seismic sources are modeled in a seismic hazard assessment with their geometric and recurrence characteristics.

Of the thirteen segments or gaps characteristic of the country, six faults were selected and characterized in this study since the interplate sources that are within a radius of influence of 500 km were analyzed. **Figure 2a** shows the six fault segments

(a) (b)

Figure 2.
(Left) Faults along the Pacific Coast [11]. (Center and Right) Seismic sources type Area affecting Mexico City [12].

described by Ref. [16] that affect the Central Mexico segment are, Oaxaca Este (OX-E), Oaxaca Central (OX-CI and OX-CII), Oaxaca Oeste (OX-O), Ometepec (OX-M), Acapulco-San Marcos (AC-SM), Guerrero Central (GC), Petatlán (PE), and Michoacán (MI). We have considered an area where earthquakes might occur, with a radius 500 km (red circle in **Figure 2b**), these areas were selected according to the faults described by [17].

Part right and center of **Figure 2** illustrate the 19 seismic sources type-Area parameters, which describe the area where seismic events are presented; these parameters can be consulted in Ref. [18], namely, the focal depth, the activity rate or number of events per year that exceed the minimum magnitude for each source, the parameters of the Richter model, and the estimated minimum and maximum seismic moment magnitudes (M_w).

We use two sets of spectral acceleration attenuation (SA) functions; the first group includes classical functions [19–23] developed with a global database, while for the second group, we use attenuation relations developed by us [18] of strong ground motion recorded in Mexico City with M > 6.0, in both groups we considered interplate and intraslab seismic sources.

For the Hazard assessment the probabilistic procedure described above was used. For the calculation of the hazard curves and the uniform hazard spectra, UHS, we used the EZ-FRISK package [24]. **Figure 3** shows the seismic hazard curves in Zone I, Zone II, and Zone IIIb, as an example of the several sites studied. In **Figure 3**, uniform hazard spectra (UHS) are included for six return periods, namely, 40, 100, 250, 475, 975, and 2475 years, which represent, respectively, 68%, 39.3%, 20% 10%, 5%, and 2% exceedance. Probability of occurring in 50 years. These spectra show the highest spectral acceleration values in Mexico City, ranging from 0.8 g to 1.05 g, indicating a high hazard in Zone III for intraplate and interplate events.

When the shape of the spectral ordinates in sites of Mexico City obtained with our functions are compared with other relationships, we observed significant differences in periods ranging between 1 and 2 s, mainly due to the incorporation of a regression model, which considers local effects, and as seen in **Figure 3**, it provides good estimates of the spectral accelerations. UHS from sites in Zone I, Zone II, and Zone IIIb cover a larger range of periods.

As an example, **Figure 4** present the displacement spectra in Zone IIIb calculated using synthetic accelerograms, which were defined from the UHS assuming a return

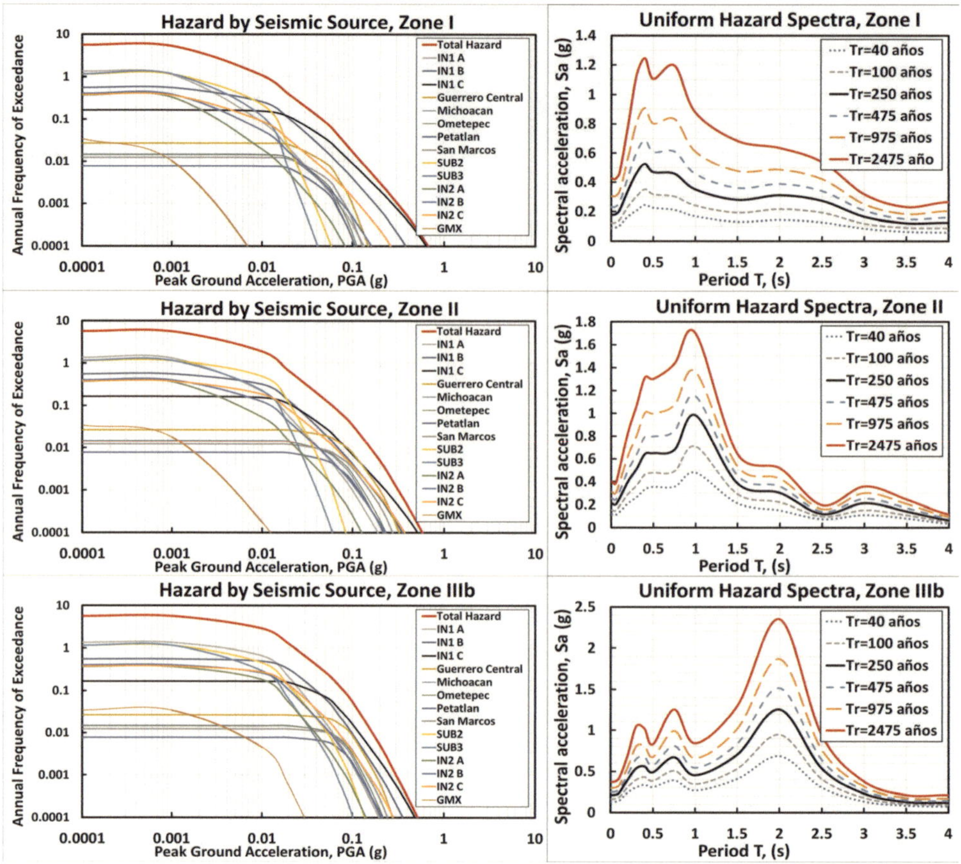

Figure 3.
Left: Hazard curves obtained in Zone I, Zone II, and Zone IIIb and Right: the corresponding uniform hazard spectra (UHS) for several return periods.

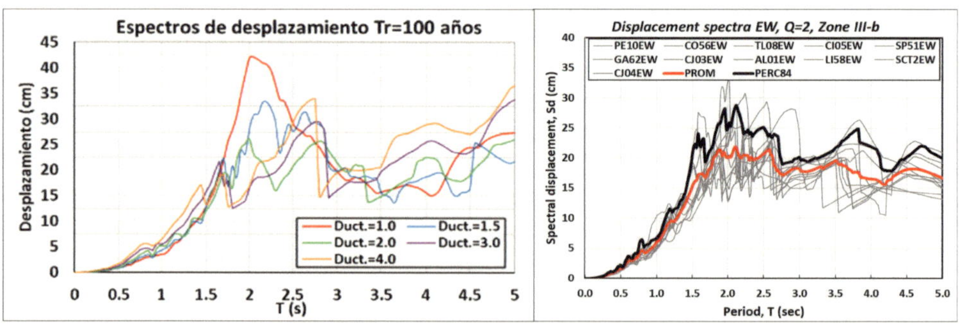

(a) (b)

Figure 4.
Left: Displacement spectra in Zone IIIb calculated from synthetic records for UHS and a return period Tr = 100 years Right: Displacement spectra at stations in Zone IIIb due to September 19, 2017, Earthquake.

period Tr = 100 years. The right part of **Figure 4** is compared this result with the displacement spectra of stations located in Zone IIIb due to the September 19, 2017, earthquake.

4. Numerical models

4.1 Buildings characteristics

The identification and selection of structural systems to be studied typical of an area is initially carried out. In this case, five buildings of a very common typology in Mexico City were selected, which has been very vulnerable to the two earthquakes of September 19 (1985 and 2017), and other seismic events.

With the purpose of studying existing buildings constructed before 1985 in Mexico City, the five medium-rise buildings damaged by the 2017 earthquake were analyzed (**Figure 5**). These buildings were built on reinforced concrete columns and reticular waffle slabs, four of them are in the Benito Juárez Sector and one more in the Cuauhtémoc Sector. All buildings are for housing use and are classified within group B according to the México City Building Code [8]. **Table 1** summarizes the main characteristics of the five buildings, including the EME-98 classification of the damage degree assigned after the 2017 earthquake.

Once the structures to be studied have been modeled, the main parameters to be considered are defined. The main function of this variation is to obtain a standard deviation from the potential capabilities of each of the analyzed models and consider them within their vulnerability assessment. The definition of the damage thresholds was made according to **Table 2**.

These buildings have long rectangular plants as shown in **Figure 3**, as can be seen, with the exception of the SE2–7 N building, the structures exceed or are equal to the value of 2.5, which is a limit value suggested in all structural recommendations because when exceeded, it has been observed that the structural response can be amplified and the torsional effects can also be increased, in addition, all these structures have no girders and the slab have relatively low depths, which generates flexible

Key	Age	Stories	Height	Bay	L_1	L_2	Zone	d slab	Cols	L_1/L_2	DD
SE2–5 N	1977	5	13.80	2	14.62	7.31	IIIa	35	70x35	2.4	DD2
SE2–7 N	1982	7	18.90	3	13.50	3.40	IIIa	35	55x45	1.5	DD3
SE2–8 N	1983	8	22.25	3	15.75	5.25	IIIa	25	65x35	5.4	DD3
SE2–10 N	1978	10	26.0	1	7.68	7.68	IIIb	35	50x80	3.3	DD3
SE2–12 N	1980	12	32.4	2	13.00	6.33	IIIa	35	145x40	2.5	DD3

Table 1.
General characteristics of the selected buildings [25].

Damage degree	ID	Damage threshold
No damage	ND	—
Sligth	DL	Sd1 = 0.7Dy
Moderate	DM	Sd2 = Dy
Severe	DS	Sd3 = Dy+0.25(Du-Dy)
Completo	DC	Sd4 = Du

Table 2.
Definition of damage threshold.

frames away from being shear systems. Another factor that increases the vulnerability of these buildings is the unfavorable orientation of the columns with their minor axis on the short side.

4.2 Capacity curves

For all buildings studied, a nonlinear static analysis (Pushover) was performed using the SAP program to define the capacity curves of each model. The structural configuration and considered loads of buildings are defined in Ref. [25]. The different capacity curves of the models analyzed are presented in part left of **Figure 5**. According to the results, it is possible to note that the model of the structure SE2–10 N is the one that provides the least resistant capacity since its last capacity is presented when a basal shear of Vu = 25.6 Ton occurs, with a roof displacement of 27 cm, while the strongest structure corresponds to the model SE2–7 N, where its basal shear is Vu = 60.31 Ton, with a roof displacement of 31 cm.

To obtain the ductility, μ, of the system it is necessary to represent the capacity curve in its bilinear form, which is obtained by defining the yield point and the ultimate capacity point of the structure. The procedure used in this study corresponds to that proposed by FEMA 356 [14], which is based on matching the energy dissipated by the structure defined by the area under the actual curve with the dissipated energy of the idealized curve. **Table 2** presents some of the most important points that define the capacity curves obtained in their bilinear form.

4.3 Capacity spectra

To compare the seismic demand with the capacity of the structure, it is necessary to transform the result of pushover to another curve that relates the spectral displacement *Sd* with the spectral acceleration *Sa*. This transformation is known as the capacity spectrum and develops by applying the dynamic characteristics of the fundamental mode. The capacity spectrum is determined using the following equations:

$$Sd_j = \frac{Dt_j}{\gamma_M \varphi_{t1}} \tag{2}$$

$$Sa_j = \frac{V_j}{M_T \alpha} \tag{3}$$

where:
D_{tj} = displacement of each point of the capacity curve [cm].
V_j = shear of each point of the capacity curve [Ton].
γ_M = participation factor of the first mode.
α = effective mass coefficient of the basal shear of the first mode.
φ_{t1} = maximum top amplitude of the structure associated with the first mode.
M_T = total mass of the structure.

The capacity spectrum is defined by two main points: the ultimate capacity point (UC) and the yield capacity point (YC), which are expressed as follows:

$$YC\left[A_y = S_{ay} = \frac{C_s SR}{\alpha_1}; Dy = S_{dy}\right] \tag{4}$$

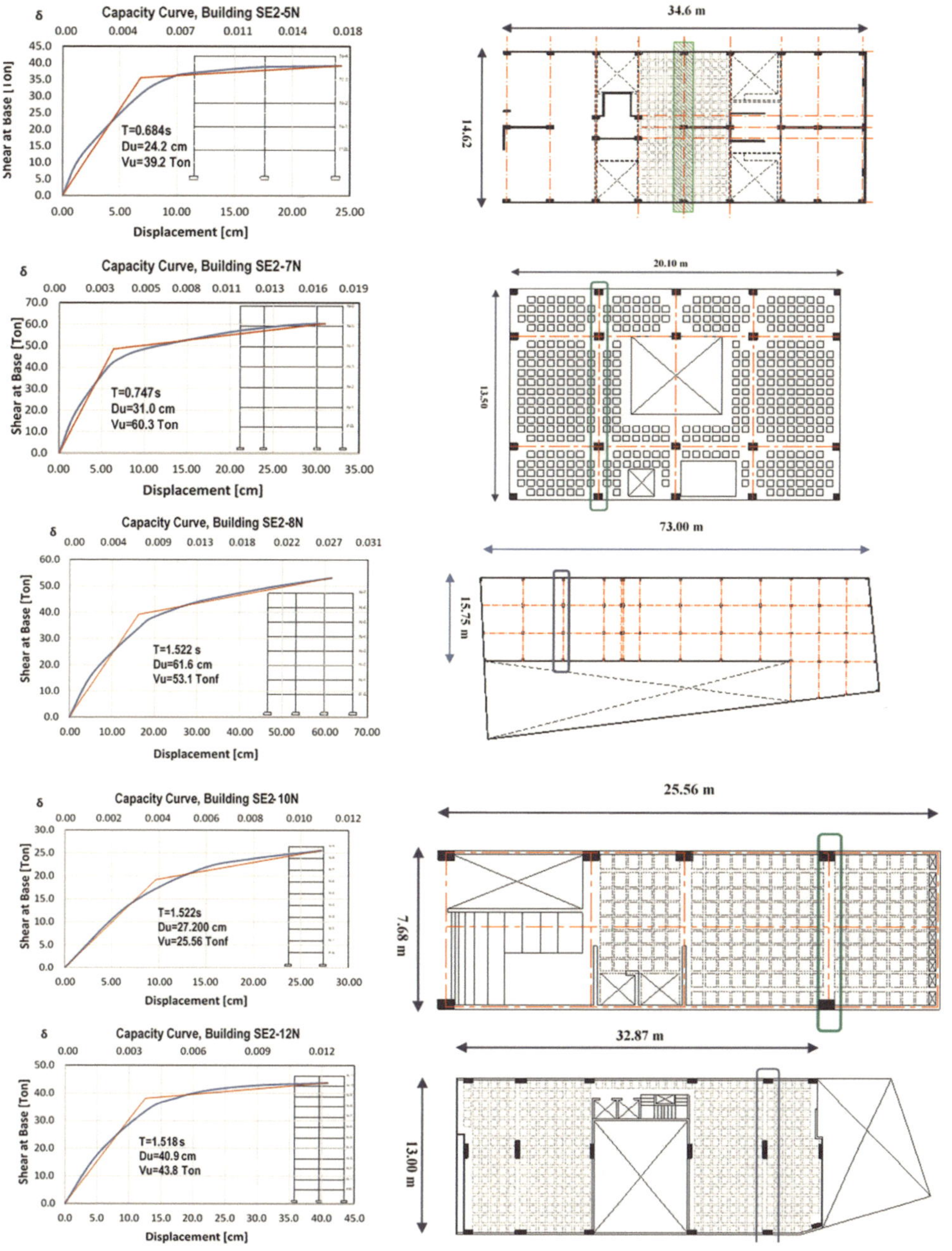

Figure 5.
Right: Structural plants of the buildings studied. SE2–5 N, SE2–7 N, SE2–8 N, SE2–10 N, SE2–12 N. Left: The corresponding capacity curves of the studied buildings.

$$UC\left[A_u = S_{au} = \lambda A_y; D_u = S_{du}\right] \tag{5}$$

where C_s represents the design seismic coefficient. This value is calculated using Eq. (4), and an approximate overstrength of the buildings analyzed is obtained.

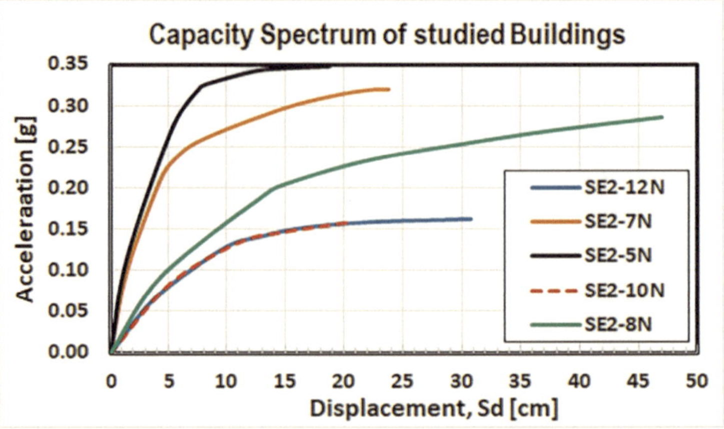

Figure 6.
Capacity spectrum, derived from the capacity curves of ***Figure*** *5 for the five selected buildings.*

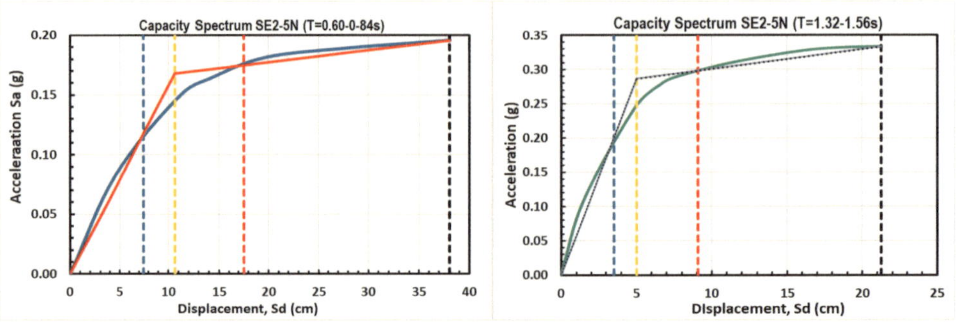

Figure 7.
Mean capacity spectra for buildings with short and long periods; damage thresholds are indicated in each case (DD2 to DD5).

As part of the methodology used in this work, the buildings analyzed were divided into two groups. Firstly, based on the characteristics and properties of each one of these structures, the dispersion of the capacity curves is greater among the buildings with five to seven levels and those from eight to 12 levels (**Figure 6**). Secondly, it is desired to reduce the uncertainty that exists due to this difference since there are only five buildings analyzed. For this reason, two mean capacity curves have been calculated, one for buildings with short periods and another for buildings with periods longer than 1 sec (**Figure 7**). This division by periods is also useful and of great interest since it allows classifying the type of buildings that were most vulnerable to the earthquake of September 19, 2017. Capacity spectra of **Figure 7** show the bilinear representation. From these capacity curves in their bilinear form, the thresholds were defined, of damage for each type (**Table 2**).

5. Fragility curves and damage probability matrices

Fragility curves are a graphical representation of the cumulative distribution function of the probability of reaching or exceeding a specific state of damage, given a

Damage degree	ID	Damage (%)	FDC (%)	SE2–5		SE2–7		SE2–8		SE2–10		SE2–12	
				Prob (%)	D. I.	Prob (%)	D.I.	Prob (%)	D.I.	Prob (%)	D.I.	Prob (%)	D.I.
No damage	DD1	0–10	5	45.7	2.3	21.4	1.1	4.6	0.2	0.4	0.0	5.5	0.3
Sligth	DD2	10–30	20	36.5	5.9	41.0	8.2	29.0	5.8	11.3	2.3	30.6	6.1
Moderate	DD3	30–60	45	11.4	8.2	25.4	11.4	43.2	19.4	41.0	18.5	42.5	19.1
Heavy	DD4	60–90	75	5.9	4.4	11.0	8.3	19.2	14.4	37.2	27.9	17.9	13.4
Complete	DD5	90–100	100	0.5	0.5	1.1	1.1	4.0	4.0	10.0	10.0	3.6	3.6
			Σ		20		30		44		59		43
Damage index, D.I. = Prob x FDC													

Table 3.
Damage degree and damage index in the five buildings.

structural response, to a given seismic action. The HAZUS methodology [11] defines fragility curves and a lognormal probability distribution defined by the following equation.

$$P[ED \geq ED_i] = \Phi\left[\frac{1}{\beta_{ED}} \ln\left(\frac{Sd}{\overline{Sd_{ED}}}\right)\right] \quad (6)$$

where $\overline{Sd}_{ED}$ is the average spectral displacement for which the probability of exceedance is 50%; $\boldsymbol{\beta_{ED}}$ is the standard deviation of the natural logarithm of spectral displacement, $\boldsymbol{\Phi}$ is the cumulative standard normal distribution function, and $\boldsymbol{S_d}$ is the spectral displacement.

In this methodology, each fragility curve is defined by the average spectral displacement value corresponding to the threshold of each damage state defined in **Table 2**. The standard deviation was calculated using two methods and adjusted to the mean of both. The first method obtains the standard deviation using the different actual capacity curves, while the second method adjusts the fragility curves with a discrete probability distribution (**Table 3**).

5.1 Fragility curves for buildings constructed before 1985

The fragility curves calculated for the first type of buildings, that is, between five and seven stories (with vibration periods between 0.60 and 0.86 s), are presented on the left side of **Figure 8**. In addition, the values of the spectral displacements S_d corresponding to the hazard spectrum for a return period Tr = 100 years in Zone IIIb, are included in each curve (see **Figure 4**). **Table 3** indicates the average damage factors and the central damage factor, FDC, for each case.

5.2 Fragility curves for buildings constructed after 1985

In order to include buildings built with recent building regulations, theoretical models of buildings were included from the design spectra of the 2004 Complementary Technical Standards for Design by Sism (GDF, 2004). **Figure 9** shows the

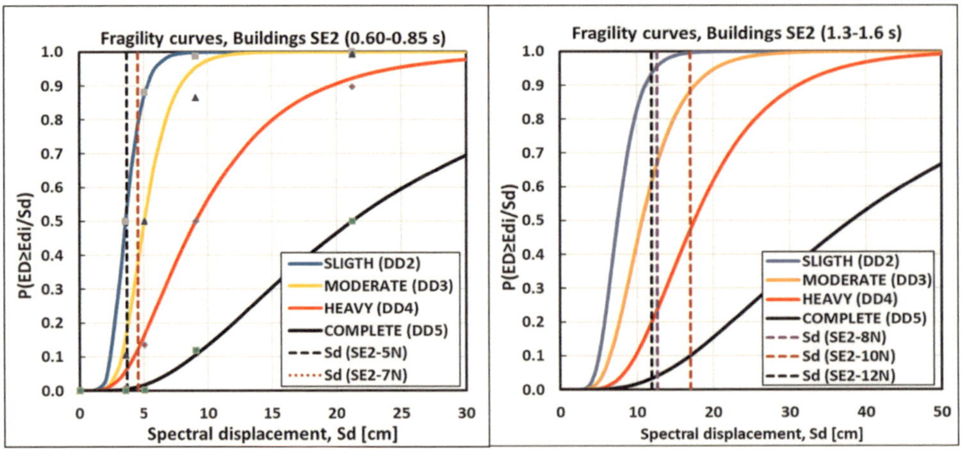

Figure 8.
Fragility curves that are indicated by dashed lines the Sd displacements for each building according to its period calculated assuming a return period Tr = 100 years.

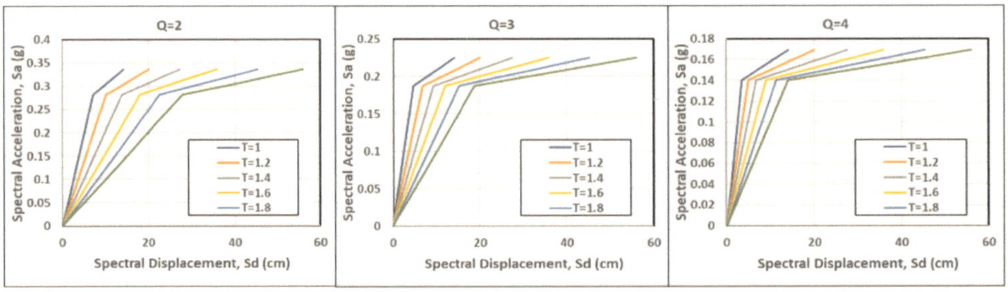

Figure 9.
Capacity spectra obtained from NTCS2004 [8] design spectra for Zone IIIb in Mexico City and using an overstrength factor of 1.5.

capacity spectra determined for three ductility reduction factors, Q = 2, Q = 3, and Q = 4, for structures built in Zone IIIb, and an overstrength factor of 1.5. These spectra are the basis for determining the fragility curves in **Figure 10**, following the same procedure used earlier. Fragility curves were calculated for six structural models with periods between 1.0 s and 2.0 s, with intervals of 0.2 s.

Figure 4b shows the displacement response spectra for 5% of critical damping at eleven stations in Mexico City's zone IIIb of September 19, 2017, which were calculated for elastic models and for models with ductility equal to two (Q-2). They are also indicated in the same **Figure 4**, the average or average response spectra and the spectrum corresponding to the 84% percentile, for all cases. It is important to note that there is little dispersion between the spectra of the different stations for displacement spectra, compared to acceleration response spectra.

In order to calibrate the fragility curves in **Figure 10**, the spectral displacement, Sd, for the return period Tr = 100 years (**Figure 4a**) associated with each natural period, is included in each of the six boxes. It can see that, in general, there is very good correlation with the behavior of the buildings of this group (built after 1985), observed during the 19 September 2017 earthquake.

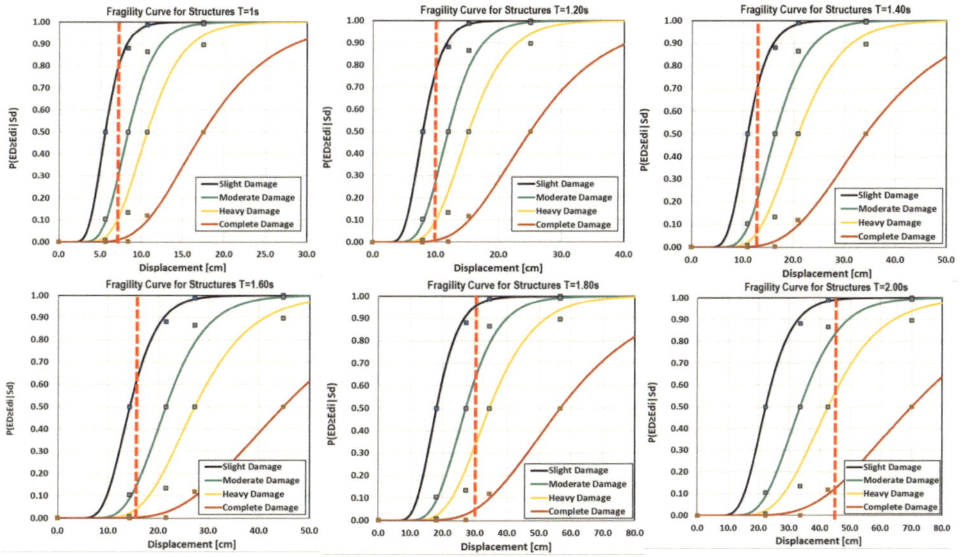

Figure 10.
Fragility curves for buildings using a coefficient of Q = 2 and with fundamental periods of 1.0, 1.2, 1.4, 1.6, 1.8, and 2.0 s.

6. Conclusions

In the first part of this article, a probabilistic seismic hazard analysis (PSHA) was carried out, which allowed the definition of seismic hazard curves as well as uniform hazard spectra for each of the six seismic zones of Mexico City; in this analysis, the seismogenic origin of the earthquakes was considered, that is, interplate or intraslab origin.

In the second part, the fragility curves that are the basis for establishing the damage index of buildings for a specific scenario were defined. Two groups of fragility curves were defined, the first for buildings built before 1985 and the second for buildings built after 1985. In the first case, nonlinear static analyzes (Pushover) of five buildings, which were performed to define the capacity curves and later the capacity spectra elements, were necessary to estimate the fragility curves. In the second case, the capacity spectra were directly defined from the design spectra of the Mexico City Building Regulations.

For the study of buildings built before 1985, five buildings were selected as representatives of this group, which have a structural system of columns of reinforced concrete with a waffle slab, and were damaged by the earthquake of September 19, 2017.

With the purpose to assign the vulnerability of residential buildings in Mexico City, the fragility curves of buildings built before 1985 were defined, which have a structural system of reinforced concrete columns with a lightened flat slab and which were damaged by the earthquake of September 19, 2017.

The findings of the probabilistic seismic hazard and probabilistic vulnerability analysis resulted in the following conclusions:

1. Uniform hazard spectra, UHS, were computed for return periods of 50, 100, 475, 975, and 2475 years. Specific attenuation relationships were developed for

Mexico City that directly includes local effects; in addition, to complement the study, classical predictive functions were used.

2. With the capacity curves of the buildings, their expected design ductility and overstrength were estimated. It was found that the average overstrength used for this type of structure (buildings built before 1985) is approximately 1.50.

3. With the uniform hazard spectra, displacement demands were defined for some return period, in this article a return period of 250 years was selected. These demands were combined with the capacity curves to obtain the percentages of damage expected for each degree of damage, and thus determine the damage index of the building.

4. The most vulnerable building due to the 19S-2017 earthquake was the 10-story building (SE2–10 N), which has an elongated floor plan with a B/A = 3.33 ratio, in addition to having only one span. This structure was assigned a damage degree due to the earthquake of 45%, while with the analytical method carried out, a global damage index of 59% was found, in both cases, it corresponds to a damage degree of 3 (DD3), or moderate damage.

5. On the other hand, the least vulnerable building turned out to be the 5-story one (SE2–5 N), since in the evaluation of the building a percentage of damage of 25% was assigned, while with the analytical method based on the fragility curves, a global damage index of 20% was determined, both cases correspond to damage degree of 2 (DD2), or slight damage.

6. For the estimation of the fragility of the buildings built with construction regulations after 1985, theoretical models of buildings were included based on the design spectra of the Mexican Construction Code, 2004. Capacity spectra were calculated for structure models in Zone IIIb; overstrength factor of 1.5 and a reduction factor of Q = 2 were used. When the results are compared with the spectral displacements registered by the 2017 earthquake, it can be verified that, in general, there is a very good correlation between results and observations.

7. In conclusion, the methodology used in this work shows that quite reasonable results are obtained, even though data from only a few buildings have been used; however, the buildings used in this work, which were damaged by the 2017 earthquake, are representative of hundreds of existing buildings in Mexico City. Besides, the study was complemented by the parametric study of the second part.

Author details

Alonso Gómez-Bernal*, Antonio Romero Peña and Jonathan de Anda Gil
Universidad Autónoma Metropolitana, Azcapotzalco, México City, Mexico

*Address all correspondence to: agb@azc.uam.mx

References

[1] Hassan AF, Sozen MA. Seismic vulnerability assessment of low-rise buildings in regions with infrequent earthquakes. ACI Structural Journal. 1997;**94**(1):31-39

[2] Lang K, Bachmann H. On the seismic vulnerability of existing buildings: A case study of the City of Basel. Earthquake Spectra. 2004;**20**(1):43-66

[3] Yücemen MS, Özcebe G, Pay AC. Prediction of potential damage due to severe earthquakes. Structural Safety. 2004;**26**(3):349-366

[4] Yakut A, Ozcebe G, Yucemen MS. Seismic vulnerability assessment using regional empirical data. Earthquake Engineering and Structural Dynamics. 2006;**35**(10):1187-1202

[5] Lautour OR, Omenzetter P. Prediction of seismic-induced structural damage using artificial neural networks. Engineering Structures. 2009;**31**(2): 600-606. DOI: 10.1016/j.engstruct.2008.11.010

[6] Villar-Vega M, Silva V, Crowley H, Yepes C, Tarque N, Acevedo AB, et al. Development of a fragility model for the residential building stock in South America. Earthquake Spectra. 2017; **33**(2):581-604

[7] Gómez Bernal A, Arellano E, González O, y Juárez, H. Características, causas, y consecuencias de los daños debidos al sismo del 19 de septiembre de 2017 (M=7.1) en México. In: XII Congreso Chileno de Sismología e Ingeniería Sísmica. Asociación Chilena de Sismología e Ingeniería Antisísmica; ACHISINA, Chile, 2019

[8] Roeslin S, Ma Q, Juárez GH, Gómez BA, Wicker J, Wotherspoon L. A machine learning damage prediction model for the 2017 Puebla-Morelos, Mexico, earthquake. Earthquake Spectra. 2020;**36**(2_suppl):314-339. DOI: 10.1177/8755293020936714

[9] Gobierno del Distrito Federal, GDF. Normas Técnicas Complementarias para Diseño por Sismo. 2017. http://www.smie.org.mx/informacion-tecnica/estados/reglamentos-construccion-ciudad-de-mexico.php

[10] Federal Emergency Management Agency FEMA. Multi-hazard loss estimation methodology. In: Earthquake Model. Washington, DC: Federal Emergency Management Agency FEMA and National Inst. of Building Sciences NIBS; 2003

[11] Federal Emergency Management Agency FEMA. HAZUS. Earthquake Loss Estimation Methodology. Washington, DC: Federal Emergency Management Agency and National Institute of Building Sciences; 1999

[12] Kramer SL. Geotechnical Earthquake Engineering. Civil Engineering and Engineering Mechanics Series. Vol. 1. Upper Saddle River, NJ: Prentice Hall; 1996

[13] Cornell CA. Engineering seismic risk analysis. Bulletin of the Seismological Society of America. 1968;**58**(5): 1583-1606

[14] McGuire RK. Seismic Hazard and Risk Analysis. Oakland: Earthquake Engineering Research Institute; 2004

[15] Aguiar R, Rivas A. Estudio Probabilístico de la Peligrosidad Sísmica de Ambato, en roca. Primera Edición: Microzonificación Sísmica de Ambato; 2018. pp. 1-22

[16] Nishenko SP, Singh SK. Conditional probabilities for the recurrence of large and great interpolate earthquakes along the Mexican Subduction Zone. Bulletin of the Seismatic Society of America. 1987;**77**:2094-2114

[17] Zúñiga R, Suárez G, Ordaz M, García-Acosta V. Peligro Sísmico en Latinoamérica y el Caribe. Mexico: Pan American Institute of Geography and History, IPGH; 1997

[18] Romero PA, Gómez BA, Arellano E. Evaluación de la Vulnerabilidad Sísmica de Edificios residenciales de la Ciudad de México. In: Memorias XXIII Congreso Nacional de Ingeniería Estructural. Sociedad Mexicana de Ingeniería Estructural SMIE, México; 2022

[19] Youngs RR, Chiou SJ, Silva WJ, Humphrey JR. Strong ground motion attenuation relationships for subduction zone earthquakes. Seismological Research Letters. 1997;**68**(1):58-73

[20] Abrahamson NA, Silva WJ, Kamai R. Summary of the ASK14 ground motion relation for active crustal regions. Earthquake Spectra. 2014;**30**(3): 1025-1055

[21] Atkinson GM, Boore DM. Empirical ground-motion relations for subduction-zone earthquakes and their application to Cascadia and other regions. Bulletin Seismatic Society of America. 2003; **93**(4):1703-1729

[22] Zhao JX, Zhang J, Asano A, Ohno Y, Oouchi T, Takahashi T, et al. Attenuation relations of strong ground motion in Japan using site classification based on predominant period. Bulletin Seismatic Society of America. 2006;**96**:3

[23] García D, Singh SK, Herráiz M, Ordaz M, Pacheco JF. Inslab earthquakes of Central Mexico: Peak ground-motion parameters and response spectra. Bulletin Seismatic Society of America. 2005;**95**:2272-2282

[24] Risk-Engineering. Ez-Frisk v.7.65. Golden, Colorado, USA: Manual Risk Engineering, Inc; 2022

[25] De Anda GJ. Vulnerabilidad sísmica de edificios construidos antes de 1985 en Ciudad de México a raíz del sismo del 19 de septiembre de 2017. Master's Thesis. Universidad Autónoma Metropolitana Azcapotzalco; 2020

Chapter 3

A Seismic Hazard Assessment of North Chhattisgarh (India)

Ashish Kumar Parashar

Abstract

Chhattisgarh, located in Central India, has been carved out of Madhya Pradesh to become the 26th state of the Indian Union. North Chhattisgarh is addressed by the tribal population. In the current study, an endeavor has been made to carry out the seismic hazard analysis for the major district headquarters of north Chhattisgarh, considering the local site effects and developing a seismic zone map for north Chhattisgarh. Seismic hazard analysis has been done for major district headquarters Ambikapur, Baikunthpur [Koria], Korba, and Jashpurnagar of north Chhattisgarh, using seismotectonic information. All earthquake sources and past seismic events have been considered within a radius of 300 km for the headquarters, applying deterministic and probabilistic seismic hazard analysis approaches. The seismic parameters and peak ground acceleration at the bedrock level for the district headquarters of north Chhattisgarh have been estimated. Using probabilistic seismic hazard analysis, hazard curves have been developed for each district headquarters. Alternatively, for peak ground acceleration of 0.05 g, 0.1 g, and 0.15 g return periods have been estimated for the study area. The probabilities of exceedance for 2% and 10% for 50 years have also been estimated for the study area. The current study throws light on the design and construction of vital civil engineering structures near and around the seismically active headquarters in northern Chhattisgarh.

Keywords: district headquarter, deterministic seismic hazard, fault, hazard curve, peak ground acceleration, probabilistic seismic hazard, probabilities of exceedance

1. Introduction

Over recent years, there has been an explosive growth of interest in the development of novel gel-phase materials based on small molecules. The occurrence of a major earthquake event is still seen as a catastrophe, with devastating consequences over human lives, buildings, and transportation networks, causing a large-scale impact on any society. Indeed, even if becoming particularly devastating to poor, underdeveloped countries, it equally affects populations and several infrastructures in modern, industrialized countries. For global picture, the occurrence of earthquakes can have an outcome, which is reflected in the deep changes in building philosophy or large investments in seismic design strategies. A number of recent earthquakes are typically referred to not due to only their destructive impact but also owing to the lessons that have been learned, leading, and sometimes to reflect changes or the establishment of turning points in seismic design philosophies. Chhattisgarh (21°17′ 42.47″ N, 81° 49′ 41.63′ E) is a newly developed state in India. North Chhattisgarh is rich in temples, natural beauty, dense forests, hill stations, water bodies, and industries. The northern portion of the state includes major districts that are Ambikapur, Baikunthpur [Koria], Korba, and Jashpurnagar. India's National Center for Seismology (NCS) and USGS reported a magnitude 4.8 earthquake near Ambikāpur, Surguja district of Chhattisgarh, in the morning on Monday, July 11th, 2022, at a moderately shallow depth. On the same date, earthquake of magnitude 4.3 is reported near Korba district. Further on an earthquake of magnitude of 4.6 on 29 July 2022 is reported by National Center for Seismology, near Baikunthpur [Koria]. Recent seismic activity around the districts of north Chhattisgarh is increased. It is essential that the current study is focused on the seismic vulnerability analysis over the major district headquarters of the north Chhattisgarh region.

2. Methodology

For estimating the seismic hazard at places of critical infrastructures, such as high-rise building, dams, bridges, and subsea tunnels, are of the utmost importance in the field of earthquake engineering, as damage to such structures results in severe economic loss and threats to the environment. Seismic hazard can be quantified by adopting two globally accepted techniques: the deterministic seismic hazard analysis (DSHA) and the probabilistic seismic hazard analysis (PSHA). Both approaches can be defined in a four-step process, and their initial steps are identical [1]. In the deterministic approach, the ground-shaking hazard at any point is evaluated based on the controlling source, which is expected to produce the maximum hazard at that point among all the potential sources [2]. On the other hand, in the probabilistic approach, the inherent uncertainties in the forecast of earthquake size, site, and earth motion parameters are explicitly combined to obtain the hazard for a given probability of exceedance in a particular time period [3]. Despite their benefits and drawbacks, these two methods can harmonize each other and give additional insights into the question of seismic hazard [4]. Moreover, a deterministic framework accounts for the hazard from each source independently; thus, it is more capable in the regions, where seismic activity is low to moderate, particularly, where limited earthquake data is available [5].

2.1 Deterministic seismic hazard analysis (DSHA)

For identification of linear seismic sources, a seismotectonic atlas developed by the Survey of India [6] has been taken as the base for the present study. The seismotectonic maps are prepared for each district headquarters of north Chhattisgarh, keeping the headquarter at the center of the circle, with a radius of 300 km, based on the latitude and longitude as shown in **Figure 1**. The faults having fault length (Li) $\geq$ 25 km coming in a 300 km radius were identified, numbered, and

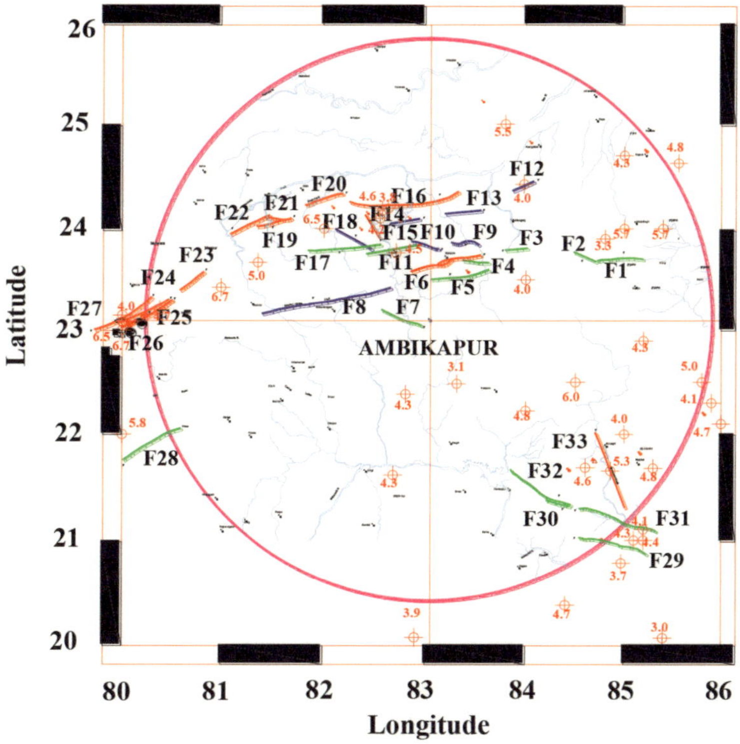
Latitude
Longitude
AMBIKAPUR
F1
F2
F3
F4
F5
F6
F7
F8
F9
F10
F11
F12
F13
F14
F15
F16
F17
F18
F19
F20
F21
F22
F23
F24
F25
F26
F27
F28
F29
F30
F31
F32
F33

(a)

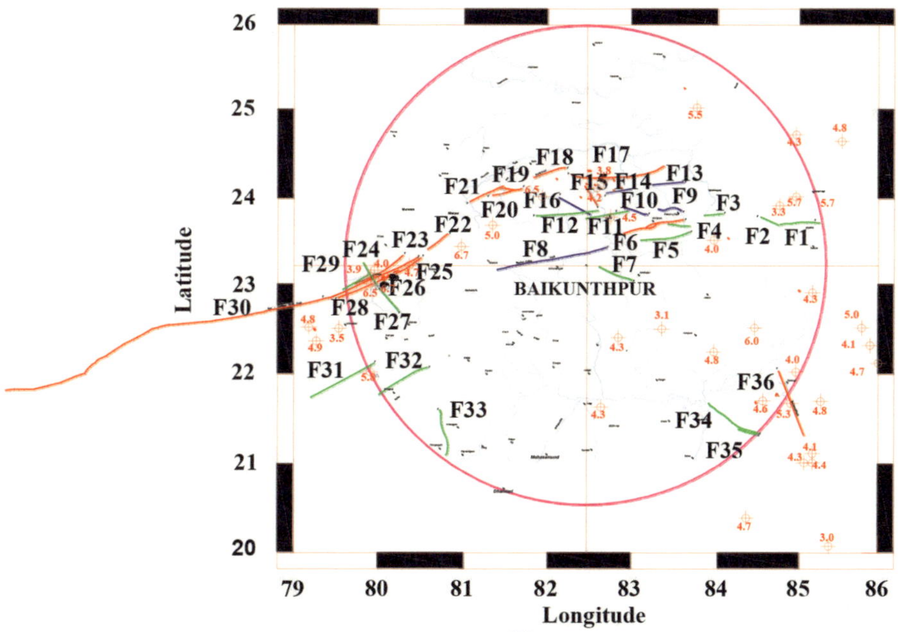
Latitude
Longitude
BAIKUNTHPUR
F1
F2
F3
F4
F5
F6
F7
F8
F9
F10
F11
F12
F13
F14
F15
F16
F17
F18
F19
F20
F21
F22
F23
F24
F25
F26
F27
F28
F29
F30
F31
F32
F33
F34
F35
F36

(b)

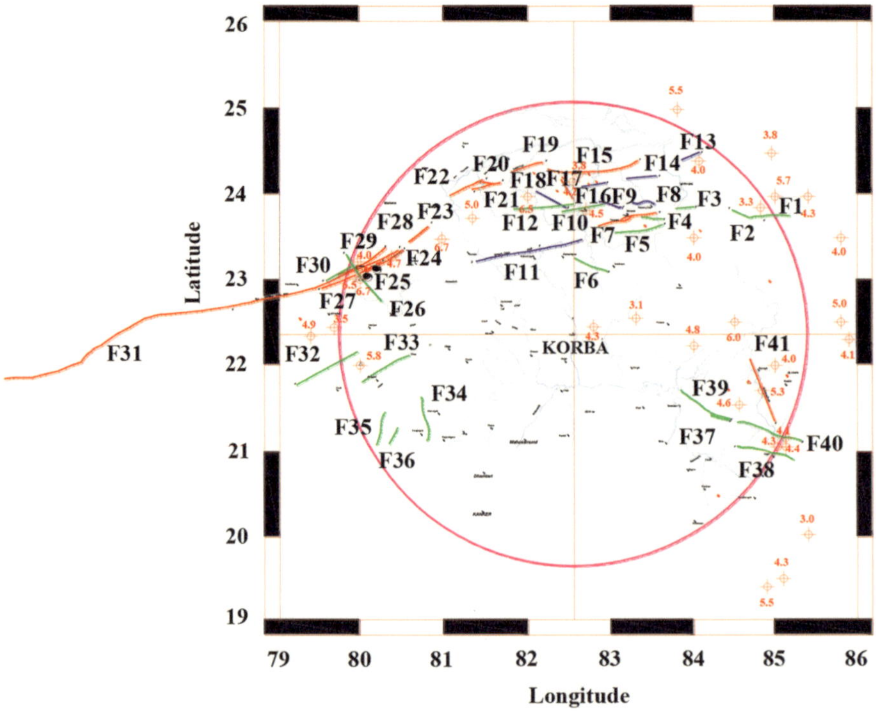

(c)

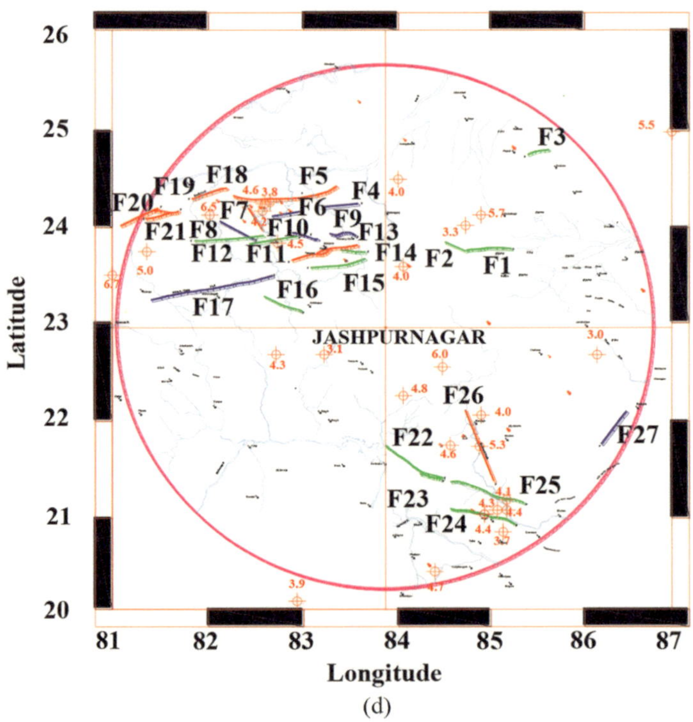

(d)

Figure 1.
Seismotectonic map of district headquarters of north Chhattisgarh: (a) Ambikapur, (b) Baikunthpur [Koria], (c) Korba, and (d) Jashpurnagar.

measured the length of the faults. The minimum map distances of identified faults were measured from the center of the circle for the study area.

The total number of linear sources as 33, 36, 41, and 27 were identified for major districts Ambikapur, Baikunthpur [Koria], Korba, and Jashpurnagar, respectively. In next step, the source-to-site distance or minimum map distance for the above linear sources for study area was measured using seismotectonic maps as shown in **Figure 1** and tabulated in **Appendix Tables A1–A4**.

For estimation of seismic parameters, the linear least-square fit method developed by Stepp [7] has been applied over the past earthquake data collected from various catalog and research agencies (USGS website and NICE) for all four-district headquarters of north Chhattisgarh. In order to assess the magnitude of completeness, time-magnitude plots of the final catalog have been generated for different time periods as shown in **Figure 2**.

The seismic activity of a region is characterized by the Gutenberg-Richter [8], recurrence relation as given below:

$$\log 10\ (\mathrm{N}) = \mathrm{a–b\ Mw} \tag{1}$$

where N is the number of earthquakes greater than or equal to magnitude m, a denotes the seismicity rate computed by the logarithm of the average number of earthquakes of magnitude M_{min} or M_{max}, and b value characterizes the proportion of large earthquakes relative to small earthquakes [9] as shown in **Figure 3**. For tectonic earthquakes, the b value is to be confined within the range $0.7018 < b < 0.8429$ as tabulated in **Table 1**.

The "b" value assesses the frequency of the occurrence of earthquakes of different sizes. The maximum magnitude is the highest potential of accumulated strain energy to be released in the region or in a seismic source [10]. The maximum probable earthquake is defined as the upper limit of earthquake magnitude for a given entire region and is synonymous with the magnitude of the largest possible earthquake in that region [11]. It assumes a sharp cut-off magnitude at a maximum magnitude, so by definition, no earthquake is to be expected with a magnitude exceeding M_{max} [12]. In the present study, M_{max} is estimated by using two methods. The first method is of Wells and Coppersmith [13] and the second method is of Gupta [14]. In Wells and Coppersmith method, a relation between M_w and surface rupture length (SRL) was developed using reliable source parameters, and this is further applicable to all types of faults, shallow earthquakes, and interplate or intraplate earthquakes.

$$\mathrm{Log\ (SRL)} = 0.57\mathrm{M_w} - 2.33 \tag{2}$$

The above equation was used to estimate the maximum magnitude (Mmax) for all sources of the study area. Gupta's method in which maximum magnitude (Mmax) was estimated from Eq. (3) as given below:

$$\mathrm{M_{max}} = \mathrm{M_{obs+}}0.5 \tag{3}$$

M_{max} = Maximum magnitude
M_{obs} = Observed magnitude

After comparing the outcome of the above two methods, maximum magnitude (M_{max}) values for seismic sources for the district headquarters of north Chhattisgarh are tabulated in Appendix A. The recurrence relation developed in the present study is

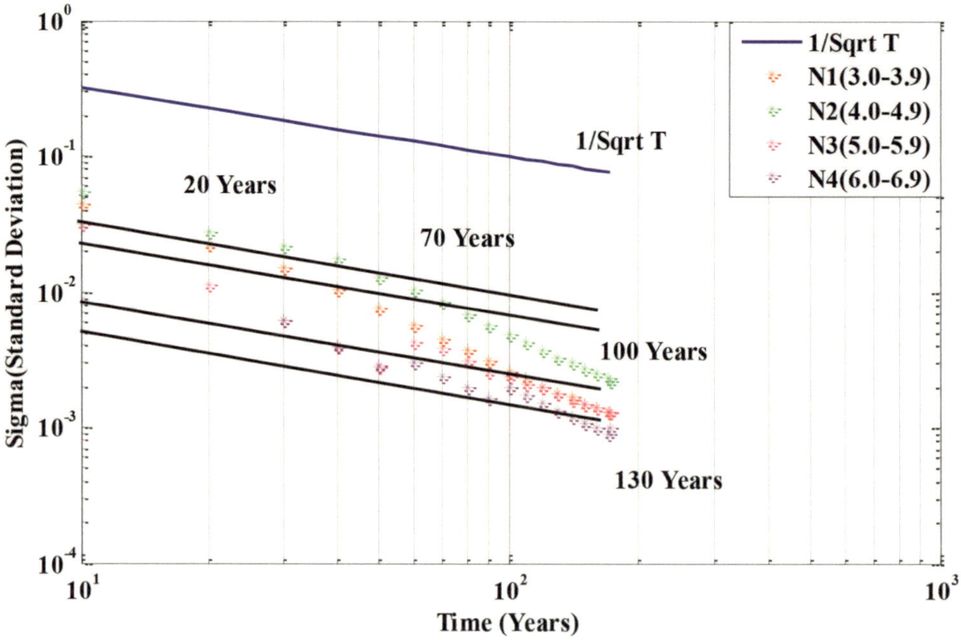
1/Sqrt T
N1(3.0-3.9)
N2(4.0-4.9)
N3(5.0-5.9)
N4(6.0-6.9)
1/Sqrt T
20 Years
70 Years
100 Years
130 Years
Sigma(Standard Deviation)
Time (Years)

(a)

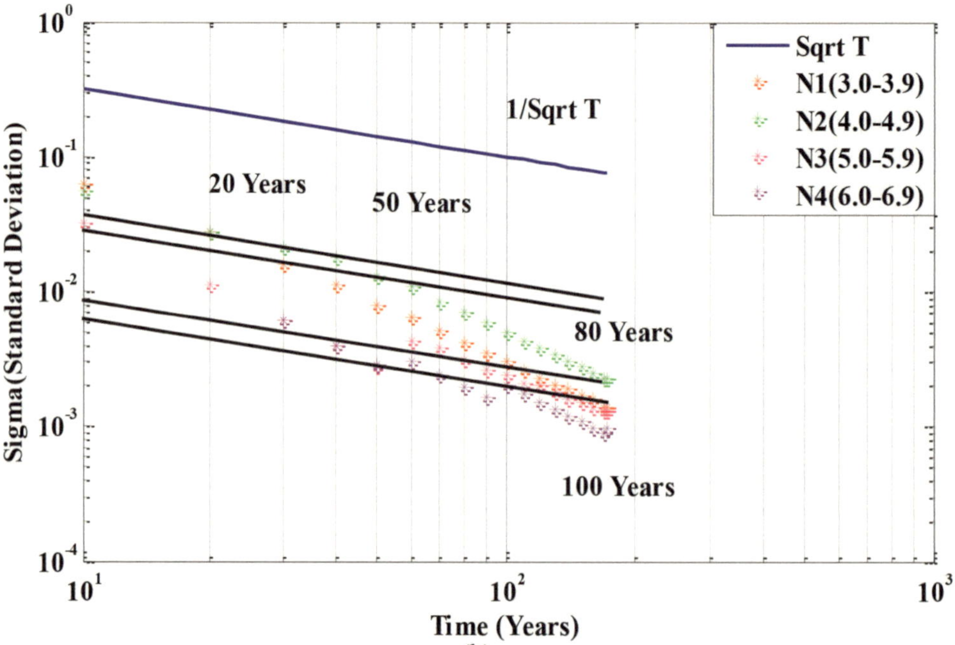
Sqrt T
N1(3.0-3.9)
N2(4.0-4.9)
N3(5.0-5.9)
N4(6.0-6.9)
1/Sqrt T
20 Years
50 Years
80 Years
100 Years
Sigma(Standard Deviation)
Time (Years)

(b)

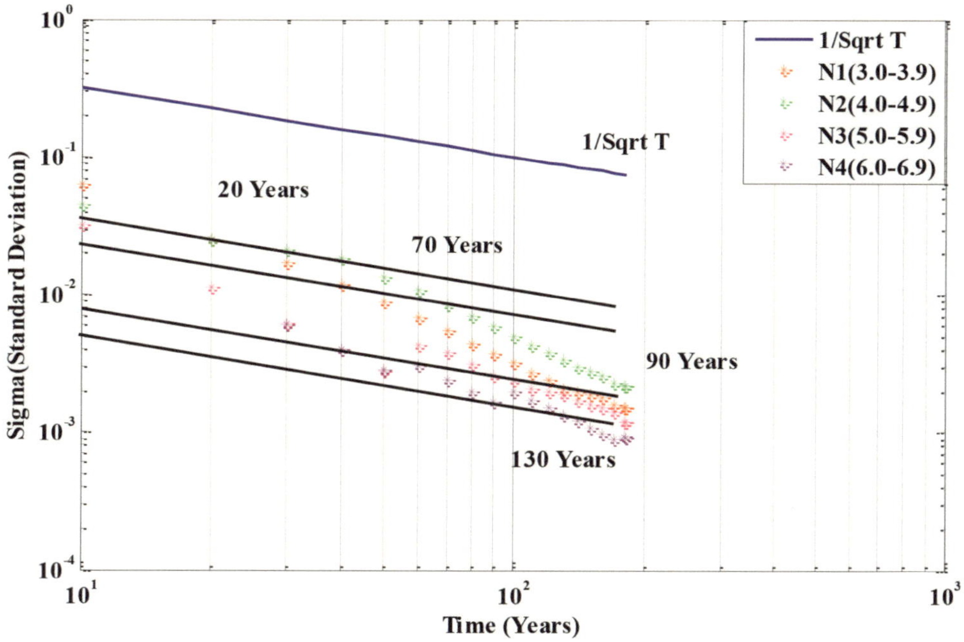

(c)

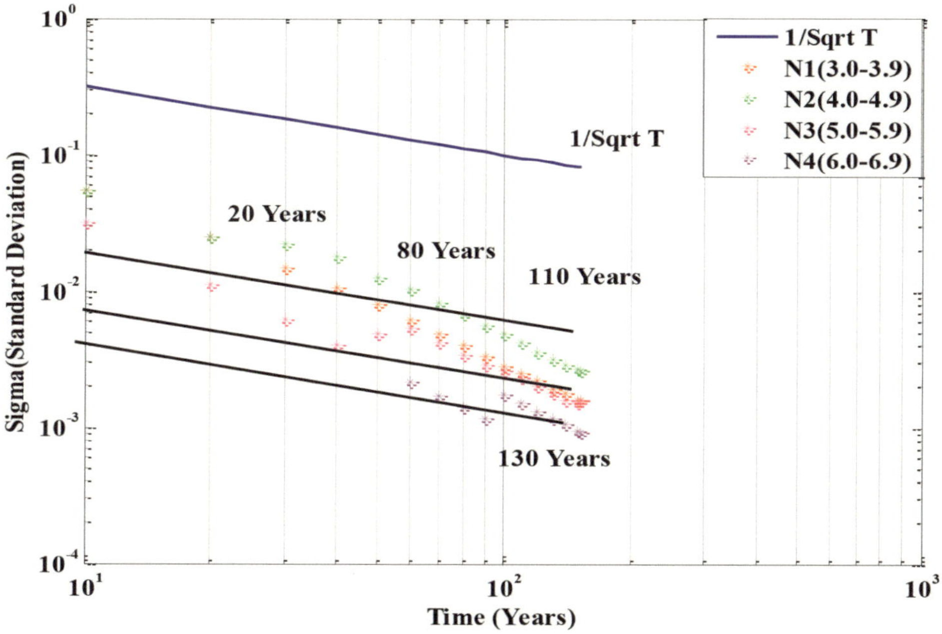

(d)

Figure 2.
Earthquake data completeness analysis for district headquarters of north Chhattisgarh: (a) Ambikapur, (b) Baikunthpur [Koria], (c) Korba, and (d) Jashpurnagar.

not for the particular seismic source but for the entire study region. It is required to differentiate the activity rates among the seismic sources and to develop the frequency magnitude relationship for an individual fault. The truncated exponential recurrence

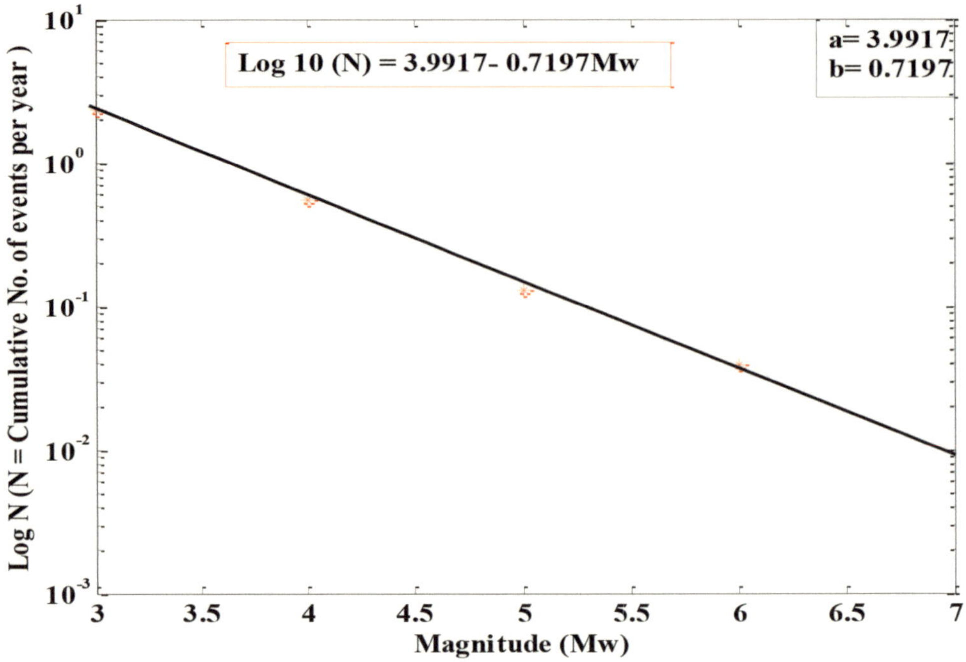
Log 10 (N) = 3.9917- 0.7197Mw
a= 3.9917
b= 0.7197
Log N (N = Cumulative No. of events per year)
Magnitude (Mw)
10^1
10^0
10^-1
10^-2
10^-3
3
3.5
4
4.5
5
5.5
6
6.5
7

(a)

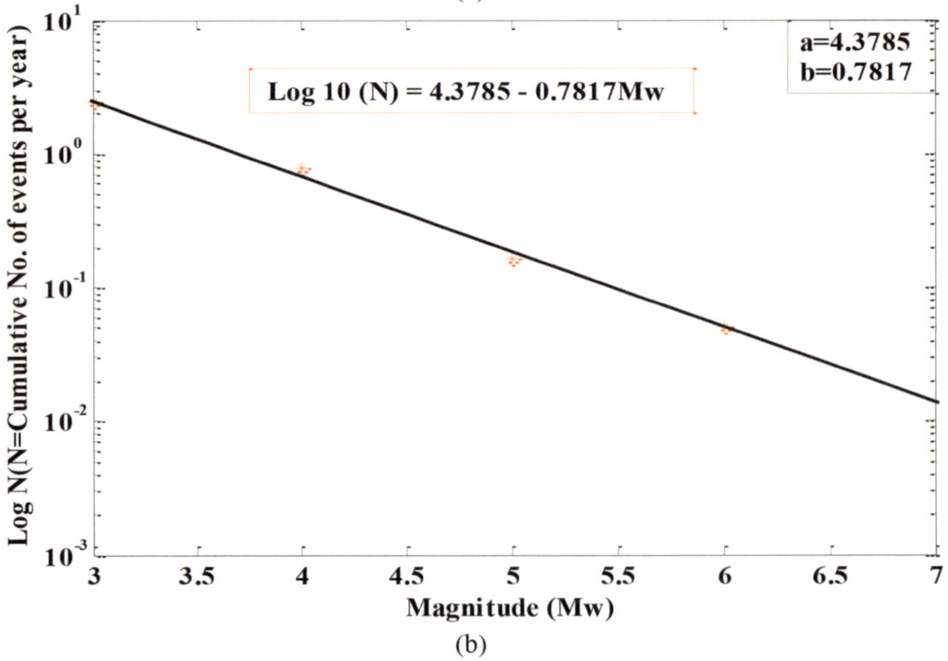
Log 10 (N) = 4.3785 - 0.7817Mw
a=4.3785
b=0.7817
Log N(N=Cumulative No. of events per year)
Magnitude (Mw)
10^1
10^0
10^-1
10^-2
10^-3
3
3.5
4
4.5
5
5.5
6
6.5
7

(b)

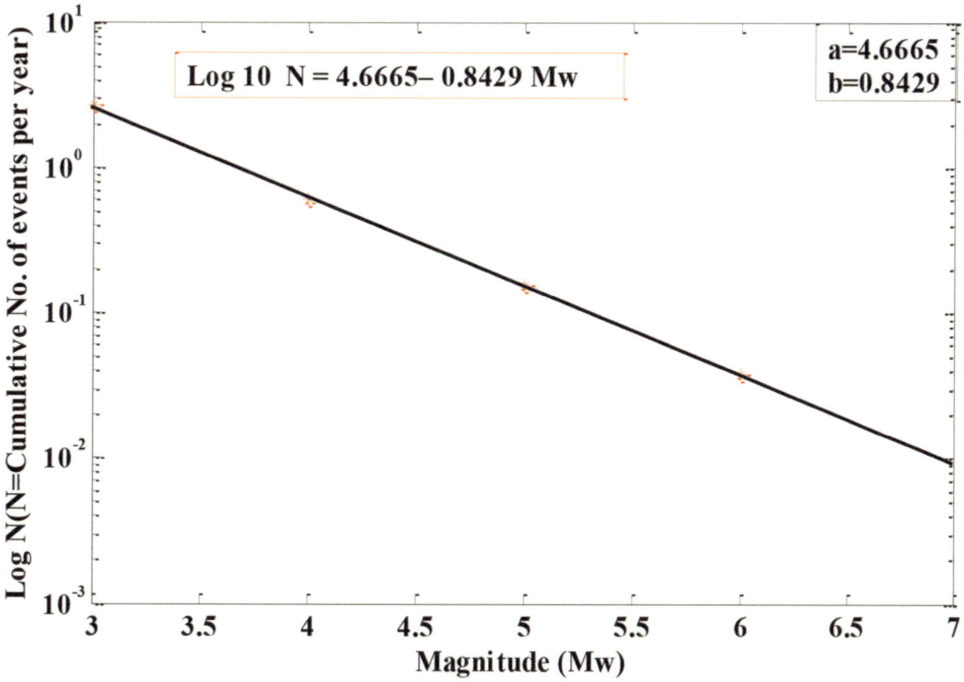

(c)

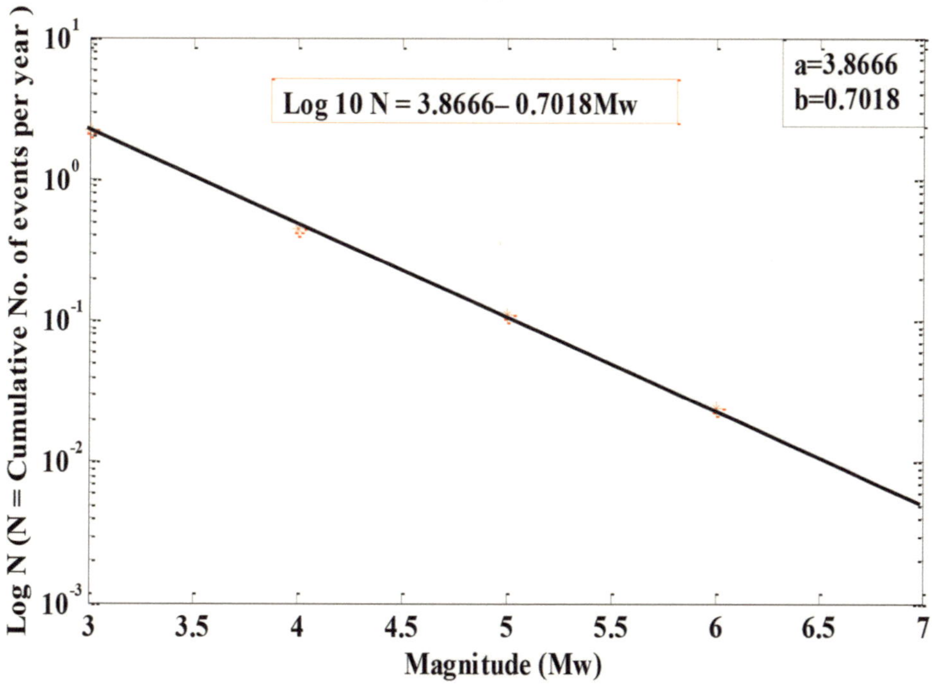

(d)

Figure 3.
Frequency-magnitude relationship for district headquarters of north Chhattisgarh: (a) Ambikapur, (b) Baikunthpur [Koria], (c) Korba, and (d) Jashpurnagar.

Name of district headquarter	regional recurrence relationship for district headquarter	b value considered for the present study
Ambikapur	Log 10 (N) = 3.9917–0.7197 Mw	0.7197
Baikunthpur [Koria]	Log 10 (N) = 4.3785–0.7817 Mw	0.7817
Korba	Log 10 (N) = 4.6665–0.8429 Mw	0.8429
Jashpurnagar	Log 10 (N) = 3.8666–0.7018 Mw	0.7018

Table 1.
Regional recurrence relationship and "b" values for district headquarters of north Chhattisgarh.

relationship is commonly used in practice: to estimate the most likely earthquake magnitude and the most likely source-site distance, for the calculation of PGA values using the ground motion records [15]. This process of disaggregation requires the mean annual rate of exceedance (λm), expressed as a function of magnitude. A MATLAB computer program has been developed to solve Eq. (4) and the graphs were plotted as shown in **Figure 4**.

$$\lambda_m = w_i * \upsilon * \frac{\exp[-\beta(m - m_0) - \exp[-\beta(m_{\max} - m_0)]}{1 - \exp[-\beta(m_{\max} - m_0)]} \tag{4}$$

where $\upsilon = \exp(\alpha - \beta * m_0)$ α = 2.303*a, β = 2.303*b, and w_i is the weight factor for a particular source.

An attenuation relationship includes the source geometry, earthquake magnitude, source-to-site distance, and site conditions. Regional geology plays an important role in the selection of an appropriate relationship in any seismic hazard study. In the recent past, devastating events have occurred in Peninsular India (PI), which is a warning about the possibility of such earthquakes in future [16]. In the present research, the study area comes under the region of PI. So, the attenuation relationship developed by Iyengar and Raghu Kanth [17] has been used. The GMPE proposed by Iyengear and Raghu Kanth is given below:

$$\ln \mathrm{Y} = \mathrm{C1} + \mathrm{C2}\ (\mathrm{M} - 6) + \mathrm{C3}(\mathrm{M} - 6)^2 - \ln(\mathrm{R}) - \mathrm{C4}(\mathrm{R}) + \ln(\varepsilon) \tag{5}$$

where Y, M, and R refer to PGA(g), moment magnitude, and hypo-central distance, respectively.

Peninsular India: C1 = 1.6858; C2 = 0.9241; C3 = −0.0760; C4 = 0.0057;

σ (ln ε) = standard deviation of error = 0 [50 Percentile, for DSHA ε = 0]

σ (ln ε) = standard deviation of error = 0.4648 [84 Percentile]

M100 = magnitude of earthquake [100 years Recurrence—period calculated by using **Figure 4**] R = hypo-central distance = $\sqrt{(D^2+F^2)}$ D = minimum map distance to the sources, and F = focal depth = 10 km.

The peak ground acceleration at bedrock level for all the sources of the study area has been estimated for a return period of 100 years, using the attenuation relationship of Iyengear and Raghu Kanth [17]. The estimated peak ground acceleration values at bedrock level of seismic sources of the district headquarters of north Chhattisgarh are tabulated in **Appendix Tables B1–B4**. It is observed that in the attenuation relationship the highest estimated PGA (g) value is found for fault no. F8 for district headquarter Baikunthpur [Koria] and tabulated in **Table 2**.

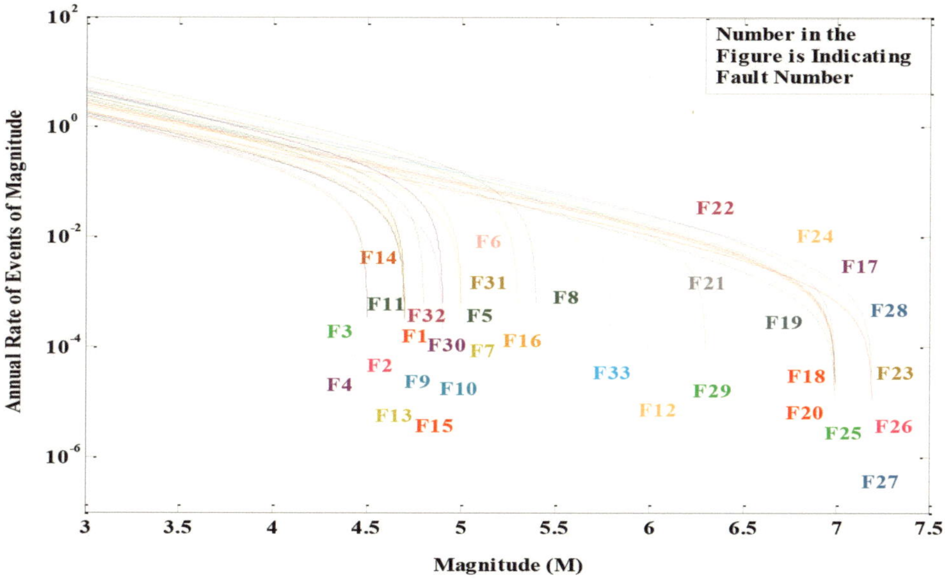
Number in the
Figure is Indicating
Fault Number
Annual Rate of Events of Magnitude
Magnitude (M)
10^{2}
10^{0}
10^{-2}
10^{-4}
10^{-6}
3
3.5
4
4.5
5
5.5
6
6.5
7
7.5
F22
F24
F17
F6
F14
F31
F21
F8
F11
F32
F5
F28
F19
F3
F1
F30
F7
F16
F2
F33
F18
F23
F4
F9
F10
F29
F12
F13
F15
F20
F25
F26
F27

(a)

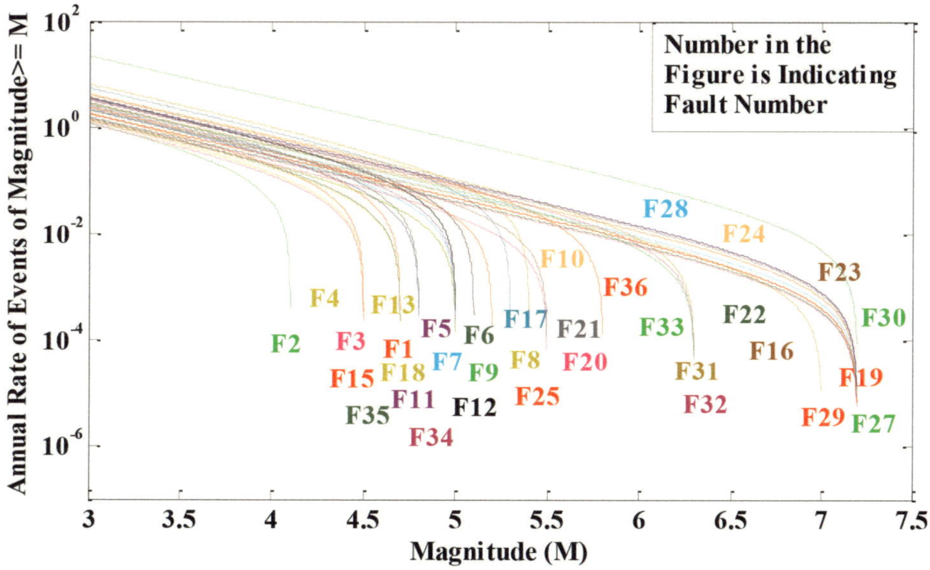
Number in the
Figure is Indicating
Fault Number
Annual Rate of Events of Magnitude>= M
Magnitude (M)
10^{2}
10^{0}
10^{-2}
10^{-4}
10^{-6}
3
3.5
4
4.5
5
5.5
6
6.5
7
7.5
F28
F24
F23
F10
F36
F4
F13
F17
F22
F30
F5
F6
F21
F33
F2
F3
F1
F16
F7
F8
F20
F15
F18
F9
F31
F19
F11
F12
F25
F32
F35
F29
F27
F34

(b)

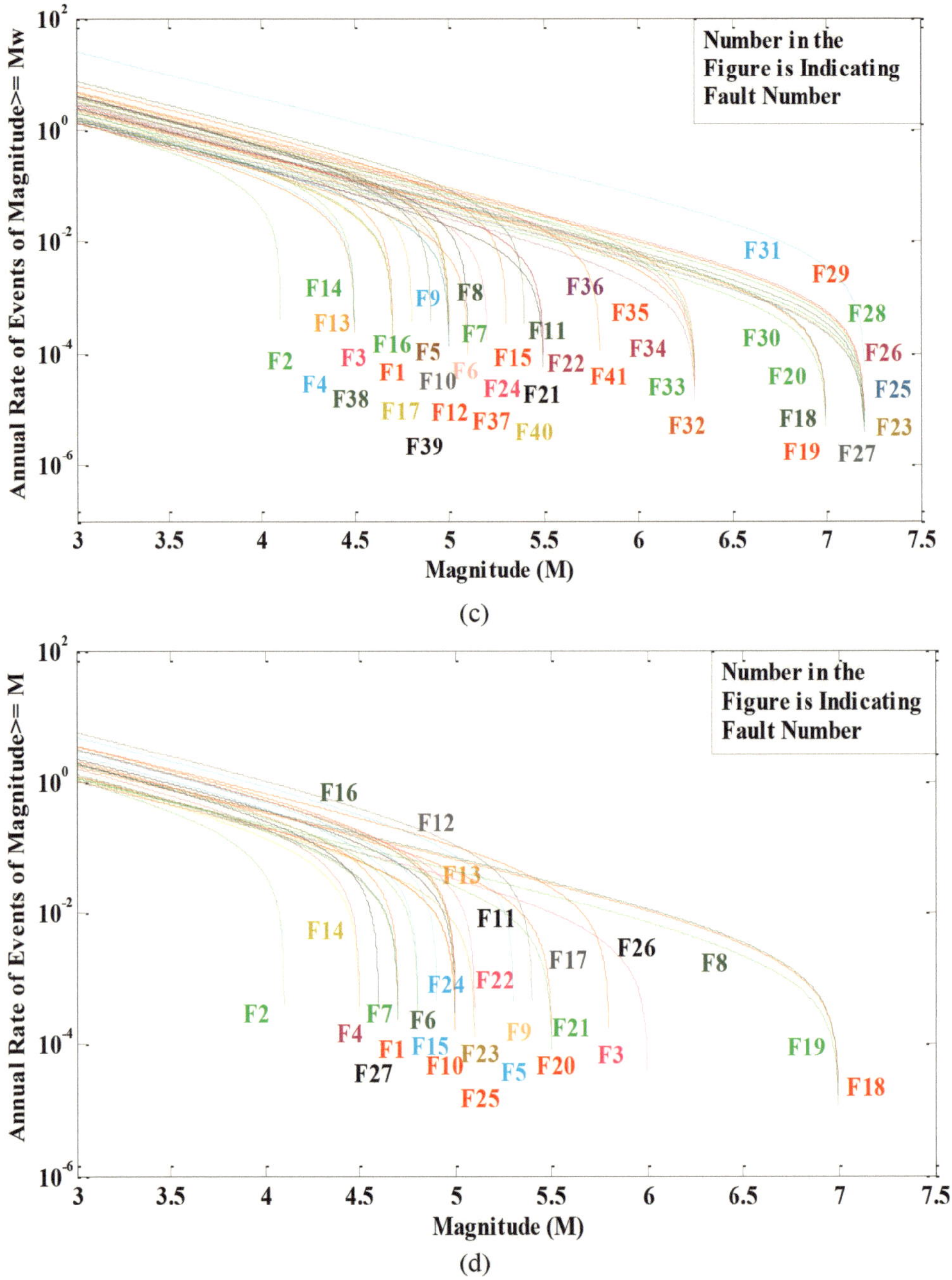

Figure 4.
Disaggregation of seismic sources near district headquarters: (a) Ambikapur, (b) Baikunthpur [Koria], (c) Korba, and (d) Jashpurnagar.

The seismic zonation map of any country acts as a guide to the seismic status of the regions and their susceptibility to earthquakes. India has been divided into five zones with respect to the severity of the earthquake. Zone V is the seismically most active zone, where an earthquake of magnitude 8 or more could occur.

Name of district headquarter	Hypo-central distance	Magnitude M100	PGA(g) value	
			50 percentile	84 percentile
Ambikapur Fault No. F7	14.366	4.741	0.09587	0.15259
Baikunthpur [Koria] Fault No. F8	17.598	5.338	0.14554	0.23165
Korba Fault No. F6	92.803	4.910	0.01144	92.803
Jashpurnagar Fault No. F9	109.015	6.709	0.04929	0.07846

Table 2.
Maximum PGA values for seismic sources for district headquarters of north Chhattisgarh.

Recent strong motion observations around the world have revolutionized the thinking on the aspect of the design of engineering structures, placing emphasis on the characteristics of the structures. BIS 1893 (Part 1): 2016 (sixth revision) [18], prepared a seismic zone map of India as shown in **Figure 5** with zone factors that are tabulated in **Table 3**.

The outcome of the present study with the recommended value for zone II is 0.1g as per IS 1893 (Part 1): 2016 (sixth revision). The maximum PGA (g) value for 50 percentile for a return period 100 years for Baikunthpur [Koria] is more and is 0.14554 g for Iyengear and Raghu Kanth's model.

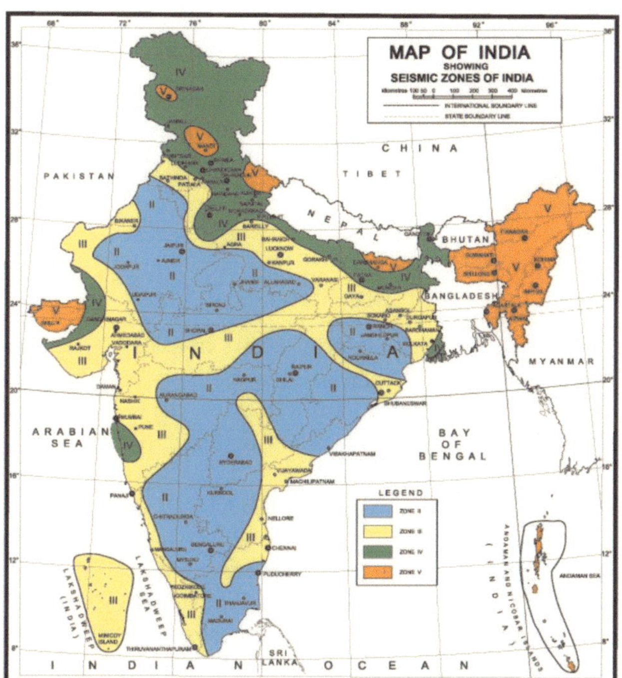

Figure 5.
Indian seismic zone map as per BIS 189 (Part 1): 2016 [Map taken from BIS 1893 (Part 1): 2016, sixth revision].

Seismic zone	II	III	IV	V
Seismic intensity	Low	Moderate	Severe	Very severe
Z	0.10	0.16	0.24	0.36

Table 3.
Seismic zone factor, Z, IS 1893 (Part 1): 2016 (Clause 6.4.2).

Figure 6.
Chhattisgarh district map.

As **Figure 6** (i) shows the district headquarters map of Chhattisgarh and the study area is highlighted by a rectangle. Using the outcome of the DSHA approach and IS 1893 (Part 1): 2016 (sixth revision) recommendation the seismic zone map for north Chhattisgarh has been developed and as shown in **Figure 7a** and **b**.

2.2 Probabilistic seismic hazard analysis (PSHA)

Thus, seismic hazard estimation has been performed considering the classical Cornell [19] approach. This approach comprises considering all potential seismic sources and their activity rate. Many researchers have considered this approach to carry out the hazard analysis of different regions of India [20–27]. The seismic hazard curves can be used to evaluate the probability of ground motion exceedance for a specified return period. Seismic hazard curves can be obtained for a particular seismic source and are combined to express aggregate hazards at a particular site. The probability of exceedance can be written as:

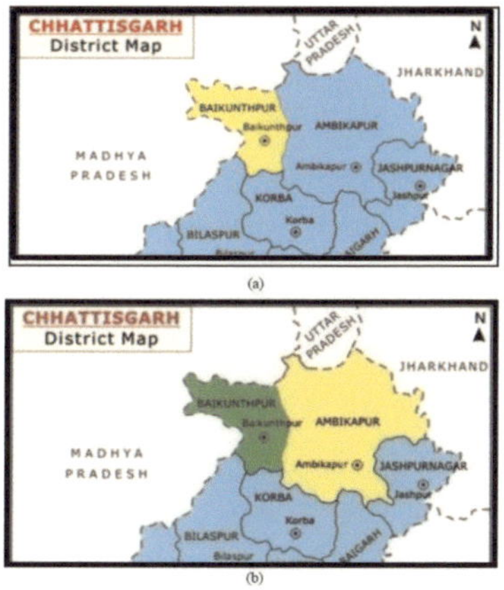

Figure 7.
Seismic zone map of north Chhattisgarh. (a) PGA (g) values for 50 percentile. (b) PGA (g) values for 84 percentile.

$$\lambda_{y*} = \sum_{i=1}^{N_s} \nu_i \int\int P[Y > y* | m, r] f_{Mi}(m) f_{Ri}(r) dmdr \tag{6}$$

The final PSHA equation is given by

$$\lambda_{y*} = \sum_{i=1}^{N_S} \sum_{j=1}^{N_M} \sum_{k=1}^{N_R} \nu_i \int\int P[Y > y* | m_j, r_k] P[M = m_j] P[R = r_k] \tag{7}$$

where P[Y > y* | m, r] is obtained from the predictive relationship and the probability density functions for magnitude and distance, respectively. The mean annual rate of exceedance λy* is computed at a site for different specified ground motion values y* in a life period of time. A computer program has been developed in MATLAB, which is used to draw the seismic hazard curves for the major district headquarters of north Chhattisgarh. The curves are depicted in **Figure 8**. Using Iyengear and Raghu Kanth's [17] attenuation relationship, the peak ground accelerations at bedrock level for 2% and 10% probability of exceedance in 50 years have been computed [28]. The aggregate value of PGA from this relationship has been worked out for each district headquarter (**Figure 9**). For 2% and 10% probability of exceedance in 50 years the PGA values for district headquarters Korba, Jashpurnagar, Ambikapur, and Baikunthpur [Koria] ranges from 0.016 to 0.086 g and from 0.008 to 0.036 g, respectively. The outcome highlighted that the seismically activity district headquarters are Ambikapur and Baikunthpur [Koria]. Thus, for PGA values 0.05 g, 0.10 g, and 0.015 g at bedrock level, the return periods have been estimated for Ambikapur and Baikunthpur [Koria] and have been tabulated in **Table 4**.

3. Conclusions

After carrying out the deterministic and probabilistic seismic hazard analysis for major district headquarters of north Chhattisgarh, considering the local site effects, the following conclusions were drawn and discussed below:

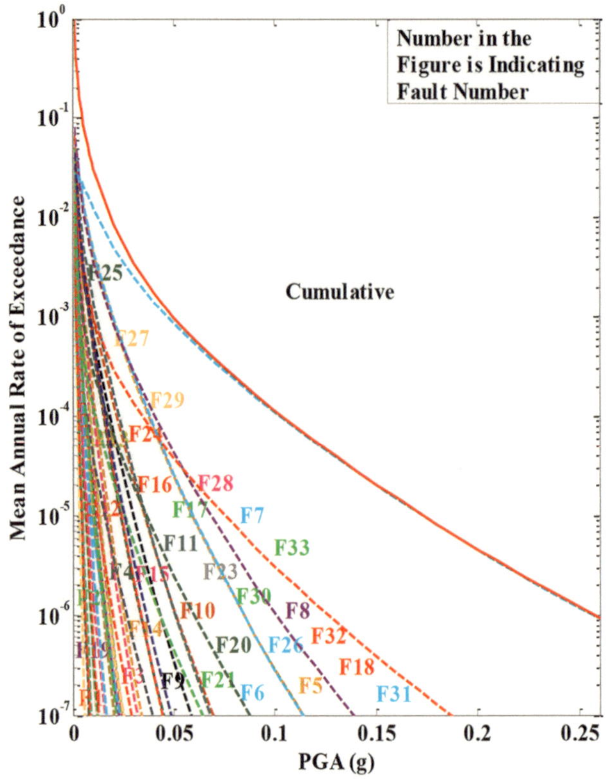
Number in the Figure is Indicating Fault Number
Cumulative
Mean Annual Rate of Exceedance
PGA (g)
F25
F27
F29
F24
F16
F28
F17
F7
F33
F11
F23
F30
F8
F10
F32
F20
F26
F18
F21
F6
F5
F31
0
0.05
0.1
0.15
0.2
0.25

(a)

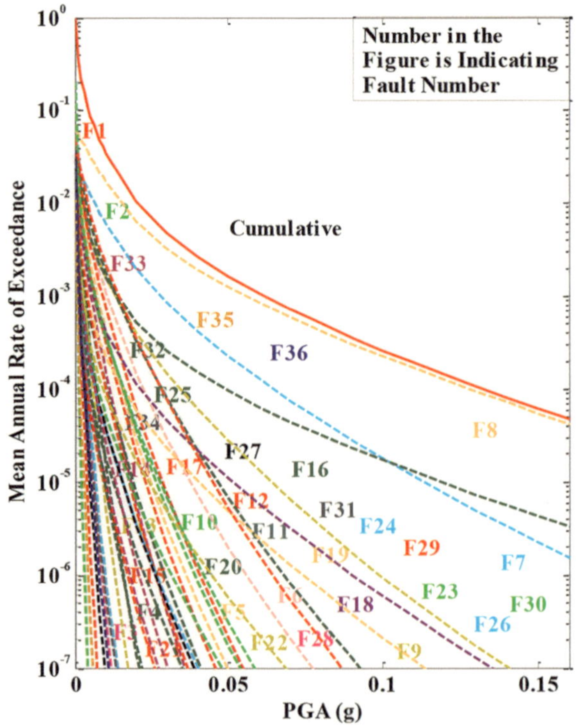
Number in the Figure is Indicating Fault Number
Cumulative
Mean Annual Rate of Exceedance
PGA (g)
F1
F2
F33
F35
F36
F32
F25
F8
F27
F16
F17
F12
F31
F24
F10
F11
F29
F7
F20
F23
F30
F18
F26
F22
F28
F9
0
0.05
0.1
0.15

(b)

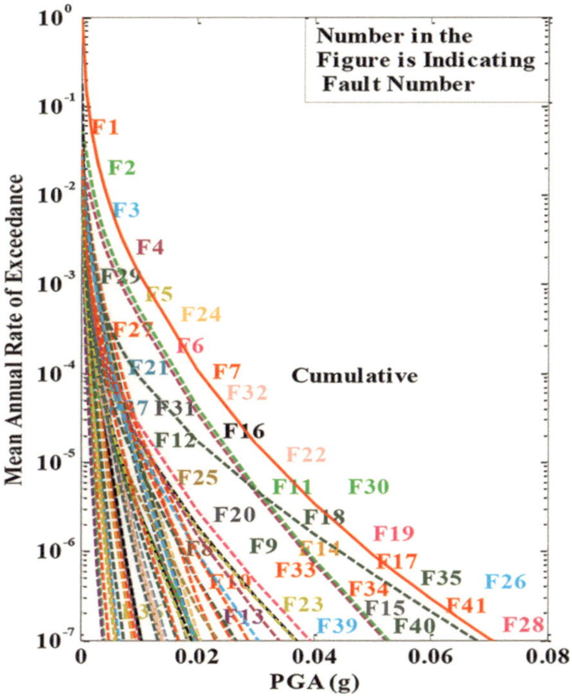

(c)

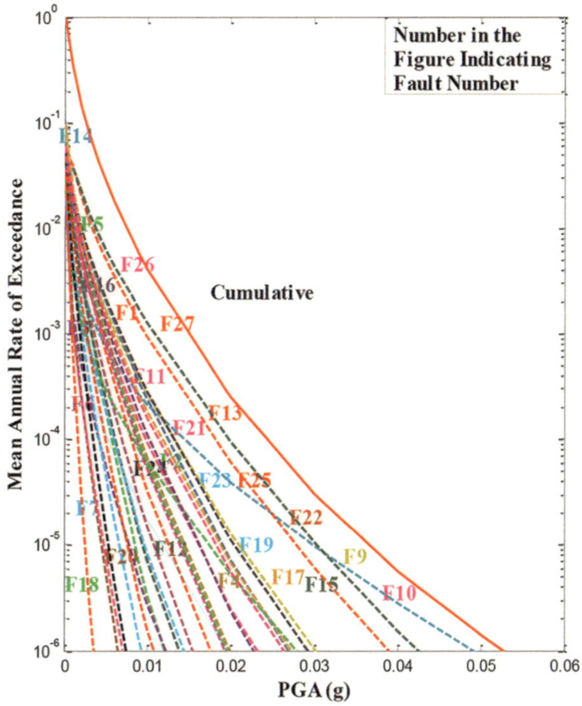

(d)

Figure 8.
Seismic hazard curves for district headquarters of north Chhattisgarh: (a) Ambikapur, (b) Baikunthpur [Koria], (c) Korba, and (d) Jashpurnagar.

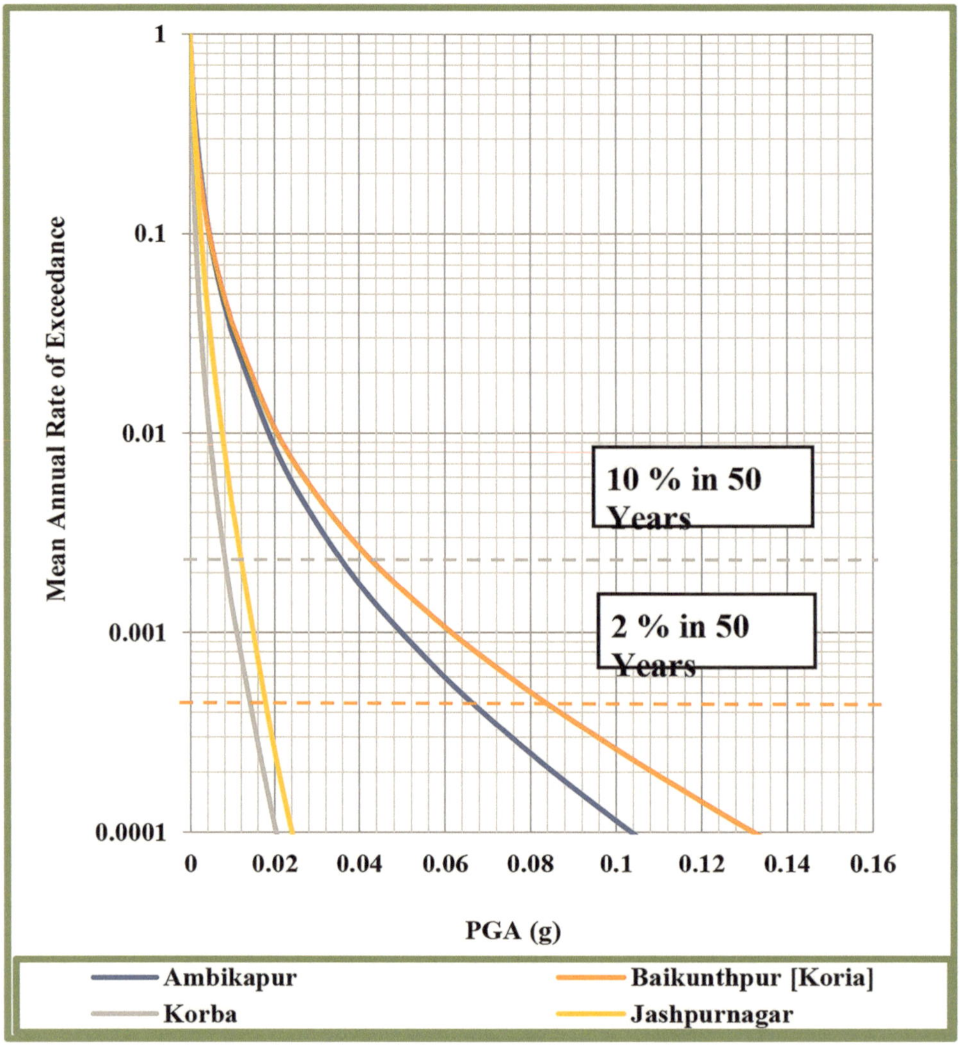

Figure 9.
Seismic hazard curves for district headquarters of north Chhattisgarh.

Name of district headquarter	For PGA(g) the return periods in years		
	0.05g	**0.10g**	**0.15g**
Ambikapur	1017	8788	48781
Baikunthpur [Koria]	614	3848	16364

Table 4.
Return periods for various PGA (g) values for district headquarters of north Chhattisgarh.

The frequency magnitude relationship was established from the research, for the study area, after carrying out the completeness analysis as per Stepp. Completeness of the data was observed for the north Chhattisgarh region and a seismic hazard parameter of "b value" estimated was found to vary from 0.7018 to

0.8429. The maximum value of peak ground acceleration (PGA) for a recurrence period of 100 years for the headquarter of Ambikapur was found to be due to fault F7 having a fault length of 46 km with a hypo-central distance of 14.366 km, and was 0.09587 g for 50 percentile and 0.15259 g for 84 percentile. On the other hand, the maximum value of peak ground acceleration (PGA), for the recurrence period of 100 years for the headquarter Baikunthpur [Koria], which was found to be due to fault F8, having fault length of 140 km, hypo-central distance of 17.598 km, and was estimated to be equal to 0.14554 g for 50 percentile and 0.23165 g for 84 percentile. For Korba and Jashpurnagar, the peak ground acceleration (PGA) for a recurrence period of 100 years found to be for fault F6, fault length of 46 km, and F9, fault length of 30 km are respectively very low. Using PSHA approach, a seismic hazard curves have been prepared for the major district headquarters of north Chhattisgarh. For Baikunthpur [Koria], the estimated return was found to be a period of 3848 year for 0.1g. The outcome of the present study is directly used for the design of earthquake resistance structures.

A. Appendix

Fault No.	Fault length (Li) in km	Mw observed in the fault	Maximum magnitude		Maximum magnitude	Weight factor $w_i = \frac{L_i}{\sum L_i}$
			Method A	Method B		
F1	51	3.3	4.7	3.8	4.7	0.0293
F2	26	3.3	4.1	3.8	4.1	0.0149
F3	25	4.0	4.1	4.5	4.5	0.0144
F4	28	4.0	4.2	4.5	4.5	0.0161
F5	62	4.0	4.8	4.5	4.8	0.0356
F6	77	4.5	5.0	5.0	5.0	0.0442
F7	46	4.3	4.6	4.8	4.8	0.0264
F8	140	4.5	5.4	5.0	5.4	0.0803
F9	30	4.5	4.3	5.0	5.0	0.0172
F10	30	4.5	4.3	5.0	5.0	0.0172
F11	55	4.5	4.7	5.0	5.0	0.0316
F12	25	4.0	4.1	4.5	4.5	0.0144
F13	39	4.0	4.4	4.5	4.5	0.0224
F14	32	4.2	4.3	4.7	4.7	0.0184
F15	30	4.2	4.3	4.7	4.7	0.0172
F16	117	3.8	5.3	4.3	5.3	0.0671
F17	78	4.5	5.0	5.0	5.0	0.0447
F18	45	6.5	4.6	7.0	7.0	0.0258
F19	28	5.0	4.2	5.5	5.5	0.0161
F20	42	6.5	4.5	7.0	7.0	0.0241
F21	28	6.5	4.2	7.0	7.0	0.0161
F22	47	5.0	4.6	5.5	5.5	0.0270

Fault No.	Fault length (Li) in km	Mw observed in the fault	Maximum magnitude		Maximum magnitude	Weight factor $wi = \frac{L_i}{\sum L_i}$
			Method A	Method B		
F23	33	6.7	4.3	7.2	7.2	0.0190
F24	60	6.7	4.8	7.2	7.2	0.0344
F25	51	4.7	4.7	5.2	5.2	0.0293
F26	31	6.7	4.3	7.2	7.2	0.0178
F27	70	6.7	4.9	7.2	7.2	0.0402
F28	70	5.8	4.9	6.3	6.3	0.0402
F29	75	4.4	4.9	4.9	4.9	0.0430
F30	26	4.6	4.1	5.1	5.1	0.0149
F31	86	4.1	5.0	4.6	5.0	0.0493
F32	75	4.6	4.9	5.1	5.1	0.0430
F33	87	5.3	5.1	5.8	5.8	0.0499

Table A1.
Estimation of maximum magnitude for sources for district headquarter Ambikapur.

Fault No.	Fault length (Li) in km	Mw observed in the fault	Maximum magnitude		Maximum magnitude	Weight factor $wi = \frac{L_i}{\sum L_i}$
			Method A	Method B		
F1	51	3.3	4.7	3.8	4.7	0.0218
F2	26	3.3	4.1	3.8	4.1	0.0111
F3	25	4.0	4.1	4.5	4.5	0.0107
F4	28	4.0	4.2	4.5	4.5	0.0120
F5	62	4.0	4.8	4.5	4.8	0.0265
F6	77	4.5	5.0	5.0	5.0	0.0329
F7	46	4.3	4.6	4.8	4.8	0.0197
F8	140	4.5	5.4	5.0	5.4	0.0597
F9	30	4.5	4.3	5.0	5.0	0.0128
F10	30	4.5	4.3	5.0	5.0	0.0128
F11	55	4.5	4.7	5.0	5.0	0.0235
F12	78	4.5	5.0	5.0	5.0	0.0333
F13	39	4.0	4.4	4.5	4.5	0.0167
F14	32	4.2	4.3	4.7	4.7	0.0137
F15	30	4.2	4.3	4.7	4.7	0.0128
F16	45	6.5	4.6	7.0	7.0	0.0192
F17	117	4.6	5.3	5.1	5.3	0.0499
F18	42	6.5	4.5	7.0	7.0	0.0180
F19	28	6.5	4.2	7.0	7.0	0.0120
F20	28	5.0	4.2	5.5	5.5	0.0120

Fault No.	Fault length (Li) in km	Mw observed in the fault	Maximum magnitude		Maximum magnitude	Weight factor $w_i = \frac{L_i}{\sum L_i}$
			Method A	Method B		
F21	47	5.0	4.6	5.5	5.5	0.0201
F22	33	6.7	4.3	7.2	7.2	0.0141
F23	60	6.7	4.8	7.2	7.2	0.0256
F24	70	6.7	4.9	7.2	7.2	0.0299
F25	51	4.7	4.7	5.2	5.2	0.0218
F26	31	6.7	4.3	7.2	7.2	0.0133
F27	76	6.7	4.9	7.2	7.2	0.0324
F28	47	6.7	4.6	7.2	7.2	0.0201
F29	38	6.7	4.4	7.2	7.2	0.0162
F30	477	6.7	6.3	7.2	7.2	0.2034
F31	91	5.8	5.1	6.3	6.3	0.0388
F32	70	5.8	4.9	6.3	6.3	0.0299
F33	58	5.8	4.8	6.3	6.3	0.0248
F34	75	4.6	4.9	5.1	5.1	0.032
F35	26	4.6	4.1	5.1	5.1	0.0111
F36	87	5.3	5.1	5.8	5.8	0.0371

Table A2.
Estimation of maximum magnitude for sources for district headquarter Baikunthpur [Koria].

Fault No.	Fault length (Li) in km	Mw observed in the fault	Maximum magnitude		Maximum magnitude	Weight factor $w_i = \frac{L_i}{\sum L_i}$
			Method A	Method B		
F1	51	3.3	4.7	3.8	4.7	0.0197
F2	26	3.3	4.1	3.8	4.1	0.0100
F3	25	4.0	4.1	4.5	4.5	0.0097
F4	28	4.0	4.2	4.5	4.5	0.0108
F5	62	4.0	4.8	4.5	4.8	0.0239
F6	46	4.5	4.6	5.0	5.0	0.0177
F7	77	4.5	5.0	5.0	5.0	0.0296
F8	30	4.5	4.3	5.0	5.0	0.0116
F9	30	4.5	4.3	5.0	5.0	0.0116
F10	55	4.5	4.7	5.0	5.0	0.0212
F11	140	4.5	5.4	5.0	5.4	0.0539
F12	78	4.5	5.0	5.0	5.0	0.0300
F13	25	4.0	4.1	4.5	4.5	0.0097
F14	39	4.0	4.4	4.5	4.5	0.0150
F15	117	4.6	5.3	5.1	5.3	0.0450

Fault No.	Fault length (Li) in km	Mw observed in the fault	Maximum magnitude		Maximum magnitude	Weight factor $wi = \frac{L_i}{\sum L_i}$
			Method A	Method B		
F16	32	4.2	4.3	4.7	4.7	0.0123
F17	30	4.2	4.3	4.7	4.7	0.0116
F18	45	6.5	4.6	7.0	7.0	0.0173
F19	42	6.5	4.5	7.0	7.0	0.0162
F20	28	6.5	4.2	7.0	7.0	0.0108
F21	28	5.0	4.2	5.5	5.5	0.0108
F22	47	5.0	4.6	5.5	5.5	0.0181
F23	33	6.7	4.3	7.2	7.2	0.0127
F24	51	4.7	4.7	5.2	5.2	0.0197
F25	31	6.7	4.3	7.2	7.2	0.0120
F26	76	6.7	4.9	7.2	7.2	0.0293
F27	47	6.7	4.6	7.2	7.2	0.0181
F28	60	6.7	4.8	7.2	7.2	0.0231
F29	70	6.7	4.9	7.2	7.2	0.0270
F30	38	6.7	4.4	7.2	7.2	0.0147
F31	477	6.7	6.3	7.2	7.2	0.1834
F32	91	5.8	5.1	6.3	6.3	0.0350
F33	70	5.8	4.9	6.3	6.3	0.0270
F34	58	5.8	4.8	6.3	6.3	0.0223
F35	45	5.8	4.6	6.3	6.3	0.0173
F36	25	5.8	4.1	6.3	6.3	0.0097
F37	26	4.6	4.1	5.1	5.1	0.0100
F38	75	4.4	4.9	4.9	4.9	0.0289
F39	75	4.6	4.9	5.1	5.1	0.0289
F40	86	4.1	5.0	4.6	5.0	0.0331
F41	87	5.3	5.1	5.8	5.8	0.0335

Table A3.
Estimation of maximum magnitude for sources for district headquarter Korba.

Fault No.	Fault length (Li) in km	Mw observed in the fault	Maximum magnitude		Maximum magnitude	Weight factor $wi = \frac{L_i}{\sum L_i}$
			Method A	Method B		
F1	51	3.3	4.7	3.8	4.7	0.0358
F2	26	3.3	4.1	3.8	4.1	0.0183
F3	25	5.5	4.1	6.0	60	0.0176
F4	39	4.0	4.4	4.5	4.5	0.0274
F5	117	4.6	5.3	5.1	5.3	0.0821
F6	32	4.2	4.3	4.7	4.7	0.0225

Fault No.	Fault length (Li) in km	Mw observed in the fault	Maximum magnitude		Maximum magnitude	Weight factor $wi = \frac{L_i}{\sum L_i}$
			Method A	Method B		
F7	30	4.2	4.3	4.7	4.7	0.0211
F8	45	6.5	4.6	7.0	7.0	0.0316
F9	30	4.5	4.3	5.0	5.0	0.0211
F10	30	4.5	4.3	5.0	5.0	0.0211
F11	55	4.5	4.7	5.0	5.0	0.0386
F12	78	4.5	5.0	5.0	5.0	0.0547
F13	77	4.5	5.0	5.0	5.0	0.0540
F14	28	4.0	4.2	4.5	4.5	0.0197
F15	62	4.0	4.8	4.5	4.8	0.0435
F16	46	4.3	4.6	4.8	4.8	0.0323
F17	140	4.5	5.4	5.0	5.4	0.0982
F18	42	6.5	4.5	7.0	7.0	0.0295
F19	28	6.5	4.2	7.0	7.0	0.0197
F20	47	5.0	4.6	5.5	5.5	0.0330
F21	28	5.0	4.2	5.5	5.5	0.0197
F22	75	4.6	4.9	5.1	5.1	0.0526
F23	26	4.6	4.1	5.1	5.1	0.0183
F24	75	4.4	4.9	4.9	4.9	0.0526
F25	86	4.1	5.0	4.6	5.0	0.0604
F26	87	5.3	5.1	5.8	5.8	0.0611
F27	47	3.0	4.6	3.5	4.6	0.0330

Table A4.
Estimation of maximum magnitude for sources for district headquarter Jashpurnagar.

B. Appendix

Fault No.	Fault length (Li) in km	Hypo-central distance (R) in km	Magnitude M100 [recurrence period—100 years]	PGA (g)	
				50 percentile	84 percentile
F1	51	188.197	4.654	0.00247	0.00393
F2	26	169.283	4.069	0.00154	0.00245
F3	25	109.722	4.436	0.00515	0.00820
F4	28	73.053	4.443	0.00962	0.01530
F5	62	43.964	4.755	0.02688	0.04279
F6	77	55.437	4.946	0.02463	0.03921
F7	46	14.366	4.741	0.09587	0.15259
F8	140	53.374	5.342	0.03930	0.06255

Fault No.	Fault length (Li) in km	Hypo-central distance (R) in km	Magnitude M100 [recurrence period—100 years]	PGA (g)	
				50 percentile	84 percentile
F9	30	85.839	4.874	0.01237	0.01969
F10	30	76.906	4.978	0.01627	0.02589
F11	55	80.845	4.926	0.01430	0.02276
F12	25	163.333	4.434	0.00255	0.00405
F13	39	114.441	4.458	0.00494	0.00785
F14	32	108.345	4.629	0.00656	0.01045
F15	30	101.339	4.625	0.00727	0.01157
F16	117	140.470	5.245	0.00823	0.01309
F17	78	95.004	4.950	0.01152	0.01834
F18	45	96.767	5.996	0.03201	0.05095
F19	28	188.56	5.340	0.00514	0.00818
F20	42	162.653	5.960	0.01266	0.02014
F21	28	181.726	5.760	0.00841	0.01339
F22	47	202.013	5.327	0.00439	0.00698
F23	32	243.519	5.848	0.00480	0.00764
F24	60	276.16	6.179	0.00477	0.00759
F25	51	272.046	5.094	0.00172	0.00273
F26	30	299.432	5.839	0.00282	0.00448
F27	70	294.673	6.255	0.00431	0.00685
F28	70	287.316	5.92	0.00340	0.00540
F29	75	282.662	4.854	0.00120	0.00191
F30	26	230.742	4.936	0.00216	0.00343
F31	86	182.893	4.954	0.00365	0.00580
F32	75	258.258	5.039	0.00184	0.00293
F33	87	213.585	5.638	0.00530	0.00844

Table B1.
PGA values for seismic sources for district headquarter Ambikapur.

Fault No.	Fault length (Li) in km	Hypo-central distance (R) in km	Magnitude M100 [recurrence period—100 Years]	PGA(g)	
				50 percentile	84 percentile
F1	51	284.311	4.654	0.00095	0.00151
F2	26	226.875	4.071	0.00083	0.00132
F3	25	159.071	4.436	0.00269	0.00427
F4	28	113.997	4.444	0.00489	0.00778
F5	62	75.268	4.753	0.01311	0.02086
F6	77	61.487	4.946	0.02146	0.03415

Fault No.	Fault length (Li) in km	Hypo-central distance (R) in km	Magnitude M100 [recurrence period—100 Years]	PGA(g)	
				50 percentile	84 percentile
F7	46	18.688	4.739	0.07174	0.11419
F8	140	17.598	5.338	0.14554	0.23165
F9	30	113.706	4.872	0.00795	0.01265
F10	30	85.684	4.784	0.01123	0.01788
F11	55	58.705	4.982	0.02374	0.03778
F12	78	70.340	4.948	0.01787	0.02845
F13	39	130.700	4.458	0.00394	0.00627
F14	32	93.850	4.629	0.00823	0.01310
F15	30	76.359	4.623	0.01110	0.01766
F16	45	63.914	5.892	0.05304	0.08443
F17	117	115.568	5.238	0.01144	0.01821
F18	42	125.292	5.852	0.01837	0.02923
F19	28	125.425	5.644	0.01501	0.02389
F20	28	130.393	5.213	0.00908	0.01445
F21	47	143.071	5.310	0.00851	0.01355
F22	32	176.807	5.764	0.00893	0.01420
F23	60	209.857	6.053	0.00817	0.01300
F24	70	228.330	6.142	0.00733	0.01166
F25	51	205.760	5.092	0.00330	0.00525
F26	31	233.794	5.743	0.00478	0.00761
F27	76	240.422	6.146	0.00652	0.01037
F28	47	266.593	5.940	0.00419	0.00667
F29	38	270.842	5.852	0.00371	0.00590
F30	477	268.928	6.860	0.00907	0.01444
F31	91	290.955	5.957	0.00340	0.00541
F32	70	235.202	5.861	0.00528	0.00840
F33	58	257.776	5.808	0.00403	0.00641
F34	75	230.631	5.035	0.00241	0.00383
F35	26	248.655	4.936	0.00181	0.00288
F36	87	275.870	5.624	0.00284	0.00452

Table B2.
PGA values for seismic sources for district headquarter Baikunthpur [Koria].

Fault No.	Fault length (Li) in km	Hypo-central distance (R) in km	Magnitude M100 [recurrence period—100 Years]	PGA(g)	
				50 percentile	84 percentile
F1	51	274.332	4.652	0.00104	0.00164
F2	26	260.081	4.074	0.00060	0.00095
F3	25	209.347	4.439	0.00154	0.00244
F4	28	174.787	4.443	0.00225	0.00358
F5	62	141.854	4.751	0.00475	0.00756
F6	46	92.803	4.910	0.01144	0.01820
F7	77	143.137	4.946	0.00579	0.00921
F8	30	186.156	4.868	0.00320	0.00509
F9	30	175.351	4.868	0.00362	0.00575
F10	55	158.859	4.922	0.00465	0.00739
F11	140	122.891	5.330	0.01135	0.01805
F12	78	168.242	4.944	0.00426	0.00678
F13	25	265.317	4.436	0.00088	0.00140
F14	39	212.380	4.458	0.00153	0.00242
F15	117	216.339	5.233	0.00343	0.00545
F16	32	190.606	4.629	0.00234	0.00371
F17	30	174.746	4.623	0.00277	0.00440
F18	45	163.464	5.788	0.01066	0.01696
F19	42	224.732	5.736	0.00520	0.00827
F20	28	217.662	5.540	0.00462	0.00734
F21	28	220.173	5.183	0.00313	0.00497
F22	47	231.025	5.277	0.00309	0.00491
F23	32	232.361	5.654	0.00445	0.00708
F24	51	244.908	5.081	0.00219	0.00348
F25	31	264.518	5.608	0.00311	0.00495
F26	76	250.588	6.045	0.00539	0.00856
F27	47	289.607	5.835	0.00307	0.00488
F28	60	249.641	5.936	0.00491	0.00781
F29	70	266.525	6.007	0.00447	0.00710
F30	38	297.601	5.717	0.00255	0.00405
F31	477	293.834	6.751	0.00660	0.01050
F32	91	278.291	5.874	0.00353	0.00562
F33	70	212.717	5.798	0.00625	0.00994
F34	58	212.942	5.739	0.00589	0.00937
F35	45	262.893	5.660	0.00333	0.00529
F36	25	256.840	5.409	0.00275	0.00436
F37	26	206.054	4.917	0.00273	0.00433
F38	75	256.393	4.853	0.00154	0.00244

Fault No.	Fault length (Li) in km	Hypo-central distance (R) in km	Magnitude M100 [recurrence period—100 Years]	PGA(g)	
				50 percentile	84 percentile
F39	75	156.850	5.029	0.00535	0.00850
F40	86	240.606	4.950	0.00199	0.00316
F41	87	230.217	5.593	0.00428	0.00681

Table B3.
PGA values for seismic sources for district headquarter Korba.

Fault No.	Fault length (Li) in km	Hypo-central distance (R) in km	Magnitude M100 [recurrence period—100 Years]	PGA(g)	
				50 percentile	84 percentile
F1	51	127.196	4.659	0.00520	0.00827
F2	26	118.412	4.071	0.00295	0.00469
F3	25	254.206	5.508	0.00311	0.00495
F4	39	143.978	4.463	0.00334	0.00531
F5	117	169.597	5.247	0.00579	0.00921
F6	32	162.409	4.634	0.00324	0.00515
F7	30	176.246	4.629	0.00274	0.00436
F8	45	176.181	6.052	0.01178	0.01874
F9	30	109.015	6.709	0.04929	0.07846
F10	30	124.640	4.888	0.00694	0.01104
F11	55	142.431	4.936	0.00578	0.00920
F12	78	170.995	4.594	0.00280	0.00445
F13	77	98.240	4.952	0.01096	0.01745
F14	28	88.029	4.447	0.00736	0.01172
F15	62	81.035	4.759	0.01186	0.01888
F16	46	94.427	4.746	0.00930	0.01480
F17	140	138.092	5.350	0.00945	0.01504
F18	42	237.152	6.016	0.00598	0.00952
F19	28	266.059	5.816	0.00375	0.00597
F20	47	287.334	5.340	0.00192	0.00306
F21	28	273.686	5.242	0.00197	0.00314
F22	75	136.807	5.043	0.00697	0.01109
F23	26	174.579	4.947	0.00397	0.00632
F24	75	222.346	4.858	0.00216	0.00343
F25	86	193.947	4.960	0.00325	0.00517
F26	87	133.076	5.654	0.01368	0.02176
F27	47	278.875	4.562	0.00090	0.00143

Table B4.
PGA values for seismic sources for district headquarter Jashpurnagar.

Author details

Ashish Kumar Parashar
Department of Civil Engineering, SoS Engineering and Technology, Guru Ghasidas Vishwavidyalaya, Bilaspur, India

*Address all correspondence to: aparashar08@gmail.com

References

[1] Zafar S, Tripathi S. Deterministic seismic hazard analysis of the region. International Journal of Engineering Research and Technolog. 2016;**5**(2): 634-636

[2] Rao D, Vansittee C. Deterministic seismic hazard analysis for the North western Part of Haryana State, India, Considering Various Seismicity Levels. Pure Applied Geophysics. 2021;**178**: 449-464

[3] Kramer S. Geotechnical Earthquake Engineering. New Jersey: Prentice Hall; 1996

[4] Mualchin L. History of modern earthquake hazard mapping and assessment in California using a deterministic or scenario approach. Pure and Applied Geophysics. 2011;**168**(3-4): 383-407

[5] Naik N, Choudhury D. Deterministic seismic hazard analysis considering different seismicity levels for the state of Goa, India. Natural Hazards. 2015;**75**(1): 557-580

[6] SEISAT: Seismotectonic Atlas of India. New Delhi: Geological Survey of India; 2000

[7] Stepp JC. Analysis of completeness of the earthquake sample in the Puget sound area and its effect on statistical estimates of earthquake hazard. In: Proceedings of the International Conference. Microzonation, Seattle; 1972. pp. 897-910

[8] Gutenberg B, Richter CF. Frequency of earthquakes in California. Bulletin of the Seismological Society of America. 1944;**34**:185-188

[9] Lindholm AC, Parvez IA, Kuhn D. Probabilistic earthquake hazard assessment for Peninsular India. Journal of Seismology. 2016;**20**(2):629-653

[10] Sunardi B, Haryoko U, Rohadi S, Shahzad S. Implications of Kendeng Fault to Seismic Hazard Potential in Malang. In: Proceeding of The 7th Annual Basic Science International Conference. 2017

[11] Anbazhagan P, Bajaj K, Patel S. Seismic hazard maps and spectrum for Patna considering region-specific seismotectonic parameters. Natural Haza rds. 2015;**78**:1163-1195

[12] Girish Joshi C, Lal M. Uncertainties in the estimation of Mmax. Journal of Earth System Science. 2008;**117**(S2): 671-682

[13] Wells LD, Coppersmith J. New empirical relationships among magnitude, rupture length, rupture width, rupture area, and surface displacement. Bulletin of the Seismological Society of America. 1994; **84**(4):974-1002

[14] Gupta I. The state-of-the-art in seismic hazard analysis. Journal of Earthquake Technology. 2002;**39**(4): 311-346

[15] Thaker TP, Rathod GW, Rao KS, Gupta K. Use of seismotectonic information for the seismic hazard analysis for Surat City, Gujarat, India: Deterministic and Probabilistic Approach. Pure Applied Geophysics. 2012;**169**:37-54

[16] Patel M, Solanki C, Thaker T. Deterministic seismic hazard analysis of Ankleshwar City, Gujarat. Advances in

Sustainable Construction and Resource Management. 2021:673-680

[17] Iyengar RN, Raghukanth ST. Attenuation of strong ground motion in peninsular India. Seismological Research Letters. 2004;**75**(4):530-540

[18] BIS: 1893. Indian Standard Criteria for Earthquake Resistant Design of Structures. Part I, General Provisions and Buildings (Sixth Revision). New Delhi: Bureau of Indian Standards; 2016

[19] Cornell C. Engineering seismic risk analysis. Bulletin of Seismological Society of America. 1968;**58**:1583-1606

[20] Iyengar RN, Ghosh S. Microzonation of earthquake hazard in greater Delhi area. Current Science. 2004;**87**(9): 1193-1202

[21] Raghukanth STG, Iyengar RN. Seismic hazard estimation for Mumbai City. Current Science. 2006;**91**(11): 1486-1494

[22] Sitharam TG, Sil A. Comprehensive seismic hazard assessment of Tripura and Mizoram States. Journal of Earth System Science. 2014;**123**:837-857

[23] Muthuganeisa P, Raghukanth STG. Site-specific probabilistic seismic hazard map of Himachal Pradesh, India, Part II. Hazard Estimation. Acta Geophysica. 2016;**64**(4):853-884

[24] Anbazhagan P, Bajaj K, Matharu K, Sayed SR, Moustafa N, Al-Arifi SN. Probabilistic seismic hazard analysis using the logic tree approach – Patna district (India). Natural Hazards Earth System Science. 2019;**19**:2097-2115

[25] San H. A probabilistic approach to the seismic hazard in Kashmir basin, NW Himalaya. Geoscience Letters. 2019; **6**(5):1-11

[26] Payal M, Thaker T. Seismic hazard analysis of Vadodara Region, Gujarat, India: Probabilistic and Deterministic Approach. Journal of Earthquake Engineering. 2020:1-24

[27] Abhik P, Pradipta C, Avijit B, Sapan K. Probabilistic Seismic Hazard Analysis of Sitamarhi near Bihar-Nepal Region2021. pp. 1-29

[28] Surve G, Jyotima K, Nitin S. Probabilistic seismic hazard assessment studies for Mumbai region. Natural Hazards. 2021;**107**:575-600

Section 2

Ground-Shaking Threats: A Closer Look at Earthquake Hazards

Chapter 4

Analysis and Prediction of the SARIMA Model for a Time Interval of Earthquakes in the Longmenshan Fault Zone

Xue Yuan, Hu Dan, Ye Qiuyin, Zeng Wenjun, Yang Jing and Rao Min

Abstract

Based on the catalog data of earthquakes with Ms $\geq$ 2.5 in the Longmenshan fault zone from January 2012 to September 2021, we establish an earthquake time interval series grouped by earthquake magnitude and then use the SARIMA model to predict the series in different periods. By analyzing the fitting effect of the models, the optimal model parameters of different magnitude sequences and the corresponding period values are obtained. Among them, the adjusted R^2 values of each model with Ms $\geq$ 2.5 and Ms $\geq$ 3.0 sequences are more than 0.86, up to 0.911; the short-time prediction effects are good, and the values of predicted RMSE are 10.686 and 8.800. The prediction results of the models show that the overall trend of the subsequent earthquake time interval in the Longmenshan fault zone is stable, and the prediction results of the Ms $\geq$ 3.0 sequence have a weak fluctuating growth trend; that is, the number of earthquakes with the Ms $\geq$ 3.0 in this area will decrease slightly, and the seismicity will decrease in a period of time. The analysis results and method can provide a scientific basis for earthquake risk management and a feasible way to predict earthquake occurrence times.

Keywords: Longmenshan fault zone, time interval of earthquakes, SARIMA model, time series analysis, magnitude grouping

1. Introduction

The Longmenshan fault zone is located on the eastern edge of the Qinghai-Tibet Plateau and the northwestern edge of the Sichuan Basin. It extends in a NE-SW direction, reaches the Qinling fault zone in the north, and ends at the Xianshuihe-Xiaojiang fault zone in the south. It is approximately 500 km long and 50 km wide and is mainly composed of four main faults, including the Houshan fault, Central fault, Qianshan fault, piedmont hidden fault, and the thrust nappe controlled by them [1, 2]. Dozens of important cities and towns, such as Dujiangyan, are distributed within its scope. In a southeastern direction, this large fault at the front of the main mountain range is close to the densely populated Chengdu Plain area and the Chengdu-Chongqing economic circle. Exploring the regularity of earthquake occurrence times along the Longmenshan fault zone can provide a scientific basis for earthquake management and decision-making for earthquake prediction in this area.

Currently, earthquake research is an interdisciplinary and comprehensive field. For earthquake time series data, many researchers have used statistical and probabilistic methods to predict important information, such as earthquake occurrence time. Some scholars have focused on the exploration of magnitude time series [3–5]. There have been studies on the time interval sequence of earthquake occurrence, most of which have analyzed the probability distribution of the earthquake time interval [6–10]. There have also been studies on the prediction of time intervals. Guo & Xu [11] made a confirmatory prediction of the Ms8.0 earthquake in northwestern China using a method of multiplied periods with golden section, which appeared to be in good agreement with the testing results. Mariani and Tweneboah [12] proposed applying the stochastic differential equation based on the superposition of independent Ornstein-Uhlenbeck processes to earthquake research and used this model to fit an earthquake sequence in South America. In addition, some scholars have expanded the prediction approach, using SVM, LSTM and regression algorithms to predict earthquake times [13–16].

The seasonal autoregressive moving average model can be used not only to predict natural phenomena such as river flow and rainfall [17–20] but also to predict social and economic phenomena such as transportation passenger flow and trade volume [21–25]. In the field of earthquake prediction, some scholars have used ARIMA to identify earthquake precursor anomalies [26–29]. However, few ARIMA or SARIMA models have been used to fit earthquake occurrence time data.

In this paper, we selected earthquake events in the Longmenshan fault zone in Sichuan Province in China since 2012 as the research object. We used the SARIMA model to fit the time interval sequence according to magnitude to predict the development trend of the earthquake occurrence time interval and the time of the next earthquake occurrence.

2. SARIMA model

2.1 Model theory

The ARIMA model was first proposed by Box and Jenkins in the 1970s. Its full name is the differential autoregressive moving average model, denoted as ARIMA (p, d, q), where p and q represent autoregressive and moving average orders, respectively; d is the difference order, and the basic model structure is as follows:

$$\varnothing(B)\nabla^d x_t = \Theta(B)\varepsilon_t \tag{1}$$

In the formula, $\varnothing(\mathrm{B}) = 1 - \varnothing_1\mathrm{B} - \dots - \varnothing_p\mathrm{B}^p\nabla^d = (1-\mathrm{B})^d$ is the autoregressive coefficient polynomial of order p, where B is the delay operator, and $\mathrm{Bx_t} = \mathrm{x_{t-1}}$; ∇^d is the d-th order difference; and $\Theta(B) = 1 - \theta_1 B - \dots - \theta_q B^q$ is the moving average coefficient polynomial of order q [30].

ARIMA can model nonstationary time series without seasonal effects. However, in real life, many time series have a certain periodicity. For the series that contains both seasonal effects and long-term trend effects and a complex interaction between them, we can use the SARIMA model.

The general expression of the SARIMA model was originally proposed by Wang et al. (Wang et al., 2018) and is recorded as ARIMA(p, d, q) × (P, D, Q)s, where P and Q are seasonal autoregressive and seasonal moving average orders, respectively; D is the order of seasonal difference; and s is the number of seasonal cycles. As an extension of the ARIMA model, this model extracts seasonal effect information from the series by seasonal difference.

For the time series x_t, the SARIMA model expression is as follows:

$$\nabla^d \nabla_s^D x_t = \frac{\theta(B)\theta_s(B)}{\varnothing(B)\varnothing_s(B)} \varepsilon_t \tag{2}$$

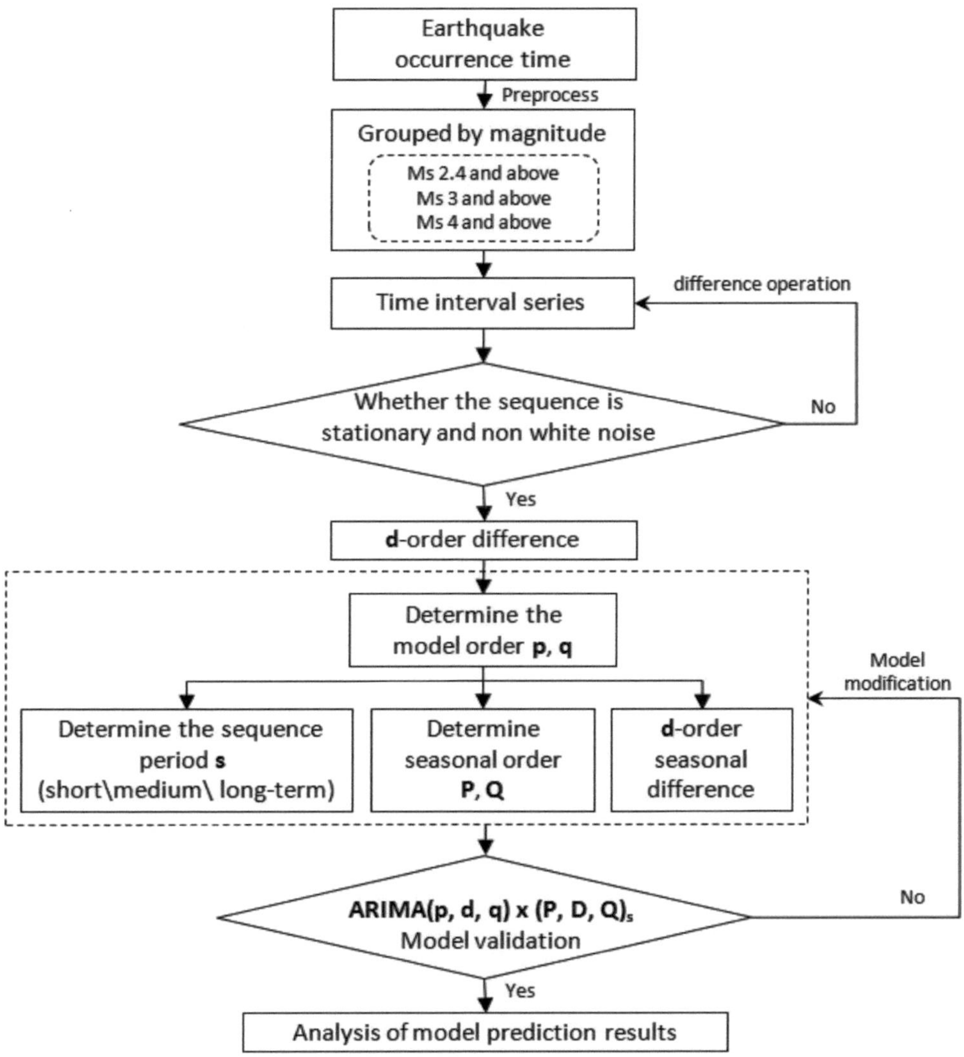

Figure 1.
SARIMA model analysis flow chart.

where ∇_s^D is the D seasonal difference in s steps; ε_t is the random interference of the error term at time t; $\varnothing_s(B)$ and $\theta_s(B)$ represent the P-order seasonal autoregressive coefficient polynomial and Q-order seasonal moving average coefficient polynomial, respectively. The meanings of other variables are shown above.

2.2 Forecast model construction

For the prediction of small sample data, the traditional time series prediction model can fully extract the information and can also avoid overfitting. After analyzing the data characteristics, this paper chooses to use SARIMA to model the earthquake time series. As a traditional statistical prediction model, the SARIMA model has a good fitting degree to small sample data. It has high prediction accuracy and fast training speed and can reflect the dynamic changes in data. While capturing the trend of the series, the model can also extract its periodic fluctuations [17, 31].

To obtain the time interval sequence, we take the difference from the earthquake sequence. Then, we use SARIMA to model the data that are grouped according to the magnitude and obtain models with different parameters, model-fitting value series Y_t' and predicted value series Y_t' . To ensure that the model can extract short-term and long-term information and improve the accuracy of prediction in each period, long, medium, and short periods with the best modeling effect are selected, expressed as s_a, s_b, and s_c. The model analysis flowchart is shown in **Figure 1**.

3. Study on the time interval of earthquakes in the Longmenshan fault zone

3.1 Data and samples

This paper uses earthquake occurrence time data from January 9, 2012, to September 24, 2021, in the earthquake catalog of the China Seismological Network (h ttp://www.ceic.ac.cn). The earthquake occurrence time, geographic coordinates, magnitude, and other factors of the earthquake event time series are characterized by volatility and nonlinearity. Therefore, we preprocess this information into data in a specific space and a certain magnitude; that is, the time series with an Ms2.4 and above in the longitude and latitude range of the Longmenshan fault zone (29.5° N-33.5° N, 102° E-107° E). By further screening whether the earthquake location is in the fault zone, 437 experimental sample data are obtained, and the earthquake occurrence time is from January 9, 2012, to August 2021, Ms. $\in$ (2.5,7).

The earthquake time interval sequence is obtained by the difference operation of the selected samples. The sample has the following characteristics:

1. The sample size is relatively small. The monitoring data in the early years are not sufficiently complete, which can lead to a significant reduction in the accuracy of calculating the time interval values. Therefore, an earthquake sequence of nearly 10 years is selected to effectively avoid this problem, but it also reduces the amount of data.

2. Large time span. The data are from January 2012 to August 2021, including the earthquake events of the Longmenshan fault zone in the past 10 years.

3. Nonstationary and strong volatility. There are many factors affecting the data, there may be complex interactions, and the earthquake sequence has complex periodic characteristics.

In addition, the 58th earthquake recorded in the sequence was the 2013 Ya'an Lushan Ms. 7.0 earthquake, followed by several very close aftershocks, resulting in a short time interval between the 58th and 203th earthquakes in the monitoring data, and the sequence approached zero in a period of time. The original sequence is represented with X_t, and the sequence diagram is shown in **Figure 2**.

3.2 Model determination

The characteristics of the data are analyzed, and its trend, seasonality, and residual series are extracted through seasonal decomposition. The results are shown in **Figure 3**, indicating that the series has a certain trend and seasonality. It is preliminarily speculated that the minimum period is approximately 5 days.

The SARIMA model requires the sequence to meet the precondition of stationarity or post-difference stationarity. After the stationarity test of X_t, we determine the number of differences d = 2. The sequence after the difference is marked as X_d, and its sequence diagram is shown in **Figure 4**. The ADF test is used to determine whether the data after the difference have a trend. The test principle is to check whether there is a unit root in the sequence: if the sequence is stable, there is no unit root. After

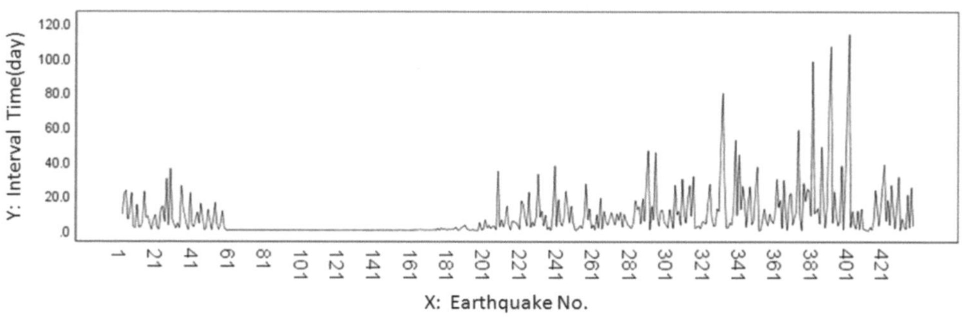

Figure 2.
Sequence diagram of earthquake time intervals.

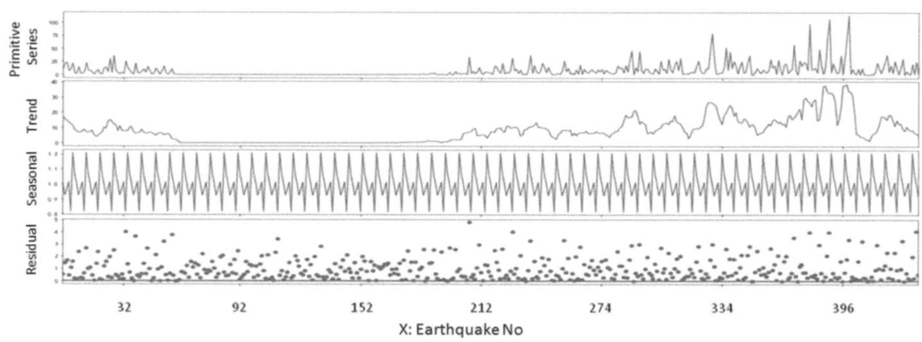

Figure 3.
Trend and seasonal decomposition of earthquake time interval series.

testing, the ADF test statistic of sequence X_d is −15.4634, and the p value is less than 0.05, indicating that the sequence is stable.

The ratio of the model training set and test set of sequence Xd is 9 to 1, and the order of the model is determined according to the ACF and PACF of the data. ACF is used to measure the correlation between the current value of the series and its lagged terms. PACF is used to measure the correlation between the current value and its lagged terms after removing the influence explained by the previous lag. The nonseasonal order of the model is preliminarily determined from the ACF and PACF charts in **Figure 5**. The PACF diagram can be regarded as a trailing or ninth order truncation.

The optimal order of the model in combination with the adjusted R^2 of the model is selected. The fitting effect of ARIMA models with different orders is shown in **Table 1**. From this, the parameters p = 9 and q = 1 in the model are determined.

Then, we analyze the ACF and PACF of integral multiple orders of s after a one-order, s-step difference of X. Taking s = 22 as an example, as shown in **Figure 6**, the ACF and PACF coefficients lagging 22 orders are outside the range of two times the standard deviation and then gradually converge. It is preliminarily assumed that P = 1 and Q = 1 in the model. We further carry out parameter tests and LB tests for the hypothetical model. A parameter test is used to verify the validity of the model; the LB test is used to check whether the residual sequence after model fitting is a white noise sequence.

The test results are shown in **Table 2**, and the above model contains an insignificant parameter (first-order seasonal autoregressive coefficient). Through comparison, the best seasonal order of the model, P = 0, and Q = 1 are determined.

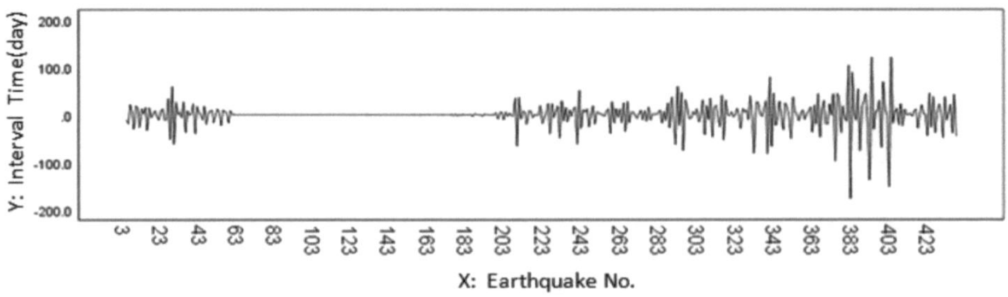

Figure 4.
Sequence diagram of the earthquake time interval after secondary difference.

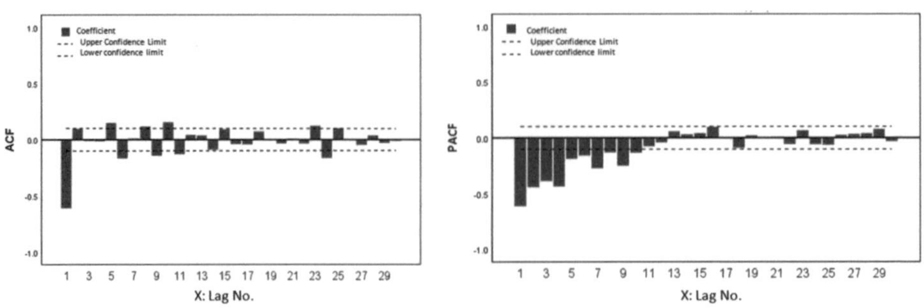

Figure 5.
ACF and PACF diagrams of the earthquake time interval sequence.

ARIMA model	Adjusted R^2	Parameter test	LB test(p value)
ARIMA (0, 2, 1)	0.643	not significant	0.0001
ARIMA (7, 2, 1)	0.803	pass	0.031
ARIMA (9, 2, 1)	0.813	pass	0.395
ARIMA (10, 2, 1)	0.813	not significant	0.305

Table 1.
Comparison of ARIMA models with different autoregressive orders.

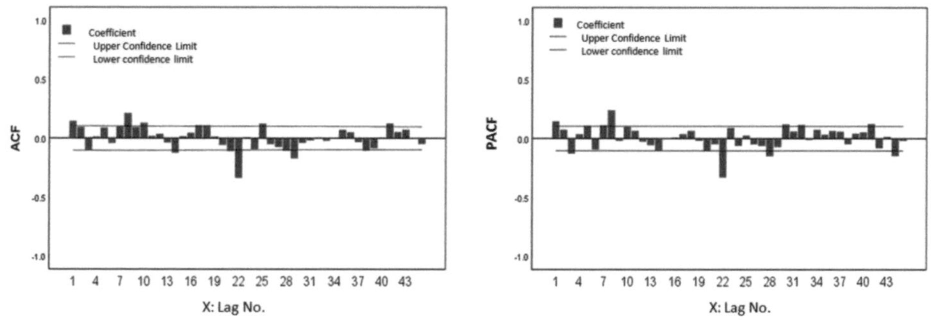

Figure 6.
ACF and PACF diagrams of the earthquake time interval sequence after secondary difference.

SARIMA model	Adjusted R^2	Parameter test	LB test(p value)
ARIMA(9, 2, 1)x(0, 1, 0)$_{22}$	0.817	pass	0.048
ARIMA(9, 2, 1)x(1, 1, 0) $_{22}$	0.846	pass	0.009
ARIMA(9, 2, 1)x(0, 1, 1) $_{22}$	0.869	pass	0.175
ARIMA(9, 2, 1)x(1, 1, 1) $_{22}$	0.869	not significant	0.115

Table 2.
Comparison of SARIMA models with different seasonal orders.

3.3 Model evaluation

The earthquake sequences used in the study have complex overlapping of different cycle periods. Therefore, when analyzing the periodicity of data, different seasonal periods are selected for short-, medium-, and long-term prediction. The center moving average of sequence X_d with span n is calculated, and the resulting sequence is marked as X_n, $n \in (4,45)$. The difference between X_n and X_d is compared, the seasonal elimination degree of each cycle is observed, and the models of $s \in (4,45)$ are fitted, respectively. To better compare the prediction effects between different models, this paper uses the RMSE to represent the fitting error of the model:

$$\text{RMSE} = \sqrt{\frac{1}{N}\sum_{t=1}^{N}(x_t - y_t)^2} \tag{3}$$

where x_t is the true value, and y_t is the fitted value. For the model passing the parameter test and LB test, we analyze its adjustment R^2 value and fitting RMSE to determine the fitting effect. The AIC is also commonly used to measure the complexity and goodness of fit of statistical models; the smaller the AIC value is, the better the model performance. The principle of model selection in this paper is to prioritize the model with a high R^2 value. When the difference between the R^2 values is not significant, the fitting RMSE of the model is compared to determine the optimal model, and the AIC value of the model is taken as an auxiliary reference. **Table 3** provides the fitting of some models with different periods that have passed the parameter test and LB test.

Therefore, we determine that the short-, medium-, and long-term prediction models of seismic data are ARIMA $(9, 2, 1) \times (0, 1, 1)_6$'ARIMA$(9, 2, 1) \times (0, 1, 1)_{22}$' and ARIMA$(9, 2, 1) \times (0, 1, 1)_{42}$. The model-fitting effect is as follows (**Figure 7**):

Period	SARIMA model	Adjusted R^2	LB test (p value)	Fitting RMSE	AIC
Short period	ARIMA$(9, 2, 1)x(0, 1, 1)_5$	0.870	0.194	12.049	2924.767
	ARIMA$(9, 2, 1)x(0, 1, 1)_6$	0.911	0.215	11.785	2914.680
Medium period	ARIMA$(9, 2, 1)x(0, 1, 1)_{22}$	0.869	0.175	13.083	2705.475
Long period	ARIMA$(9, 2, 1)x(0, 1, 1)_{32}$	0.872	0.655	13.133	2579.307
	ARIMA$(9, 2, 1)x(0, 1, 1)_{40}$	0.871	0.369	13.791	2473.724
	ARIMA$(9, 2, 1)x(0, 1, 1)_{42}$	0.881	0.298	13.212	2435.224

Table 3.
Comparison of SARIMA models with different periods.

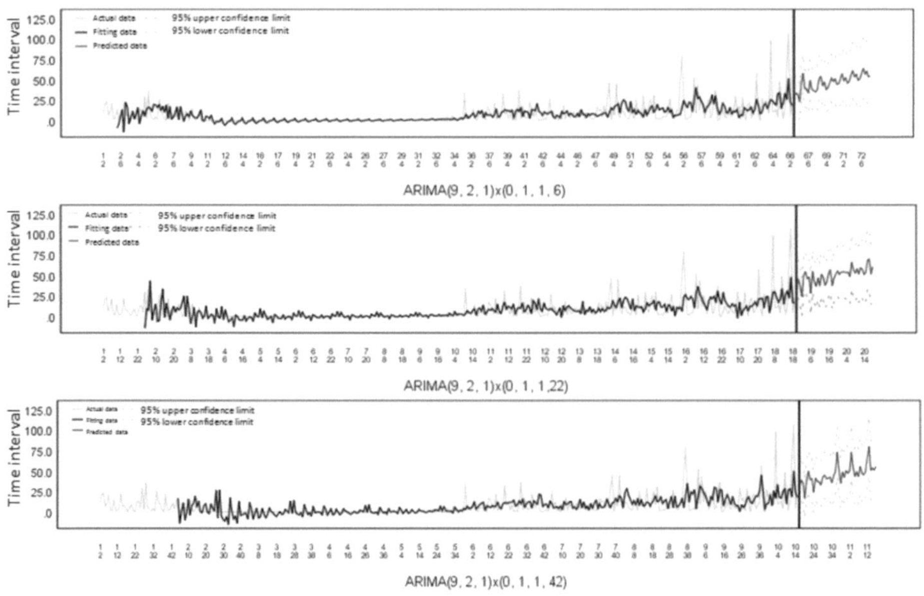

Figure 7.
Fitting effect of models with different periods.

4. Magnitude grouping

By grouping the original sequence by magnitude, 390 points of data for the Ms. ≥ 3.0 sequence and 58 points of data for the Ms. ≥ 4.0 sequence are obtained. After analyzing and modeling each sequence according to the above steps, we consider fitting X_{d3} with model ARIMA(7,2,1) × $(0,1,1)_s$ and X_{d4} with model ARIMA (3,2,1) × $(0,1,0)_s$. The comparison of each group's models in different periods is shown in **Tables 4** and **5**.

According to the above principles, we use s = 7, s = 16, and s = 32 to make short-term, medium-term, and long-term predictions for Xd3, respectively. The prediction model is ARIMA(7,2,1) × $(0,1,1)_7$'ARIMA(7,2,1) × $(0,1,1)_{16}$'and ARIMA(7,2,1) × $(0,1,1)_{32}$. For X_{d4}, due to the small amount of earthquake sequence data with an Ms. ≥ 4.0, it is difficult to capture the long period changing trend, and the fitting effect of those long period models is poor. In addition, the fitting effect of the short-period models is unsatisfactory. The X_{d4} sequence only shows an obvious medium periodic. Therefore, the medium-term prediction of X_{d4} is made with s = 11, and the prediction model is ARIMA (3,2,0) × (0,1,0)11. The fitting effects of different models of sequences are shown in **Figures 8** and **9**.

The predicted results are compared with the true values, as shown in **Figures 10** and **11**. The long-term prediction of the X_d and X_{d3} series shows that the model can capture their periodic changes, but the overall prediction value is higher than the real value.

Short-term prediction of X_{d4} shows that ARIMA(3,2,0) × $(0,1,0)_{11}$ can predict well the development trend of this series in recent few times and does not show a high prediction result. The predicted RMSE of each model is shown in **Table 6**.

Period	SARIMA model	Adjusted R^2	LB test (p value)	Fitting RMSE	AIC
Short period	ARIMA(7, 2, 1)x$(0, 1, 1)_7$	0.872	0.058	13.901	2705.256
Medium period	ARIMA(7, 2, 1)x$(0, 1, 1)_{14}$	0.881	0.061	14.059	2604.135
	ARIMA(7, 2, 1)x$(0, 1, 1)_{16}$	0.891	0.201	13.970	2577.105
	ARIMA(7, 2, 1)x$(0, 1, 1)_{23}$	0.849	0.179	17.243	2442.787
Long period	ARIMA(7, 2, 1)x$(0, 1, 1)_{32}$	0.867	0.383	15.112	2349.765
	ARIMA(7, 2, 1)x$(0, 1, 1)_{35}$	0.863	0.074	15.041	2294.26

Table 4.
Comparison of different periodic models of sequence X_{d3}.

Period	SARIMA model	Adjusted R^2	LB test (p value)	Fitting RMSE	AIC
Short period	ARIMA(3, 2, 0)x$(0, 1, 0)_7$	0.809	0.108	73.242	472.971
Medium period	ARIMA(3, 2, 0)x$(0, 1, 0)_{10}$	0.708	0.106	76.754	443.281
	ARIMA(3, 2, 0)x$(0, 1, 0)_{11}$	0.775	0.672	60.594	414.368
	ARIMA(3, 2, 0)x$(0, 1, 0)_{15}$	0.77	0.165	59.875	372.062
Long period	ARIMA(3, 2, 0)x$(0, 1, 0)_{20}$	0.781	0.321	59.076	318.484

Table 5.
Comparison of different periodic models of sequence X_{d4}.

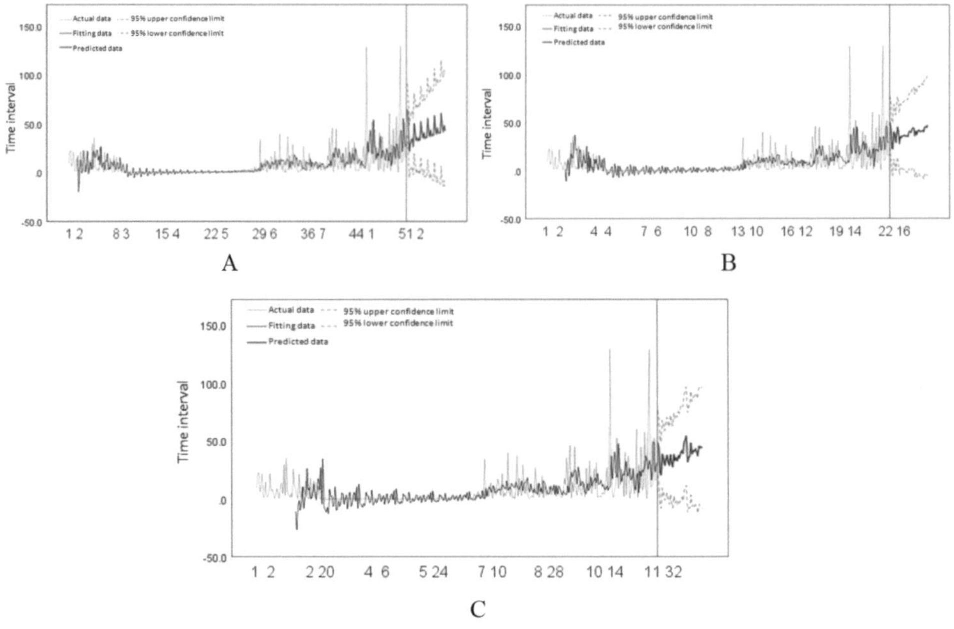

Figure 8.
Fitting effect of the optimal model with different periods of sequence X_{d3}. A. ARIMA$(7,2,1) \times (0,1,1)_7$. B. ARIMA$(7,2,1) \times (0,1,1)_{16}$. C. ARIMA$(7,2,1) \times (0,1,1)_{32}$.

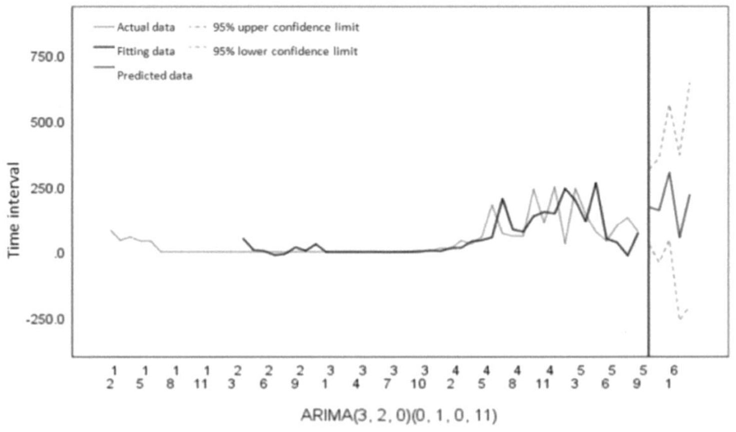

Figure 9.
Fitting effect of the optimal model of sequence X_{d4}.

The average level of the true and predicted values of the series X_d and X_{d3} is calculated. The average error of the models is 17.578 days at the lowest and 32.967 days at the highest. With reference to the general accuracy of the prediction of the earthquake occurrence time, the prediction error can be considered to be within an acceptable range. We use the predicted value to subtract the average error to correct the series predicted value. Taking ARIMA(9, 2, 1) x $(0, 1, 1)_{22}$ for X_d and ARIMA(7, 2, 1) x $(0, 1, 1)_{32}$ for X_{d3} as examples, the correction effect is shown in **Figure 12**, which shows that the models have good prediction performance for trend and periodicity.

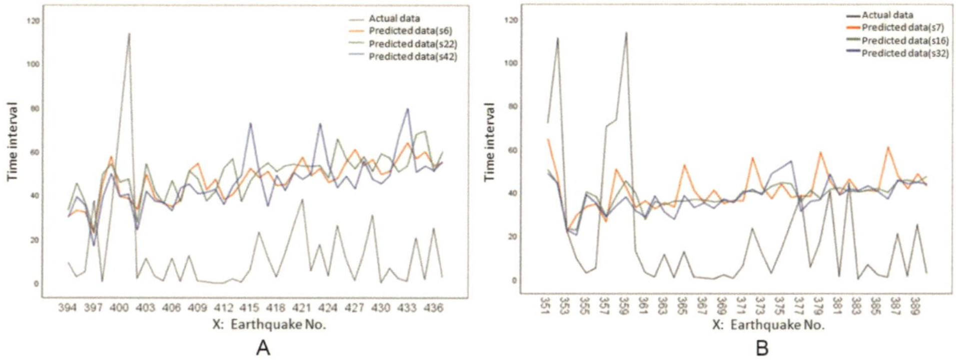

Figure 10.
Comparison of prediction effects of optimal models with different periods of sequences X_d and X_{d3}. A. Prediction of X_d. B. Prediction of X_{d3}.

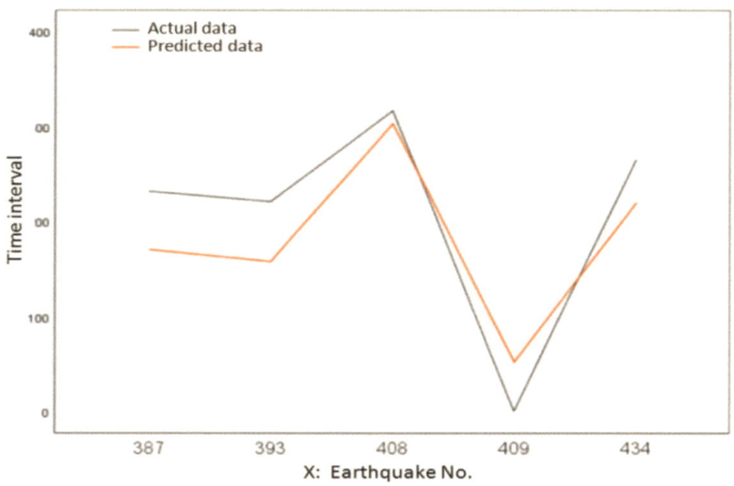

Figure 11.
Prediction effect of the optimal model of sequence X_{d4}.

Series	Optimal model	Predicted RMSE
X_d	ARIMA(9, 2, 1)x(0, 1, 1)$_6$	35.546
	ARIMA(9, 2, 1)x(0, 1, 1)$_{22}$	35.437
	ARIMA(9, 2, 1)x(0, 1, 1)$_{42}$	35.206
X_{d3}	ARIMA(7, 2, 1)x(0, 1, 1)$_{42}$	37.385
	ARIMA(7, 2, 1)x(0, 1, 1)$_{16}$	37.571
	ARIMA(7, 2, 1)x(0, 1, 1)$_{32}$	37.060
X_{d4}	ARIMA(3, 2, 0)x(0, 1, 0)$_{11}$	56.091

Table 6.
Prediction RMSE of different periodic models for each magnitude series.

Since the prediction step of sequences X_d and X_{d3} is more than 40, which may reduce the prediction accuracy, we consider making 15-step predictions for sequence X_d and 5-step predictions for sequence X_{d3}, which has a relatively small data volume. Taking the above two models as examples, the prediction results are shown in **Figure 13**. The predicted RMSEs are 10.6860 and 8.8009, respectively. There is no

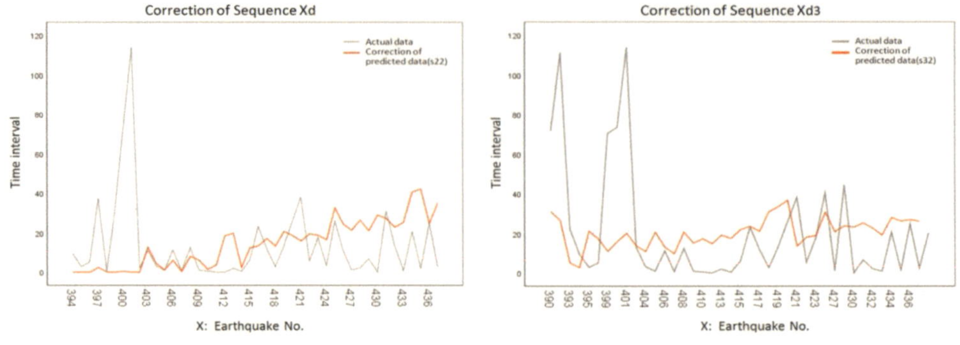

Figure 12.
Prediction effects after correction of sequences X_d and X_{d3}.

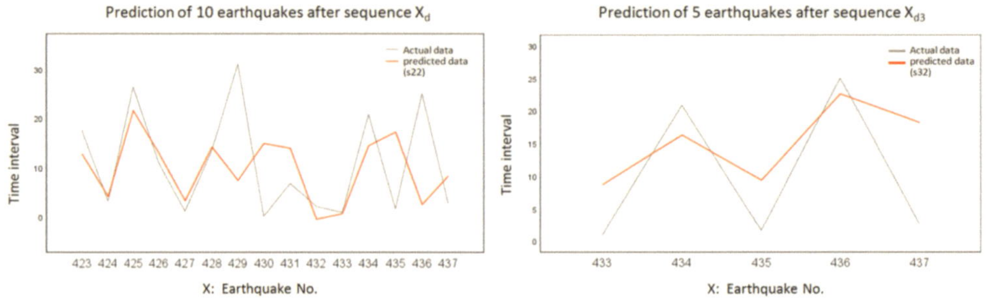

Figure 13.
Prediction effects of sequences X_d and X_{d3} after reducing prediction times.

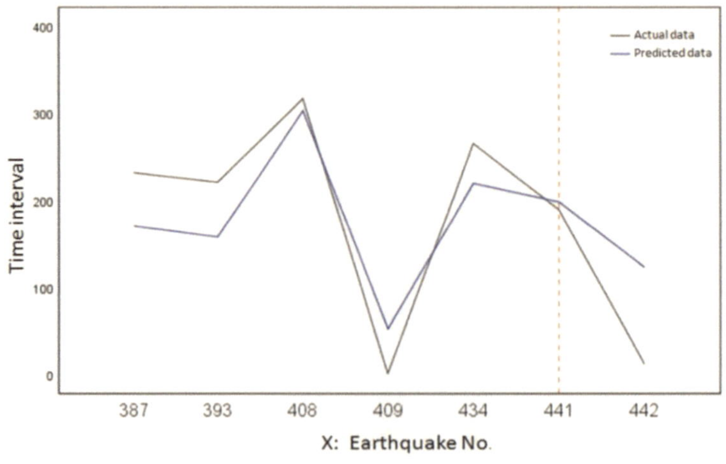

Figure 14.
Prediction results of new data ($Ms \geq 4.0$) in the Longmenshan fault zone in 2022.

obvious trend of the earthquake occurrence interval predicted by sequence X_d, while the prediction sequence of X_{d3} has a slightly increasing trend.

Two new datasets for an earthquake sequence with an Ms. ≥ 4.0 in the Longmenshan fault zone as of March 6, 2022 are added, and the ARIMA(3, 2, 0) x $(0, 1, 0)_{11}$ model is used to predict the time interval series. Comparing the predicted value with the true value, as shown in **Figure 12**, the predicted RSME is 55.7112. It is found that the trend is still captured well, and the model accuracy can be further improved on this basis (**Figure 14**).

5. Conclusion

This paper proposes to use SARIMA to model the time series from the perspective of the time interval of the Longmenshan fault zone, analyze the hidden information of the earthquake time series, and predict the next earthquake occurrence time. According to the model analysis and prediction results, the following conclusions are drawn: (1) The SARIMA model is applicable to the analysis and prediction of earthquake time interval series. The optimal model adjusted R^2 value of each series is above 0.86, up to 0.911. (2) For long-term prediction, the models of series X_d and X_{d3} have higher prediction values than the true values, and the prediction performance for lower values (time interval approaching 0) is relatively poor. (3) In short-term prediction, the optimal models of sequences X_d and X_{d3} have good prediction effects and can predict the sequence periodicity well. The prediction result of X_{d3} shows a slight fluctuation growth trend; that is, the number of earthquakes with an Ms. ≥ 3.0 in the Longmenshan fault zone decreases slightly. The periodicity of sequence X_{d4} is obvious, and in short-term prediction, the model can capture well its development. The prediction trend is consistent with the real situation. The prediction accuracy of the model can be further improved.

Author details

Xue Yuan[1,2], Hu Dan[1,3]*, Ye Qiuyin[1,2], Zeng Wenjun[1,2], Yang Jing[1,2] and Rao Min[1,2]

1 Geomathematics Key Laboratory of Sichuan Province, Chengdu University of Technology, Chengdu, Sichuan, People's Republic of China

2 College of Mathematics and Physics, Chengdu University of Technology, Chengdu, Sichuan, People's Republic of China

3 Office of Academic Affairs, Chengdu University of Technology, Chengdu, Sichuan, People's Republic of China

*Address all correspondence to: 443185980@qq.com

References

[1] Kangsheng X, Weihuan W, Ying L, Ju P, Xiaolong C. Evolution of microseismic activity in Longmenshan area from 2016 to 2017: Based on fuzzy clustering analysis of microseismic catalogue. Progress in Geophysics. 2022; **37**(01):69-77 (in Chinese)

[2] Yihai Y, Xuemei Z, Hua Xi S, Lina FC, Yurong Q, Liang Chuntao S, et al. Segmentation characteristics of Longmenshan fault — Constraints from dense focal mechanism data. Chinese Journal of Geophysics. 2021;**64**(04): 1181-1205 (in Chinese)

[3] Hongwei H. Application of threshold autoregressive model in earthquake prediction. Journal of Seismological Research. 1992;**02**:154-161 (in Chinese)

[4] Salamat M, Zoller G, Zare M, et al. The maximum expected earthquake magnitudes in different future time intervals of six seismotectonic zones of Iran and its surroundings. Journal of Seismology. 2018;**22**:1485-1498

[5] Kundu S, Opris A, Yukutake Y, et al. Extracting correlations in earthquake time series using visibility graph analysis. Frontiers in Physics. 2021;**9**:1-13

[6] Xuan W. Statistical analysis and probability prediction of M5.0 and above earthquakes in Sichuan. Earthquake Research in Sichuan. 2021;**04**:40-45 (in Chinese)

[7] Sadeghian R. Finding a probability density function for time intervals between earthquake occurrences. Arabian Journal of Geosciences. 2018; **11**(17):1-10

[8] Liu GL. A new physical model for earthquake time interval distribution. Physica a-statistical mechanics and its applications. 2016;**465**:62-65

[9] Khan MY, Iqbal T, Shah MA. Probabilistic modeling of earthquake Interevent times in different regions of Pakistan. Pure and Applied Geophysics. 2020;**177**(12):5673-5694

[10] Sil A, Sitharam T, Haider ST. Probabilistic models for forecasting earthquakes in the northeast region of India. Bulletin of the Seismological Society of America. 2015;**205**:2910-2927

[11] Guo A, Xu A. Analyzing the Time Interval of Great Earthquakes (M≥8) in Northwest China and Making Prediction by the Method of Multiplied Period with Golden Section. Northwestern Seismological Journal. 2005;(S1):138-139 (in Chinese)

[12] Mariani MC, Tweneboah OK. Stochastic differential equation of earthquakes series. Pure and Applied Geophysics. 2016;**173**(07):2357-2364

[13] Hanlin C, Jin'an C, Wei H. Earthquake Time Prediction Experiment in Xinjiang Based on LSTM Method. Abstracts of the Third Symposium on Geophysical Information Frontier Technology. 2021:81 (in Chinese)

[14] Jia Y, Xin Y, Yidan L, Peihuang H. Earthquake prediction and visualization based on grey prediction and SVR algorithm. Journal of Minjiang University. 2021;**42**(05):1-7 (in Chinese)

[15] Haining Z. Research on Earthquake Prediction Method Based on Improved Support Vector Machine Regression. Hefei, Anhui, China: Anhui University; 2016 (in Chinese)

[16] Asencio-Cortes G, Morales-Esteban A, Shang X, et al. Earthquake prediction in California using regression algorithms and cloud-based big data infrastructure.

Computers & Geosciences. 2018;**115**: 198-210

[17] Kun W, Jinmei R, Ni D. Prediction of Qujing air quality index based on SARIMA model. Journal of Qujing Normal University. 2018;**37**(03):25-29 (in Chinese)

[18] Dastorani M, Mirzavand M, Dastorani MT, et al. Comparative study among different time series models applied to monthly rainfall forecasting in semi-arid climate condition. Natural Hazard. 2016;**81**(3):1811-1827

[19] Moeeni H, Bonakdari H. Monthly reservoir inflow forecasting using a new hybrid SARIMA genetic programming approach. Journal of Earth System Science. 2017;**126**(02):18

[20] Ray S, Das S, Mishra P, et al. Time series SARIMA modelling and forecasting of monthly rainfall and temperature in the south Asian countries. Earth Systems and Environment. 2021;**5**:531-546

[21] Chen Y, Yuhui F, Yifei Z, Chao W. Ship track prediction based on ARIMA-BIGRU. Ship & Ocean engineering. 2021;**50**(06): 147-152 (in Chinese)

[22] Jingtao Y, Huan Y. Research on foreign trade risk prediction and early warning based on wavelet decomposition and ARIMA-GRU hybrid model. China Management Science. 2022:1-11. DOI: 10.16381/j.cnki.issn1003-207x.2021.1174 (in Chinese)

[23] Yue S, Xiaoyu S, Liting J, Tong L. Railway passenger flow forecast based on ARMA-LSTM combined model. Computer Applications and Software. 2021;**38**(12):262-267 + 273 (in Chinese)

[24] Milenkovic M, Svadlenka, Melichar V, et al. SARIMA modelling approach for railway passenger flow forecasting. Transport. 2015;**33**(5): 1113-1120

[25] Farhan J, Ong GP. Forecasting seasonal container throughput at international ports using sarima models. Maritime Economics & Logistics. 2018; **20**(1):131-148

[26] Jiyan Z, Xin'an W, Shanshan Y, Jipan H. Precursor anomaly detection of Jiuzhaigou M7.0 earthquake based on ARIMA model. North China Earthquake Sciences. 2019;**37**(01):28-33 (in Chinese)

[27] Lei L, Ning Z, Shuhui Y, Yizhuo W. ARMA residual correction model for detecting ionospheric TEC anomalies before Ms 6.4 earthquake. Journal of Geodesy and Geodynamics. 2021; **41**(04):382-386 (in Chinese)

[28] Saqib M, Senturk E. Comparisons of autoregressive integrated moving average (ARIMA) and long short term memory (LSTM) network models for ionospheric anomalies detection: A study on Haiti (M-w=7.0) earthquake. Acta Geodaetica et Geophysica. 2022; **57**(01):195-213

[29] Saqib M, Senturk E, Sahu SA, et al. Ionospheric anomalies detection using autoregressive integrated moving average (ARIMA) model as an earthquake precursor. Acta Geophysica. 2021;**69**(04):493-1507

[30] Danhui Y, Yan W. Appliied Time Series Analysis. Beijing, China: China Renmin University Press; 2019. p. 7 (in Chinese)

[31] Yanqun S, Shougang Z, Moyuan L, Yan Z, Yanyu P, Chong W, et al. Prediction of mosquito invasion in Nanjing based on SARIMA model. Journal of Nanjing Medical University (Natural Sciences). 2022;**42**(01):108-111 (in Chinese)

Chapter 5

Urban Damage Assessment after the *Mw* 5.8 Silivri Earthquake: The Case of Istanbul City

Oğuzhan Çetindemir and Abdullah Can Zülfikar

Abstract

This chapter presents the results of an urban damage assessment after a moderate seismic event, the *Mw* 5.8 Silivri earthquake, which is the most significant earthquake to have struck the region since two major catastrophic earthquakes, the *Mw* 7.6 Kocaeli and the *Mw* 7.1 Düzce earthquakes. First, distribution maps for earthquake parameters and building damages using an appropriate ground motion prediction equation are created for İstanbul. Then, near-real-time hazard and damage distribution maps are generated using the data recorded during the event by the ground motion network established in Istanbul. Comparing the results of the two analyses reveals that the ground motion and damage distributions generated by the selected ground motion prediction equations (GMPEs) are more conservative than those generated by the network, and this is because the actual station data surpass the GMPE's projections. This research concludes by emphasizing the significance of both GMPEs and densely installed ground motion station networks that capture real-time data during earthquakes and providing motivations for constructing or expanding such systems.

Keywords: earthquake damage assessment, NGA GMPE, near-real-time strong motion network, earthquake hazard, earthquake risk mitigation

1. Introduction

Earthquakes are one of the most destructive natural disasters in history. They are a potential cause of fatalities as well as structure and infrastructure damage in densely inhabited seismic-prone regions. In recent years, earthquakes have wreaked havoc on numerous cities around the world, causing a variety of issues [1–3]. Numerous scholars investigated seismic hazards to human lives and their economic effects [4–7]. Seismic activity is a severe natural force to which civil engineering structures are exposed and poses grave risks to human life [8, 9]. In seismically sensitive regions, constructing structures to resist this force becomes an economic burden. Previous research has examined the detrimental consequences of earthquakes on urban infrastructures [10–13]. This is why earthquake damage assessment is crucial for emergency response, disaster management, and seismic risk mitigation.

IntechOpen

Seismic hazard analysis needs the use of region-specific attenuation relations. Since a large number of ground motion prediction equations (GMPEs) can be used to assess a region's seismic hazard, selecting proper GMPEs can significantly impact design and safety evaluation. However, determining an acceptable GMPE has proven to be somewhat challenging [14]. Therefore, establishing strong motion networks is a crucial initiative for seismic risk mitigation. Strong motion networks can provide recorded ground motions to near-real-time seismic damage assessment networks, such as Prompt Assessment of Global Earthquakes for Response [15] and Real-time Earthquake Damage Assessment using City-scale Time History Analysis [16], enabling a robust, accurate assessment. Crucially, the sensor density effect in a network on seismic damage assessment was examined as a detailed case study for Zeytinburnu District, Istanbul, Turkey [17]. They found that the sensor density has a proportional effect on the regional-scale seismic damage assessment accuracy. Today, strong motion networks are operating in Italy [18], Taiwan [19], India [20], Japan [21], Iran [22], Greece [23], Romania [24], the US [25], and so on.

Regarding Turkey, Kandilli Observatory and the Earthquake Research Institute (KOERI) developed Istanbul's Earthquake Rapid Response and Early Warning System (IERREWS) in 2002, which is one of the most advanced strong motion networks in the world. In addition, in July 2008, a 16-station strong motion network was installed in İzmir. Earthquake Research and Implementation Center (ERIC-DAUM) of Dokuz Eylül University (DEU, Izmir), Earthquake Research Department (ERD) of the General Directorate of Disaster Affairs (GDDA, Ankara), Izmir Metropolitan Municipality, and Ministry of Public Works and Settlement collaborated in this network's establishment. The Scientific and Technological Research Council of Turkey (TUBITAK) funded the project to obtain strong motion data for earthquake hazard assessment and to establish a real-time monitoring system in Turkey to address public safety concerns [26].

This work focuses on a case study regarding urban damage assessment in metropolitan Istanbul, Turkey, after a recent moderate earthquake (the *Mw* 5.8 Silivri Earthquake, 2019), as one of the most significant seismic events in the region since two major earthquakes (the Kocaeli and Düzce earthquakes) that struck the region in 1999. The utilization of real-time data obtained from a densely deployed strong motion network and comparison with a GMPE that provides empirical results is one of the most significant merits of this research.

The remaining sections of this chapter are organized as follows: Section 2 represents a literature survey on seismicity in the Marmara Region, earthquake early warning and rapid response system (EEWRRS) in Istanbul, structural health monitoring systems, and next generation attenuation (NGA), and ground motion prediction equations (GMPE) models. Section 3 provides insight near-real-time strong motion network incorporated into the earthquake early warning system in Istanbul. In addition, near-real-time hazard (ground motion distribution) and damage maps are generated using data recorded by the Istanbul Natural Gas Distribution Network Seismic Risk Reduction Project (IGRAS) system. Section 4 introduces a case study regarding a recent offshore earthquake that hit metropolitan Istanbul. In the case study, earthquake hazard maps are created using a proper GMPE, and building damage distribution is estimated. In Section 5, distribution maps were generated based on chosen GMPE following bias correction of phantom stations using data from the strong motion network. Hazard maps using an appropriate GMPE and near-real-time data during the event are compared and discussed in Section 6. Finally, Section 7 concludes the paper by presenting significant findings in two major remarks.

2. Literature survey

This section mainly focuses on a survey regarding four different aspects. First, high seismicity in Turkey, particularly in the Marmara region, is briefly explained. Secondly, two initiatives functioning seismic risk mitigation measures, earthquake early warning, and structural health monitoring systems in the region, are introduced. Finally, next generation attenuation (NGA) models are discussed.

2.1 Seismicity in Marmara region

Turkey is one of Europe's most seismically active countries, and it has been struck by multiple catastrophic earthquakes throughout its history. These earthquakes (*Mw* 6.7 1992 Erzurum, *Mw* 6.1 1995 Dinar, *Mw* 6.3 1998 Adana-Ceyhan, *Mw* 7.6 1999 Kocaeli, *Mw* 7.2 1999 Düzce, *Mw* 6.4 2003 Bingöl, *Mw* 7.2 2011 Van, *Mw* 7.0 2020 İzmir, and *Mw* 6.7 2020 Elazığ) resulted in severe economic losses and physical damage, as well as a large number of fatalities. Ambraseys and Finkel [27] documented 32 devastating earthquakes that hit the larger Marmara Sea region between the fourth and nineteenth centuries, affecting Istanbul. Following the devastation caused by the two major earthquakes that struck the Marmara region in 1999, governmental, non-governmental, and academic organizations in Turkey have acknowledged the need for extensive response planning based on rigorous risk analyses of likely seismic hazards in Turkey and Istanbul in particular [5].

Le Pichon, Şengör [28] examined an earthquake scenario and provided a projection for two active fault lines located 20–30 kilometers southwest and south of central Istanbul: (1) the Central Marmara Basin (CMB) and (2) the North Boundary Fault (NBF), respectively. Some researchers found that within the next 30 years, the likelihood of an earthquake of magnitude seven or greater occurring in the Marmara Sea is around 65% Parsons [29, 30]. Considering more than two decades have passed after two severe earthquakes, a possible Istanbul earthquake may be in the near future.

Ansal, Akinci [31] provided significant results on earthquake loss assessment in Istanbul. Based on the 1999 Kocaeli earthquake investigation, they discovered critical parameters for the seismic ground motion simulation. In terms of the loss estimate, the CMB fault section is more critical due to the higher seismic excitation intensity it may cause. Okay, Kaşlılar-Özcan [32] indicated that the CMB segment has not raptured since 1766 and poses an urgent threat to the region. Ergintav, Reilinger [33] determined that the Princes' Island region of the Marmara Sea is most likely to cause the next Mw > 7 earthquakes along the NAFZ. Recent studies were conducted by Woessner, Laurentiu [34] to construct a reference hazard model for Europe. They reported that the mean peak ground acceleration (PGA) with a 10% probability of exceedance in 50 years provides the highest seismic hazard estimate (PGA $\geq$ 0.25 g) along the NAFZ, with values up to 0.75 g in the northern Aegean and the Marmara Sea.

2.2 Earthquake early warning and rapid response system (EEWRRS) in Istanbul

Ground motion estimation without earthquake source parameters aims to provide near-real-time (within a few seconds of the event origin time) assessments of earthquake-induced ground motions and associated building damage in Istanbul. It is based on the fact that a dense accelerometric network (approximately 2–3 kilometers between adjacent instruments) is installed in the southern part of the city, where both population density and earthquake hazards are highest. No source parameters (magnitude, rupture length, or mechanism) are required to compute the earthquake-generated ground motion distribution. PGAs and spectral accelerations (SAs) at various periods are interpolated using two-dimensional splines to derive the ground motion values at the center of each 0.01×0.01° geo-cell (1120×830 m grid size). Spectral displacements are used to calculate the seismic demand in the center of each geo-cell. Coherency functions are being used to assess shaking maps created by interpolating discrete ground motion measurements. The ELER hazard module executes the following tasks: (1) using a regional seismotectonic database, determine the most likely site of the earthquake's origins, (2) the spatial distribution of chosen ground motion parameters at engineering bedrock is estimated using region-specific ground motion prediction models, (3) a bias correction of ground motion estimations using strong motion data where available, (4) estimation of the spatial distribution of chosen site-specific ground motion parameters by the use of a regional geology database and appropriate amplification models, (5) alternatively, ground motion parameters can be estimated site-specifically using Next Generation Attenuation (NGA) models.

Due to the complicated segmentation of the Marmara fault line and its proximity to the city, the Istanbul earthquake rapid response system (IEEWS) designed a robust and straightforward earthquake early warning algorithm based on the exceeding threshold values. To trigger, at least three stations must surpass the threshold level within a 5-second interval. Böse [35] stochastically simulated 280 earthquake scenarios in the Marmara Sea ranging from *Mw* 4.5 to *Mw* 7.5 and determined that the average early warning time spans between 8 and 15 seconds, depending on the event's source location.

The data transmission between the remote stations and the KOERI processing hub is accomplished via fiber optic cable with redundancy provided by a satellite system. The time required to transmit data from remote stations to the KOERI data center is a few milliseconds via fiber optic connections and less than a second via satellites. The continuous online data from these stations is processed at the hub, and subsequent alerts of emerging potentially disastrous ground motions provide real-time warning to vital infrastructures, allowing for the activation of shut-off mechanisms before the damaging waves reach the location.

At the moment, the IEEWS has not issued a public alert. Only the Istanbul Natural Gas Distribution Company (IGDAS) and the Marmaray Tube Tunnel actively utilize the EEW alert to activate automated shut-off systems in these facilities [36]. Both end users manage their own networks, with strong motion stations co-located at high-pressure district gas regulators in the case of IGDAS and spaced throughout the tunnel in the case of Marmaray. Regarding the IGDAS, gas flow is automatically halted at the district regulator level in response to IEEWS alarms and the local site exceeding ground motion parameter threshold values. Local threshold values are determined on a case-by-case basis based on the local building stock.

The Marmaray Tube Tunnel, completed in 2013, is operated by the Turkish State Railways (TCDD). Train operations within a 1.4-kilometer-long tunnel beneath the Bosphorus, which connects the city's European and Asian sides, can be halted based on a combination of IEEWS alerts and a local threshold exceedance detected by its 26 tunnel sensors. Although IEEWS signals were sent to these vital facilities in recent seismic events such as the *Mw* 3.8 Yalova on 13 August 2015 and the *Mw* 4.2 Marmara Sea on 16 November 2015, *Mw* 4.5 Silivri on 24 September 2019, and the *Mw* 5.8 on 26 September 2019, no action was taken because the local threshold values were not surpassed.

Along with IEEWS, KOERI has implemented the regional EEW algorithms VS(SC3) and Probabilistic and Evolutionary Early Warning System (PRESTo) as part of the REAKT project. The Virtual Seismologist (VS) algorithm [37] is a network-based Bayesian approach to EEW, and the SED group integrated an operational VS into the evolving Californian EEW prototype system [38] with support from the USGS ShakeAlert project. In 2013, VS was integrated as a set of self-contained modules into the open-source and widely distributed SeisComP3 (SC3) [39, 40] earthquake monitoring software, incorporating an EEW algorithm in the same system that many seismic networks use on a regular basis [41, 42]. This solution is referred to as "VS" (SC3). Further detail can be found in Clinton, Zollo [43]. These applications take advantage of KOERI's Marmara regional seismic network, which consists of 40 broad-band and 30 strong motion seismic stations. PRESTo is currently monitoring 18 of the regional network's strong motion stations. Scenarios for various seismic events, including the 1999 *Mw* 7.4 Kocaeli Earthquake, show that a recurrence of this event would provide Istanbul with around 11 seconds of early warning. It is intended to expand the number of stations, including those utilized by PRESTo. Regional EEW algorithms VS (SC3) and PRESTo are not integrated with the existing IEEWS. Therefore, in their current configuration, the VS (SC3) and PRESTo systems would not provide warning in advance of strong motions associated with near-source seismic occurrences such as those in the Marmara Sea. On the other hand, the regional early warning system (EWS) is meant to be used in conjunction with the threshold-based IEEWS to provide warning for remote events that could be catastrophic for tall buildings and long-span bridge structures. On May 24, 2014, the *Mw* 6.9 Northern Aegean earthquake was intensely felt across Istanbul's high-rise buildings and was accurately characterized by VS (SC3) within 36 seconds of the origin time. The *Mw* 6.9 earthquake in Northern Aegean illustrated the importance of merging regional and threshold-based techniques. With the IEEWS and regional EWS algorithms, Istanbul has on-site structural monitoring efforts for historical buildings, high-rise buildings, and suspension bridges. These initiatives are not currently integrated into early warning systems.

2.3 Structural health monitoring systems

Instrumentation of major structures using strong ground motion recorders monitors the vibrations during an earthquake. Comparing these recordings with those made prior to the earthquake can reveal differences in structural response that have relevance for structural damage and a loss in seismic resistance. Therefore, these systems can be integrated into EEW systems to take critical measures for mitigating seismic risk. The structural health of significant structures in the Marmara region, particularly in Istanbul, has begun to be examined due to the region's high seismicity. Especially after the strong earthquakes in 1999, the public began to demand SHM systems, and the relevant government officials began to meet these demands.

SHM systems are currently monitoring five different long-span cable-supported bridges in the Marmara region. These are the 15 July Martyrs Bridge (the First Bosphorus Bridge), the Fatih Sultan Mehmet Suspension Bridge (the Second Bosphorus Bridge), the Yavuz Sultan Selim Bridge (the Third Bosphorus Bridge), the Osman Gazi Bridge (Izmit Bay Bridge), and the 1915 Çanakkale Bridge (Dardanelles). Regarding their SHM systems, design considerations and further information can be found in [44]. The following are some examples of currently monitored facilities in the Marmara region: Hagia Sophia, The Maltepe Mosque, Sapphire Tower, Kanyon Building, Isbank Tower, Polat Tower, and the Marmaray Tube Tunnel [45].

2.4 Next generation attenuation and ground motion prediction equations models (NGA GMPE)

The initial stage is to mitigate the potential seismic damage by developing earthquake-resistant structures. Additionally, seismic hazard assessments of a region are critical for preventing earthquake-related destruction. Establishing an efficient seismic hazard and a comprehensive seismic risk assessment is crucial for the nation's sustainable growth.

Seismic hazard analyses demand the application of region-specific attenuation relations. Ground motion prediction equations (GMPEs) are used to determine ground motion parameters required for designing and evaluating vital structures [46]. Since a large number of GMPEs can be used to assess a region's seismic hazards and risks, selecting proper GMPEs can significantly impact design and safety evaluation [14]. Attenuation relationships allow for the accurate estimation of ground motion parameters (macroseismic intensity, PGA, PGV, and SAs) in terms of magnitude and source-to-site distance, and site features [47].

GMPEs for shallow crustal earthquakes were recently created in the Next Generation Attenuation (NGA). PEER created a number of GMPE models based on the NGA-West1 database (original NGA project). Following that, GMPEs were updated using the NGA-West2 database. The average horizontal component of shallow crustal earthquakes in active tectonic zones is estimated empirically using the PEER NGA-West2 database [48]. While it was designed for tectonically active locations, it is also applicable to other regions. NGA-West2 investigated the regionalization of ground motion properties, including an elastic attenuation, site response, and within-event standard deviation. GMPEs are frequently used to predict ground motions using deterministic and probabilistic seismic hazard analyses. The results of the seismic hazard analysis are used to conduct site-specific seismic analysis and design of facilities, to estimate social and financial losses, and to generate regional seismic hazard maps for use in building standards and financial estimation, among other purposes [47].

The NGA-West2 incorporates regionalization to account for regional changes in far-source distance attenuation and soil response. This is accomplished by using recordings from severe earthquakes in other active tectonic zones and California data [49]. As part of the NGA-West2 project, five GMPEs were developed: Abrahamson, Silva, and Kamai (ASK2014) [48], Boore, Stewart, Seyhan, and Atkinson (BSSA) [50], Campbell and Bozorgnia (CB) [51], Chiou and Youngs (CY2014) [52], and Idriss (I) [53].

The NGA-West2 project developed GMPEs for the purpose of calculating the medians and standard deviations of average horizontal component intensity measures (IMs) for shallow crustal earthquakes in active tectonic zones [50]. The equations were developed using data from a global database containing M 3.0-7.9 events. NGA-West2 GMPEs have a general limit of M8.5 for strike-slip faults, M8.0 for reverse faults, and M7.5 for normal faults, and a rupture distance R_{rup} or Joyner & Boore distance, R_{jb} of 0–300 km [53]. The ground motion IMs that comprise the GMPEs' dependent variables include the horizontal components PGA, PGV, and 5%-damped SAs [50]. These IMs were calculated using the RotD50 parameter [54], which represents the median horizontal single-component ground motion over all non-redundant azimuths. ASK2014 is applicable to magnitudes 3.0–8.5, distances from 0 to 300 km, and spectral periods ranging from 0 to 10 seconds. Regional differences were incorporated into the ASK2014 model based on distances. The ASK2014 model assumes that in active crustal zones, median ground motions at distances smaller than around 80 km are similar worldwide. The model confirmed that median stress dips are similar to those experienced during earthquakes in many active crustal locations such as California, Alaska, Taiwan, Japan, Turkey, Italy, Greece, New Zealand, and Northwest China [48]. BSSA2014 GMPE was developed to be generally applicable for earthquakes with magnitudes of M 3.0 to 8.5 (except for the lack of constraint for M > 7 normal slip events), at distances ranging from 0 to 400 km, at locations with Vs30 values ranging from 150 to 1500 m/s, and for spectral periods ranging from 0.01 to 10 sec [50]. The BSSA2014 model includes regional variation in source, path, and site effects. The CY2014 GMPE model is appropriate for predicting horizontal ground motion amplitudes associated with earthquakes in active tectonic zones that meet the following criteria [49]: (1) $3.5 \leq M \leq 8.5$ for strike-slip earthquakes, (2) $3.5 \leq M \leq 8.0$ for reverse and normal faulting earthquakes, (3) $Z_{TOR} \leq 20$ km, (4) $0 \leq R_{RUP} \leq 300$ km and $180 \leq V_{S30} \leq 1500$ m/s.

Akcan et al. [55] evaluated the next generation attenuation (NGA) ground motion models that would be used in the region by comparing them to ground motion recordings from the Silivri earthquake in 2019. Among the NGA-West2 models, they found that ASK2014 and CY2014 provide the best fit for local PGA datasets. In conclusion, the high performance of these models proves that they can be used to estimate ground motions in metropolitan Istanbul.

3. Near-real-time strong motion network in Istanbul

Istanbul has one of the most sophisticated strong motion networks in the world, integrated with Istanbul's Earthquake Rapid Response and Early Warning System (IERREWS), constructed in 2002 by Kandilli Observatory and the Earthquake Research Institute (KOERI) [56]. In this section, Istanbul's strong motion network is introduced. Later, the near-real-time hazard maps created by the officials using an earthquake parameter (PGA) recorded from the network are presented.

3.1 Introduction to the network

Apart from structures, a strong earthquake in Istanbul could damage infrastructures like the natural gas distribution network and subsequent damage from gas leaks from damaged pipelines (e.g., building damage, deaths, and injuries from fire and explosions), as well as production failures due to the natural gas cut-off for industrial subscribers. As a result, it is vital to determine the natural gas distribution system's reliability in the event of an earthquake to calculate the total economic losses and develop appropriate emergency response measures and plans. Following the loss and damage caused by the August 17th, and November 12th, 1999 earthquakes, the local administration, governmental agencies, non-governmental organizations, and scholars recognized the importance of developing elaborate earthquake response plans based on detailed earthquake analyses in Istanbul. This resulted in the Istanbul Metropolitan Municipality commissioning the research as "Updating the Possible Earthquake Losses in Istanbul" [57]. As a result of these investigations, IGDAS and the KOERI conducted a study to design a real-time risk mitigation system for the gas distribution network.

The strong motion network consists of 832 strong motion stations for use in automatic shut-off mechanisms when a ground motion parameter threshold is exceeded. The network's accelerometer stations were created in collaboration with the

Turkish Scientific and Technological Research Center (TUBITAK). KOERI provides firmware for real-time ground motion parameter calculation. As shown in **Figure 1**, these accelerometers were installed at 832 selected district regulator stations to enable real-time shut-off of gas valves using the parametric data received online from the stations. The network was also built in the city's south, where population and seismic risk are highest. **Figure 2** depicts Istanbul's night-time population density.

The accelerometer stations were configured to compute the following real-time strong motion parameters to initiate the shut-offs: (1) Peak Ground Acceleration (PGA) and Peak Ground Velocity (PGV), (2) Arias intensity for ground motions in the x-direction (lax), (3) Cumulative Absolute Velocity integrated with a 5 cm/s^2 lower threshold (CAV5) appears to adequately represent the ground motion's longer-period (lower-frequency) components, (4) As response spectral parameters, Pseudo-Spectral acceleration (PSA), Pseudo-Spectral velocity (PSV), and Spectral displacement (Sd), (5) As initially stated by [58], Spectral Intensity (SI) is defined as the average spectral velocity over the vibration period range [0.1, 2.5] s [59].

The KOERI earthquake rapid response system's real-time ground motion data and the strong motion network's accelerometric data are integrated to estimate earthquake hazards (real-time ground motion distribution) and risks (damage distribution). For rapid response purposes, a total of 220 strong motion stations (110 from KOERI and 110 from IGDAS) were first integrated in 2013 [60]. Data integration is carried out using the **E**arthquake **L**oss **E**stimation **R**outine (ELER)-earthquake shaking map algorithm. The data is merged at the KOERI and IGDAS Scada Centers via two servers. The data is exchanged via a fiber optic virtual private network (VPN).

The Istanbul Natural Gas Distribution Network Seismic Risk Reduction Project (IGRAS) earthquake risk mitigation system comprises three major modules: a scenario earthquake module, a real-time earthquake module, and a map management

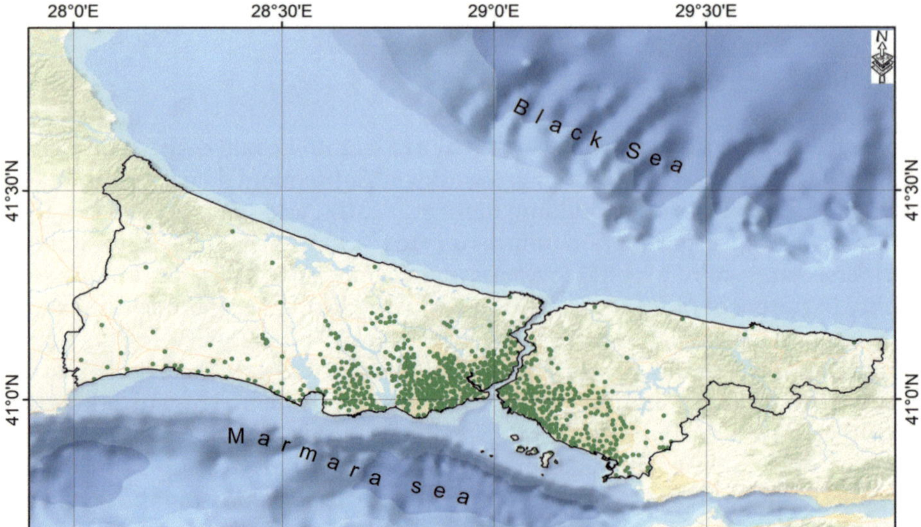

Figure 1.
The location of the strong motion network in Istanbul.

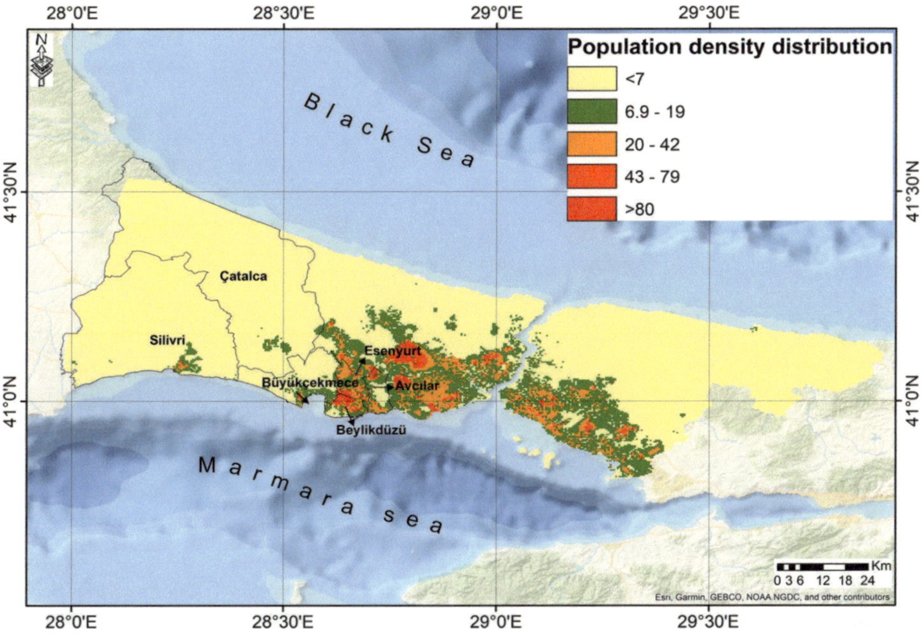

Figure 2.
Night-time population density distribution for metropolitan Istanbul.

module. These primary modules are subdivided into the following: (a) an earthquake hazard map module; (b) an IGDAS pipeline infrastructure damage assessment module; and (c) a service box damage assessment module.

The earthquake hazard map module generates grid-based distributions of ground motion parameters such as macroseismic intensity, PGA, PGV, and spectral acceleration for various vibration periods based on the magnitude and location of the earthquake. The module makes use of GMPEs that are appropriate for Istanbul, as well as grid-based local site information such as V_{s30} and a fault database. The recorded ground motion parameters from KOERI and IGDAS strong motion networks are used for GMPE bias correction, and integrated final earthquake hazard maps are delivered. The IGDAS pipeline infrastructure damage assessment module associates the earthquake hazard information generated by the earthquake hazard map module with the natural gas infrastructure elements and assesses the damage to the natural gas infrastructure elements analytically. A module for calculating natural gas building damage first connects the earthquake hazard and the building inventory and then calculates the building damages. Then, it correlates the findings of the building damage analysis with the locations of natural gas service boxes and determines the estimated number of damaged ones. Additionally, a GIS-based Web application for the IGRAS system is designed, which allows for online monitoring of scenario-based and real-time earthquake hazard, damage, and loss estimation findings.

3.2 Earthquake hazard maps created by the IGRAS system during the *Mw* 5.8 Silivri earthquake

During the *Mw* 5.8 Silivri earthquake, the real-time earthquake ground motion data was obtained by the rapid response and risk mitigation system in İstanbul.

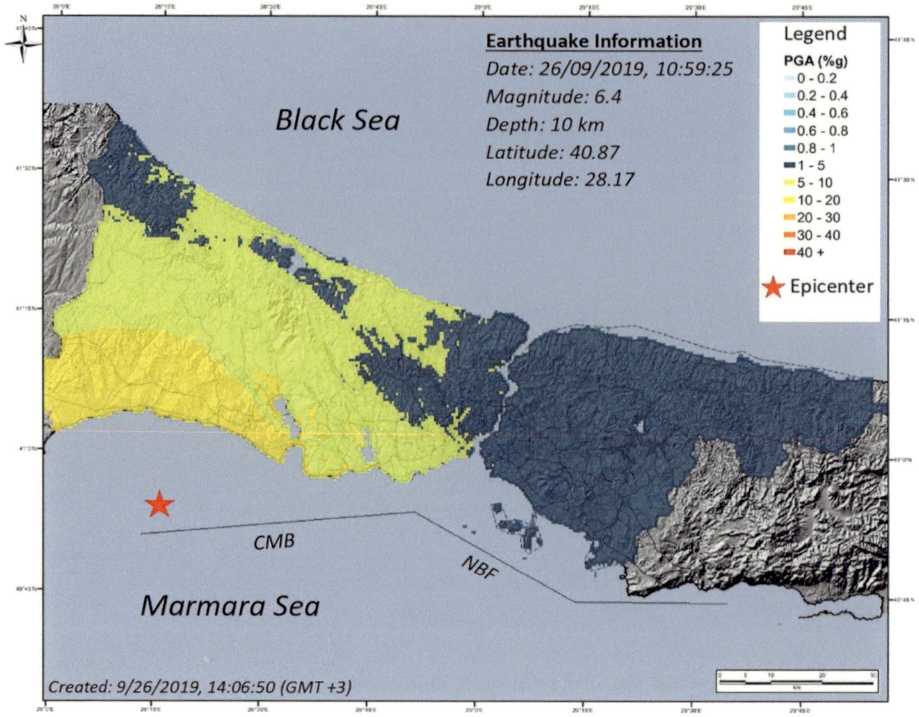

Figure 3.
PGA distribution map created at 14:06:50.

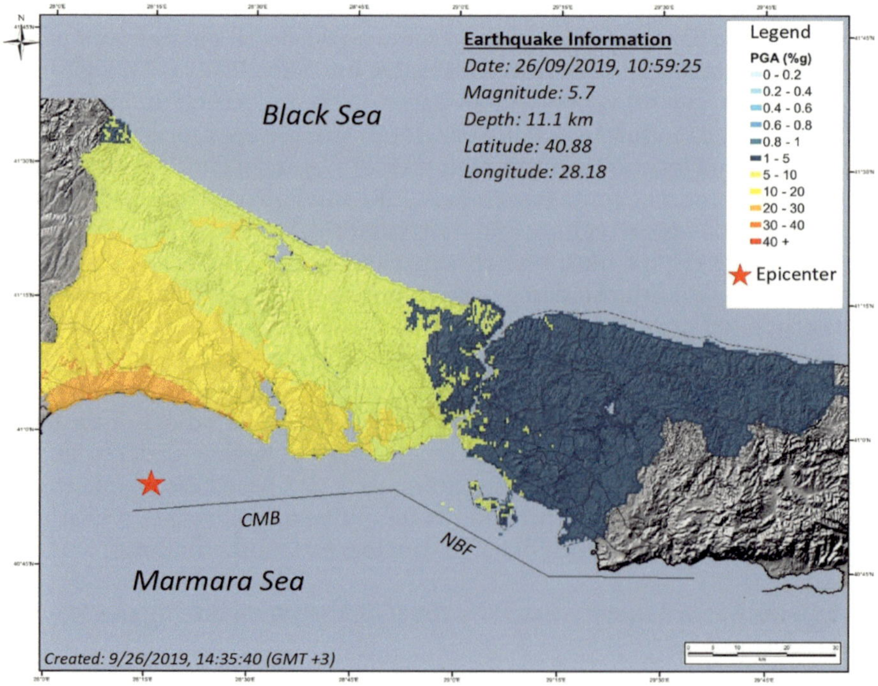

Figure 4.
PGA distribution map created at 14:35:40.

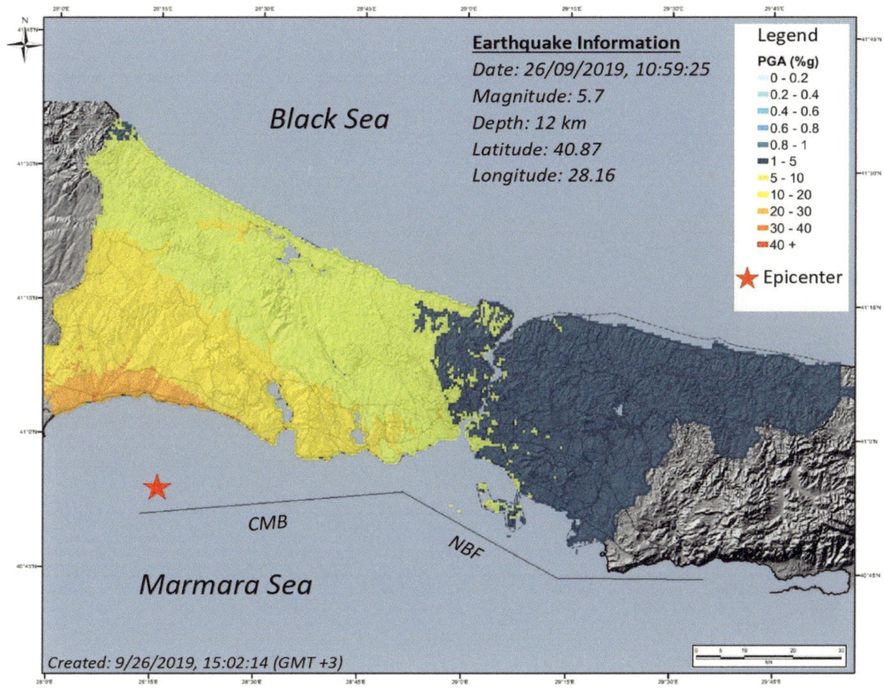

Figure 5.
PGA distribution map created at 15:02:14.

Figures 3–5 illustrate PGA hazard maps automatically generated by the network at intervals of about 30 minutes after the earthquake. As the network updates the data collected after the event, the PGA distribution and earthquake information change for each map, especially between the first and the last two maps created. Figures show that the final PGA is between 0.3 and 0.40 g.

4. Earthquake hazard maps and damage assessment: aftermath of the *Mw* 5.8 Silivri earthquake

This section will share earthquake hazard maps and results of building damage distribution after the Silivri Earthquake (*Mw* 5.8). First, a brief introduction about the Silivri Earthquake will be given. Then, seismic hazard maps created using ELER will be presented. Finally, the damage estimations will be depicted based on hazard maps. Details regarding adopted methods for urban damage assessment can be found in the ELER V3.0 Manual [61].

4.1 A brief information on the *Mw* 5.8 Silivri earthquake

On September 26, 2019, at 13:59 Turkish time (GMT +3), an *Mw* 5.8 offshore earthquake struck the Marmara Region, Turkey. This earthquake occurred at a depth of 6.99 kilometers from the ground and is located 21.99 kilometers from the nearest settlement, Silivri, in Istanbul. **Figure 6** depicts the epicenter of the earthquake. 150 aftershocks with magnitudes ranging from 1.0 to 4.1 were recorded between the mainshock and 07:40 on September 27, 2019 [62]. The Silivri earthquake is significant

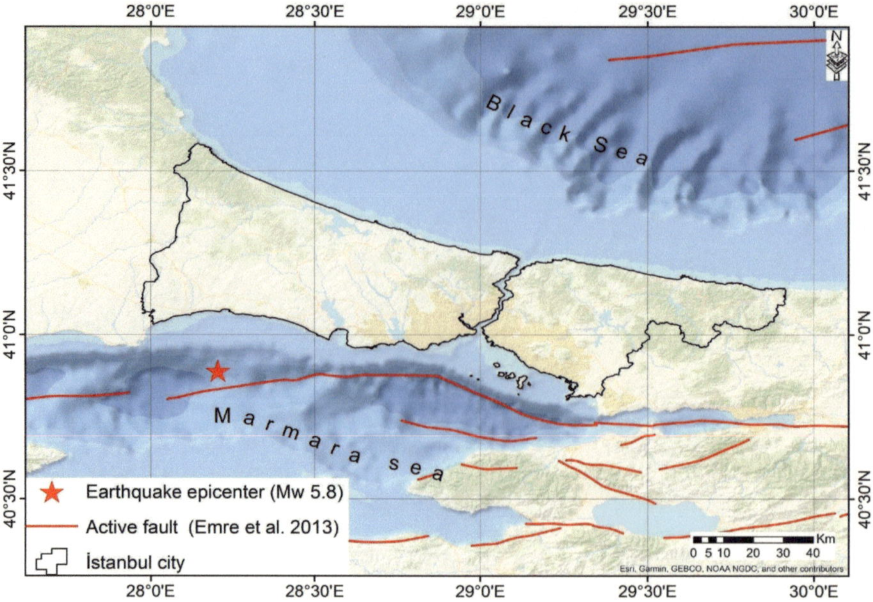

Figure 6.
The epicenter of the Mw 5.8 Silivri earthquake.

since it is the region's largest earthquake since the 1999 Kocaeli (*Mw* 7.6) and 1999 Düzce (*Mw* 7.1) earthquakes and their aftershocks. For further information, readers are advised to refer a recent paper by [63].

4.2 Earthquake hazard maps during the Silivri earthquake

The earthquake hazard maps are generated by grid-based distributions of ground motion parameters such as macroseismic intensity, PGA, and spectral acceleration for various vibration periods (SA02, SA10) based on the magnitude and location of the earthquake. This is achieved by making use of ground motion prediction equations (GMPEs) that are appropriate for Istanbul, as well as grid-based local site information such as V_{s30} and a fault database. **Figure 7** illustrates the V_{s30} map for metropolitan Istanbul. The choice of proper GMPEs can have a significant impact on design and safety evaluation as mentioned before (see Section 2.4). It is well known that attenuation relationships offer a satisfying estimation of ground motion parameters (PGA and SAs) depending on the magnitude and source-to-site distance.

The NGA2014's relationship with regional factors affecting Turkey in the aftermath of the mid-sized Marmara Sea earthquake was examined by [64]. According to their study, it appears that Chiou and Youngs 2014 (CY2014) provides the best fit to local PGA datasets recorded during the 26 September 2019 Silivri earthquake.

The CY2014 GMPE model is appropriate for predicting horizontal ground motion amplitudes associated with earthquakes in active tectonic zones that meet the following criteria [49]: (1) $3.5 \leq M \leq 8.5$ for strike-slip earthquakes, (2) $3.5 \leq M \leq 8.0$ for reverse and normal faulting earthquakes, (3) $Z_{TOR} \leq 20$ km, (4) $0 \leq R_{RUP} \leq 300$ km • $180 \leq V_{S30} \leq 1500$ m/s. Therefore, in this work, CY2014 GMPE model is used for generating the earthquake hazard maps.

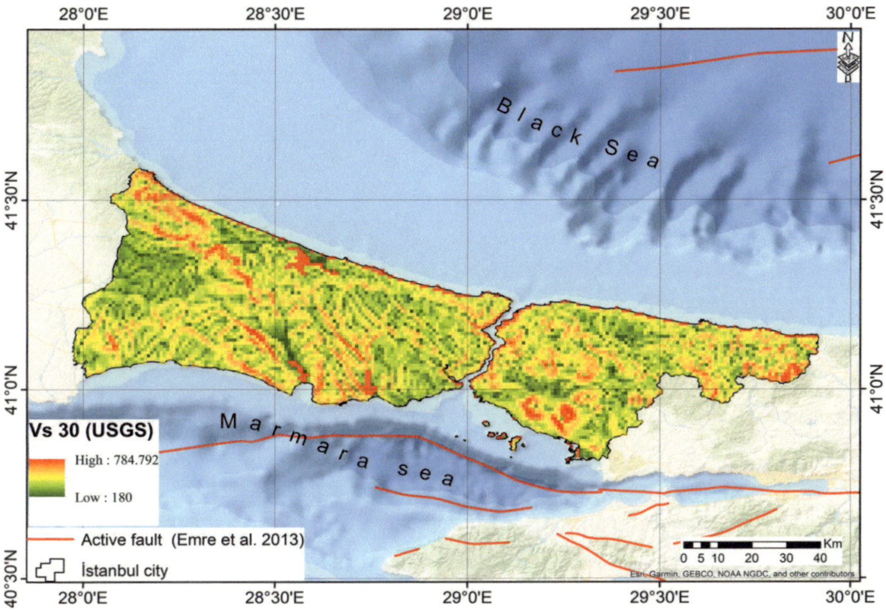

Figure 7.
V_{s30} map for metropolitan Istanbul.

The Modified Mercalli (MMI) conversion has been achieved through the regression relationships developed by [65]. The distribution maps for earthquake parameters (MMI, PGA, SA02, and SA10) are depicted in **Figures 8** and **9**. The European coastal part of Istanbul represents the highest intensity, MMI V (**Figure 7a**). PGA is found 0.1529 g, as seen in **Figure 8b**. Spectral accelerations descend from the coastline to the midland, as seen in **Figure 9**.

4.3 Building damage assessment

The loss estimation engine for the Level 1 (Loss assessment) module in ELER is based on macroseismic damage estimation tools and aims to analyze both building damage and casualties. In Level 1, the intensity-based empirical vulnerability relationships and casualty vulnerability models based on multiple approaches can be

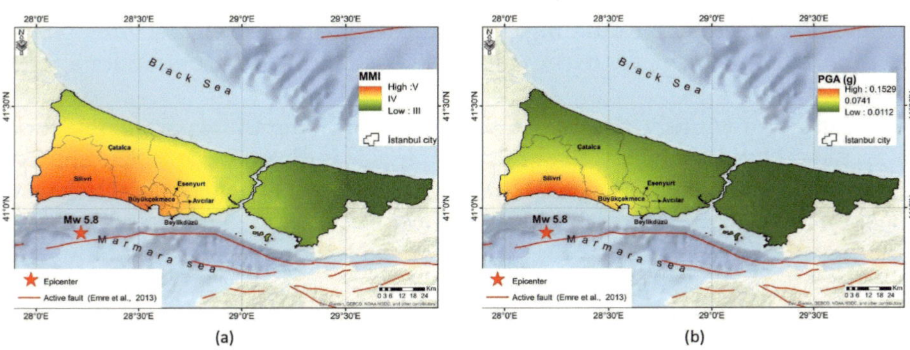

Figure 8.
Earthquake hazard maps (ground motion distribution) using CY14 GMPE model: (a) MMI, (b) PGA.

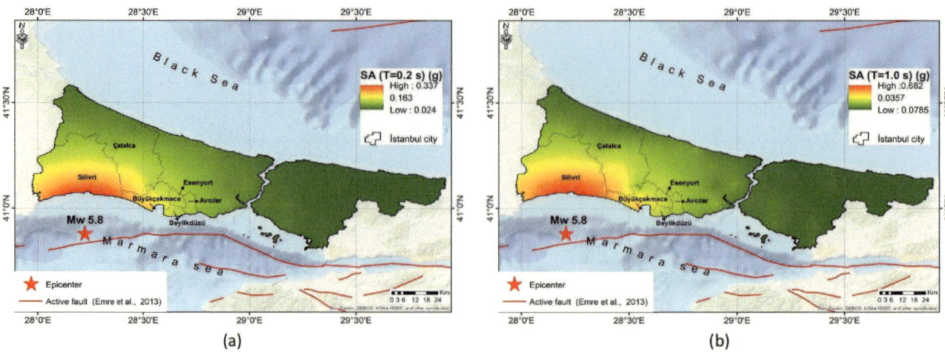

Figure 9.
Earthquake hazard maps (ground motion distribution) using CY14 GMPE model: (a) SA02 (b) SA10.

utilized. The intensity grid should be a MATLAB (.mat) file containing a grid matrix and a reference vector created in the Hazard module. The building database file is essentially a Shapefile (.shp) containing the distribution of buildings within each cell. This file may also include the population of each cell for the purpose of computing casualties. In the absence of a population field in the building database, casualty estimates are approximated using the regional population (obtained from the Land Scan population distribution). The vulnerability-ductility as a MATLAB (.mat) file including a table describing the vulnerability, ductility, t parameter, and replacement cost of each building type. ELER scans the building database for each building type specified in the vulnerability-ductility table and, if found, calculates the damage.

For Level 1 analysis, the building inventory and population data consist of grid-based (geocell) building and population distribution. Based on the Risk UE building taxonomy, the building distribution for the Marmara region is used as sample data. In ELER, Turkey's data is supplied as a model for other regions and nations to develop/incorporate their inventory data.

Damage estimation involves obtaining a cumulative damage probability using a normal distribution for each building type in Level 1. The damage probability distribution is dependent on the vulnerability and ductility factors of each building [66]. They derived the observed damage-based vulnerability approach known as the macroseismic method from the definition supplied by the European Macroseismic Scale [67] utilizing classical probability theory and fuzzy-set theory. As the purpose of a Macroseismic Scale is to acquire a measure of the earthquake's severity based on the damage sustained by the buildings, the scale itself can be used as a vulnerability model for predicting purposes to provide the likely damage distribution for a given intensity.

As discussed previously, the vulnerability table provides these parameters for each building type. The probability of cumulative damage is discretized to yield the five damage states as D1-slight damage, D2-moderate damage, D3-substantial to heavy damage, D4-very heavy damage, D5-destruction.

Since more significant damage is expected in coastal areas close to the epicenter, only Silivri, Çatalca, Büyükçekmece, Beylikdüzü, Esenyurt, and Avcılar are considered for damage distribution plots. **Figure 10** illustrates building damage distribution created employing the MMI results provided from Hazard Module in ELER. Slightly higher distribution levels of damage are seen in Esenyurt, Avcılar, and Beylikdüzü.

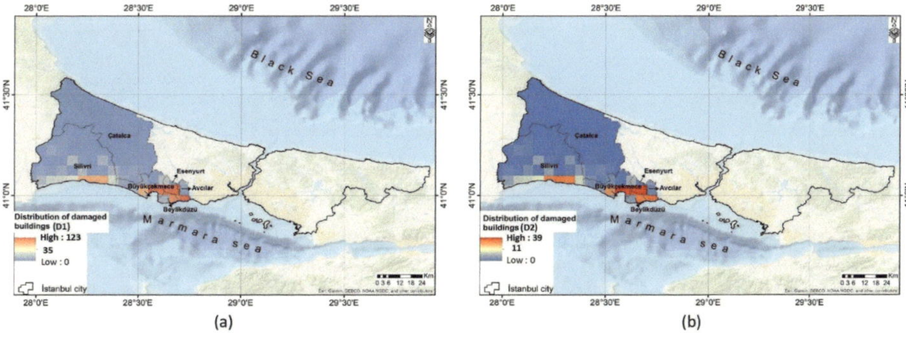

Figure 10.
Building damage distribution obtained for the Mw 5.8 Silivri earthquake: (a) D1-slight damage, (b) D2-moderate damage.

5. Near-real-time hazard and damage distribution maps generated using the data recorded by the strong motion network in Istanbul

The Hazard module estimates spatially distributed intensity and ground motion parameters (PGA, PGV, Sa, and Sd) using region-specific ground motion prediction equations and gridded shear wave velocity data for a given earthquake magnitude and epicenter. In this section, near-real-time earthquake parameters (PGA, SA02, and SA10) and damage (slight and moderate) distribution maps are generated by utilizing data from Istanbul's strong motion network. The data collected from triggered stations during the Silivri earthquake were used as event data, and distribution maps were plotted based on the GMPE of CY2014 after bias correction of phantom stations by the data utilized from the strong motion network in İstanbul. The source type panel defines the source mechanism associated with the event. For small-magnitude events, the source can be given as a point; for large-magnitude events, the user can specify the source type as a finite fault. The source type was assigned as "point source" since we simulated a moderate earthquake in this case. ELER can use ground motion prediction equations with V_{s30} [52, 68–70] as an input parameter to directly calculate the ground motion values at the surface. This feature was employed because the network recorded the event at the surface level. Custom site condition map is used as in the form of V_{s30} grids.

Figures 11–13 show the PGA, SA02, and SA10 distributions, respectively. Of 832 stations, 116 detected ground vibrations during the Silivri earthquake. Some data collected from acceleration stations is seen as a clear difference from the estimation of phantom stations. These abrupt and apparent variations in the distribution maps are shown as circles with a dashed line. As expected, the higher ground accelerations are seen in Silivri and Büyükçekmece's coastal region. This condition is also the same for SA02. However, Beylikdüzü also has a high SA10 spectral acceleration as other specified coastal counties. PGA is found to be 0.3496 g, as seen in **Figure 11**.

The building damage was estimated using the spectral acceleration-displacement-based vulnerability assessment methodology. ELER employs a variety of strategies for achieving this goal in the Level 2 (Loss assessment) module. There are currently two options for the seismic demand spectrum: (1) Euro Code 8 spectrum and (2) IBC 2006 spectrum. In this work, the construction of the 5%-damped elastic response spectra is selected as IBC (International Building Code) 2006. [71] provides a general

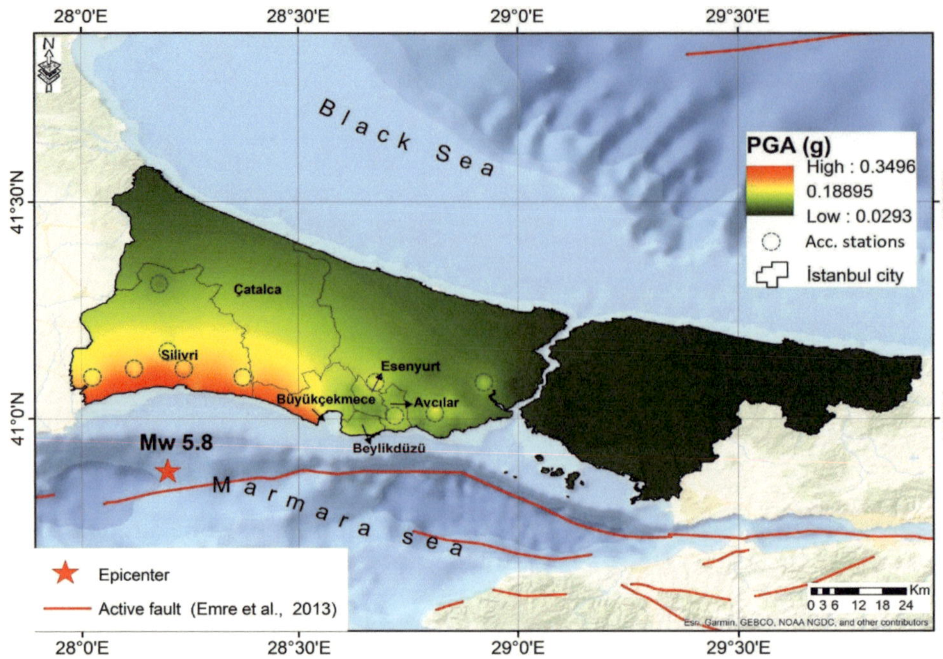

Figure 11.
PGA distribution maps obtained for the Mw 5.8 Silivri earthquake based on the GMPE of CY2014 after bias correction of phantom stations by the strong motion network in İstanbul.

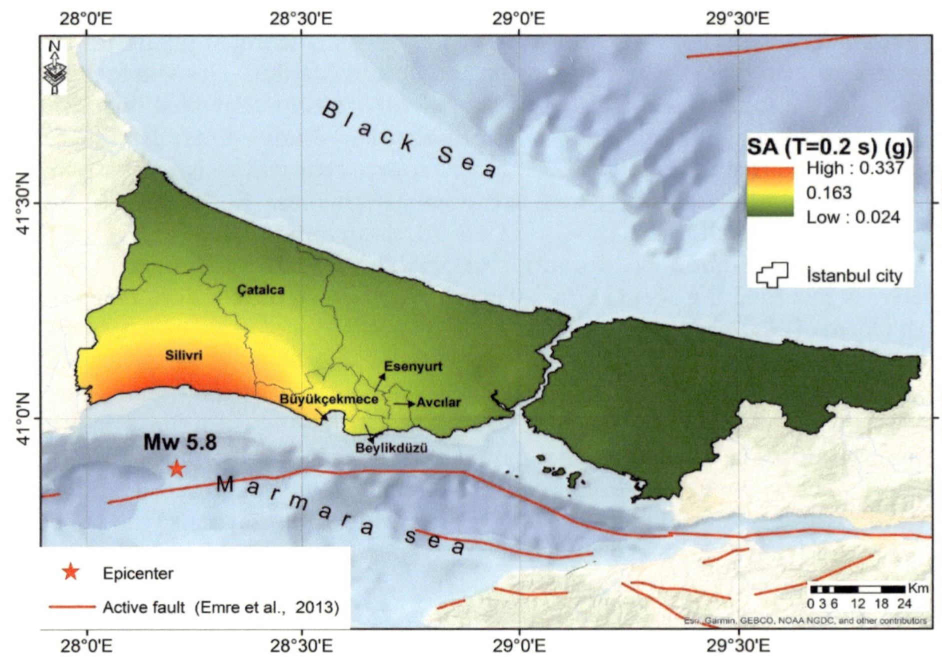

Figure 12.
SA02 spectral acceleration distribution maps obtained for the Mw 5.8 Silivri earthquake based on the GMPE of CY2014 after bias correction of phantom stations by the strong motion network in İstanbul.

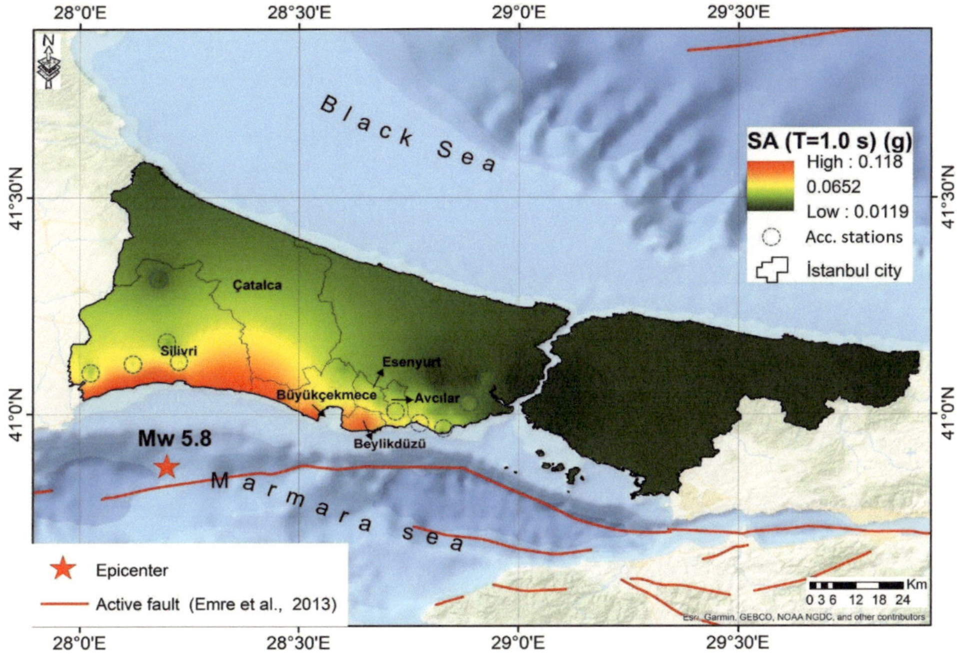

Figure 13.
SA10 spectral acceleration distribution maps obtained for the Mw 5.8 Silivri earthquake based on the GMPE of CY2014 after bias correction of phantom stations by the strong motion network in İstanbul.

horizontal elastic acceleration response spectrum. It is defined by (1) spectral accelerations at short period and 1-sec period, respectively, (2) short period and 1-sec period design response spectral accelerations adjusted for the specified site class and damping value (S_{DS}, S_{D1}), (3) corner periods of the constant spectral acceleration region given by $T_0 = 0.2T_S$ and $T_S = S_{D1}/S_{DS}$, and (4) long-period transition period. It is a regional-dependent parameter, and it is assumed that $T_L = 5$ s herein. In IBC 2006 and NEHRP 2003 Provisions [72], the recommended values for site and damping adjustments are stated. Spectral acceleration values at 0.2 and 1 sec periods are required to generate the IBC-2006 demand spectrum for each geographical unit. The user provides two MATLAB (.mat) files including a grid matrix and a reference vector of 0.2 and 1.0 sec spectral accelerations.

These two spectral acceleration files can be obtained from the previous calculations in Hazard Module or it can be developed by the user in a proper format for Level 2. In this section the spectral accelerations recorded from the strong motion network in İstanbul during the Silivri earthquake were used.

For the estimation of building damage in a Level 2 module, analytical fragility relationships and spectral acceleration-displacement-based vulnerability assessment approaches are applied. In the current case study, Capacity Spectrum Method (CSM-ATC 40) [73] was employed. The CSM is an approximate heuristic method that essentially assumes that a complex non-linear multi-degree-of-freedom system, such as a multi-story building experiencing severe plastic deformations during an earthquake, can be modeled as an equivalent single-degree-of-freedom system with an appropriate level of inelasticity. The simplicity of the procedure is the advantage of the method since no time history analysis is required.

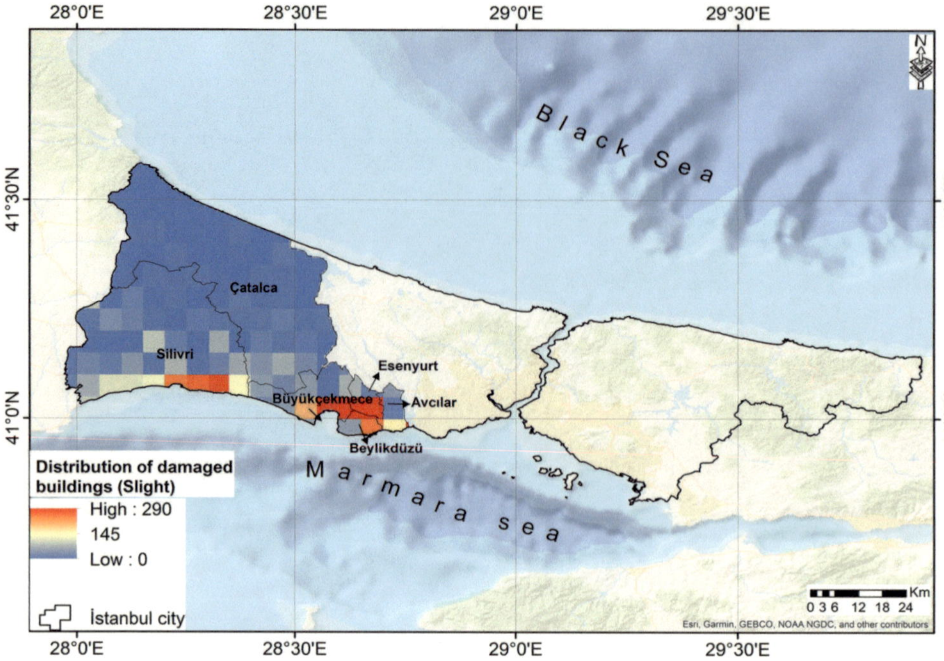

Figure 14.
Slight damage distribution map of buildings obtained for the Mw 5.8 Silivri earthquake.

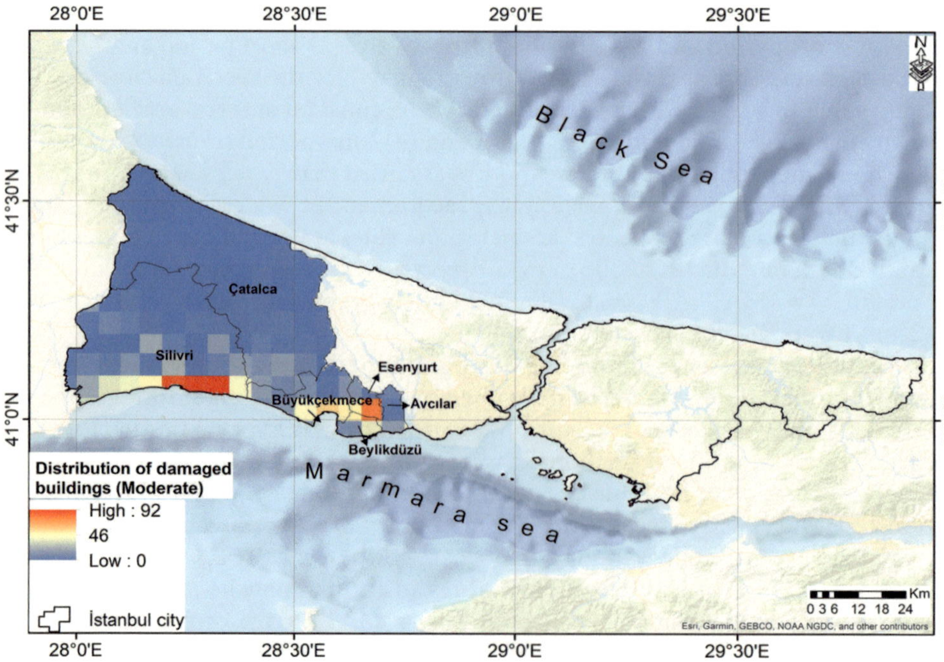

Figure 15.
Moderate damage distribution map of buildings obtained for the Mw 5.8 Silivri earthquake.

The building and population data for Level 2 analysis are grid-based (geocell) urban building and demographic inventory. For building grouping, the European building taxonomy created as part of the EU-FP5 RISK-UE project [66, 74].

Since more significant damage is expected to occur in coastal areas close to the epicenter, only Silivri, Çatalca, Büyükçekmece, Beylikdüzü, Esenyurt, and Avcılar are considered for damage distribution. **Figures 14** and **15** show slight and moderate damage distribution maps of buildings obtained for the *Mw* 5.8 Silivri earthquake using the strong motion network data, respectively. The higher levels of damage are seen in Silivri, partially in the coastal regions of Büyükçekmece, Beylikdüzü, and Avcılar. In addition, a higher damage distribution is observed in Esenyurt than in other hinterland regions. Because extreme and complete damage are not observed, their maps were not created.

6. Discussion

The results in Sections 4 and 5 are discussed in terms of PGA, SA02, SA10 (hazard maps), and damage distribution maps (loss assessment) of buildings, with comparisons between attenuation relationships (CY2014) and the distribution maps generated from data recorded by the strong motion network in İstanbul during the *Mw* 5.8 Silivri earthquake.

- PGA distribution maps created automatically at Istanbul Natural Gas Distribution Company (IGDAS) Scada Centers by IGDAS strong motion stations require a specific amount of time until the final map is made, providing reliable results of PGA values and earthquake information. This is because earthquake parameters are continuously updated after the event. After corrections are made to earthquake parameters, the map is updated accordingly, and very comparable results are obtained with the map created by using data acquired from the near-real-time ground motion network and the map created at Technical Safety and SCADA Directorate Disaster Management Chiefdom by IGDAS (see **Figures 5** and **11**).

- It was determined that the maximum acceleration value of the PGA distribution map drawn with the real station network was roughly 2.3 times higher than the maximum acceleration value obtained in the PGA distribution map drawn with the attenuation relationship utilizing the Level 1 analysis module (see **Figures 8b** and **11**).

- In Section 5, seismic hazard maps (PGA, SA02, and SA10) were created using the records obtained from the strong motion station network and calculated using an appropriate attenuation relationship (CY2014) after bias correction of phantom stations using the data utilized from the strong motion network in İstanbul. In **Figure 13**, it is observed that some PGA values recorded by the network are not compatible with the values obtained from the bias correction of phantom stations using CY2014.

- As mentioned in Section 5, currently, out of the 832 strong motion stations constructed by IGDAS in Istanbul, 116 stations were able to record the offshore event since most of the stations were located in highly populated counties in which seismic risk was considered to be higher. Because most stations were away from the event epicenter, 716 strong motion stations were not triggered.

- Although the event was felt in many cities in the region (such as Yalova, Kocaeli, Bursa, Kırklareli, and Tekirdag) and its destructive significance was moderate, the significant effects of the earthquake were observed more on the European side of Istanbul, where the epicenter was closer to the region. This can be detected from the PGA distribution maps created using recorded stations right after the earthquake (see **Figures 11–13**).

- In Section 4, damage analysis was conducted in Level 1 using the MMI results of GMPE-CY2014 obtained from the Hazard Module in ELER. Section 5 presents the analysis results of Level 2 using SA02 and SA10 recorded from the station network in İstanbul. It is seen that the damage distributions corresponding to the highest damage per cell obtained from Level 2 (Section 5) are approximately 2.4 times more than those obtained from Level 1 (Section 4). The reason for this is that the actual station recordings exceed the CY2014 predictions.

7. Conclusions

This chapter mainly focuses on urban damage assessment using data acquired from a real-time ground motion network during a recent moderate offshore event, the Silivri earthquake (*Mw* 5.8), which struck the Istanbul metropolitan area. Earthquake early warning and rapid response systems in İstanbul are introduced as one of the most sophisticated measures to mitigate seismic risk in the world. This highlights the significance of such systems' beneficial effects in earthquake-prone regions. The major findings of this paper are listed in two aspects as follows:

Urban damage assessment after a moderate earthquake event:

1. Using data from 116 stations recorded in real-time, seismic hazard (ground motion distribution) maps are created for earthquake parameters (such as PGA, SA02, and SA10). To the authors' best knowledge, this is the first research conducted for the region using real-time strong motion data collected from densely installed stations to generate a hazard map of a metropolitan area like Istanbul. In addition, it is believed that this is a convenient method to validate the results obtained from attenuation relationships (GMPEs).

2. Regarding European counties, the highest PGA is monitored in the coastal regions of Silivri and Buyukcekmece, which are the two counties that are nearest to the epicenter (see **Figure 11**).

3. In the case that the expected Istanbul earthquake occurs near the Central Marmara Basin (CMB) fault, where the Silivri earthquake occurred, significant damage is also expected in Silivri and along with its coastal line. Higher damage distribution will be seen in some part of the coastal regions of Büyükçekmece, Beylikdüzü, and Avcılar. In addition, Esenyurt may have a greater damage distribution than other hinterland regions mentioned.

4. Comparing the results of the two analyses reveals that Level 1 damage distributions are more conservative than Level 2 outcomes, and this is because the actual station recordings exceed the CY2014 predictions.

5. The recorded ground motion parameters from the strong motion networks are used for GMPE bias correction, and integrated final earthquake hazard maps are delivered. In this study, as in previous research stated, CY2014 provides one of the best fits to local PGA datasets recorded during the earthquake, CY2014 was chosen as GMPE for attenuation relationship. However, as stated in the Discussion Section, since the results are more conservative than the actual results, a study with different attenuation relationships as future work will contribute to the literature.

6. Only the counties nearest to the epicenter are considered for damage distributions map plotting. In addition, extensive and complete damage maps were not plotted because only slight and moderate damage was detected due to the Silivri earthquake as a result of analyses.

7. Although Level 2 damage distribution results per cell were much higher than Level 1, similar damage distribution patterns were obtained as a result of both analyses (see **Figures 10**, **14** and **15**).

8. Damage assessments are conducted using the Marmara region's building inventory provided by ELER as sample data. Utilizing an updated building inventory and V_{s30} values will be beneficial for future endeavors.

9. During the on-site inspection, it was stated that 320 extremely damaged structures were identified. It is believed that there are two reasons behind the difference between analysis results and field inspections. First, the sample inventory data used for the analyses might not be detailed enough to represent the current building stock in the region. Secondly, the ground motion distribution estimated using the GMPE for Level 1 and Level 2 (for phantom stations) is more conservative than actual values. Therefore, it is recommended to use either more compatible GMPEs or sensor density might be considered to be increased, especially on southwest coast of Istanbul.

Beneficial effects of earthquake damage mitigation systems and methods:

1. Structural health monitoring systems can provide insight into structures' conditions during and after a seismic event. Comparing the recordings with those made prior to the earthquake can reveal a shift in structural response that have relevance for structural damage and a reduction in seismic resistance. Moreover, these technologies can be linked to early warning and rapid response systems to mitigate seismic risk.

2. Through an early warning and rapid response system network, besides the physical destructive effects of an earthquake, ensure that the gas systems are shut down in advance to prevent the destructive effects that the big explosions may cause.

3. GMPEs are prediction equations established based on numerous factors, such as the site's ground condition, the magnitude of the earthquake, and the type of fault. Consequently, their precision is mostly dependent on these characteristics. Utilizing such equations in seismic hazard analyses necessitates a high level of competence in this discipline. In addition, since the parameters to be used for attenuation relationships calculated with GMPEs will already be inherently present in the recorded acceleration data obtained from the densely established ground motion station network, these experimentally recorded data will yield the most accurate results in terms of ground motion and building damage distributions. However, it is vital to emphasize that the appropriate distribution of the stations provides better resolution for the hazard maps that may be created.

4. The estimations made with empirical GMPEs will be validated by comparing them with the experimentally recorded data in the field. Thus, the most realistic results in the field during the seismic risk mitigation process will be available to stakeholders for consideration.

In conclusion, this study emphasizes the importance of both GMPEs and densely installed ground motion station networks that collect real-time data during earthquakes and provides motivation for establishing or further developing similar systems.

Conflict of interest

The authors declare no conflict of interest.

Author details

Oğuzhan Çetindemir* and Abdullah Can Zülfikar
Civil Engineering Department, Gebze Technical University, Kocaeli, Turkey

*Address all correspondence to: ocetindemir@gtu.edu.tr

References

[1] Ali Z, Qaisar M, Mahmood T, Shah MA, Iqbal T, Serva L, et al. The Muzaffarabad, Pakistan, earthquake of 8 October 2005: Surface faulting, environmental effects and macroseismic intensity. Geological Society, London, Special Publications. 2009;**316**(1):155-172

[2] Vina A, Chen X, McConnell WJ, Liu W, Xu W, Ouyang Z, et al. Effects of natural disasters on conservation policies: The case of the 2008 Wenchuan earthquake, China. Ambio. 2011;**40**(3):274-284

[3] Tang B, Liu X, Liu Y, Xue C, Zhang L. A meta-analysis of risk factors for depression in adults and children after natural disasters. BMC Public Health. 2014;**14**(1):1-12

[4] Noy I, Yonson R. Economic vulnerability and resilience to natural hazards: A survey of concepts and measurements. Sustainability. 2018;**10**(8):2850

[5] Erdik M. Earthquake risk, earthquake rapid response system, and early warning system in Istanbul. Macalester International. 2005;**15**(1):12

[6] Anbarci N, Escaleras M, Register CA. Earthquake fatalities: The interaction of nature and political economy. Journal of Public Economics. 2005;**89**(9-10):1907-1933

[7] Nanto DK. Japan's 2011 Earthquake and Tsunami: Economic Effects and Implications for the United States. Darby, PA: Diane Publishing Co; 2011

[8] Elbaz K, Shen JS, Arulrajah A, Horpibulsuk S. Geohazards induced by anthropic activities of geoconstruction: A review of recent failure cases. Arabian Journal of Geosciences. 2016;**9**(18):1-11

[9] Lyu H-M, Shen S-L, Yang J, Zhou A-N. Risk assessment of earthquake-triggered geohazards surrounding Wenchuan, China. Natural Hazards Review. 2020;**21**(3):05020007

[10] Al-Hussein A, Haldar A. Condition assessment of civil structures under earthquake excitation. Journal of Aerospace Engineering. 2019;**32**(1):04018119

[11] Kelly DJ, Zona J, editors. Current practice and future trends for seismic design for buildings in new england. In: Structures Congress 2014. Reston, VA: American Society of Civil Engineers (ASCE). 2014. pp. 2330-2341

[12] Rosyidi SAP, Aldiamar F, Faizal N, Taha MR. Lesson learned from road infrastructure deterioration caused by Earthquake 7.6 and 6.2 MW Padang, September 30, 2009. In: Advances in Unsaturated Soil, Geo-Hazard, and Geo-Environmental Engineering. Reston, VA, United States: American Society of Civil Engineers; 2011. pp. 180-187

[13] Abbas M, Elbaz K, Shen S-L, Chen J. Earthquake effects on civil engineering structures and perspective mitigation solutions: A review. Arabian Journal of Geosciences. 2021;**14**(14):1-17

[14] Douglas J. Ground motion prediction equations 1964-2021. 2021

[15] Wald D, Jaiswal K, Marano K, Bausch D, Hearne M. PAGER–Rapid assessment of an earthquakes impact. US Geological Survey. Report No.: 2327-6932; 2010

[16] Lu X, Cheng Q, Xu Z, Xu Y, Sun C. Real-time city-scale time-history analysis and its application in resilience-oriented earthquake emergency responses. Applied Sciences. 2019;**9**(17):3497

[17] Cheng Q, Fei Y, Lu X, Liao W, Zhang W, Chen PY, et al. Influence of sensor density on seismic damage assessment: A case study for Istanbul. Bulletin of the Seismological Society of America. 2022;**112**(4):2156-2169. DOI: 10.1785/0120220005

[18] Gorini A, Nicoletti M, Marsan P, Bianconi R, De Nardis R, Filippi L, et al. The Italian strong motion network. Bulletin of Earthquake Engineering. 2010;**8**(5):1075-1090

[19] Liu K-S, Shin T-C, Tsai Y-B. A free-field strong motion network in Taiwan: TSMIP. Terrestrial, Atmospheric and Oceanic Sciences. 1999;**10**(2):377-396

[20] Mittal H, Kumar A, Ramhmachhuani R. Indian National strong motion instrumentation network and site characterization of its stations. International Journal of Geosciences. 2012;**3**(6A):1151-1167. DOI: 10.4236/ijg.2012.326117

[21] Aoi S, Kunugi T, Fujiwara H. Strong-motion seismograph network operated by NIED: K-NET and KiK-net. Journal of Japan Association for Earthquake Engineering. 2004;**4**(3):65-74

[22] Shahvar MP, Farzanegan E, Eshaghi A, Mirzaei H. i1-net: The Iran strong motion network. Seismological Research Letters. 2021;**92**(4):2100-2108

[23] Skarlatoudis A, Margaris B, Papazachos C, editors. Recent advances in Greece on strong-motion networking and data processing. In: The Consortium of Organizations for Strong-motion Observation Systems (COSMOS) Invited WORKSHOP. California: COSMOS Publications; 2004

[24] Craifaleanu I-G, Borcia I, Praun I. Strong-motion networks in Romania and their efficient use in the structural engineering applications. In: Earthquake Data in Engineering Seismology. Dordrecht: Springer; 2011. pp. 247-259

[25] Wald DJ, Quitoriano V, Heaton TH, Kanamori H, Scrivner CW, Worden CB. TriNet "ShakeMaps": Rapid generation of peak ground motion and intensity maps for earthquakes in southern California. Earthquake Spectra. 1999;**15**(3):537-555

[26] Polat O, Ceken U, Uran T, Gok E, Yilmaz N, Beyhan M, et al. IzmirNet: A strong-motion network in metropolitan Izmir, Western Anatolia, Turkey. Seismological Research Letters. 2009;**80**(5):831-838

[27] Ambraseys N, Finkel C. Long-term seismicity of Istanbul and of the Marmara Sea region. Terra Nova. 1991;**3**(5):527-539

[28] Le Pichon X, Şengör A, Demirbağ E, Rangin C, Imren C, Armijo R, et al. The active main Marmara fault. Earth and Planetary Science Letters. 2001;**192**(4):595-616

[29] Parsons T. Recalculated probability of M≥ 7 earthquakes beneath the Sea of Marmara, Turkey. Journal of Geophysical Research: Solid Earth. 2004;**109**(B5):B05304. DOI: 10.1029/2003JB002667

[30] Erdik M, Demircioglu M, Sesetyan K, Durukal E, Siyahi B. Earthquake hazard in Marmara region, Turkey. Soil Dynamics and Earthquake Engineering. 2004;**24**(8):605-631

[31] Ansal A, Akinci A, Cultrera G, Erdik M, Pessina V, Tönük G, et al. Loss estimation in Istanbul based on deterministic earthquake scenarios of the Marmara Sea region (Turkey). Soil Dynamics and Earthquake Engineering. 2009;**29**(4):699-709

[32] Okay AI, Kaşlılar-Özcan A, İmren C, Boztepe-Güney A, Demirbağ E, Kuşçu İ. Active faults and evolving strike-slip basins in the Marmara Sea, northwest Turkey: A multichannel seismic reflection study. Tectonophysics. 2000;**321**(2):189-218

[33] Ergintav S, Reilinger R, Çakmak R, Floyd M, Cakir Z, Doğan U, et al. Istanbul's earthquake hot spots: Geodetic constraints on strain accumulation along faults in the Marmara seismic gap. Geophysical Research Letters. 2014;**41**(16):5783-5788

[34] Woessner J, Laurentiu D, Giardini D, Crowley H, Cotton F, Grünthal G, et al. The 2013 European seismic hazard model: Key components and results. Bulletin of Earthquake Engineering. 2015;**13**(12):3553-3596

[35] Böse M. Earthquake early warning for Istanbul using artificial neural networks: Karlsruhe, Univ., [Diss., 2006]. 2006

[36] Zulfikar C, Pinar A, Tunc S, Erdik M, editors. Real-time earthquake risk mitigation of infrastructures using Istanbul earthquake early warning and rapid response network. In: EGU General Assembly Conference Abstracts. Geophysical Research Abstracts. Vienna, Austria. 2014;**16**:EGU2014-16149

[37] Cua G, Heaton T. The virtual seismologist (VS) method: A Bayesian approach to earthquake early warning. In: Earthquake Early Warning Systems. Berlin, Heidelberg: Springer; 2007. pp. 97-132

[38] Böse M, Allen R, Brown H, Gua G, Fischer M, Hauksson E, et al. CISN shakealert: An earthquake early warning demonstration system for California. In: Early Warning for Geological Disasters. Berlin, Heidelberg: Springer; 2014. pp. 49-69

[39] Hanka W, Saul J, Weber B, Becker J, Harjadi P. Real-time earthquake monitoring for tsunami warning in the Indian Ocean and beyond. Natural Hazards and Earth System Sciences. 2010;**10**(12):2611-2622

[40] Olivieri M, Clinton J. An almost fair comparison between earthworm and SeisComp3. Seismological Research Letters. 2012;**83**(4):720-727

[41] Behr Y, Clinton J, Kästli P, Cauzzi C, Racine R, Meier MA. Anatomy of an earthquake early warning (EEW) alert: Predicting time delays for an end-to-end EEW system. Seismological Research Letters. 2015;**86**(3):830-840

[42] Behr Y, Clinton JF, Cauzzi C, Hauksson E, Jónsdóttir K, Marius CG, et al. The virtual seismologist in SeisComP3: A new implementation strategy for earthquake early warning algorithms. Seismological Research Letters. 2016;**87**(2A):363-373

[43] Clinton J, Zollo A, Marmureanu A, Zulfikar C, Parolai S. State-of-the art and future of earthquake early warning in the European region. Bulletin of Earthquake Engineering. 2016;**14**(9):2441-2458

[44] Memisoglu Apaydin N, Zulfikar A, Cetindemir O. Structural health monitoring systems of long-span bridges in Turkey and lessons learned from experienced extreme events. Journal of Civil Structural Health Monitoring. 2022;**12**(6):1375-1412

[45] Çaktı E, Şafak E. Structural health monitoring: Lessons learned. In: Seismic Isolation, Structural Health Monitoring, and Performance Based Seismic Design in Earthquake Engineering. Cham, Switzerland: Springer; 2019. pp. 145-164

[46] Harith N, Ramadhansyah P, Adiyanto M, Ramli N, editors. Ground

motion observation of Sabah earthquakes on the use of next generation attenuation (NGA) ground-motion models. In: IOP Conference Series: Earth and Environmental Science. Bristol: IOP Publishing; 2021;**682**(1):012050

[47] Shiuly A, Roy N. A generalized VS–N correlation using various regression analysis and genetic algorithm. Acta Geodaetica et Geophysica. 2018;**53**(3):479-502

[48] Abrahamson NA, Silva WJ, Kamai R. Summary of the ASK14 ground motion relation for active crustal regions. Earthquake Spectra. 2014;**30**(3):1025-1055

[49] Chiou BS-J, Youngs RR. Update of the Chiou and Youngs NGA model for the average horizontal component of peak ground motion and response spectra. Earthquake Spectra. 2014;**30**(3):1117-1153

[50] Boore DM, Stewart JP, Seyhan E, Atkinson GM. NGA-West2 equations for predicting PGA, PGV, and 5% damped PSA for shallow crustal earthquakes. Earthquake Spectra. 2014;**30**(3):1057-1085

[51] Campbell KW, Bozorgnia Y. NGA-West2 ground motion model for the average horizontal components of PGA, PGV, and 5% damped linear acceleration response spectra. Earthquake Spectra. 2014;**30**(3):1087-1115

[52] Chiou B-J, Youngs RR. An NGA model for the average horizontal component of peak ground motion and response spectra. Earthquake Spectra. 2008;**24**(1):173-215

[53] Bozorgnia Y, Abrahamson NA, Atik LA, Ancheta TD, Atkinson GM, Baker JW, et al. NGA-West2 research project. Earthquake Spectra. 2014;**30**(3):973-987

[54] Boore DM. Orientation-independent, nongeometric-mean measures of seismic intensity from two horizontal components of motion. Bulletin of the Seismological Society of America. 2010;**100**(4):1830-1835

[55] Akcan SO, Zülfikar A, Bıyıkoğlu H. Evaluation of strong ground motion records of 2019 *Mw* 5.8 Silivri Earthquake with NGA-WEST2 GMPEs

[56] Erdik M, Fahjan Y, Ozel O, Alcik H, Mert A, Gul M. Istanbul earthquake rapid response and the early warning system. Bulletin of Earthquake Engineering. 2003;**1**(1):157-163

[57] Report. IMMI. Updating of the Possible Earthquake Losses in Istanbul. Istanbul: IMM; 2009

[58] Housner GW. Spectrum Intensities of Strong-motion Earthquakes. In: Proceedings of the Symposium on Earthquake and Blast Effects on Structures: Los Angeles, California, Earthquake Engineering Research Institute, Los Angeles. Pasadena CA, United States: CalTech Library. June 1952. pp. 20-36. Available from: https://resolver.caltech.edu/CaltechAUTHORS:20161010-155126031

[59] Zulfikar C, Erdik M, Safak E, Biyikoglu H, Kariptas C. Istanbul natural gas network rapid response and risk mitigation system. Bulletin of Earthquake Engineering. 2016;**14**(9):2565-2578

[60] IGRAS. Istanbul natural gas distribution network seismic risk reduction project. 2013

[61] ELERv3.0. Earthquake loss estimation routine. 2010

[62] AFAD. The disaster and emergency management presidency. 2019. Available from: https://en.afad.gov.tr/

[63] Malcioglu FS, Süleyman H, Çaktı E. Seismological and engineering characteristics of strong motion data from 24 and 26 September 2019 Marmara Sea earthquakes. Bulletin of Earthquake Engineering. 2022;**20**(11):5567-5599. DOI: 10.1007/s10518-022-01422-y

[64] Zülfikar AC, Tekin S, Akcan SO, Gök MG. 26 Eylül 2019 Silivri açıkları (Marmara Denizi) depreminin kuvvetli yer hareketi verilerinin değerlendirilmesi. Journal of the Institute of Science and Technology. 2020;**10**(3):1720-1736

[65] Wald DJ, Quitoriano V, Heaton TH, Kanamori H. Relationships between peak ground acceleration, peak ground velocity, and modified Mercalli intensity in California. Earthquake Spectra. 1999;**15**(3):557-564

[66] Lagomarsino S, Giovinazzi S. Macroseismic and mechanical models for the vulnerability and damage assessment of current buildings. Bulletin of Earthquake Engineering. 2006;**4**(4):415-443

[67] Grünthal G, Levret A. L'echelle macrosismique européenne. In: European Macroseismic Scale 1998:(EMS-98). Luxembourg: Centre Européen de Géodynamique et de Séismologie Musée National d'Histoire Naturelle, Luxembourg Section Géophysique et Astrophysique; 2001

[68] Campbell KW, Bozorgnia Y. NGA ground motion model for the geometric mean horizontal component of PGA, PGV, PGD and 5% damped linear elastic response spectra for periods ranging from 0.01 to 10 s. Earthquake Spectra. 2008;**24**(1):139-171

[69] Boore DM, Atkinson GM. Ground-motion prediction equations for the average horizontal component of PGA, PGV, and 5%-damped PSA at spectral periods between 0.01 and 10.0 s. Earthquake Spectra. 2008;**24**(1): 99-138

[70] Boore DM, Joyner WB, Fumal TE. Equations for estimating horizontal response spectra and peak acceleration from western North American earthquakes: A summary of recent work. Seismological Research Letters. 1997;**68**(1):128-153

[71] IBC I. International building code. International Code Council, Inc.: (formerly BOCA, ICBO and SBCCI). 2006;**4051**:60478-65795

[72] Council BSS. NEHRP Recommended Provisions for Seismic Regulations for New Buildings and Other Structures. United States: FEMA; 2003

[73] Comartin C, Niewiarowski R, Rojahn C. ATC-40 Seismic evaluation and retrofit of concrete buildings. SSC 96. 1996;1.

[74] Mouroux P, Le Brun B. RISK-UE project: An advanced approach to earthquake risk scenarios with application to different European towns. In: Assessing and Managing Earthquake Risk. Springer; 2008. pp. 479-508

Chapter 6

Crustal Faults Reactivated during 2010 Mw = 8.8 Maule Earthquake in South Chile

Jorge Quezada and Arturo Belmonte

Abstract

On February 27th 2010 occurred the Mw = 8.8 Maule subduction earthquake, filling a seismic gap of south Chile. The uplift trend is mostly typical for subduction earthquakes with decreasing uplift trend from trench to arc in Andes Cordillera. However local perturbations occurred due to the reactivations of crustal faults occurred such as Pichilemu fault (normal), Santa María fault (normal) and Tirua-Mocha fault (reverse). Different kind of faults and seismic behavior evidence complex stress distribution at the overriding South American Plate. In this paper, the activity and seismicity linked of some crustal faults at Maule earthquake rupture área are considered, and the related seismic potential that can increase the seismic hazard. Some questions are the bigger magnitude that can generate these faults and if their activity is related to the interseismic or coseismic phases of the subduction seismic cycle.

Keywords: earthquakes, crustal, faults, Chile, seismicity

1. Introduction

Chile is a seismic country due to its tectonic framework, dominated by the subduction of Nazca Plate beneath South American Plate at approximately 7 cm/yr at N77°E trend. The Wadati Benioff zone has N10°E strike and 20°E dip. Geological hazard consequences of this are earthquakes, tsunami, and volcanic activity. Subduction earthquakes of thrust focal mechanism included the biggest ever recorder, the Mw = 9.5 Valdivia earthquake that occurred on May 22, 1960 and the sixth, the February 27, 2010 Mw = 8.8 Maule earthquake. In a same segment of the subduction zone, Mw = 8 earthquakes occurred every 100 years and an Mw = 9 earthquake every 300 years or more [1–9]. **Figure 1** shows some of the main subduction earthquakes and rupture lengths that occurred in central and south Chile, divided in Central Chile, Concepción, and Valdivia segments. The epicenters of the subduction seismicity are located from 10 km east to the trench to 160 km eastward in the littoral zones and hypocenters between 10 and 55 km being deeper eastward [1, 9]. Other Nazca intraplate earthquakes are due to slab pull with normal mechanism [10–13]. Epicenters of these earthquakes are located 160–270 km from the trench eastward and hypocenters between 100 and 200 km. The Chillán Mw = 8 earthquake of January 25, 1939 [10]

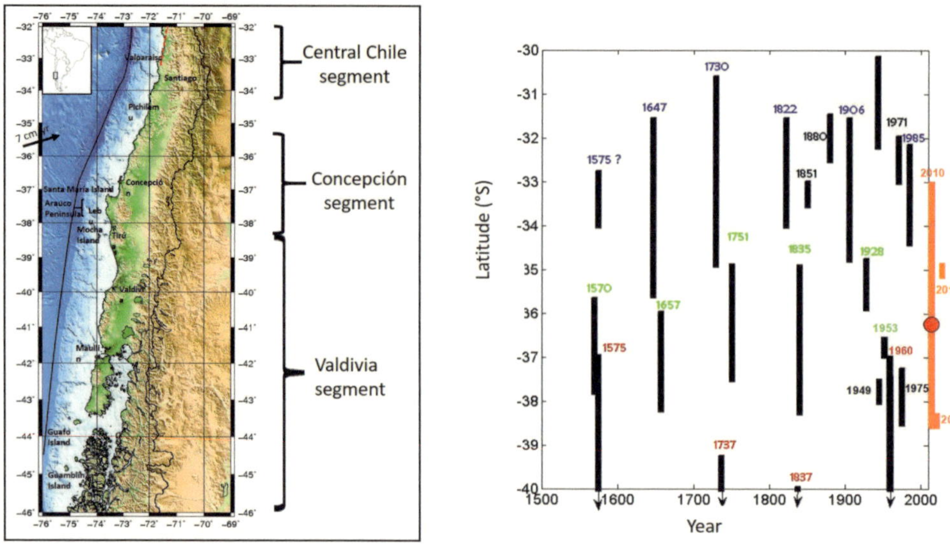

Figure 1.
Significant earthquakes in central and south Chile divided in three segments: Central Chile, Concepcion, and Valdivia. Right: years and extension of the ruptures of the earthquakes. Blue years: Central Chile earthquakes; green years: Concepcion earthquakes; and red years: Valdivia earthquakes.

was the deadliest earthquake that occurred in Chile with more than 25,000 persons killed. Other significant slab pull earthquakes are the Punitaqui 1997 Mw = 7.1 [12] and Tarapaca 2005 Mw = 7.8 [14]. Other Intraplate earthquakes are located at South American Plate with strike slip mechanism. These events are concentrated in crustal faults along the axis of Andes Cordillera of N-S strike and vertical dip, right lateral movement [15–18]. Significant earthquakes of such kind are 2001 Mw = 6.3 Aroma earthquake, 2004 Mw = 6.4 Curicó earthquake, and 2007 Mw = 6.2 Aysén earthquake. Due to the shallow hypocentral depths (≤10 km), in mountain areas, these earthquakes generate significant damages due to the landslides triggered.

However, crustal faults can also generate significant earthquakes or being activated or reactivated during significant subduction earthquakes. The 2010 Maule earthquake activated two crustal faults: the Pichilemu fault of normal mechanism [19–21], the normal Santa Maria fault [22, 23], and the Tirúa-Mocha fault of thrust mechanism [5]. In this paper, we assess the seismic potential of such faults.

2. Maule earthquake

Maule earthquake affected an area of more than 650 km length between 33 and 38.6°S. Two main asperities are identified, the northern one with main slip of 18 m between 34 and 35°S with depth between 20 and 40 km, and the southern one with main slip of 10 m between 37 and 37.6°S [3, 24–26]. Before Maule earthquake, a seismic gap between 34 37° S, named the Pichilemu-Concepcion seismic gap [1, 2] was identified. An earthquake of Mw = 8.5 was expected in this area, and the previous big earthquake located in such area occurred in 1835. This seismic gap is characterized by a lack of coastal seismicity in this area (Campos et al. [1]. About vertical changes, an interseismic subsidence is noted, mainly in littoral zones located close to the trench such as Arauco Peninsula and Santa María Island between 37 and 37.7°S [5, 21].

When finally, Maule earthquake filled this gap, several anomalies that are not expected occurred. The Mw = 8.8 magnitude was bigger than the Mw = 8.5 predicted [1, 2], overlapping rupture zones of recent earthquakes [3, 5]. In the north, the 1985 Central Chile Mw = 8 earthquake ruptured also the zone between 33 and 34°S. Uplift of 50 cm was observed in the littoral zone that contrasted with the subsidence in the same zone, which was observed in 2010 Maule earthquake. The southern rupture zone of Maule earthquake was affected also by the Concepcion Mw = 8.1 earthquake of May 21st 1960, the first earthquake of the great 1960 seismic sequence. The place of Lebu (37.6°S) was uplifted 1.2 m during 1960 earthquake, such uplift was recovered between 1960 and 2010 earthquakes and after Maule earthquake, was uplifted again 2 meters. Mocha island (38.3°S) was uplifted 1.5 m during 1960 earthquake, but unlike Arauco Peninsula and Santa María Island (37–37.7°S), the 1960 coseismic uplift remained [5].

The Maule earthquake and related tsunami provoked around 570 inhabitants killed and hundreds injured with extensive damage. Land changes included coastal uplift and subsidence, being decreasing uplift arcward (toward east). The hinge zone between uplift and subsidence was around 123 km in the area ruptured by the northern asperity and around 135 km in the southern asperity, indicating the shallower position of the northern asperity [5].

3. Crustal faults activity triggered during and following Maule Earthquake

3.1 Pichilemu Fault

On March 11, 2010, a series of earthquakes started at Pichilemu zone (34.3°S/72°W) with an Mw = 6.9 event, followed by several events of Mw = 5.5–6.5 in the next minutes. Then a crustal seismic activity continued in such zone to date. The focal mechanism of this earthquake was normal, with strike NW-SE, and the seismic activity is located in an alignment of such orientation with 50 km length [19, 20, 27]. Farias et al. [20] had fieldwork at Pichilemu after Maule earthquake but before March 11 event, indicating 0.2 m uplift. Quezada et al. [21] visited Pichilemu littoral zone in April 2010, and it exists at that date at a subsidence of around 50 cm. For this reason, the true dip of the NW-SE fault is toward SW. The littoral subsidence of the Pichilemu fault is bigger than the former uplift generated by Maule earthquake. The aftershocks of Pichilemu fault are separated from the events of the Wadati-Benioff plane [20, 27, 28]. This fault is explained because their strike is normal to the maximum extensional axis (T axis) of the rupture in Wadati-Benioff zone at this latitude [19].

3.2 Santa María Fault

A normal fault was activated close to the northern tip at Santa María Island (37°S) of NE-SW strike and visible open cracks and scarp until 30 cm height. The length of the trace is 600 m. Not significant seismicity or aftershocks are recognized near to this fault [27, 28]. One explanation of this fault is that this corresponds to a shallow extension due to a deeper splay fault of thrust mechanism [23]. This hypothesis was refused by Allmendinger et al. [22] that considers that this fault was reactivated due to the orientation of such fault normal to the maximum extensional axis (T axis) of Maule earthquake main rupture, at Wadati-Benioff zone in this latitude. In subduction earthquakes, the distribution of the T axis changes being normal to a curve trenchward concave between both tips of the main fault. It is the same explanation of the Pichilemu fault [19].

3.3 Tirua Mocha Fault

The vertical movements at the southern zone of Maule earthquake (38.3–38.4°S) are not typical trend with decreasing uplift from trench to arc. In fact, Mocha Island experienced an uplift of 0.25–0.3 m, whereas Tirua place, located at the continental margin in front of Mocha Island, experienced 0.9 m uplift. This anomaly was explained by the existence of a splay fault between Tirua and Mocha Island [5]. The activity of such fault began after 1960 earthquake due to the development of a strong asperity at the updip of Wadati-Benioff zone. This splay fault is synthetic, west vergent, also thrust movement. The main deformation due to the interplate convergence was in such fault explaining the permanence of the 1960 coseismic uplift at Mocha Island and the big interseismic uplift at Tirúa. The strong asperity remained even during Maule earthquake, and the main slip occurred along Tirua-Mocha splay fault explaining the abnormal uplift trend. Significant earthquakes of Mw = 6 magnitude occurred in such zone in the following months, evidencing the instability conditions, and finally, on January 2, 2011, an Mw = 7.1 earthquake occurred in such area that generated 0.5 m uplift at Mocha island and 0.2 m subsidence at Tirúa, indicating the final removal of the asperity at the updip of Wadati Benioff zone. Hicks and Rietbrock [29] indicated hybrid mechanism with thrust movement along Wadati-Benioff zone and a crustal normal fault. Quezada et al. [5] indicated that such normal fault was the Tirua-Mocha fault that experienced tectonic inversion because the movement is in a shallower position at the Wadati Benioff zone, being favorable conditions of such opposite movement.

3.4 Seismicity linked to the crustal faults

From the Centro Sismológico Nacional-U de Chile, we collected those seismic events occurred between January 2010 and November 2022 around Pichilemu (34°S), Santa María Island (37°S), and Tirúa-Mocha (38.3°S) areas. Around each of these locations, seismicity is exposed on a map view and along an ENE-WSW profile: in both cases, a subduction azimuth of 82° is considered (**Figures 2–7**). Distribution of seismicity is shown in circles according to its magnitude range: M < 3.5 cyan color, 3.5 < M < 4.5 yellow color, 4.5 < M < 5.5 red color, 5.5 < M < 6.5 green color, and 6.5 < M < 7.5 blue color. Distribution of seismicity and magnitudes in time are exposed in graphs. Here it can be seen that this data set is fully influenced by aftershocks linked

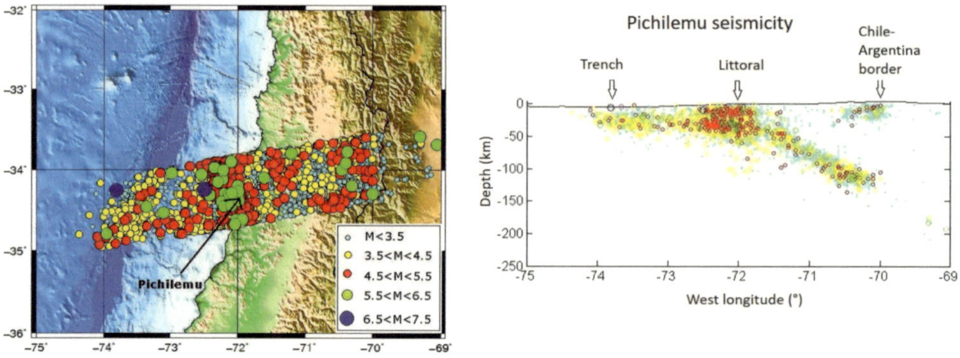

Figure 2.
Seismicity at Pichilemu. It is noted in the profile the crustal seismicity in the littoral zone related to Pichilemu fault.

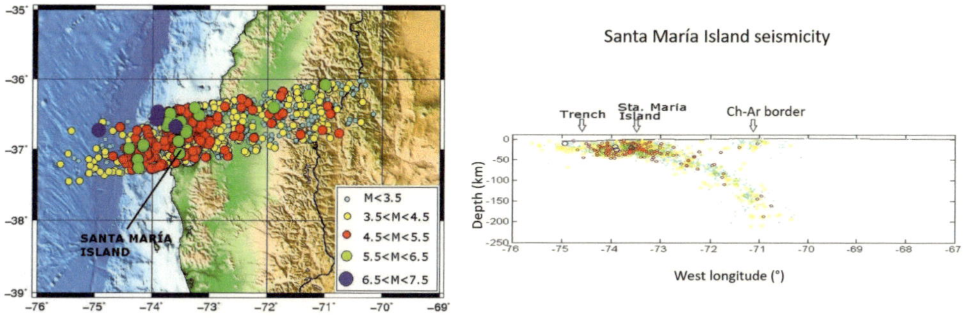

Figure 3.
Seismicity at Santa Maria Island. Shallow seismicity exists between the trench and littoral, but it is difficult to separate crustal from interplate events.

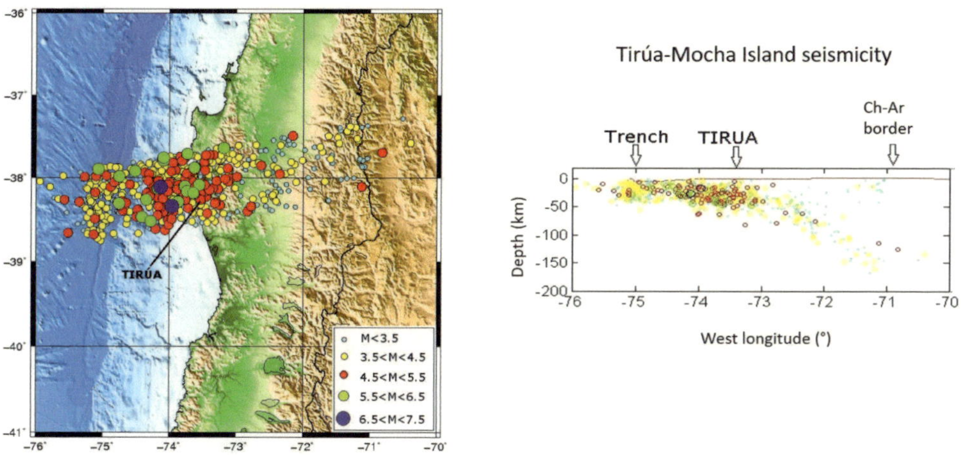

Figure 4.
Seismicity at Tirúa-Mocha Island. Shallow seismicity exists close to Tirúa that could be related to Tirúa-Mocha fault.

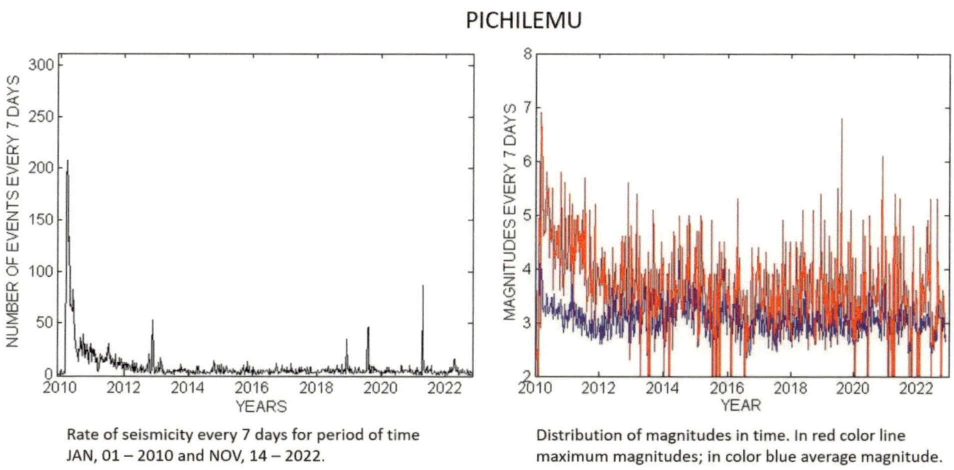

Figure 5.
Frequency of events and magnitude at Pichilemu.

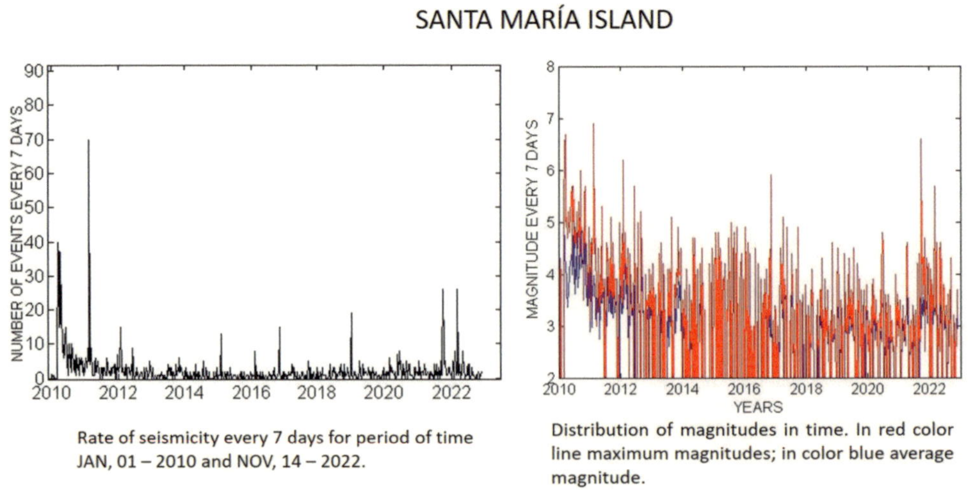

Figure 6.
Frequency of events and magnitude at Santa Maria Island.

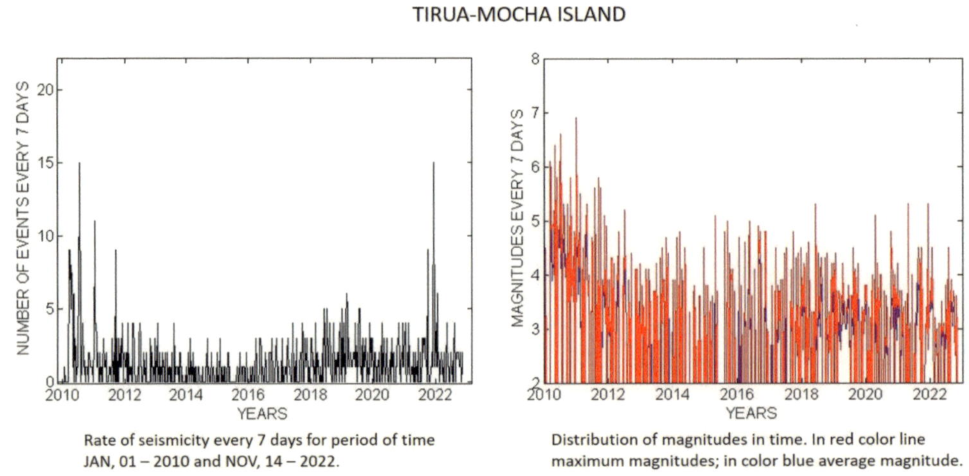

Figure 7.
Frequency of events and magnitude at Tirúa-Mocha Island.

with 2010 Maule Earthquake (Mw = 8.8). Error location for seismic events can be considered up to 10 km around determined hypocenter.

3.5 Seismicity linked to the crustal faults

As it can be seen in Pichilemu, Santa Maria Island, and Pichilemu areas, it exists as a continuous seismicity from trench to arc (**Figures 2–4**). However, the vertical profiles show separated clusters. The slab is clearly defined. Crustal seismicity occurs at Andean Cordillera. This seismicity must be related to the strike slip partitioning and compressive crustal faults linked to the growing of this relief. The magnitudes of these events are decreasing from north to south. At Pichilemu latitude (34°S, **Figure 2**), the biggest magnitude reached Mw = 5.5, at Santa Maria

Island latitude (37°S, **Figure 3**), the magnitude reached Mw = 4.5, and at Tirúa-Mocha Island latitude (38.3°S, **Figure 4**), the magnitude reached Mw = 3.5. Slab pull seismicity is visible in all profiles with depth hypocenters bigger than 50 km increasing depth eastward illuminating the slab. The number of events and bigger magnitude also occur at the northern profile at Pichilemu.

Between the trench and the littoral zone, important seismic activity being interplate and crustal exists. The Pichilemu profile shows a big amount of seismicity in the littoral zone linked to the Pichilemu fault with magnitudes reaching Mw = 5.5. West of this cluster until the trench, lesser shallow seismicity is noted. This light is better than the littoral seismic zone linked to the Pichilemu fault. The Santa María Island profile also displays shallow seismicity between the trench and littoral zone, but separate crustal seismicity to interplate one is more difficult, but crustal seismicity exists. Magnitudes are lesser than Mw = 5.5. At Tirúa-Mocha Island profile, crustal activity also exists between the trench and Tirúa, but with fewer events. Despite this, the hypocenters are located close to Tirúa and may illuminate the Tirúa-Mocha fault with some events of magnitude reaching Mw = 5.5.

About the frequency of the seismicity, it displayed the number of events every 7 days and the magnitude every 7 days between 2010 and 2022 (**Figures 5–7**). At Pichilemu (**Figure 5**), during all 2010 year occurred until 200 events every 7 days decreasing later stabilizing in 10 events since 2013, with some specific days with more activity. Average magnitude of these events is 3, but the maximum magnitudes decreased from Mw = 7 in 2010 to Mw = 5.5 since 2012 with some isolated events later.

The Santa María profile (**Figure 6**) shows a maximum of 40 events per 7 days in 2010 decreasing to 5 events, but since 2017, a small increase of the amount of events with some days with a big amount of seismicity exists. It is typically when occurred an event of magnitude between Mw = 5–6, then an aftershock sequence takes place related to the first event. Between 2010 and 2012, the average magnitude was between Mw = 3.5–4.5; then until 2021, it was Mw = 3, and since 2022, it increases.

The Tirúa-Mocha profile (**Figure 7**) shows important seismic activity between 2010 and 2012. Also occurred significant events of magnitudes of Mw = 6–7. The fourth peak corresponds to the Mw = 7.1 Araucania earthquake of January 2, 2011. Then the average magnitude of the events stabilized at Mw = 3.5, and since 2020, the amount of seismicity increased slowly.

4. Discussion

Maule earthquake triggered the rupture of three crustal faults: Pichilemu, Santa María, and Tirúa-Mocha, all with different behaviors. Pichilemu and Santa María faults with normal mechanism can be explained better due to their orientation normal to the maximum T axis of the rupture along Wadati-Benioff zone. We agree with Allmendinger et al. [22] that Santa María fault is not a splay fault. By contrast, Tirúa-Mocha is a splay fault with thrust mechanism. Such fault can be explained due to the development of a strong asperity in the interplate zone after 1960 earthquake, which remained even during Maule earthquake, being the movement accommodated in that zone at the Tirúa-Mocha splay fault, and the asperity at the updip of Wadati-Benioff zone was removed finally during Araucania Mw = 7.1 earthquake of January 2, 2011. It is not easily distinguished seismic activity linked to Santa María fault and Tirúa-Mocha fault. But the large number of events linked to Pichilemu fault are noted. Many of them are felt by the population of Pichilemu and surrounding areas.

It is not clear if these tree faults can generate their own earthquakes, not related with the subduction seismic cycle. These faults appear to be activated or reactivated due to the stress changes of South American plate, after Maule earthquake. The interplate border is irregular, so, during some earthquakes, different asperities can be developed in successive seismic cycles, some at the updip, some at the down dip of Wadati-Benioff zone. The position of these asperities, especially if such faults are located at the updip, can trigger the development or the reactivation of a splay fault. Such asperities can be of different sizes, and depending on this, the position of the T axis can vary in successive earthquakes, and some rupture lengths can generate a geometry of T axis that can be favorable to the normal reactivation of crustal faults [19].

About seismic potential, Pichilemu and Tirua Mocha fault proved to trigger earthquakes of magnitude Mw = 7. Pichilemu is located 100 km from places in which is located half of the population of Chile, including the capital Santiago and Valparaiso and Viña del Mar big cities. The Tirua- Mocha fault is located less than 200 km from two main cities of south Chile: Concepcion and Temuco.

5. Conclusions

Crustal faults can be reactivated during some subduction seismic cycles. Their activity is poorly constrained. Some faults have persistent seismic activity, and others appear to be reactivated only after a major subduction earthquake. Some are normal, others are reverse splay faults. The irregular shape in the interplate zone can develop different asperities in successive subduction cycles, and in some of them, crustal faults are reactivated. It is necessary to know the existence of these faults and the seismic potential. Due to their shallow depth, these crustal faults can be another source of seismic hazard, even the traces can be across populated areas. It is necessary to improve the knowledge of the activity of such faults.

Author details

Jorge Quezada[1*] and Arturo Belmonte[2]

1 Departamento de Ciencias de la Tierra, Facultad Ciencias Quimicas, Universidad de Concepcion, Concepción, Chile

2 Departamento de Geofísica, Facultad de Ciencias Físicas y Matemáticas, Universidad de Concepción, Concepción, Chile

*Address all correspondence to: jquezad@udec.cl

References

[1] Campos J, Hatzfeld D, Madariaga R, López G, Kausel E, Zollo A, et al. A seismologic study of the 1835 eismic gap in south-central Chile. Physics of the Earth and Planetary Interiors. 2002;**132**:177-195

[2] Ruegg J-C, Rudloff A, Vigny C, Madariaga R, De Chabalier J, Campos J, et al. Interseismic strain accumulation measured by GPS in the seismic gap between Constitución and Concepción in Chile. Physics of the Earth and Planetary Interiors. 2009;**175**:78-85

[3] Lay T. Earthquakes: A Chilean surprise. Nature. 2011;**471**:174-175

[4] Farías M, Vargas G, Tassara A, Carretier S, Baize S, Melnick D, et al. Land-level changes produced by the Mw 8.8 2010 Chilean Earthquake. Science. 2010;**329**:916

[5] Quezada J, Jaque E, Catalán N, Belmonte A, Fernández A, Isla F. Unexpected coseismic surface uplift at Tirúa-Mocha Island area of south Chile before and during the Mw 8.8 Maule 2010 earthquake: A possible upper plate splay fault. Andean Geology. 2020;**47**(2):295-315. DOI: 10.5027/andgeoV47n2-3057

[6] Vargas G, Farías M, Carretier S, Tassara A, Baize S, Melnick D. Coastal uplift and tsunami effects associated to the 2010 Mw 8.8 Maule earthquake in Central Chile. Andean Geology. 2011;**38**(1):219-238

[7] Cisternas M, Atwater BF, Torrejón F, Sawai Y, Machuca G, Lagos M, et al. Predecessors of the giant 1960 Chile earthquake. Nature. 2005;**437**:404-407

[8] Melnick D, Moreno M, Quinteros J, Baez J. The super-interseismic phase of the megathrust earthquake cycle in Chile: The super-interseismic earthquake cycle. Geophysical Research Letters. 2017;**44**(2). DOI: 10.1002/2016GL071845

[9] Ruiz S, Madariaga R. Historical and recent large megathrust earthquakes in Chile. Tectonophysics. 2018;**733**:37-56

[10] Beck S, Barrientos S, Kausel E, Reyes M. Source characteristics of historic earthquakes along the central Chile subduction zone. Journal of South American Earth Sciences. 1998;**11**:115-129

[11] Lemoine A, Madariaga R, Campos R. Slab-pull and slab-push earthquakes in the Mexican, Chilean and Peruvian subduction zones. Physics of the Earth and Planetary Interiors. 2002;**132**(1-3):157-175. DOI: 10.1016/S0031-9201(02)00050-X

[12] Pardo M, Comte D, Monfret T, Borosheck R. The October 15, 1997 Punitaqui earthquake (Mw=7.1): A destructive event within the subducting Nazca plate in the Central Chile. Tectonophysics. 2002;**345**(1-4):199-210

[13] Peyrat S, Campos J, de Chabalier J, Perez A, Bonvallot S, Bouin M. The Tarapaca intermediate-depth earthquake (Mw 7.7, 2005, Northern Chile): A slabpull event with horizontal fault plane constrained from seismologic and geodetic observations. Geophysical Research Letters. 2006;**33**(22):1-6. DOI: 10.1029/2006GL027710

[14] Kuge K, Kase Y, Urata Y, Campos J, Perez A. Rupture characteristics of the 2005 Tarapaca, northern Chile, intermediate-depth earthquake: Evidence for heterogeneous fluid distribution across the subducting

oceanic plate? Journal of Geophysical Research Solid Earth. 2010;**115**(B9):1-15. DOI: 10.1029/2009JB007106

[15] Legrand D, Delouis B, Dorbath L, David C, Campos J, Marquez L, et al. Source parameters of the Mw = 6.3 Aroma crustal earthquake of July 24, 2001 (northern Chile), and its aftershock sequence. Journal of South American Earth Sciences. 2007;**24**:58-68

[16] Cembrano J, Lara L. The link between volcanism and tectonics in the southern volcanic zone of the Chilean Andes: A review. Tectonophysics. 2009;**471**:96-113

[17] Vargas G, Rebolledo S, Sepulveda S, Lahsen A, Thiele R, Townley B, et al. Submarine earthquake rupture, active faulting and volcanism along the major Liquiñe-Ofqui Fault Zone and implications for seismic hazard assessment in the Patagonian Andes. Andean Geology. 2013;**40**(1):141-171

[18] Agurto H, Rietbrock A, Barrientos S, Bataille K, Legrand D. Seismo-tectonic structure of the Aysen Region, Southern Chile, inferred from the 2007 Mw = 6.2 Aysen earthquake sequence. Geophysical Journal International. 2007;**190**:116-130

[19] Aron F, Allmendinger R, Cembrano J, Gonzalez G, Yañez G. Permanent fore arc extensión and seismic segmentation: Insigts from the 2010 Maule earthquake, Chile. Journal of Geophysical Research. 2013;**118**:1-16. DOI: 10.1029/2012JB009339

[20] Farias M, Comte D, Roecker S, Carrizo D, Pardo M. Crustal extensional faulting triggered by the 2010 Chilean earthquake: The Pichilemu Seismic Sequence. Tectonics. 2011;**30**:TC6010

[21] Quezada J, Jaque E, Belmonte A, Fernández A, Vásquez D, Martínez C. Movimientos cosísmicos verticales y cambios geomorfológicos generados durante el terremoto Mw=8,8 del 27 de Febrero de 2010 en el centro-sur de Chile. Revista Geográfica del Sur. 2010;**2**:11-44

[22] Allmendinger R, Gonzalez G, Cembrano J, Aron F, Yañez G. Splay fault slip during the Mw 8.8 2010 Maule Chile earthquake. Geology. 2013;**41**(12):e309

[23] Melnick D, Moreno M, Motagh M, Cisternas M, Wesson RL. Splay fault slip during the Mw 8.8 2010 Maule chile earthquake. Geology. 2013;**40**(3):251-254

[24] Hayes G, Bergman E, Johnson K, Benz H, Brown L, Meltzer A. Seismotectonic framework of the 2010 February 27Mw 8.8 Maule Chile earthquake sequence. Geophysical Journal International. 2013;**195**(2):1034-1051. DOI: 10.1093/gji/ggt238

[25] Lay T, Ammon C, Kanamori H, Koper K, Sufri O, Hutko A. Teleseismic inversion for rupture process of the 27 February 2010 Chile (Mw 8.8) earthquake. Geophysical Research Letters. 2010;**37**(13):1-5. DOI: 10.1029/2010GL043379

[26] Lorito S, Romano F, Atzori S, Tong X, Avallone A, McCloskey J, et al. Limited overlap between the seismic gap and coseismic slip of the great 2010 Chile earthquake. Nature Geoscience. 2011;**4**:173-177

[27] Lieser K, Grevemeyer I, Lange D, Flueh E, Tilmann F, Contreras-Reyes E. Splay fault activity revealed by aftershocks of the 2010 Mw 8.8 Maule earthquake, central Chile. Geology. 2014;**42**(9):823-826

[28] Rietbrock A, Ryder I, Hayes G, Haberland C, Comte D, Roecker S, et al. Aftershock seismicity of the 2010 Maule Mw=8.8, Chile, earthquake: Correlation

between co-seismic slip models and aftershock distribution? Geophysical Research Letters. 2012;**39**:L08310. DOI: 10.1029/2012GL051308

[29] Hicks S, Rietbrock A. Seismic slip on an upper-plate normal fault during a large subduction megathrust rupture. Nature Geoscience. 2015;**8**:955-960

Chapter 7

The "Natural Time" Method Used for the Potential Assessment for Strong Earthquakes in China Seismic Experimental Site

Shengfeng Zhang and Yongxian Zhang

Abstract

Due to the direct achieving for the state of stress or the strain along the earthquake fault which is quite essential in the assessment for the potential of strong earthquakes, the method of nowcasting earthquakes using the 'natural time' concept has been used in several locations worldwide and shown significant result. In this work, the Earthquake Potential Score (EPS) was determined using the nowcasting approach before several earthquake cases in the China Seismic Experimental Site (CSES) and analyze the consistency with the observation to evaluate its effectiveness. Firstly, with the importance of the data quality to this statistical method, we describe the background seismicity of the CSES area. Secondly, ergodicity research demonstrates the differences that exist in sub-regions such as Sichuan and the Yunnan region, mainly due to the simultaneous impact with the 2008 Wenchuan 8.0 earthquake. In the end, the strong earthquake potential prior to four earthquakes with magnitude larger than 6.0 was ultimately determined using the nowcasting method, which has EPS above 0.8. This may give support for the interpretation of EPS in earthquake nowcasting and will serve as a key reference for the ongoing development of this technology.

Keywords: China seismic experimental site (CSES), "natural time", nowcasting method, earthquake potential, strong earthquakes

1. Introduction

The North-South Seismic Region in central China is one of the focal regions that pays special attention to decreasing the threat of earthquakes in China. A number of devastating earthquakes took place in this area, containing the 1920 Haiyuan M8.5 earthquake, the 1970 Tonghai M_S7.8 earthquake, the 2008 Wenchuan M_S8.0 earthquake, and 2013 Lushan M_S7.0 earthquake. This seismic zone was encircled by the Ordos block, the Sichuan Basin, and the Tibetan Plateau in the tectonic backdrop [1]. The Indian plate colliding with the Eurasia plate, which caused the Tibetan plateau to rise and expand, is represented by the region's variation in crust thickness, which ranges from 30 to 46 km in the east to 46–74 km in the west [2]. There exist two experiment sites in history, one is the West-Yunnan Earthquake Prediction Experiment Site in 1980, and another is the National Experiment Site for Earthquake Monitoring and Forecast in 2014, have both been established there because of the region's reality [3]. Following the Regional Earthquake Likelihood Models (RELM) workshop, SCEC suggested that other regions take part in the testing process for various prediction models using religious assessment methodologies. The Collaboratory for the Study of Earthquake Predictability is offering this location as one of the first testing sites in China [4, 5].

The China Seismic Experimental Site (CSES) was launched on May 12, 2018 [6–8]. Given that CSES only covers the Chinese provinces of Sichuan and Yunnan, its geographic scope is less extensive than that of the field site before it. It builds on ideas from ecology and environmental science to propose Coordinated Distributed Experiments (CDEs), a novel collaborative research strategy that follows the design for CSES [9]. As CESE work has progressed in recent years, the concepts of retrospective and prospective [10], start and trial [7], planning and test [11], earthquake forecasting, and system design [12] have been revised. At the same time, the testing facility will be built in the CSES area as part of CSEP2.0, which was also announced in 2019 [13–17]. This study will provide an overview of seismicity analysis using historical and contemporary catalogs, the nowcasting experiment, and ergodicity feature, particularly for the potential assessment prior to strong events like August 3, 2014, Ludian M_S6.5 earthquake; October 7, 2014, Jinggu M_S6.6 earthquake; May 21, 2021, Yangbi M_S6.4 earthquake; and September 5, 2022, Luding M_S6.8 earthquake. **Figure 1** depicts the geographical distribution of earthquakes in the CSES area with a magnitude greater than 6.0 from 700 B.C. to A.D. 2022.

2. The earthquake catalog used

The earthquake catalog we utilized was integrated with the historical catalog for occurrences before 1970/01/01 and the contemporary catalog from 1970/01/01 to 2022/11/20 given by the China Earthquake Networks Center (CENC). **Figure 2** illustrates the capacity to record earthquake events in the CSES area, where ancient Chinese recording mentions some 8+ earthquakes, by varying the event number plot inside each magnitude bin and the magnitude-sequence number plot [18, 19]. As a consequence of the construction of the seismic station, and the disturbing occurrence of a few big earthquakes, the result indicates that there are certain peculiar time nodes that disclose the variation of monitoring capability in this area.

To obtain the completeness distribution of the modern earthquake catalog after 1970, many statistical methods based on the earthquake catalog [20] are usually used. **Figure 3** shows the assessment of the completeness state using the Best Combination method (Mc95-Mc90-Max curvature) [20], which reveals the temporal variation mainly depending on the development of observational facilities. The results for the catalog before 1970 give a completeness magnitude of around 5.0–6.0, however, there is a significant error since there are so few records, compared with that of the modern earthquake catalog. In general, the magnitude threshold of the historical earthquake catalog should be chosen as 4.0–4.5 to ensure the completeness level and enough sample size if a statistical algorithm would be used to analyze the sequence. For the catalog after 1970, 3.0–4.0 can be determined to be the completeness magnitude level [5]. On the other hand, strong earthquakes, such as the seismic sequence of Wenchuan 8.0 in 2008 and Lushan 7.0 in 2013, have a substantial influence on the catalog's

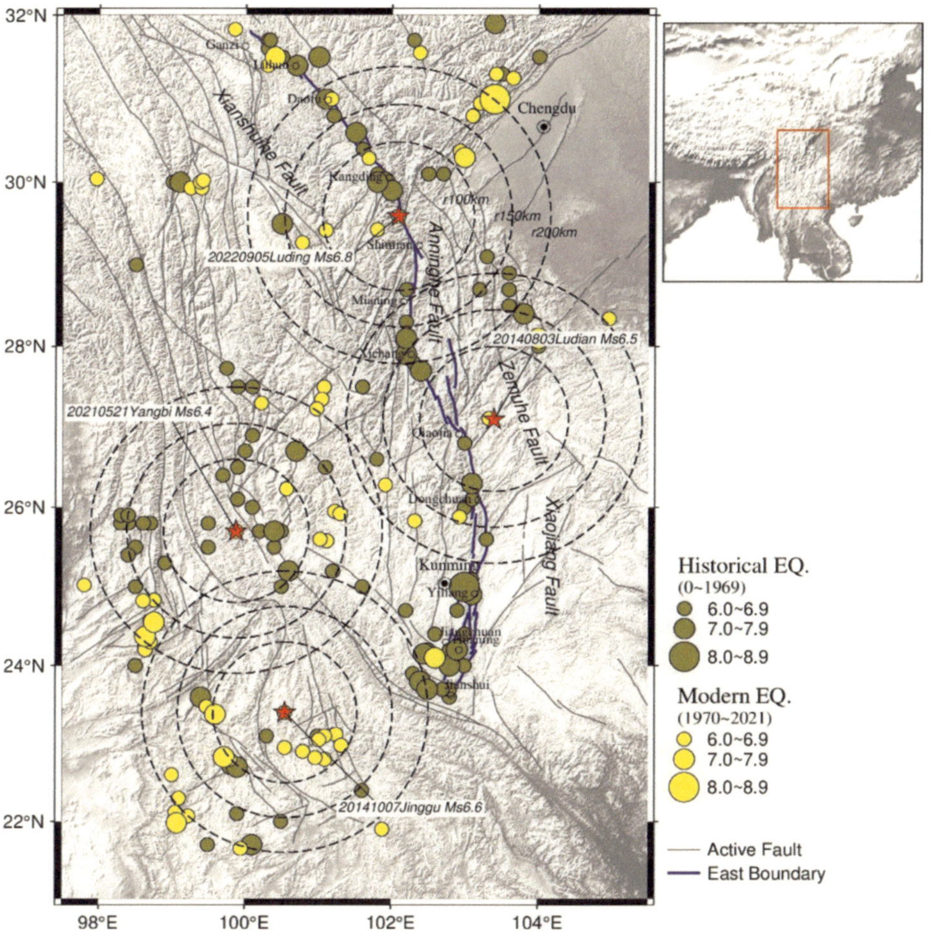

Figure 1.
Spatial distribution of earthquakes with magnitude above 6.0 for the period from 700 B.C. to A.D. 2022 in CSES region. The dots in dark yellow indicate the strong earthquakes up to 1970. Earthquakes that occurred since 1970/01/01, are shown in light yellow. The area being studied is shown in the indexing graphic at the upper right. The three dashed circles indicate the region with a radius of 100, 150, and 200 km, respectively, used in the nowcasting analysis. Center of these circles (the star in red) is the epicenter of the September 5, 2022, Luding M_S*6.8 earthquake; May 21, 2021, Yangbi* M_S*6.4 earthquake; October 7, 2014, Jinggu* M_S*6.6 earthquake; and August 3, 2014, Ludian* M_S*6.5 earthquake.*

completeness owing to observational limitations and regular seismological interpretation, which should be addressed in the future. The cut-off magnitude of 4.0 may be interpreted as the magnitude threshold in the following statistical models as a global estimate.

A "mixed" magnitude system is used in the earthquake catalog since 1970. The local type of magnitude (M_L) is utilized for events with magnitudes under 4.0. Surface wave type of magnitude (M_S) is utilized for earthquakes greater than magnitude 5.5. The change from M_L to M_S is not well defined for the events in between. Only "magnitude" is used in this research to indicate the size of events due to the magnitude uncertainty. The outcomes of the numerical calculations conducted in this chapter, such as the determination of the Earthquake Potential Score using the nowcasting technique, are unaffected by the magnitude uncertainty [21–24]. In line with Rundle's research [24], we used a threshold magnitude of 4.0 and a target magnitude of larger

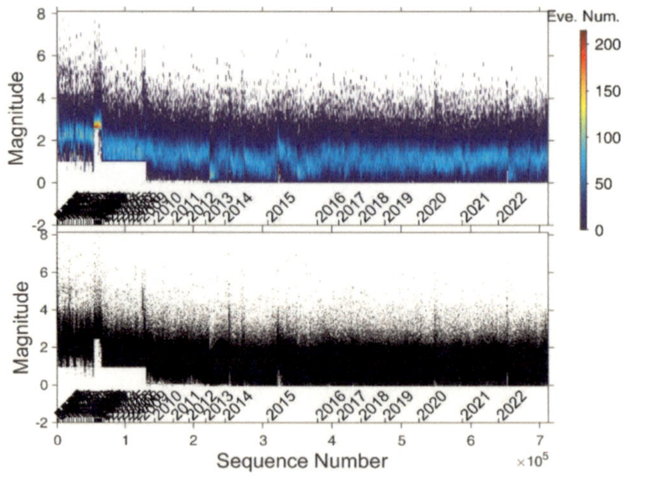

Figure 2.
Variation of event number within each magnitude bin (top) and the magnitude-sequence number plot (bottom) of the catalog of CSES region from 700 B.C. to A.D. 2022/11/20. In the event number plot, the number window of the x-axis is chosen as 500. The text above the x-axis shows the calendar time. In the magnitude-sequence plot, to avoid the difficulty to observe the little distribution of magnitude, a random of [−0.05, 0.05] is added into the fluctuation of the magnitude bin.

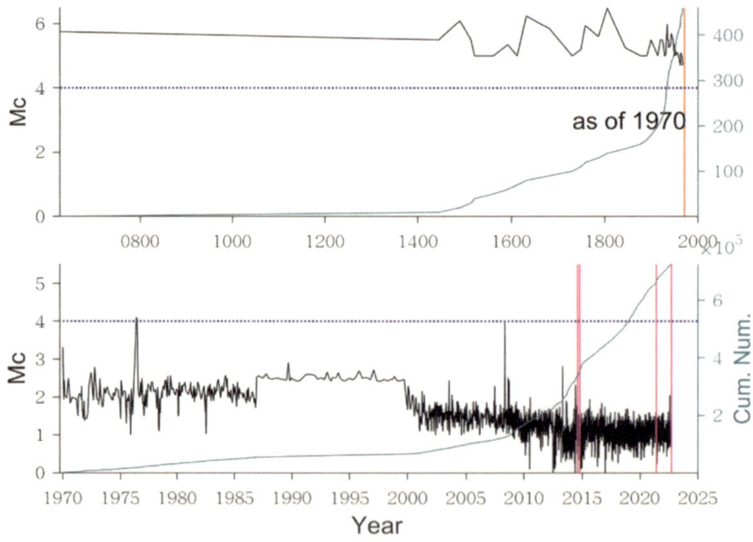

Figure 3.
Completeness magnitude and cumulative number analysis for the earthquake catalog of CSES. The above and bottom plots show the result of 700 B.C. to 1970 and 1970 to 2022/11/20. The black line in the plot indicates the completeness magnitude calculated using the Best Combination method (Mc95-Mc90-Max curvature) [20]. The blue lines show the cut-off magnitude level considering the global temporal variation of completeness magnitude. The dark green line shows the cumulative event number as the time. The vertical pure light red line marks the time of 1970/01/01, which separates the whole catalog into above and bottom plots. The vertical purplish red lines mark the occurrence time of four target strong earthquakes in this study.

than 6.0 in our investigation. However, for alternative methods that evaluate not only the frequency of earthquakes but also their size, concerns of magnitude transition, and magnitude uncertainty must be considered.

3. Ergodicity analysis

Some studies have suggested that driven mean-field systems may often display significant ergodic behavior. Moreover, Egolf [25] and Tiampo et al. [26, 27] gave the study that statistically stationary models have the propensity to live in a set of physical states resembling equilibrium. Large events, however, usually reveal the potential to temporarily throw the study target out of balance before it returns to its original state. We apply the method of Tiampo et al. [27, 28] to quantify the base level of heterogeneity and, as a consequence, the predictability of target earthquakes due to the temporal complexity of the small events in the CSES zone, as shown in **Figure 2**. According to the method of Thirumalai et al. [29] and Thirumalai and Mountain [30], it is possible to evaluate ergodicity behavior generally using the Thirumalai-Mountain plot (the TM metric). Originally, the TM metrics may be used to assess effective ergodicity or the discrepancy between the time average of a quantity, commonly associated with energy, at every cell or grid of the system. A simple statistical algorithm, such as the pattern of Informatics (PI) algorithm, may be determined by the TM metrics, not by looking at energy levels, but rather at the frequency of observations [31].

The beginning of the earthquake catalog corresponds to the starting time t = 0 and the function $TM_n(t)$, which is computed in the analysis at extensive time t, is defined as

$$TM_n(t) = \frac{1}{L}\sum_{i=1}^{L}[n_i(t) - \overline{n}(t)]^2 \tag{1}$$

$n_i(t)$ is the count of events located within the ith grid in the period of $0 \sim t$, L is the number of grids in the spatial range, and $\overline{n}(t)$ is the average count of events across all cells during the period $0 \sim t$,

$$n_i(t) = \frac{1}{t}\int_0^t n_i(t')dt' \tag{2}$$

$$\overline{n}(t) = \frac{1}{L}\sum_{i}^{L} n_i(t) \tag{3}$$

Tiampo et al. [27, 28] found that, as shown in Eq. (1), all grids or cells in the system are comparable in properties, especially in the case of physical characteristics, and the deviation of the average quantity in temporal from grouped mean number is diminishing. The statement states that if the system exhibits the behavior of "effective ergodic" over an extended length of time, time t will have a direct relationship with the function $1/TM$. With a grid size of 0.2° × 0.2° and a cut-off magnitude of 3.0, **Figure 4** depicts the $1/TM$ metrics of the seismicity from 1970 to 2022 in the CSES area, Sichuan and Yunnan portions. Before 1980, we can observe from the instance of including and not containing the Wenchuan aftershocks that both the Yunnan and Sichuan regions had weak ergodicity and display unique ergodic tendencies. After accounting for changes in the earthquake catalog caused by technology and network problems, Tiampo et al. [27, 28] found that natural seismicity exhibited ergodicity. In this study, when taking into account the condition of China's seismic networks, the regionalized aspects of seismological observation and artificial processing can be responsible for the behavior of ergodicity. The seismicity across the entire CSES area and two sub-regions exhibits excellent ergodicity since about 1980. Taking the Sichuan region as an example, where the Wenchuan earthquake's aftershocks mostly contributed to the disruption of the 1/*TM* metric in 2008.

4. Nowcasting method

The term "nowcast" has existed for a very long period in the fields of meteorology and economics. It has now been applied to a variety of fields, including stock market trend forecasting, displaying the cloud movement in real-time, and application in hazard assessment of strong earthquakes under the name "Nowcasting Earthquakes" [21–24]. In contrast to "forecasting," which was usually used to produce a probabilistic estimate of future events in the seismic cycle, it often focuses on the identification of the current state of a system using indirect approaches. In the field of seismology research, this strategy has traditionally been utilized in regions with a long history of earthquake records and reasonably high seismic activity. The technique has recently been used to estimate the seismic risk for a number of places, including California [24, 32], Tokyo [22], the Himalayas [33], and New Zealand [34], among others. To compute the Earthquake Potential Score (EPS), the nowcasting technique made use of the frequency of little events that occurred between larger ones in the same or a nearby study location with a similar dynamic development history. The magnitude threshold for small events should be chosen so that all events in the database are a completeness sequence. The key benefit of this technique is that it is easier to use than the direct method of hazard identification in earthquake forecasting and prediction which seems to be difficult to implement in reality.

The study of Varotsos et al. [35–37] suggested that the "natural time" in the nowcasting method is calculated by counting the number of little events that have occurred in a studied location since the previous strong earthquake. Target events are indicated by M_λ, whereas little events are indicated by M_σ (4.0), which accounts for the completeness of the historical and present catalog. The Gutenberg-Richter magnitude-frequency relationship may characterize the average frequency of small events larger than M_σ but with magnitude under M_λ [38]. The relation of Gutenberg-Richter relation is used here to obtain the average count of events larger than M,

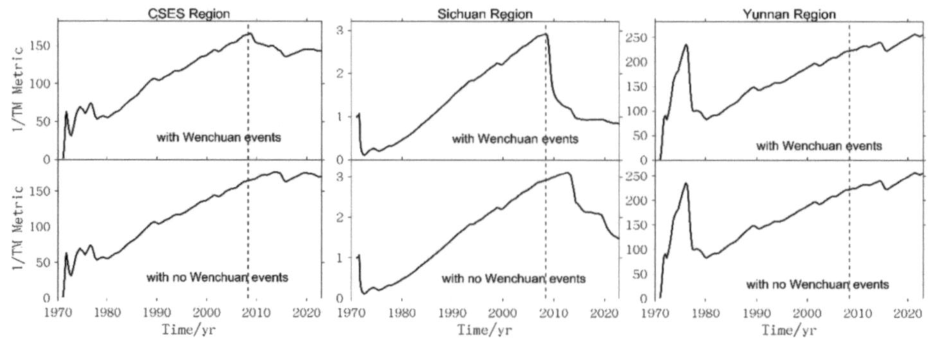

Figure 4.
The 1/TM metric indicates the seismic ergodicity in the CSES region. The left, middle and right plots show the entire CSES, Sichuan, and Yunnan region, respectively. The top and bottom show the case computed with and without aftershocks of the Wenchuan 8.0 earthquake. The vertical dashed lines reflect the occurrence time of the 2008 Wenchuan earthquake.

$$N_{avg} = 10^a 10^{-bM} \tag{4}$$

where b is normally close to 1, a value indicates the background level of seismicity.

If we use N_σ to represent the average number of little events more than M_σ, N_λ the mean frequency of events greater than M_λ, respectively,

$$N_\sigma = 10^a 10^{-bM_\sigma} \tag{5}$$

$$N_\lambda = 10^a 10^{-bM_\lambda} \tag{6}$$

then we can determine how many little earthquakes there are on average between large earthquakes:

$$N = \frac{N_\sigma - N_\lambda}{N_\lambda} = 10^{b(M_\lambda - M_\sigma)} - 1 \tag{7}$$

N in this case is independent of a value.

By calculating the cumulative distribution function (CDF) with the help of small events with magnitudes in $[M_\sigma, M_\lambda]$, the mathematical process of nowcasting computation may be stated in terms of the potential for massive earthquakes (target events). The probability density function (PDF) and cumulative distribution function of small events with magnitudes in $[M_\sigma, M_\lambda]$ within every larger cycle are calculated using the scientific mathematical approach of Bevington and Robinson [39]. The current CDF may be calculated using the present frequency of little events, $n(t)$, where t is the time from the most recent large event. The earthquake potential score (EPS) at time t may be used to define this value:

$$\text{EPS} = p[n \le n(t)] \tag{8}$$

The likelihood that the next significant earthquake in our research zone with a magnitude larger than M_λ will occur is determined by the EPS. According to Eqs. (7) and (8), the EPS value will rise over time after the most recent significant earthquake before abruptly returning to zero when the next significant earthquake strikes. It will then begin to rise once again until the next significant event happens. Therefore, in this process, there is just one straightforward method of interpreting the earthquake data throughout the whole procedure, with no fitting calculations for the model parameters. This approach may be used for both a broad seismic zone and a small area, such as a regional area of a city, as there is no connection between the EPS definition and the rate of earthquakes in the area under study.

5. Earthquake potential before target earthquakes

In the previous study as Rundle et al. [24], a circle area around the city was usually used when we want to assess the potential to occur next strong earthquake, and the seismicity in a relatively large region will be selected to build the background database to describe the statistical characteristic. Alternately, here we choose a circle area around the epicenter of four earthquakes, that is, August 3, 2014, Ludian $M_S6.5$

earthquake; October 7, 2014, Jinggu M_S6.6 earthquake; May 21, 2021, Yangbi M_S6.4 earthquake; and September 5, 2022, Luding M_S6.8 earthquake, to assess the potential before the occurrence of these four events, and used the seismicity of CSES as the large region to build the background database. We investigate circular regions with radii of 100, 150, and 200 km to test the analysis's resilience to parameter fluctuation. **Figure 5** shows the EPS result (case of radii 200 km) calculated using the nowcasting technique for these four events. For instance, **Figure 5a** shows that since the previous significant event with a magnitude over 6.8 on 2013/4/20, there have been 113 small events within 200 km of the epicenter of the Luding M_S6.8 earthquake, and the EPS value has reached 81%. The EPS value for the remaining three earthquakes approaches 98, 87, and 88%, respectively, according to **Figure 5b–d**. **Table 1** lists the results obtained using various radii. We can observe that the EPS was high but did not reach 100% prior to the occurrence of these four events.

As a result of the big events' ability to segregate smaller events or samples, the amount of larger events will have an impact on how smoothly the EPS curve behaves in each plot. Even yet, the outcome of EPS results before these earthquakes might still

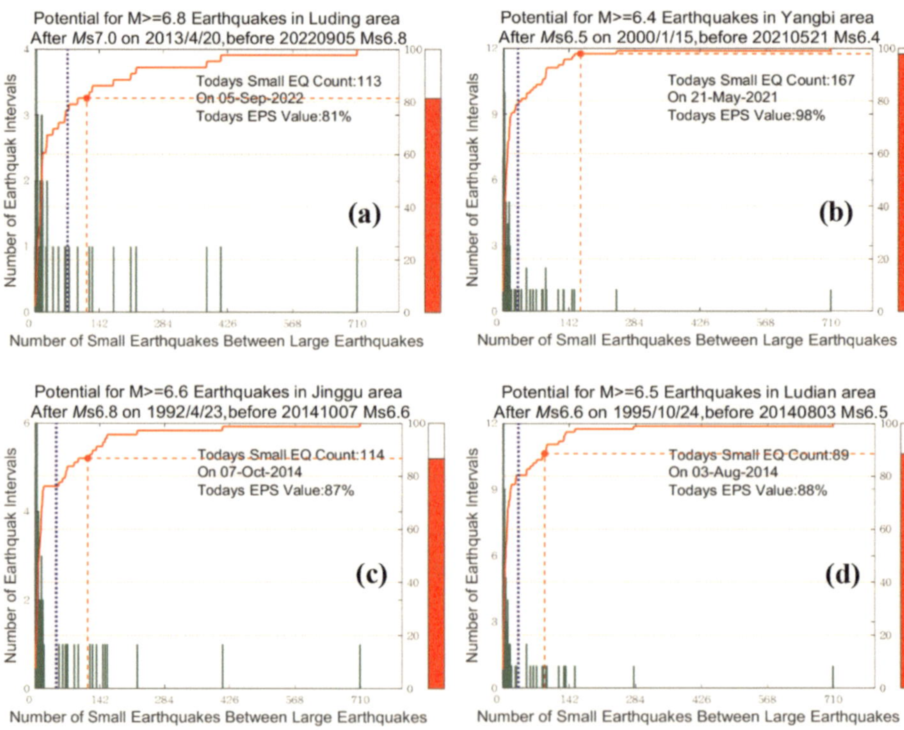

Figure 5.
The nowcasting approach was used to compute the Earthquake Potential Score (EPS) up to the occurrence of certain target events in the CSES zone. (a) Assessment for M ≥ *6.8 before the occurrence of Luding* M_S*6.8 on September 20, 2022; (b) assessment for* M ≥ *6.4 before the occurrence of Yangbi* M_S*6.4 on May 21, 2021; (c) assessment for* M ≥ *6.6 before the occurrence of Jinggu* M_S*6.6 on October 7, 2014; and (d) assessment for* M ≥ *6.5 before the occurrence of Ludian* M_S*6.5 on August 3, 2014. The green bars in each plot show the number of small events (*$M_\sigma \geq 4.0$*) in the interval of strong earthquakes. The red curve shows the CDF of earthquake intervals varied with the number of small events. The red points on the curve show the current EPS value since the occurrence of the last strong earthquake and the EPS state is shown on the right with a "hot bar." The dashed vertical line in blue indicates the mean count of small events in the intervals according to the statistical analysis for the database.*

Target events	M_λ	EPS result		
		R = 100 km	R = 150 km	R = 200 km
2014-08-03 Ludian M_S6.5	6.5	86%	85%	88%
2014-10-07 Jinggu M_S6.6	6.6	84%	91%	87%
2021-05-21 Yangbi M_S6.4	6.4	96%	96%	98%
2022-09-05 Luding M_S6.8	6.8	73%	79%	81%

Table 1.
EPS before different target events with different circular regions with radii R.

indicate a rather high likelihood of subsequent large earthquakes. How to comprehend the high EPS with the possibility of the next large earthquake is one issue that has persisted from earlier investigations. Because the EPS is high in this research but does not reach 100%, it is not required to wait for the EPS to reach 100% before the next significant earthquake occurs. When interpreting the EPS using the nowcasting technique, it is useful to keep in mind that the area under investigation is nearing another large earthquake when the EPS is rising high.

6. Summary and discussion

To characterize the fundamental aspects of seismicity, we examined the earthquake catalog in the CSES area. We determine the cut-off magnitude with 4.0 in this study according to the temporal distribution of completeness magnitude and the features of the database containing historical and modern earthquake catalog. Further research showed that ergodicity persists over the vast majority of the time from 1980 to the present, presenting evidence for the meta-stable equilibrium hypotheses. The clustering of Wenchuan aftershocks makes the Yunnan area's ergodicity superior to that of Sichuan's, which means Sichuan's influence on the ergodicity of the whole CSES is dominant. Four earthquakes, that is, August 3, 2014, Ludian M_S6.5 earthquake; October 7, 2014, Jinggu M_S6.6 earthquake; May 21, 2021, Yangbi M_S6.4 earthquake; and September 5, 2022, Luding M_S6.8 earthquake, whose potential was assessed using the nowcasting approach. To consider the influence of chosen parameters, we select radii 100, 150, and 200 km in the nowcasting method to analyze the EPS before the occurrence time of them. The outcome indicates that the EPS is high but does not achieve 100%, suggesting a scientific interpretation of the EPS when the next significant earthquake in this area is about to occur.

It was revealed in this research that the nowcasting approach may be used to define the earthquake potential after the geographical area has been identified. How far the prediction window from now can be given by this approach, in reality, is one of the concerns that need further debate. For this problem, combining the Annual Consultation Conference in China with the nowcasting approach may be a viable option, which may determine its relevance to the particular earthquake work in different locations. Previous research has shown that such yearly forecasts perform better than random guesses [40]. However, for the nowcasting technique, the present EPS is a crucial and fundamental piece of knowledge when we want to understand the potential in the sphere of catastrophe preparation.

Acknowledgements

We are grateful to Prof. Wu Zhongliang in the Institute of Earthquake Forecasting, China Earthquake Administration, Prof. John B. Rundle in UC Davis, and Jiang Changsheng in Institute of Geophysics, China Earthquake Administration for their assistance and guidance with the nowcasting method and earthquake catalog analysis. The earthquake catalog utilized in this investigation was given by the China Earthquake Networks Center (CENC). This study is also sponsored by the National Natural Science Foundation of China (42004038, U2039207), the National Key Research and Development Program of China (2018YFE0109700) and the Special Fund of the Institute of Earthquake Forecasting, China Earthquake Administration (CEAIEF2022030206).

Author details

Shengfeng Zhang* and Yongxian Zhang
Institute of Earthquake Forecasting, China Earthquake Administration, Beijing, China

*Address all correspondence to: 085012104@163.com

References

[1] Zhang PZ, Deng QD, Zhang GM, Ma J, Gan WJ, Min W, et al. Active tectonic blocks and strong earthquakes in the continent of China. Science in China Series D: Earth Sciences. 2003;**46**: 13-24. DOI: 10.1360/03dz0002

[2] Li SL, Mooney WD, Fan JC. Crustal structure of mainland China from deep seismic sounding data. Tectonophysics. 2006;**420**:239-252. DOI: 10.1016/j.tecto.2006.01.026

[3] Sun QZ, Wu SG. Development of the Earthquake Monitoring and Prediction in China during 1966~2006. Beijing: Seismological Press; 2007

[4] Jordan TH. Earthquake predictability, brick by brick. Seismological Research Letters. 2006;**77**:3-6. DOI: 10.1785/gssrl.77.1.3

[5] Mignan A, Jiang CS, Zechar JD, Wiemer S, Wu ZL, Huang Z. Completeness of the Mainland China earthquake catalog and implications for the setup of the China Earthquake Forecast Testing Center. Bulletin of the Seismological Society of America. 2013; **103**:845-859. DOI: 10.1785/0120120052

[6] Wu ZL. Seismic experimental sites: Challenges and opportunities. Journal of Geological Society of India. 2020;**95**: 113-116. DOI: 10.1007/s12594-020-1400-9

[7] Wu ZL, Li Q, Zhang XD, Li L, Tang Y, Che S, et al. China seismic experimental site (CSES): Start and trial. Reviews of Geophysics and Planetary Physics. 2021;**52**(6):675-678. DOI: 10.19975/j.dqyxx.2021-012

[8] Wu ZL, Zhang XD, Sun K. China seismic experiment site: Scientific challenges. Acta Geologica Sinica-English Edition. 2019;**93**:273-273. DOI: 10.1111/1755-6724.14084

[9] Wu ZL, Zhang Y, Li JW. Coordinated distributed experiments (CDEs) applied to earthquake forecast test sites. In: Li Y-G, editor. Earthquake and Disaster Risk: Decade Retrospective of the Wenchuan Earthquake. Singapore: Higher Education Press and Springer Nature Singapore Pte Ltd.; 2019. pp. 107-115

[10] Wu ZL, Ding ZF, Zhang XD, Li L, Shao ZG, Li Y, et al. China seismic experimental site: Retrospective and prospective. Reviews of Geophysics and Planetary Physics. 2021;**52**(2):234-238. DOI: 10.19975/j.dqyxx.2020-023

[11] Wu ZL, Wang L, Che S, Li L, Zhang XD, Shao ZG, et al. China seismic experimental site (CSES): Planning and test. Reviews of Geophysics and Planetary Physics. 2021;**52**(3):348-352. DOI: 10.19975/j.dqyxx.2021-010

[12] Wu ZL, Wang L, Li L, Zhang XD, Shao ZG, Li Y, et al. China seismic experimental site (CSES): Earthquake forecast and system design. Reviews of Geophysics and Planetary Physics. 2021; **52**(6):679-683. DOI: 10.19975/j.dqyxx.2021-028

[13] Michael AJ, Werner MJ. Preface to the focus section on the Collaboratory for the Study of Earthquake Predictability (CSEP): New results and future directions. Seismological Research Letters. 2018;**89**:1226-1228. DOI: 10.1785/0220180161

[14] Savran WH, Maechling PJ, Werner MJ, Jordan TH, Schorlemmer D, Rhoades DA, et al. The Collaboratory for the Study of Earthquake Predictability

version 2.0 (CSEP2.0): New capabilities in earthquake forecasting and testing. In: Poster Presentation at 2018 SCEC Annual Meeting. Los Angeles. 2018

[15] Zhang YX, Wu ZL, Zhang XD, Yu HZ, Zhang S, Zhang XT, et al. CSEP2 from China perspective. In: Poster Presentation at 2019 SCEC Annual Meeting. Los Angeles. 2019

[16] Zhang SF, Zhang YX. Collaborratory for the Study of Earthquake Predictability working philosophy and ten years' achievements of CSEP1.0. Earth. 2021;**41**(4):203-217. DOI: 10.12196/j.issn.10003274.2021.04.016

[17] Zhang YX, Wu ZL, Zhang XT, Li G. Annual earthquake potential consultation: A real forward prediction test in China. In: Li Y-G, editor. Earthquake and Disaster Risk: Decade Retrospective of the Wenchuan Earthquake. Singapore: Higher Education Press and Springer Nature Singapore Pte Ltd.; 2019. pp. 117-134. DOI: 10.1007/978-981-13-1085-0_5

[18] Department of Earthquake Disaster Prevention of the State Seismological Bureau. The Catalogue of Chinese Historical Strong Earthquakes. Beijing: Seismological Press; 1995

[19] Department of Earthquake Disaster Prevention of the China Earthquake Administration. The Catalogue of Chinese Modern Earthquakes. Beijing: China Science and Technology Press; 1999

[20] Woessner J, Wiemer S. Assessing the quality of earthquake catalogues: Estimating the magnitude of completeness and its uncertainty. Bulletin of the Seismological Society of America. 2005;**95**:684-698. DOI: 10.1785/0120040007

[21] Rundle JB, Luginbuhl M, Giguere A, Turcotte DL. Natural time, nowcasting and the physics of earthquakes: Estimation of seismic risk to global megacities. Pure and Applied Geophysics. 2017;**175**:647-660. DOI: 10.1007/s00024-017-1720-x

[22] Rundle JB, Luginbuhl M, Giguere A, Turcotte DL. Natural time, nowcasting and the physics of earthquakes: Estimation of risk to global megacities. Pure and Applied Geophysics. 2018;**175**:647-660. DOI: 10.1007/s00024-017-1720-x

[23] Rundle JB, Luginbuhl M, Giguere A, Turcotte DL. Natural Time, Nowcasting and the Physics of Earthquakes: Estimation of Seismic Risk to Global Megacities. In: Williams C, Peng Z, Zhang Y, Fukuyama E, Goebel T, Yoder M, editors. Earthquakes and Multi-hazards Around the Pacific Rim, Vol. II. Pageoph Topical Volumes. Cham: Birkhäuser. 2019. DOI: 10.1007/978-3-319-92297-3_10

[24] Rundle JB, Turcotte DL, Donnellan A, Grant Ludwig L, Luginbuhl M, Gong G. Nowcasting earthquakes. Earth and Space Science. 2016;**3**:480-486. DOI: 10.1002/2016ea000185

[25] Egolf DA. Equilibrium regained: From nonequilibrium chaos to statistical mechanics. Science. 2000;**287**(5450): 101-104. DOI: 10.1126/science.287.5450.101

[26] Tiampo KF, Rundle JB, Klein W, Martins JSS, Ferguson CD. Ergodic dynamics in a natural threshold system. Physical Review Letters. 2003;**91**(23): 238501. DOI: 10.1103/PhysRevLett.91.238501

[27] Tiampo KF, Rundle JB, Klein W, Holliday J, Sá Martins JS, Ferguson CD.

Ergodicity in natural earthquake fault networks. Physical Review E. 2007;**75**: 066107. DOI: 10.1103/PhysRevE. 75.066107

[28] Tiampo KF, Rundle JB, Klein W, Martins JSS. Ergodicity in natural fault systems. Pure and Applied Geophysics. 2004;**161**:1957-1968. DOI: 10.1007/ s00024-004-2542-1

[29] Thirumalai D, Mountain RD, Kirkpatrick TR. Ergodic behavior in supercooled liquids and in glasses. Physical Review A. 1989;**39**:3563-3574. DOI: 10.1103/PhysRevA.39.3563

[30] Thirumalai D, Mountain RD. Activated dynamics, loss of ergodicity, and transport in supercooled liquids. Physical Review E. 1993;**47**:479-489. DOI: 10.1103/PhysRevE.47.479

[31] Tiampo KF, Rundle JB, McGinnis S, Gross SJ, Klein W. Mean-field threshold systems and phase dynamics: An application to earthquake fault systems. Europhysics Letters. 2002;**60**(3):481-487. DOI: 10.1209/epl/i2002-00289-y

[32] Rundle JB, Donnellan A. Nowcasting earthquakes in southern California with machine learning: Bursts, swarms, and aftershocks may be related to levels of regional tectonic stress. Earth and Space Science. 2020;**7**:e2020. DOI: 10.1029/ 2020EA001097

[33] Pasari S, Sharma Y. Contemporary earthquake hazards in the west-northwest Himalaya: A statistical perspective through natural times. Seismological Research Letters. 2020;**91**: 3358-3369. DOI: 10.1785/0220200104

[34] Pasari S, Neha. Nowcasting-based earthquake hazard estimation at major cities in New Zealand. Pure and Applied Geophysics. 2022;**179**:1597-1612. DOI: 10.1007/s00024-022-03021-z

[35] Varotsos PA, Sarlis NV, Skordas ES. Long-range correlations in the electric signals that precede rupture. Physical Review E. 2002;**66**:011902. DOI: 10.1103/PhysRevE.66.011902

[36] Varotsos PA, Sarlis NV, Skordas ES. Natural Time Analysis: The New View of Time. Berlin: Springer; 2011

[37] Varotsos PA, Sarlis NV, Tanaka HK, Skordas ES. Some properties of the entropy in the natural time. Physical Review E. 2005;**71**:032102. DOI: 10.1103/ PhysRevE.71.032102

[38] Scholz CH. The Mechanics of Earthquakes Faulting. Cambridge, UK: Cambridge Univ. Press; 1990

[39] Bevington PR, Robinson DK. Data Reduction and Error Analysis for the Physical Sciences. 3rd ed. Boston (Mass.): McGraw-Hill; 2003

[40] Shi YL, Liu J, Zhang GM. An evaluation of Chinese annual earthquake predictions, 1990-1998. Journal of Applied Probability. 2001;**38A**:222-231. DOI: 10.1239/jap/1085496604

Chapter 8

Seismic Conditions Required to Cause Structural Failures in Tectonic Earthquakes

Tse-Shan Hsu

Abstract

It is found that the failure of the structure conforming to the current seismic design code can only occur in the shear band of the tectonic earthquake, and with the increase in the amount of shear banding, the boundary conditions of the structure gradually deviate from the original design ones; therefore, the seismic insufficiency of structures therefore continues to increase. With the increase of the seismic insufficiency of the structure, the structure will appear more and more serious damage such as cracking, tilting and subsidence, and collapse. Based on the findings of this study, the author suggests that the main task of the seismic design of structures is to prevent the shear banding of tectonic earthquakes from extending to each element of the structure, rather than continuously increasing the level of vibration fortification of each structural element. Only in this way can it be ensured that the structures complying with the seismic design specifications will not be damaged such as cracks, tilting and subsidence, and collapse due to the deviation of the boundary conditions from the original design ones in shear banding.

Keywords: tectonic earthquake, shear banding, ground vibration, seismic conditions, boundary conditions, structural failures

1. Introduction

Although Zhang Heng invented the first ground vibration measuring instrument as early as 132 AD, and subsequent to that the seismic design codes of countries located within earthquake zones have specified increasingly strict vibration fortification standards for structures, it can be seen from **Table 1** that the death toll of tectonic earthquakes remains high.

It has been established that traditional earthquake disaster mitigation scholars, in the absence of a clear definition of the seismic conditions that result in structural failure during tectonic earthquakes, have attributed such failures (illustrated in **Figures 1–4**) to insufficient protection against ground vibration resistance, and thus the vibration fortification specifications for structures have continued to be increased after tectonic earthquakes that have had a large impact. The main effect of a tectonic earthquake however results from shear banding, which accounts for more than 90%

No.	Year	Country	Death
1	1556	China	830,000
2	1976	China	655,000
3	2010	Haiti	316,000
4	2004	Indonesia	280,000
5	2010	Haiti	316,000
6	526	Turkey	300,000
7	1920	China	273,400
8	1923	Japan	142,800
9	1948	Russia	110,000
10	1922	Philippines	100,000

Table 1.
The 10 major earthquakes that have caused the highest number of deaths around the world [1].

(b)

Figure 1.
Damage sustained by the Zhenong Building in Taipei during the 418 Hualien earthquake of 2019 [2]: (a) building appearance; (b) cracks in the floor slab.

Figure 2.
Damage sustained by the Yutai Building in Taipei during the 418 Hualien earthquake of 2019 [3]: (a) failure of tilting and subsidence; (b) close-up of damage caused by tilting and subsidence.

of the total energy dissipated by a seismic event; the secondary effect is ground vibration, whose energy accounts for less than 10% of the total seismic energy dissipation [6]. In fact, the earthquake disaster prevention methods proposed, based on this flawed understanding, including vibration isolation and vibration reduction, actually only increase the vibration resistance of structures under earthquake conditions that would not result in failure in any case. Since the seismic design specifications do not properly reflect the seismic conditions that result in structural failure during tectonic earthquakes, it is not possible to ensure that structures that fully comply with the seismic design specifications will not fail during earthquakes. In view of this, this chapter will first define the seismic conditions that do result in structural failure during tectonic earthquakes and then analyze the reasons why the traditional methods of earthquake disaster prevention cannot perform their desired functions.

Figure 3.
The Dongshing Building in Taipei collapsed during the 921 Jiji earthquake of 1999 [4].

(a)

(b)

Figure 4.
During the 921 Jiji earthquake of 1999: (a) Sanmin Junior High School in Hualien, Taiwan remained stable [5]; (b) Guangfu Junior High School in Taichung, Taiwan suffered severe damage. Note: The red line in (b) indicates the position of the first floor.

2. Seismic conditions that result in structural failure

Building seismic design codes are formulated and revised by experts in structural dynamics and soil dynamics and thus only pertain to ground vibration fortification. Hsu *et al.* [2] showed (using **Figures 1–3**) that the increase in the amount of shear banding during tectonic earthquakes in Taipei caused problems such as the cracking of the floor slab of the Zhenong Building, tilting and subsidence of the Yutai Building and collapse of the Dongshing Building.

As illustrated in **Figures 1–3**, it was determined that the seismic conditions that resulted in structural failure during tectonic earthquakes were the boundary conditions at the bottom ends of a part or all of the columns of a building gradually deviating from being a fixed end as specified in the original design, due to an increase in the accumulated amount of shear banding. The degree of damage sustained by the building increases with an increase in the extent of this deviation, *i.e.*, it increases with an increase in the cumulative amount of shear banding.

Secondly, by comparing **Figures 4a** and **b** we see that during the 921 Jiji earthquake the school building located in a non-shear banding area (**Figure 4a**) remained stable and sustained no damage; however, the school building located in a shear banding area (**Figure 4b**) could not maintain stability and suffered serious failure.

As illustrated in **Figures 1–4**, it has thus been determined that (1) the seismic conditions that result in structural failure during earthquakes are the boundary conditions at the bottom ends of a part of the columns or all the columns of a building being unable to maintain the fixed end conditions of the original design; (2) the seismic conditions that do not result in structural failure during earthquakes are the boundary conditions at the bottom ends of all columns of the building maintaining the fixed end conditions of the original design.

2.1 Traditional pushover analysis and test results

Figure 5 shows the hyperbolic deformation mechanism after the bottom end of a column was set as the fixed end in a traditional pushover analysis and test, and **Figure 6** shows the deformed mesh of a school building model obtained using traditional pushover analysis.

Figure 7a shows that before the traditional pushover test was conducted, the boundary conditions at the bottom end of each column of the selected physical school building had the fixed end conditions of the original design after the 921 Jiji earthquake; in other words, the school building had the required seismic conditions that would not result in structural failure during a tectonic earthquake. **Figure 7b** also shows that after the traditional pushover test of the physical school building was conducted, the boundary conditions at the bottom ends of all columns maintained the fixed end conditions of the original design.

From **Figure 4b**, it can be seen that after the 921 Jiji earthquake, the boundary conditions of each column base of the physical school building of Guangfu Junior High School had deviated from the fixed end conditions of the original design, which is the seismic condition that results in structural failure during tectonic earthquakes, thus suffering serious damage. From **Figures 6** and **7b**, it can be deduced that whether traditional pushover analysis or traditional pushover testing is conduced, the boundary conditions of each column bottom in the structural analysis model maintained the fixed end conditions of the original design. Therefore, it is clear that the results obtained using these methods are not valid.

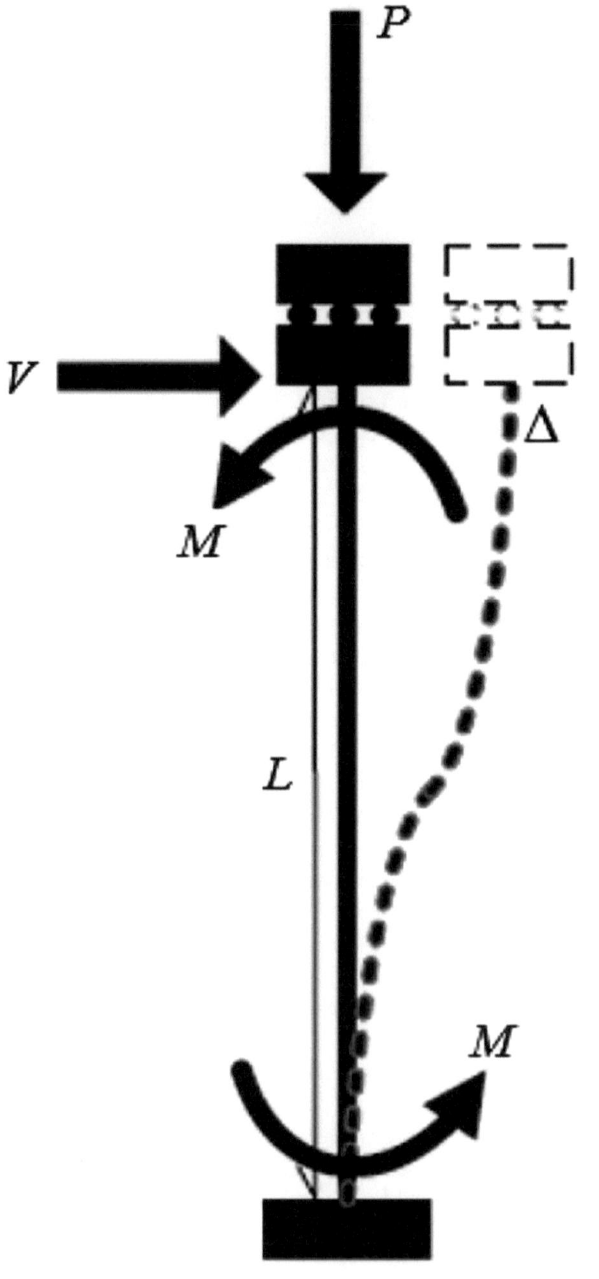

Figure 5.
The deformation mechanism generated after the bottom end of a column was set as the fixed end in a traditional pushover analysis and test [7].

2.2 Shaking table test results

A shaking table is a rigid thick plate capable of vibrating according to an input ground vibration acceleration history. Before a shaking table test is conducted, the bottom ends of all the columns of the building model are fixed to the shaking table; in other words, the building model is set up prior to testing under the seismic conditions

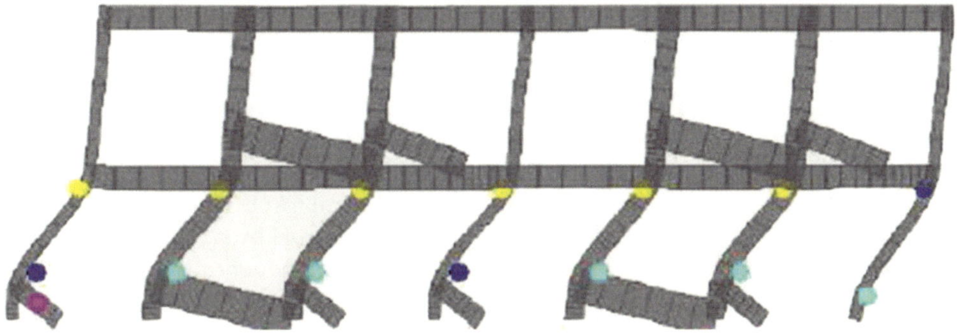

Figure 6.
The deformed mesh for the model of Kouhu Elementary School building, Yunlin, Taiwan, for a traditional pushover analysis [8].

(a)

(b)

Figure 7.
The traditional pushover test for the physical Kouhu Elementary School building, Yunlin, Taiwan [9]: (a) before test; (b) after test.

in which structural failure does not actually occur. Secondly, during the shaking table test, the model is pushed down under these same seismic conditions. Therefore, the failure pattern of the building model after the shaking table test has the same flaws as the pushover analysis result shown in **Figure 6** and the pushover test result shown in **Figure 7b**, and therefore the results of the shaking table test are not valid.

It can be seen from **Figures 6** and **7b** that the failure mode of the building model during a shaking table test is that of collapse due to lateral forces under the condition of weak columns and strong beams. Since buildings designed by structural engineers must meet the requirements of the building seismic design code, this problem of weak columns and strong beams does not occur in practice; however, because the results of the shaking table test directly show that when the model fails, the bottom ends of its columns still maintain their fixed end boundary condition and fail in the weak column and strong beam mode; thus, the validity of the shaking table test results is further undermined.

2.3 Shear banding table test results

Figure 8 shows a building model placed on a shear banding table. When the left side of the shear banding table was lifted up obliquely relative to the right, the failure pattern of the building model was similar to that of the Guangfu Junior High School building in Taichung, Taiwan during the 921 Jiji earthquake (**Figure 4b**). Therefore, it is known that in the failure process of the building model during the earthquake, the building model first changed to have the seismic conditions that resulted in structure failure caused by shear banding. With the increase in the amount of shear banding, the degree of inability of the building model increases, and finally fails.

Figure 9a shows a bridge model placed on a shear banding table. When the left side of the shear banding table was tilted and uplifted relative to the right side, the bridge model also underwent tilting and uplift. The failure pattern was similar to that of the Wuxi Bridge in Taichung, Taiwan due to the 921 Jiji earthquake as shown in **Figure 9b**. Therefore, it is known that during tectonic earthquakes, when a bridge is deformed by tilting and uplift, the seismic conditions shift from those that do not result in failure to those that do result in failure due to shear banding.

Figure 8.
The shear banding table test results of a building model [10].

(a)

(b)

Figure 9.
Comparison of the tilting uplift failure of a model bridge and a real bridge [10]: (a) model bridge after the shear banding table test; (b) Wuxi Bridge in Taichung, Taiwan after the 921 Jiji earthquake.

Figure 10 shows another example of bridge failure by further uplift and tilting due to shear banding. **Figure 10a** shows a bridge model placed on the shear banding table. When the right side of the shear banding table was further tilted and uplifted relative to the left side, the failure pattern of the falling bridge model was similar to that of the Shiwei Bridge in Taichung, Taiwan during the 921 Jiji earthquake as shown in **Figure 10b**. Therefore, before the bridge fell during the tectonic earthquake, the initial seismic conditions that would not have resulted in failure are

(a)

(b)

Figure 10.
Comparison of falling bridge phenomena [10]: (a) model bridge after a shear banding test; (b) the Shiwei Bridge in Taichung, Taiwan after the 921 Jiji earthquake.

changed to those that would result in failure due to shear banding. Ultimately, the bridge collapsed after the amount of shear banding had increased significantly.

Figure 11a shows a model of a weir placed on a shear banding table. When the right side of the shear banding table was tilted and uplifted relative to the left side, the failure pattern of the weir model was similar to that of the Shigang weir in Taichung, Taiwan during the 921 Jiji earthquake shown in **Figure 11b**. Therefore, before the weir failed during the tectonic earthquake, the initial seismic conditions that would not have resulted in weir failure were changed to those that would result in failure because of shear banding. Ultimately, the weir failed as the amount of shear banding was greatly increased.

(a)

(b)

Figure 11.
Comparison of the collapse of a model weir and a real weir [10]: (a) the model weir after the shear banding table test; (b) the weir in Shigang, Taichung, Taiwan after the 921 Jiji earthquake.

3. Seismic failure of a caisson pier

Since the traditional soil liquefaction safety factor, FS_L, is defined such that the ratio of the cyclic stress ratio of liquefaction resistance (CSRRL) to the period stress ratio of an earthquake (CSRE) be less than 1.0 [11], from the point of view of plastic mechanics this actually indicates soil yield, and it is thus quite easy to misjudge the cause of soil yield as soil liquefaction. Taking the 921 Jiji earthquake as an example, the rear line region of the caisson pier in Taichung Port, Taiwan (**Figure 12**) showed three different types of failure: subsidence, a pothole, and the spouting of silt, sand, and gravel.

Figure 12.
Different types of seismic failures at the rear line region of the caisson pier of Taichung port [12].

After the 921 Jiji earthquake, most traditional scholars believed that all three types of failure were caused by soil liquefaction. The reason they provided was that under the ground vibration of a tectonic earthquake and the densification process of the saturated loose sand below the groundwater table, the effective stress would have been reduced to zero or less than zero with the increase in the excess pore water pressure; therefore, soil liquefaction, by definition, would have occurred. Thus, it is believed by traditional scholars that the significant subsidence, the large pothole, and the ejection of silt, sand, and gravel shown in **Figure 12** were all caused by soil liquefaction.

Only after identifying the major causes of the three different types of failures that occurred at Taichung Port during the 921 Jiji earthquake can we develop effective suppression methods to prevent such failures from taking place in the rear line region of other caisson piers.

In 2017, Hsu *et al.* [13] presented evidence that soil liquefaction occurs only in shear banding or shear texturing zones induced by the plastic strain softening of dense soil, but not in the yield failure of loose soil. In addition, the compacted soil of the rear line region of the caisson pier had already passed the field density quality test. According to the contract of the entrusted project, the degree of compaction was greater than 90%, and the corresponding relative density was greater than 70%. Therefore, the compacted soil was dense; however, the compacted soil of the rear line region of the caisson pier was misjudged as loose by conventional scholars in Taiwan.

Figure 13 shows a schematic profile of a caisson pier. It can be seen that the caisson pier is placed directly on the cobble base of the seabed. At low tide and in the absence of embedded depth, D_f when depth h_2 from the seawater table to the seabed is less than depth h_1 from the groundwater table of the rear line region to the seabed and the hydraulic gradient i of the water exit point E (shown in **Figure 13**) is greater than the critical hydraulic gradient i_c, then some sandy gravel backfill in the rear line region of the caisson pier flows out to the seabed below the seawater table due to piping failure (as shown schematically in **Figure 14**), and then large piping holes appear such as those shown in **Figures 12** and **14**, respectively.

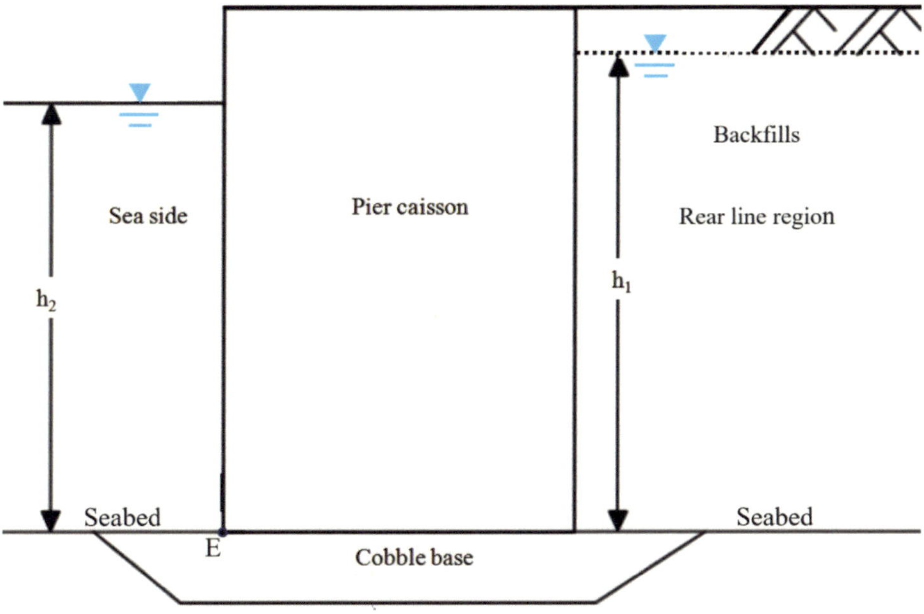

Figure 13.
Schematic diagram of the caisson pier profile [14].

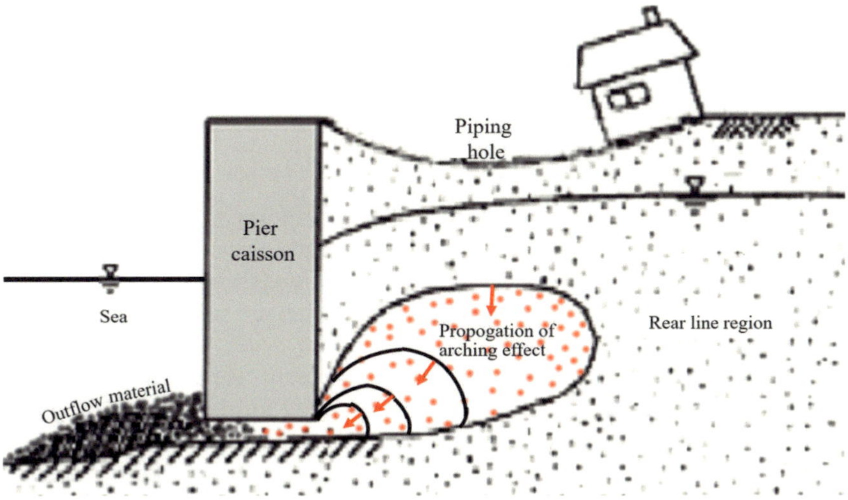

Figure 14.
Schematic diagram of the propagation of the arching effect and the appearance of a piping hole in the rear line region of a caisson pier (reproduced from [12]).

Furthermore, **Figure 15a** shows the test model of the caisson pier with very dense backfill in its rear line region. **Figure 15b** shows that when the caisson pier is turned seawards, its rear line region appears to have shear textures with different strikes, resulting in substantial subsidence. The shear textures shown in **Figure 15c** include the principal shear D, thrust shear P, and Riedel shear R. During tectonic earthquakes, when shear textures of different strikes are dislocated, high excess pore water pressure will locally appear in the shear textures, and so the groundwater in the shear

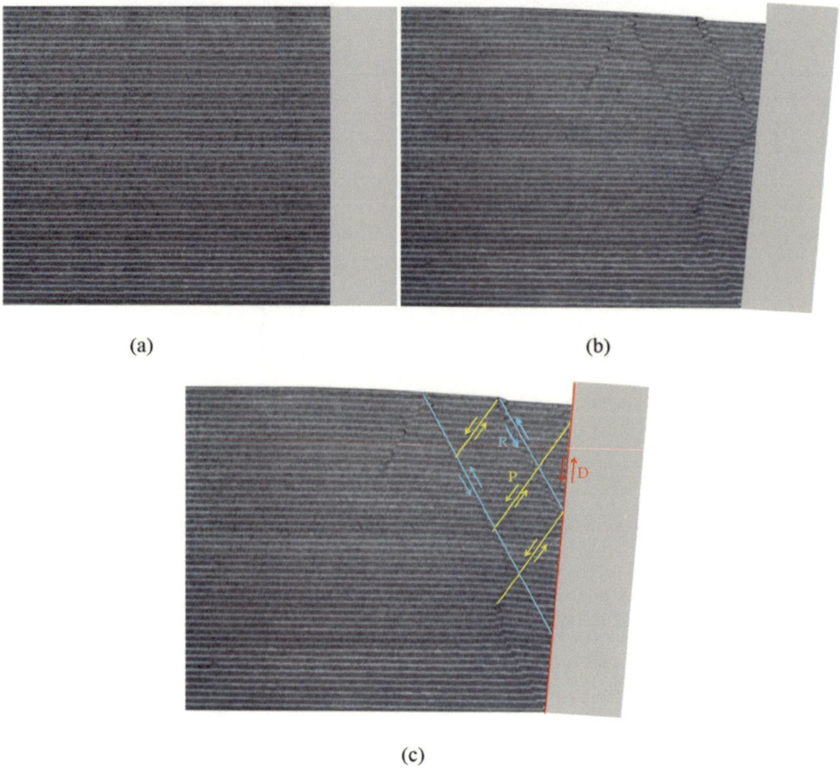

Figure 15.
Test results of shear textures and subsidence induced by the inclined caisson pier [12]: (a) before test; (b) after test; (c) the induced shear textures and uneven subsidence.

textures will entrain the brittle fractured silt, sand, and gravel and eject upward (see **Figure 12**) along the outlet tunnel formed by the pore space of the shear texture.

The test results in **Figure 15** show that the subsidence shown in **Figure 12** was caused by the caisson pier turning seawards, rather than by soil liquefaction as claimed in the past.

Hsu *et al*. [13] proposed the three constituent elements that must exist in order to induce soil liquefaction based on real-world data: (1) local shear banding or shear texturing during tectonic earthquakes; (2) high excess pore water pressure localized in the shear band or in the shear texture; (3) groundwater entrained silt, sand, and/or gravel ejection upward along the outlet tunnel formed by the pore space of the brittle fractured shear band or shear texture. For the ejected silt, sand, and gravel shown in **Figure 12**, the cause was soil liquefaction since the three elements required for soil liquefaction clearly existed.

4. Methods of earthquake-resistance and building reinforcement

Traditional building earthquake-resistance and reinforcement methods include the installation of vibration isolation pads, the installation of dampers, and increasing the stiffness of superstructural elements such as columns, beams, panels, and walls, thereby inducing vibration isolation, vibration reduction, and vibration resistance. Therefore, traditional earthquake-resistance and reinforcement methods only improve the vibration fortification level of a building under the seismic conditions required to prevent structural failure, but do not have an impact on the conditions in which structural failure does in fact occur due to shear banding during tectonic earthquakes. Therefore, buildings will still fail under the shear banding effect in future tectonic earthquakes.

The school building earthquake-resistance and reinforcement methods provided by the National Center for Research on Earthquake Engineering of Taiwan after the 921 Jiji earthquake included the reinforced concrete (RC) jacketing retrofit method, the RC wing wall retrofit method, the shear wall retrofit method, and the composite column retrofit method. It turns out that these methods can only improve the fortification level of a building against seismic vibration conditions required to prevent structural failure, but do not protect the building from failing under the shear banding effect of future tectonic earthquakes.

Since the failure of buildings during tectonic earthquakes is mainly caused by the shear band or shear texture extending into the area surrounding the shear failure plane induced in the foundation soil under the ultimate load, when the safety factor of the earthquake bearing capacity of a foundation is less than 1.0, earthquake subsidence will be induced; after the foundation loses its stability, an asymmetrical general shear failure plane will be generated, which will lead to the failure of the foundation and building [15]. Therefore, in order to effectively prevent the collapse of buildings during tectonic earthquakes, earthquake-resistance and reinforcement methods need to restrain the shear band or the shear texture that is induced by the ultimate load of a tectonic earthquake from extending into the area surrounding the shear failure plane in the foundation soil (see **Figure 16**). Only in this way can it be ensured that buildings complying with the seismic design specifications for ground vibration fortification will not be affected by shear banding or shear texturing, thereby ensuring the stability and safety of buildings during tectonic earthquakes.

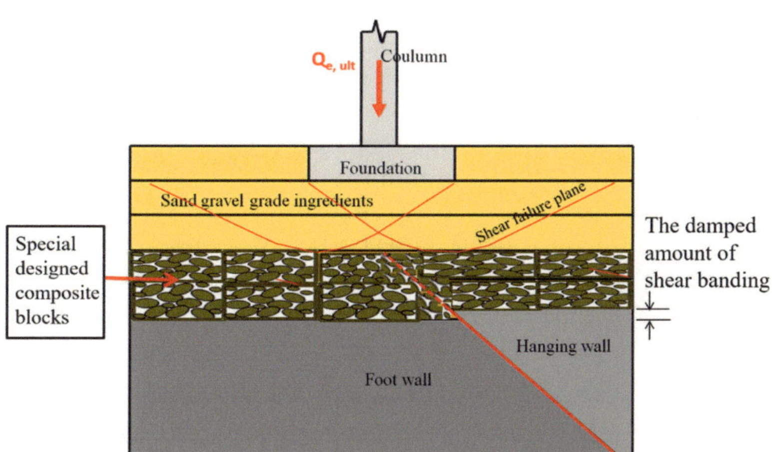

Figure 16.
Schematic diagram of effective structural earthquake-resistance and reinforcement methods for restraining the propagation of shear banding or shear texturing [16].

5. Conclusions

It has been 1890 years since the invention of the first ground vibration measuring instrument. During the whole of recorded history, the vibration fortification level specified by earthquake-resistant design codes has continued to increase, but the associated death toll due to earthquakes is still high. In view of this, the author has outlined the seismic conditions in which structural failure does in fact occur during tectonic earthquakes so as to explore the causes of failure of different structures and thereby reached the following six conclusions:

1. During a tectonic earthquake, the failure of the structure is caused by the shear banding or shear texturing induced by the tectonic earthquake, resulting in the seismic conditions changing to those that do cause structural failure.

2. In the pushover test and shaking table test, the failure of the building model occurs under the seismic conditions that do not, in reality, result in structural failures. Therefore, the failure modes of the building models are completely different to the failure modes of the physical buildings in actuality; the results of the pushover tests and the shaking table tests are therefore unrealistic.

3. In the shear banding table test, the failure of the building model occurs when the shear banding of the tectonic earthquake induces the seismic conditions to change from those that do not result in structural failure to those that do, and the failure of the physical buildings also follow such a pattern. In other words, the failure mode obtained from these tests are similar to those of actual buildings under the shear banding effect of tectonic earthquakes. Therefore, the results of the shear banding test are realistic.

4. Traditional scholars define the yield failure of saturated loose sand as soil liquefaction, without including the three constituent elements required for soil liquefaction of dense sand produced in the process of strain softening caused by the shear banding or shear texturing of tectonic earthquakes.

5. Seismic failure of the rear line region of a particular caisson pier included great subsidence, a large pothole, and the spouting of silt, sand, and gravel. The great subsidence was caused by the rotation or movement of the pier towards the sea; the cause of the large pothole was a piping failure, and the cause of the ejection of silt, sand, and gravel was soil liquefaction induced by shear texturing. These causes are different from the conclusions drawn by traditional scholars.

6. After the 921 Jiji earthquake, Taiwan completed the earthquake-resistance and reinforcement of school buildings at a total cost of more than NT$40 billion. The methods provided by the National Centre for Research on Earthquake Engineering only increased the fortification level of the columns, beams, and slabs under seismic vibration conditions to prevent structural failure, and so the school building will still collapse from the shear banding effect of future tectonic earthquakes.

Acknowledgements

The previous work of this research is a part of the preliminary studies on the stability of existing earth structures as part of the Repair, Evaluation, Maintenance, and Rehabilitation Research Program (REMR). Financial support provided by the office of Chief of Engineers, U. S. Army is acknowledged with thanks. Close consultation with Dr. S. K. Saxena and Dr. J. F. Peters was of great benefit in determining the results of the computer analyses. The chance to use the computer facilities in the U. S. Army Engineer, Waterways Experiment Station to finish most of the computer work in this research is highly appreciated.

Author details

Tse-Shan Hsu
Department of Civil Engineering, Institute of Mitigation for Earthquake Shear Banding Disasters, Feng-Chia University, Taiwan

*Address all correspondence to: tshsu@fcu.edu.tw

References

[1] Wikipedia, The Free Encyclopedia. List of Natural Disasters by Death toll. 2022. Available from: https://en.wikipedia.org/wiki/List_of_natural_disasters_by_death_toll

[2] Hsu TS, Wu ZL, Wu YC, Chuang TF. Influence of shear banding on building safety after urban renewal. International Journal of Organizational Innovation. 2022;**15**(2):165-184

[3] Hsu SE. This Building in Taipei Repeats its History of Tilting 33 Years ago during the Hualien Earthquake. Chungshi News Network; 2019. Available from: https://www.chinatimes.com/realtimenews/20190418002921-260405?chdtv

[4] Zhao YT. The Dongshing Building in Taipei Collapsed during the 921 Jiji Earthquake of 1999. ETtoday Net News; 2020. Available from: https://www.ettoday.net/news/20200921/1810526.htm

[5] Google Earth 2020. Available from: https://www.google.com.tw/intl/zh-TW/earth/download/gep/agree.html

[6] Coffey J. What Are the Different Types of Earthquakes? Universe Today, Space and Astronomy News; 2019. Available from: https://www.universetoday.com/82164/types-of-earthquakes/

[7] Huang SJ. E-paper of seismic evaluation and reinforcement for school buildings. National Center for Research on Earthquake Engineering. 2014;**2014**:17

[8] Huang SJ. E-paper of seismic evaluation and reinforcement for school buildings. National Center for Research on Earthquake Engineering. 2009;**2009**:2

[9] Chung LL, Ku TY, Huang CT, Yang YS, Lai YA, Wu LY. Simplified pushover analysis for in-situ test of existing school buildings. Structural Engineering. 2013;**28**(2):3-29

[10] Huang SX. Shear Banding Proving Test Necessary for Earthquake Disasters Prevention. Taiwan: Feng-Chia University; 2018

[11] Seed HB, Idriss IM. Simplified procedure for evaluating soil liquefaction potential. Journal of the Soil Mechanics and Foundations Division, ASCE. 1971;**97**(9):1249-1273

[12] Chang HG. A Study of the Major Cause for the 2018 Hualien Earthquake Disasters. Taiwan: Feng-Chia University; 2019

[13] Hsu TS, Tsao CC, Lin CT. Localizations of soil liquefactions induced by tectonic earthquakes. The International Journal of Organizational Innovation. 2017;**9**(3):110-131

[14] Lai JY, Lai SY, Hsieh MC, Lin YW. Study of the Influence of Liquefaction on the Stability of Harbor Structures. Institute of Transportation; 2007

[15] Richards R, Elms D, Mudhu M. Seismic bearing capacity and settlements of foundations. Journal of Geotechnical Engineering Division (ASCE). 1993;**199**(4):662-674

[16] Ho CC. Tectonic Earthquake Disasters and their Mitigation Methods. Taiwan: Feng-Chia University; 2017

Chapter 9

New Insights into Major Seismic Events by Coulomb Stress Change Pattern and Aftershock Distributions – Implication for Active Tectonics

Mahnaz Nedaei

Abstract

Identification of the fault plane of earthquakes can be a critical contribution of seismology to regional tectonic studies and assessment of expected deformation and damage patterns. A fundamental ambiguity in the representation of an earthquake with a focal mechanism is to recognize the causative fault plane accommodating the slip during the event among the two nodal planes. The Coulomb static stress has been commonly used to determine the stress distribution induced by an event. However, for the first time in this research, the Coulomb regional stress was resolved on nodal planes to realize the optimally oriented plane for failure having maximum Coulomb stress on which the regional stress triggers an event. The method has been conducted for the April 5th, 2017 Sefidsang earthquake in NE Iran. The results reveal that the earthquake-triggering fault is a northeast-dipping listric fault with dextral reverse movement. The identified structural aspects subjected to active deformation in the area have crucial implications for seismic hazard assessment of the region and potential future failure areas.

Keywords: Coulomb stress change, causative fault geometry, active tectonics, earthquake, aftershock distributions

1. Introduction

According to the Andersonian faulting theory, the relative magnitude of the principle stresses can be simply related to the faulting style currently active in region. Information on the state of stress in the lithosphere comes from a variety of sources: earthquake focal mechanisms, young geologic data on fault slip and volcanic alignments, *in situ* stress measurements, and tensile fractures. The most simple and straightforward information about the state of stress is obtainable from focal mechanisms. However, a fundamental ambiguity of the most general point-source representation of an earthquake, the centroid moment tensor (CMT), is that it does not specify which of the two nodal planes is the actual fault plane [1]. There are many approaches that give the seismic moment tensor (e.g., [2–4]) but the fault plane cannot be identified among the two nodal planes. On the other hand, when surface ruptures are not observed, as in the case of too small events or blind faults, the fault plane may be undetermined. Then, the determination of focal mechanisms and the identification of the fault plane of earthquakes can be a critical contribution of seismology to regional tectonic studies and to the assessment of expected deformation and damage patterns. The research aims to take advantage of the Coulomb stress change to discriminate between the fault and auxiliary planes of earthquakes.

Stress in brittle upper crust builds up during the interseismic period until it ruptures seismically. An earthquake occurs when the volume close to a fault moves. Earthquakes, as a result of an instability in faulting, are so pervasive that on many faults most slip occurs during them. The seismicity is an extremely important resource to studies of both continental tectonics and seismic hazard assessment in deforming regions. Stress fields and optimally oriented active faults reveal seismic potential around the cities that improves future earthquake scenarios and are helpful for disaster mitigation (e.g., [5]). The Coulomb failure stress changes caused by mainshock rupture effectively explain the aftershock distributions for the earthquakes. Not only do aftershocks appear to be triggered by such stress changes but moderate events prior to the earthquake increased the potential for failure along most of the future rupture zone, perhaps controlling the location of the later rupture.

The Coulomb stress change is one of a number of models relating to the stress triggering in earthquakes [6]. Since the 1990s, the Coulomb stress change has been widely employed to probe the cause of triggering in several general types of studies: a) main shock-main shock triggering, b) main shock-aftershock triggering, and c) faults optimally oriented for failure. A large number of studies have investigated the Coulomb stress changes and earthquake interactions (e.g., [6–24]). There are many recent works that applied the Coulomb stress change on the causative fault parameter determined by the other approaches to investigate the stress distribution after the earthquake triggering the second mainshock or the aftershocks (the “a” and “b” categories).

Methods generally used to select the fault plane of earthquakes are based on the observation of directivity effects associated with finite ruptures (e.g., [25–27]). The method presented in this article includes Coulomb stress change and its interaction with the surrounding area. Coulomb stress changes induced by regional stress (before the earthquake) are implemented to determine the optimally oriented nodal plane for failure. Our research lays in the “c” category of Coulomb stress studies mentioned above. The correlation between the epicentral distribution of aftershocks and the increased Coulomb stress area promoting failure is investigated. Our motivation for this investigation is the recognition of seismic faults, in particular, when there are uncertainties in location and fault plane orientations, and the ambiguity of which nodal plane is the right one. It could advance out knowledge of the pattern and kinematics of active faulting, and understanding the regional active deformation. This research aims to improve our understanding of kinematics and has important implications for seismic hazard assessment.

2. Method

The slip that occurs on faults during earthquakes (referred to as “source faults”) deforms the surrounding area and changes the stress field. A measure of this change is the Coulomb static stress [15, 28]. The strain created by the displacement of a source fault results in the Coulomb stress changes. Therefore, the stress change can be resolved on an area or specified "receiver" fault planes. The stresses imparted by the source faults or tectonic regime affect faults with a specified strike, dip, and rake namely receiver faults. In other representation, the faults with an optimal orientation, concerning the regional (also called "tectonic") stress or the stress imparted by the source fault, and the assumed friction coefficient are suitable for sliding and could be as a receiver fault. The Coulomb failure stress change, ΔCFS, is defined as [10, 29, 30]:

$$\Delta\sigma_f(\Delta CFS) = \Delta\tau_s + \mu' \Delta\sigma_{n'} \quad (1)$$

where $\Delta\tau_s$ is the change in shear stress on the receiver fault (positive in the direction of fault slip), $\Delta\sigma_{n'}$ is the change in normal stress acting on the target fault (positive for unclamping), and μ' is the effective coefficient of friction [31, 32]. The shear stress change is dependent on the position, geometry, and slip of the source fault and the position, geometry, and rake of the receiver fault. The normal stress increase or decrease is independent of the receiver fault rake. The parameter μ' is often called the apparent coefficient of friction and includes the effects of pore pressure changes and the material properties of the fault zone [10]. Fault friction μ' is often inferred to be 0.4–0.8 for faults with little cumulative slip, which tend to be rough, and 0–0.4 for faults with great cumulative slip, which tend to be smooth [33]. This parameter is typically found to be around 0.4 for strike-slip faults or faults with unknown orientation [34, 35].

The positive ΔCFS represents that the desired fault is likely closer to failure; the negative ΔCFS indicates that the desired fault is away from failure. Both increased shear and unclamping of faults promote failure [36–40]. The Coulomb stress change depends on the geometry and slip of the earthquake, the geometry and sense of slip of the fault, and the effective coefficient of friction [20].

There are two principal considerations of the Coulomb stress changes on receiver faults: stress changes on the specified receiver fault and stress changes on an optimally oriented receiver fault. The specified receiver faults rely on resolving stress changes on faults with known geometry [20, 33, 41]. The optimally oriented receiver faults are determined by assuming that the earthquakes will be triggered only on those planes with maximum total Coulomb stress [7, 12, 13, 42].

The Coulomb static stress has been commonly used to determine the distribution of stress induced by an event. However, for the first time, it is applied to realize the distribution of regional stress triggering an event on appropriate fault. Determination of the principal regional stress directions and a measure of relative stress magnitudes are possible by a variety of different focal mechanisms within a region of uniform stress [43]. Theoretically, the earthquakes will be triggered only on the fault planes with maximum total Coulomb stress. It means that the earthquake-triggering faults might have an optimal orientation with maximum Coulomb stress imparted by the regional stress (e.g., [44]). Therefore, the earthquake causative faults can be determined from nodal planes by resolving Coulomb stress on them with respect to the regional stress and selecting those on which the imparted Coulomb stress is maximum.

3. Results

The Sefidsang earthquake occurred on April 5th, 2017 Mw 6.15 at a depth of about 11.5 km. Its epicenter was about 35.776 degrees of latitude and 60.436 degrees of longitude, southeast of Mashhad and east-northeast of Fariman, NE Iran (**Figure 1**). The maximum acceleration of this earthquake recorded by the Iranian Strong Motion Network was about 120 cm/s at the station 38 km from the epicenter. Based on the GSI field observations, the most destructive effects of the earthquake were observed near one of the Kashafrud fault segments. The Kashafrud reverse fault closely parallels the boundary between the Binalud and the Kopeh Dagh mountains. East of Mashad, the Kashafrud fault system runs to the Binalud mountains south of the Cimmerian arc-related Fariman complex (**Figure 2**).

According to the focal mechanism solutions presented by some global seismological centers, the mechanism of the Sefidsang earthquake is reversed with a minor strike slip on northwest-southeast oriented nodal planes. The characteristic of the two nodal planes and the event are presented in **Table 1**.

Although historical seismicity in the northeast of Iran is mostly associated with the Baghan, Quchan, and Neyshabur faults [45–48], the Sefidsang earthquake occurred in a region that has faced almost no historic or recent seismicity and has been characterized by low deformation rate concerning Eurasia [49].

Though in such a moderate earthquake that the causative fault is not exposed at the surface, theoretically, each nodal plane has the same possibility for failure, however, following the rules of focal mechanism inversion [50], one of them must be selected for each earthquake. The seismic fault is determined by assuming that the earthquake will be triggered on the plane with maximum total Coulomb stress. In other words, the earthquake-triggering fault should have maximum Coulomb stress with respect to

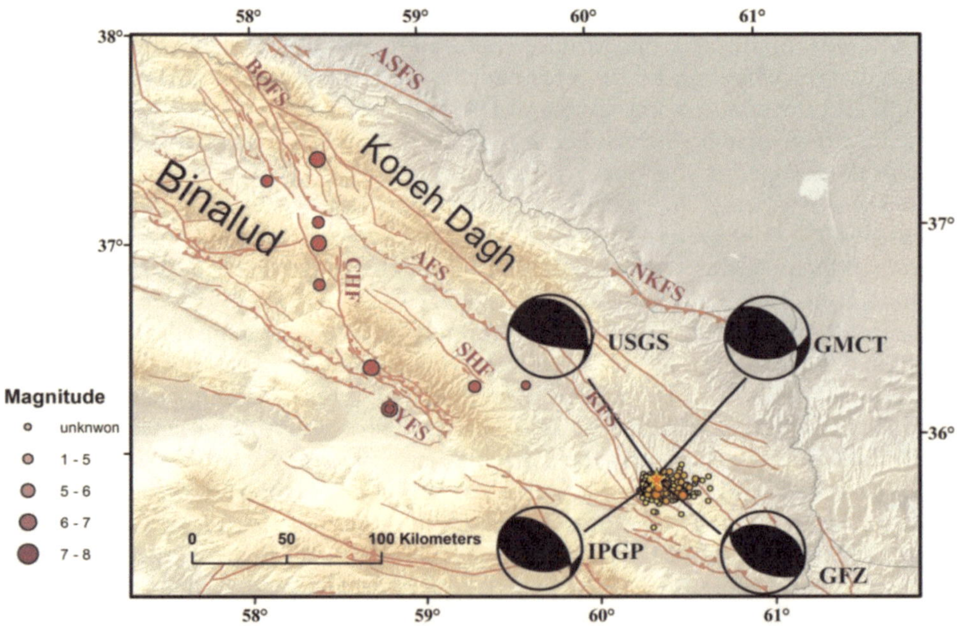

Figure 1.
The epicenter and focal mechanism solutions processed by international seismological centers of the Sefidsang earthquake (2017.04.05) plotted over the SRTM 30 m DEM. The red star presents the mainshock of the Sefidsang earthquake. The epicenters of historical earthquakes with magnitude above 6 from 1500 to 1965, modified after Refs. [45, 46], are shown as violet circles. Abbreviations for active faults are ASFS: Ashkabad fault system, NKFS: North Kopeh Dagh fault system, AFS: Atrak fault system, BF: Bijvard fault, BQFS: Baghan-Quchan fault system, CHF: Chakaneh fault, DBF: Dasht-e-Bayaz fault, DFS: Doruneh fault system, HF: Herat fault, KFS: Kashafrud fault system, JF: Jangal fault, KHF: Khaf fault, SHF: Shandiz fault, and NYFS: Neyshabur fault system.

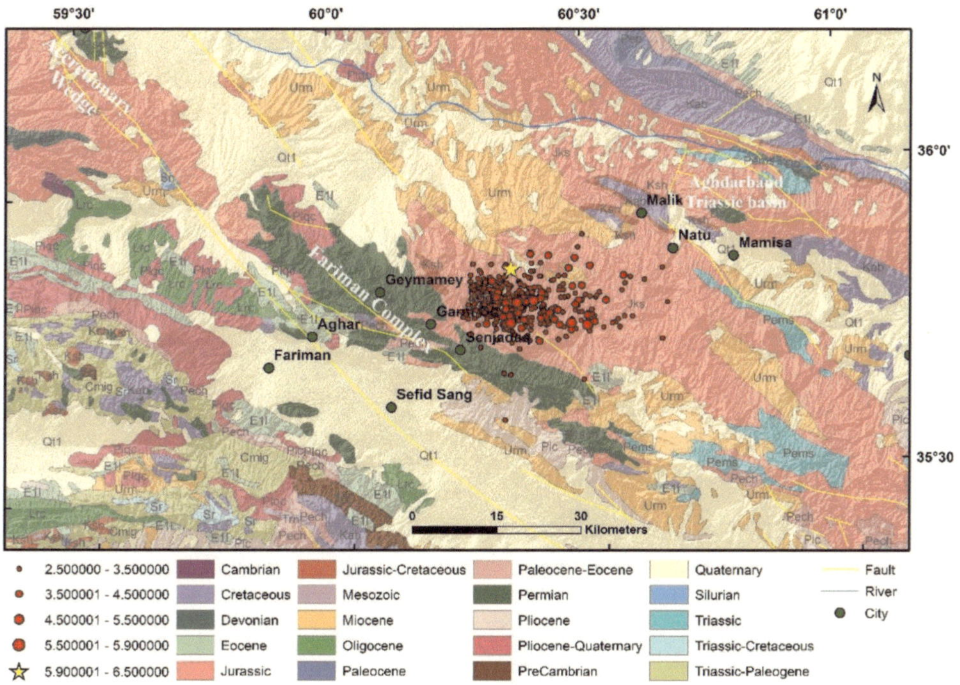

Figure 2.
Simplified geological map of the area. The Sefidsang earthquake sequence is represented by circles (the greater earthquakes, the greater circles). The location of the Sefidsang mainshock is marked by the yellow star.

Date	Time (UTC)	Lat (°N)	Lon (°E)	Depth (km)	Mw	Mechanism		
						Strike (°)	Dip (°)	Rake (°)
2017-04-05	6:9	35.776	60.436	11.5	6.15	316	20	120
						105	73	80

Table 1.
The geometry of the Sefidsang earthquake nodal planes taken from GCMT and ISC.

regional stress. The best-fitting regional stress tensor compatible with the majority of ISC collected earthquake focal mechanisms over the period of years was defined by the Win-Tensor program (**Table 2**). Then, we should select the nodal plane on which imparted Coulomb stress is maximum.

The stress change calculations were performed using the software Coulomb 3.3 [44]. For all the calculations of the Coulomb stress change, the shear modulus (G), 32×10^5 bar, Poisson ratio (ν) 0.25, Young modulus (E) 8×10^5 bar, and effective coefficient of friction (μ') 0.6 were used. The nodal plane located in the red region on

	σ_1	σ_2	σ_3
Trend	216	120	317
Plunge	8	35	54

Table 2.
Principle stress direction.

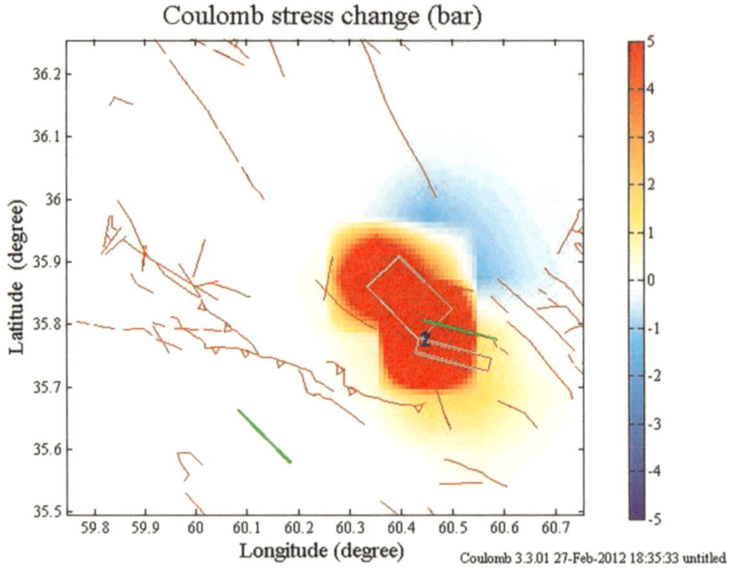

Figure 3.
The regional Coulomb stress change has been calculated for two nodal planes of the Sefidsang earthquake. The maximum Coulomb stress region is shown in red. White rectangles show the nodal plane geometry. Green lines represent the assumptive fault traces on the ground. The yellow star represents the epicenter of the Sefidsang earthquake.

the Coulomb map has the most consistency with the inferred Coulomb stress field and should be selected as the earthquake-triggering fault. However, the result represents both nodal planes, more or less, are situated in the positive ΔCFS area (**Figure 3**). The calculated positive ΔCFS area has exactly placed the bend of the Kashafrud fault proximity in the Fariman complex.

The aftershocks preferentially occur in the calculated stress increase and are less likely in the calculated stress decrease areas [51]. Therefore, the aftershock distributions control which nodal plane results in more matching with high Coulomb stress. The epicentral distribution of aftershocks by $M_N \geq 2$ and ≥ 3 is mostly situated in the maximum (most positive) Coulomb stress area (**Figure 4**) and is more consistent with the first nodal plane in **Table 1**. The general pattern of aftershocks brings up that the geometry of the Sefidsang earthquake sequence causative faults is in listric form.

Factors such as unknown stress concentration before the mainshock, crustal heterogeneity, and the existence of small faults with different orientations [15, 52, 53] may play a role in perturbing the stress field and thus it is not surprising that all of the aftershocks have not been completely matched with regions of increased Coulomb stress.

4. Discussion

The regional inferred Coulomb stress changes triggering the April 5th, 2017 Mw 6.15 Sefidsang earthquake (pre-event stress; **Table 2**) was resolved on two nodal planes. The status of ΔCFS distribution was investigated to identify which nodal plane has been more consistent with high Coulomb stress. The nodal plane has the most consistency with the regional inferred Coulomb stress field (situated in the red region

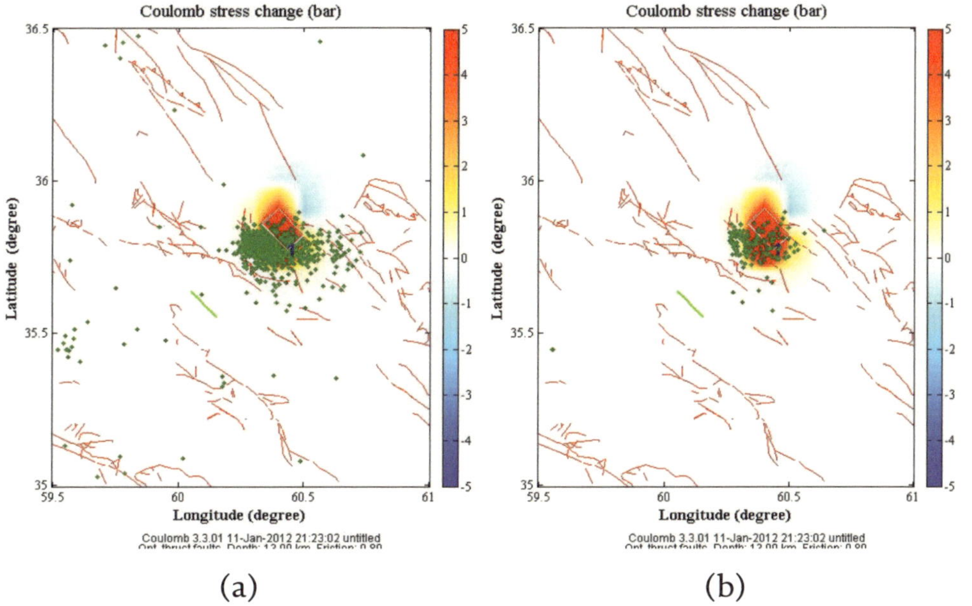

Figure 4.
The green dots represent the aftershock distribution. It is more compatible with the first geometry of nodal planes in ***Table 1***. *The epicentral distribution of aftershocks is limited to the area behind the assumptive fault trace (the green line on the coulomb map), suggesting a listric form of the causative fault. a) The magnitude of aftershocks is greater than 2. b) The magnitude of aftershocks is greater than 3. Most of the aftershocks have well matched with the maximum Coulomb stress area.*

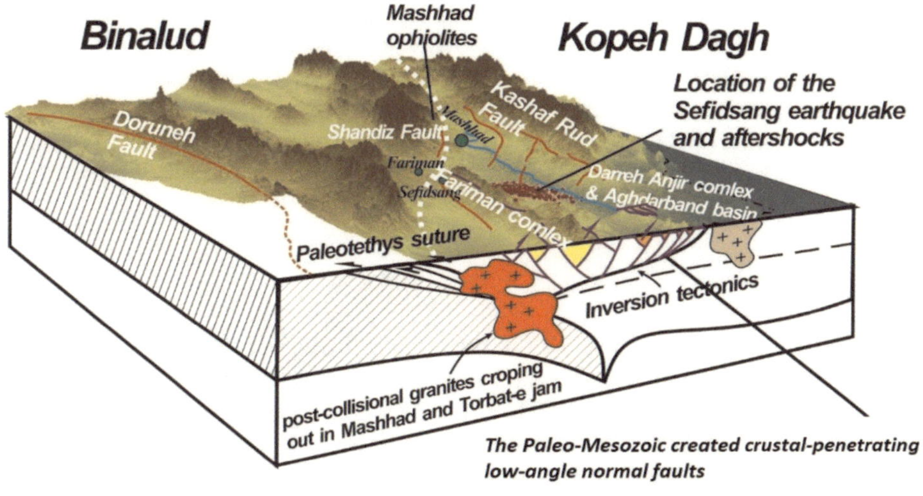

Figure 5.
Conceptual model with 3D topographic relief of SRTM 30m DEM on it represents the preexisting normal faults that were reactivated as reverse faults in current tectonic regime, viewed from S70°E. The Sefidsang earthquake sequence is shown by the red circles on the top. The structural pattern of the area illustrates thick-skin tectonics including crustal-penetrating low-angle faults reactivated in the current stress field coupled with mafic–ultramafic magmatism in the lower crust.

on the Coulomb map), on one hand, and the epicentral distribution of aftershocks, on the other hand, is closer to failure and should be selected as the earthquake-triggering fault. Although the regional inferred Coulomb stress field is consistent with both nodal planes, the epicentral distribution of aftershocks confirmed the nodal plane with 316° strike, 20° dip to the north, and 120° rake geometry as the earthquake-triggering fault (**Figure 4**). From the general pattern of aftershocks distribution, it is also obvious that the geometry of the Sefidsang earthquake sequence causative faults is in the form of listric. Considering the proximity of the Sefidsang sequence to the Fariman complex (**Figure 1**) as an arc-related basin formed by extensional faults with low initial dips [54] reveals that reversal slips occurred on preexisting normal faults. Therefore, it is very likely that the Sefidsang earthquake-triggering fault is the inverted arc-related structures reactivated in the recent tectonic regime. A kinematic model of fault planes reactivated during the Sefidsang earthquake has been represented in **Figure 5**.

5. Conclusion

The most important question concerning earthquakes is about the geometry and kinematics of the causative faults. In other words, we like to know which nodal planes would be the causative fault plane accommodating the slip during seismic activation, especially in the case that there is no rupture evidence at the surface. The Coulomb static stress has been commonly used to determine the stress distribution induced by an event. However, for the first time in this research, the Coulomb regional stress was resolved on nodal planes to realize the optimally oriented one for failure. The seismic fault is determined by assuming that the earthquake will be triggered on the plane with maximum total Coulomb stress. In other words, the earthquake-triggering fault should have maximum Coulomb stress concerning regional stress. The method has been conducted for the April 5th, 2017 Sefidsang earthquake in NE Iran. The results reveal that the earthquake occurred on a northeast-dipping listric fault with dextral reverse movement. Recognition of the pattern and kinematics of active deformation in addition to paleo-structures can help us to shed light on structural aspects subjected to active deformation in the area. The results of this study have crucial implications for seismic hazard assessment of the region and potential future failure areas.

Acknowledgements

I thank the Iranian Seismological Center (IRSC). I utilized Win-Tensor and Coulomb software. I am grateful to the International Seismological Centre (ISC), Global Centroid Moment Tensor (GCMT), and USGS for open access to their online Bulletin. I also used SRTM DEM derived from the USGS/NASA SRTM data.

The author has no relevant financial or nonfinancial interests to disclose.

Author details

Mahnaz Nedaei
Department of Geology, Payame Noor University (PNU), Tehran, Iran

*Address all correspondence to: m.nedaei@pnu.ac.ir

References

[1] Aki K, Richards PG. Quantitative seismology. 2nd ed. University Science Books; 2002. p. 704

[2] Dziewonski AM, Chou TA, Woodhouse JH. Determination of earthquake source parameters from waveform data for studies of global and regional seismicity. Journal of Geophysical Research - Solid Earth. 1981; **86**(B4):2825-2852

[3] Kawakatsu H. Automated near-realtime CMT inversion. Geophysical Research Letters. 1995;**22**(19):2569-2572

[4] Sipkin SA. Estimation of earthquake source parameters by the inversion of waveform data: synthetic waveforms. Physics of the Earth and Planetary Interiors. 1982;**30**(2-3):242-259

[5] Hori T, Yalciner A, Ozel N, Kilic I, Si M, Hyodo M. Earthquake generation cycles and tsunami simulations providing possible scenarios for Turkey (Marmara sea) and Japan (Nankai trough and Japan trench). In: EGU General Assembly Conference Abstracts. April 2015. p. 7566

[6] Hardebeck JL, Nazareth JJ, Hauksson E. The static stress change triggering model: Constraints from two southern California aftershock sequences. Journal of Geophysical Research - Solid Earth. 1998;**103**(B10):24427-24437

[7] Catalli F, Chan C-H. New insights into the application of the Coulomb model in real-time. Geophysical Journal International. 2012;**188**(2):583-599

[8] Deng J, Sykes LR. Stress evolution in southern California and triggering of moderate-, small-, and micro-size earthquakes. Journal of Geophysical Research - Solid Earth. 1997;**102**(B11): 24411-24435

[9] Hainzl S, Moradpour J, Davidsen J. Static stress triggering explains the empirical aftershock distance decay. Geophysical Research Letters. 2014; **41**(24):8818-8824

[10] Harris RA. Introduction to special section: Stress triggers, stress shadows, and implications for seismic hazard. Journal of Geophysical Research - Solid Earth. 1998;**103**(B10):24347-24358

[11] Harris RA, Simpson RW, Reasenberg PA. Influence of static stress changes on earthquake locations in southern California. Nature. 1995; **375**(6528):221-224

[12] King GC, Stein RS, Lin J. Static stress changes and the triggering of earthquakes. Bulletin of the Seismological Society of America. 1994;**84**(3):935-953

[13] Lin J, Stein RS. Stress triggering in thrust and subduction earthquakes and stress interaction between the southern San Andreas and nearby thrust and strike-slip faults. Journal of Geophysical Research: Solid Earth. 2004;**109**(B2). pp. B02303. DOI: 10.1029/2003JB00 2607

[14] Lin J, Stein RS, Meghraoui M, Toda S, Ayadi A, Dorbath C, et al. Stress transfer among en echelon and opposing thrusts and tear faults: Triggering caused by the 2003 Mw= 6.9 Zemmouri, Algeria, earthquake. Journal of Geophysical Research: Solid Earth. 2011; **116**(B3). pp. B03305. DOI: 10.1029/ 2010JB007654.hal-00747421

[15] Mitsakaki C, Rondoyanni T, Anastasiou D, Papazissi K, Marinou A, Sakellariou M. Static stress changes and fault interactions in Lefkada Island. Western Greece. Journal of Geodynamics. 2013;**67**:53-61

[16] Parsons T. Global Omori law decay of triggered earthquakes: Large aftershocks outside the classical aftershock zone. Journal of Geophysical Research: Solid Earth. 2002;**107**(B9):ESE 9-1-ESE 9-20

[17] Sarkarinejad K, Ansari S. The coulomb stress changes and seismicity rate due to the 1990 M W 7.3 Rudbar earthquake. Bulletin of the Seismological Society of America. 2014;**104**(6):2943-2952

[18] Sarkarinejad K, Ansari S. Did the 1983 Charazeh earthquake trigger the destructive 1990 Rudbar earthquake? International Journal of Earth Sciences. 2015;**104**(1):309-319

[19] Steacy S, Gomberg J, Cocco M. Introduction to special section: Stress transfer, earthquake triggering, and time-dependent seismic hazard. Journal of Geophysical Research: Solid Earth. 2005;**110**(B5). pp. B05S01. DOI: 10.1029/2005JB003692

[20] Stein RS, King GC, Lin J. Stress triggering of the 1994 M= 6.7 Northridge, California, earthquake by its predecessors. Science. 1994;**265**(5177): 1432-1435

[21] Wang J, Xu C, Freymueller JT, Li Z, Shen W. Sensitivity of Coulomb stress change to the parameters of the Coulomb failure model: A case study using the 2008 Mw 7.9 Wenchuan earthquake. Journal of Geophysical Research - Solid Earth. 2014;**119**(4):3371-3392

[22] Wang J-C, Shieh C-F, Chang T-M. Static stress changes as a triggering mechanism of a shallow earthquake: case study of the 1999 Chi–Chi (Taiwan) earthquake. Physics of the Earth and Planetary Interiors. 2003;**135**(1):17-25

[23] Wang W-H, Chen C-H. Static stress transferred by the 1999 Chi-Chi, Taiwan, earthquake: Effects on the stability of the surrounding fault systems and aftershock triggering with a 3D fault-slip model. Bulletin of the Seismological Society of America. 2001;**91**(5): 1041-1052

[24] Yadav R, Gahalaut V, Chopra S, Shan B. Tectonic implications and seismicity triggering during the 2008 Baluchistan, Pakistan earthquake sequence. Journal of Asia Earth Sciences. 2012;**45**:167-178

[25] Dreger D. Empirical Green's function study of the January 17, 1994 Northridge, California earthquake. Geophysical Research Letters. 1994; **21**(24):2633-2636

[26] Li Y, Doll C Jr, Toksöz M. Source characterization and fault plane determination for MbLg= 1.2 to 4.4 earthquakes in the Charlevoix Seismic Zone, Quebec, Canada. Bulletin of the Seismological Society of America. 1995; **85**(6):1604-1621

[27] Mori J, Hartzell S. Source inversion of the 1988 Upland, California, earthquake: Determination of a fault plane for a small event. Bulletin of the Seismological Society of America. 1990; **80**(3):507-518

[28] Meier MA, Werner M, Woessner J, Wiemer S. A search for evidence of secondary static stress triggering during the 1992 Mw7. 3 Landers, California, earthquake sequence. Journal of Geophysical Research - Solid Earth. 2014;**119**(4):3354-3370

[29] Das S, Scholz CH. Off-fault aftershock clusters caused by shear stress increase? Bulletin of the Seismological Society of America. 1981;**71**(5):1669-1675

[30] Stein RS, King GC, Lin J. Change in failure stress on the southern San

Andreas fault system caused by the 1992 magnitude= 7.4 Landers earthquake. Science. 1992;**258**(5086):1328-1332

[31] Stein RS. The role of stress transfer in earthquake occurrence. Nature. 1999; **402**(6762):605-609

[32] Toda S. Coulomb stresses imparted by the 25 March 2007 M w= 6.6 Noto-Hanto, Japan, earthquake explain its 'butterfly'distribution of aftershocks and suggest a heightened seismic hazard. Earth, Planets and Space. 2008;**60**(10): 1041-1046

[33] Ma KF, Chan CH, Stein RS. Response of seismicity to Coulomb stress triggers and shadows of the 1999 Mw= 7.6 Chi-Chi, Taiwan, earthquake. Journal of Geophysical Research: Solid Earth. 2005; **110**(B5)

[34] Parsons T, Stein RS, Simpson RW, Reasenberg PA. Stress sensitivity of fault seismicity: A comparison between limited-offset oblique and major strike-slip faults. Journal of Geophysical Research - Solid Earth. 1999;**104**(B9): 20183-20202

[35] Sumy DF, Cochran ES, Keranen KM, Wei M, Abers GA. Observations of static Coulomb stress triggering of the November 2011 M5. 7 Oklahoma earthquake sequence. Journal of Geophysical Research - Solid Earth. 2014;**119**(3):1904-1923

[36] Aron A, Hardebeck JL. Seismicity rate changes along the central California coast due to stress changes from the 2003 M 6.5 San Simeon and 2004 M 6.0 Parkfield earthquakes. Bulletin of the Seismological Society of America. 2009; **99**(4):2280-2292

[37] Harris RA, Simpson RW. Suppression of large earthquakes by stress shadows: A comparison of Coulomb and rate-and-state failure. Journal of Geophysical Research - Solid Earth. 1998;**103**(B10): 24439-24451

[38] Parsons T. Significance of stress transfer in time-dependent earthquake probability calculations. Journal of Geophysical Research - Solid Earth. 2005;**110**(B5)

[39] Stein RS, Barka AA, Dieterich JH. Progressive failure on the North Anatolian fault since 1939 by earthquake stress triggering. Geophysical Journal International. 1997; **128**(3):594-604

[40] Toda S, Lin J, Meghraoui M, Stein RS. 12 May 2008 M= 7.9 Wenchuan, China, earthquake calculated to increase failure stress and seismicity rate on three major fault systems. Geophysical Research Letters. 2008;**35**(17)

[41] Hainzl S, Steacy D, Marsan S. Seismicity models based on Coulomb stress calculations. Community Online Resource for Statistical Seismicity Analysis. 2010

[42] Ishibe T, Shimazaki K, Tsuruoka H, Yamanaka Y, Satake K. Correlation between Coulomb stress changes imparted by large historical strike-slip earthquakes and current seismicity in Japan. Earth, Planets and Space. 2011; **63**(3):12

[43] Bott MHP. The mechanics of oblique slip faulting. Geological Magazine. 1959; **96**(2):109-117. DOI: 10.1017/S0016756800059987

[44] Toda S, Stein RS, Sevilgen V, Lin J. Coulomb 3.3 Graphic-rich deformation and stress-change software for earthquake, tectonic, and volcano research and teaching—user guide. US

Geological Survey Open-file Report. 2011;**1060**(2011):63

[45] Ambraseys NN, Melville CP. A History of Persian Earthquakes. Cambridge University Press; 2005. p. 219

[46] Tchalenko J. Seismicity and structure of the Kopet Dagh (Iran, USSR). Philosophical Transactions of the Royal Society of London Series A, Mathematical and Physical Sciences. 1975;**278**(1275):1-28

[47] Berberian M, Yeats RS. Patterns of historical earthquake rupture in the Iranian Plateau. Bulletin of the Seismological Society of America. 1999; **89**(1):120-139

[48] Berberian M, Yeats RS. Contribution of archaeological data to studies of earthquake history in the Iranian Plateau. Journal of Structural Geology. 2001;**23**(2-3):563-584

[49] Masson F, Anvari M, Djamour Y, Walpersdorf A, Tavakoli F, Daignieres M, et al. Large-scale velocity field and strain tensor in Iran inferred from GPS measurements: new insight for the present-day deformation pattern within NE Iran. Geophysical Journal International. 2007;**170**(1): 436-440

[50] Carey-Gailhardis E, Mercier JL. A numerical method for determining the state of stress using focal mechanisms of earthquake populations: application to Tibetan teleseisms and microseismicity of Southern Peru. Earth and Planetary Science Letters. 1987;**82**(1-2):165-179

[51] Parsons T, Segou M, Sevilgen V, Milner K, Field E, Toda S, et al. Stress-based aftershock forecasts made within 24 h postmain shock: Expected north San Francisco Bay area seismicity changes after the 2014 M= 6.0 West Napa earthquake. Geophysical Research Letters. 2014;**41**(24):8792-8799

[52] Kozłowska M, Orlecka-Sikora B, Kwiatek G, Boettcher MS, Dresen G. Nanoseismicity and picoseismicity rate changes from static stress triggering caused by a Mw 2.2 earthquake in Mponeng gold mine, South Africa. Journal of Geophysical Research - Solid Earth. 2015;**120**(1):290-307

[53] Xie C, Lei X, Wu X, Hu X. Short-and long-term earthquake triggering along the strike-slip Kunlun fault, China: insights gained from the Ms 8.1 Kunlun earthquake and other modern large earthquakes. Tectonophysics. 2014;**617**: 114-125

[54] Zanchetta S, Berra F, Zanchi A, Bergomi M, Caridroit M, Nicora A, et al. The record of the Late Palaeozoic active margin of the Palaeotethys in NE Iran: constraints on the Cimmerian orogeny. Gondwana Research. 2013;**24**(3-4): 1237-1266

Section 3

Water's Wrath: Examining Flooding as a Natural Hazard

Chapter 10

Hydrological Extremes in Western Himalayas-Trends and Their Physical Factors

Nischal Sharma, Rohtash Saini, Sreehari K, Akash Pathaikara, Pravin Punde and Raju Attada

Abstract

Recent exacerbation of extreme precipitation events (EPEs) and related massive disasters in western Himalayas (WH) underpins the influence of climate change. Such events introduce significant losses to life, infrastructure, agriculture, in turn the country's economy. This chapter provides an assessment of long-term (1979–2020) as well as recent changes (2000–2020) in precipitation extremes over WH for summer (JJAS) and winter (DJF) seasons. Different high-resolution multi-source climate datasets have been utilized to compute the spatiotemporal trends in intensity and frequency of EPEs. The hotspots of rising extremes over the region have been quantified using the percentile approach where daily precipitation exceeds the 95th percentile threshold at a given grid. The findings reveal geographically heterogeneous trends among different datasets; however, precipitation intensity and frequency show enhancement both spatially and temporally (though insignificant). For both seasons, dynamic and thermodynamic parameters highlight the role of increased air temperatures and, as a result, available moisture in the atmosphere, signifying the consequences of global warming. Rising precipitation extremes in summer are sustained by enhanced moisture supply combined with increased instability and updraft, due to orography, in the atmosphere whereas winter atmosphere is observing an increase in baroclinicity, available kinetic energy, vertical shear and instability, contributing to a rise in precipitation extremes.

Keywords: extreme precipitation events, western Himalayas, summer monsoon, winter season, climate factors, physical factors

1. Introduction

The Western Himalayas (WH) are a strong modulator of weather and climate patterns over northern India and surrounding regions. WH are highly rich in biodiversity and covered with forests, agricultural landscapes, glaciers, wetlands and urbanized land, which underlines the significance of the region. The regional spatio-temporal distribution of precipitation over WH depicts high variability [1, 2] which can be partly attributed to influences from the atmosphere-land surface exchange processes over the region, keeping in mind the diverse land surface characteristics and large geographical variability [3, 4]. Additionally, complex interplay of regional topography with moist airflow and temperature gradient magnifies this variability further [5–7].

WH receives precipitation during both summer and winter monsoons [8]. The Indian summer monsoon (ISM), spanning through June-September (JJAS), contributes about 67–75% to the annual precipitation received over WH [2]. These mountainous environments have a significant impact on the spatiotemporal distribution of precipitation [1, 9]. Majorly, precipitation over the region during the ISM is contributed by convection followed by an orographically locked system, with Himalayas as barriers forcing moisture-laden southwest monsoon winds to dissipate moisture. Additionally, strong Tibetan high combined with the monsoon trough in northern India, creates a strong moist flow from Bay of Bengal and Arabian Sea into the Himalayas [10, 11]. Occasionally, interaction of tropical monsoon depressions and extratropical disturbances leads to the formation of heavy precipitation over the region during summer [12].

The region also receives a significant amount of precipitation during the winter season (December to February), primarily through extra-tropical cyclonic systems called Western Disturbances (WD; e.g. [1, 13]). WDs, embedded in upper tropospheric sub-tropical westerly jet stream, propagate eastward towards WH carrying moisture mainly from Mediterranean sea, Caspian sea, and Black sea (e.g. [13–15]). The interaction of these disturbances with elevated topography of WH results in their intensification and subsequent precipitation [16]. This precipitation holds key significance for sustenance of regional glaciers through snow accumulation and agricultural activities. The glacial mass balance is particularly crucial for regional river runoff and flows [17]. Any major precipitation variations in these glaciers can lead to adverse consequences on freshwater availability in downstream areas (e.g. [18]). Moreover, the vast river basin of WH acts as a watershed for a large population, assists in sustaining the regional biodiversity and provides various key ecosystem services to the surrounding north Indian plains. However, changing climate and expected hydroclimatic variability raises serious concerns related to impacts on this richly biodiverse and fragile mountainous landscape.

WH is highly prone to extreme precipitation events (EPEs) due to its intricate topography and altitude-dependent climate [10, 19, 20], which can give rise to surface runoff during such events, causing additional natural hazards such as landslides and floods (e.g. [21, 22]). Sharp regional weather fluctuations over the Himalayas makes this region unpredictable leading to sudden occurrences of heavy precipitation events. Various states in WH including Uttarakhand (UK), Jammu and Kashmir (J&K), Himachal Pradesh (HP), often face the problem of river flooding and landslides due to torrential downpours and localized occurrences of intense precipitation along the southern slopes of the Himalayas. Several case studies of EPEs over WH have highlighted massive losses through cloudbursts triggered by terrain-locked deep convective systems in valleys, as well as flash floods triggered by extratropical disturbances [23, 24]. Many scientific reports suggest an enhancement of precipitation in the order of 5–20% in the Himalayas in the 21st century (e.g. [25]). Moreover, the changes in the intensity and frequency of EPEs may vary seasonally. The increased susceptibility of WH to heavy precipitation during ISM has been discussed in various studies (e.g., [22, 26–32]). The variability aspects of WH winter precipitation under climate change scenarios, including a possibility of enhancement in precipitation extremes has also been frequently highlighted [14, 33, 34]. An increase in avalanche activity over WH slopes related to enhanced frequency of wet-snow conditions during recent decades has also been reported by [35]. Being thickly populated, WH and surrounding regions are highly vulnerable to climate extremities. Such extreme precipitation events (EPEs) can affect both natural and anthropogenic ecosystems through damage to life, infrastructure, agriculture, energy sectors, etc.

Many massive disasters related to unexpected heavy precipitation such as Leh flood (August 2010), Kedarnath disaster (June 2013), Chamoli river floods in Uttarakhand (July 2016), Nadum disaster (August 2018), Bilaspur and Shimla floods (August 2019), Jammu & Kashmir floods in Kishtwar district (July 2021) and Dharmsala floods (August 2022) have been reported in the recent times during the ISM, leading to immense life and economic losses. Increased frequency of summer EPEs (see [10, 36]) has been attributed to a number of possible causes including enhanced water vapor transport in the northern Hindu Kush Himalayan region [37], formation and movement of local deep convective systems along the orography [10, 23, 24], convergence of low-level monsoon westerly winds and northeasterly winds along the foothills combined with enhanced vertical wind shear and interaction of tropical systems with extratropical disturbances (see [9, 10, 12, 23, 26, 34]). Furthermore, elevated Tibetan plateau plays a key role during ISM by producing mesoscale precipitation through small-scale circulations and enhancing the synoptic weather conditions, leading to extreme precipitation in the WH [19, 38, 39]). Additionally, [12] found that EPEs over the Himalayas are associated with a southward extension of the western upper trough and a simultaneous northward migration of the lower monsoon trough towards the foothills of the Himalayas. These two systems amplify the low-level moisture flux from the Arabian Sea and Bay of Bengal towards the foothills of the Himalayas and contribute to cloud development and heavy precipitation.

Studies pertaining to the wintertime trends of extreme precipitation intensity and frequency are comparatively fewer. Shekhar et al. [40] reported a significantly increasing trend for heavy precipitation events (>70 mm) in the Pir-Panjal range of WH, however, other altitudes and ranges did not portray any clear trends. A significantly increasing trend for winter to early spring precipitation extremes intensity (exceeding 90th percentile threshold) has also been observed by [14] between 1900 and 2011, attributable to higher baroclinicity and in turn enhanced variability of WD activity during the recent decades. Krishnan et al. [33] also studied the impact of climate change on WD activity over WH and reported an increasing trend of winter precipitation extremes in the recent decades appertaining largely to anthropogenic forcings in conjunction with natural factors. Increasing trends of different extreme precipitation indices (exceeding 90th and 95th percentiles) using station-based records were further observed by [41] for all western Himalayan ranges during the month of February. Further, [42] demonstrated an increased frequency of atmospheric rivers (ARs) over the Himalayas during winter and underlined the association of intense ARs with EPEs over the Ganga and Indus basins.

Weather and climate extremes are generally a result of variations in different atmospheric dynamic and thermodynamic variables, as well as of some surface properties or states. Several studies, involving observations or climate model simulations, indicate that the frequency of these events would intensify with global warming due to an increment in the atmospheric moisture holding capacity as per Clausius Clapeyron relationship [43–46]. Additionally, the influence of local thermodynamics and orographic forcing in the WH produces abrupt changes in synoptic circulation, which have the potential to produce EPEs that can last for a few days [12, 47].

However, certain limitations are associated with the study of precipitation extremes over complex topographic regimes of WH. The remoteness of the region and the sparse coverage of rain gauges and automatic weather stations in mountainous areas makes precipitation monitoring in this region quite difficult [10, 48, 49]. The lack of availability of data directly affects the research studies investigating EPEs to under-perform in the Himalayan region. As a result, studies of extreme precipitation patterns, trends and possible causes in the WH remain limited and insufficient. Furthermore, very few studies have been conducted on the spatial distribution of EPE trends over WH. The inadequate representation of regional orography in the available coarser resolution datasets adds to the uncertainties associated with assessment of EPEs over any region [50]. Most of the past studies for EPEs over WH utilize comparatively coarser resolution datasets and there is a lack of high-resolution data-based studies for these extremes. Thus, it becomes important to utilize high-resolution datasets for a finer and accurate understanding of extreme precipitation trends and their possible causes over WH. The latest advancements in meteorological satellites and reanalysis products at finer resolutions with improved precipitation estimation algorithms have further facilitated the research on extreme weather events.

Our study focuses on changes in the spatial and temporal distributions of extreme precipitation events over WH during 1979–2020 in various high-resolution multi-source climate datasets, including the potential of recently released high-resolution regional reanalysis, Indian Monsoon Data Assimilation and Analysis (IMDAA). We evaluate extreme precipitation distribution and associated changes in key atmospheric parameters of EPEs over WH. Such knowledge about the climatological features of precipitation extremes and their associated dynamical and thermodynamic changes is crucial to interpret how precipitation patterns are changing in a varying climate scenario, further giving way to carry out vulnerability impact assessment studies.

2. Data and methods

2.1 Data used

In this study, we analyze daily summer (JJAS) and winter (DJF) precipitation extremes using various fine resolution multi-source gridded datasets including a gauge-based, a satellite dataset, a recently released regional reanalysis as well as a global reanalysis over WH (29°N-37.5°N and 72.5°E-80.5°E, see **Figure 1a**) within the time period 1979–2020. We have used India Meteorological Department's (IMD) daily rainfall data available at 0.25° × 0.25°, interpolated from 6955 rain gauge stations throughout the Indian subcontinent [51]. However, comparatively less stations are available over WH. We have also used, Integrated MultisatellitE Retrievals (V3) for Global Precipitation Measurement (GPM-IMERG), a merged high-resolution satellite product. Precipitation estimates are produced using Day-1 IMERG algorithm through intercalibration, merging, and interpolation of microwave and infrared records from GPM satellite constellation with gauge-based observations [52]. The regional reanalysis-IMDAA, is a recently released high resolution (12 km) product over the South Asian domain, generated by National Centre for Medium Range Weather Forecasting in collaboration with UK Met Office and IMD using a unified atmospheric model and the four-dimensional variational (4D-Var) data assimilation technique [53]. The dataset provides advantages in better representation of orographic features owing to its high spatial resolution [1, 2, 54]. Finally, we have also utilized the

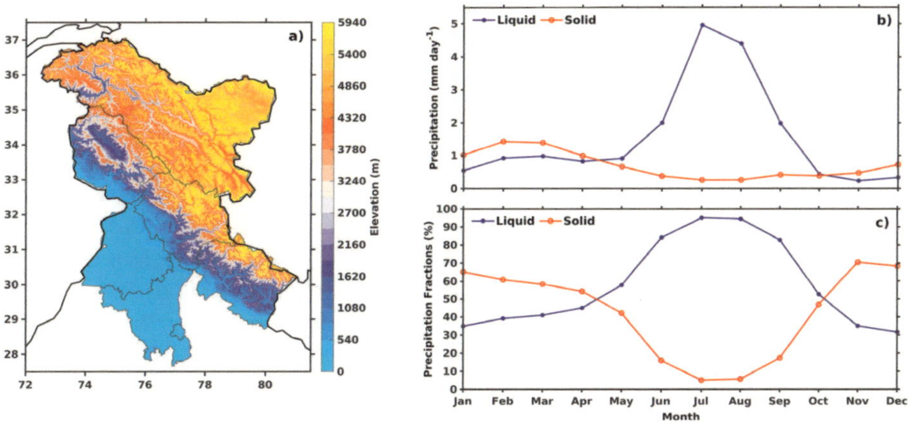

Figure 1.
(a) Elevation (in meters) of Western Himalayan region. Subplots (b) represent annual distribution of liquid (blue line) and solid (red line) precipitation and (c) indicates same as (b) in terms of precipitation fraction over WH during 1979–2020.

state-of-the-art global reanalysis ERA5, developed by European Centre for Medium-Range Weather Forecasts [55] available at a resolution of 0.25° × 0.25°. Further, daily values of different meteorological variables, including air temperature, specific humidity, vorticity, three-dimensional wind components, etc. at various pressure levels from ERA5 have also been considered.

2.2 Methodology

2.2.1 Identification of precipitation extremes

Keeping in mind the complexity of the study region and high spatio-temporal variability of precipitation over the regime, we have identified extreme precipitation thresholds at each grid point using percentile approach. This helps in describing intensity of extreme precipitation without having a stringent threshold for such varying terrains. Extremes are considered when the daily precipitation amount from the entire time series of precipitation exceeds the 95th percentile threshold at a particular grid point. The accumulated precipitation exceeding these thresholds have the potential to expedite floods in the downstream regions and affect the agricultural crops sown in the region during summer and winter seasons. Thus, we have chosen the 95th percentile as the threshold for our study for identifying EPEs.

2.2.2 Trends for intensity and frequency of EPEs

WH is known to be a complex and topographically heterogeneous regime. A wide discrepancy in precipitation patterns is observed among different datasets over this region [1]. Thus, we focus on analyzing precipitation extremes in various datasets to understand how different datasets depict precipitation extremes over the region. Datasets from different sources have been selected on the basis of the availability of long-term (~20 years) daily precipitation records, and having a relatively finer resolution. Considering that the selected datasets are generated with different input data and dissimilar developmental methods, the presence of any similar signals are strong indicators of real situation [1, 56]. Collaterally, we also explore the fidelity of the newly developed high resolution IMDAA reanalysis in representing WH precipitation extremes during the summer and winter season. IMDAA's high resolution offers significant potential in better resolution of orography [1, 2, 54], thus, a relatively better depiction of precipitation extremes is a key possibility. IMDAA's potential in the representation of climatological winter and summer precipitation characteristics over WH has been described thoroughly by [1, 2], respectively. The present study provides additional characterization of precipitation extremes in IMDAA.

Here, we investigate the intensity and frequency of EPEs in the respective datasets for summer and winter seasons after performing a bilinear interpolation to a common spatial resolution of $0.25° \times 0.25°$ for a fair comparison purpose. Further, a common period 2000–2020 has also been considered for extreme precipitation intensity and frequency trend analysis to understand how EPEs are changing in the recent decades. Spatial trends at decadal scale for intensity and frequency of EPEs are calculated at each grid point from the daily time series of precipitation. The magnitude of trend to the whole seasonal extreme precipitation at individual grid points is computed using a non-parametric Mann-Kendall Test [57, 58]. This is a rank-based method and widely used in hydrometeorological data studies [59]. Temporal trends for the frequency of EPEs are the seasonal average of the total number of grid points on each day that satisfies the EPE criteria specified earlier.

2.2.3 Trends for dynamical and thermodynamic variables

In addition, the climatological trends of key atmospheric thermodynamic variables such as air temperature and specific humidity at different pressure levels and hydrometeors in the total atmospheric column are investigated as these parameters strongly influence the extreme precipitation over any region. We further estimated climatological temporal trends for different derived variables such as moist static energy (MSE), vertically integrated moisture transport (VIMT), eady growth rate (EGR) and eddy kinetic energy (EKE) over WH to explore the impact of synoptic signatures associated with summer and winter precipitation extremes during the period 1979–2020.

3. Results and discussion

3.1 Distribution of liquid and solid precipitation over WH

The study area includes highly rugged mountains and comparatively gentler foothills as well as surrounding plain regions of Punjab and Haryana (see **Figure 1a**). WH experiences precipitation in different seasons through different weather systems. Moreover, the effect of regional orography as well as seasonal variations is predominant in the distribution of liquid and solid precipitation over different sub-regimes of WH (e.g. [3]). The intra-annual cycle of climatological mean precipitation averaged over the WH region from ERA5 reanalysis (**Figure 1b**) shows dominant contributions of liquid form of precipitation (rainfall) during summer season (JJAS) whereas solid precipitation (snowfall) is observed to be the primary form of precipitation during winter months (DJF). This implies that a large fraction (about 61% or more) of the received total precipitation during the DJF months comes from frozen hydrometeors (**Figure 1c**). The maximum precipitation over WH is received during the month of

July followed by August (**Figure 1b**), of which rainfall contributes almost 95% and 94% respectively (**Figure 1c**). During DJF months, highest amounts of total precipitation are observed during February followed by January and December, respectively. Similar results were reported for winter precipitation amounts at sub-seasonal scale by [1, 54].

Looking into the segregated precipitation fractions, December observes almost 68% of total monthly precipitation in solid form and the rest 32% as rainfall. Approximately 65% and 61% of the total monthly precipitation during January and February respectively are observed as snowfall. Although the individual fractions of monthly solid precipitation are lesser for February compared to December, it is significant to note that the total monthly precipitation observed during February is nearly more than double compared to December, implying that total snowfall amounts observed in February are higher.

3.2 Extreme precipitation intensity, frequency and trends

3.2.1 Summer season

We have examined spatiotemporal changes in the intensity and frequency of precipitation extremes over the WH using the daily gridded precipitation data (IMD), satellite-based data (IMERG), IMDAA regional reanalysis and global reanalysis ERA5. **Figure 2** shows the precipitation intensity exceeding the 95th percentile at each grid point for the summer monsoon and trends in the intensity and frequency of EPEs. The distribution of grid-wise precipitation intensity exhibits heterogeneity in precipitation amounts over the WH (**Figure 2a–d**). However, the Himalayan foothill belt shows the highest precipitation intensity of EPEs. All datasets show that the spatial distribution of precipitation intensity over the WH is characterized by low precipitation at higher altitudes (the northern part of the WH) and high precipitation over low-altitude regions, albeit with varying magnitudes. Satellite-based data (IMERG) and reanalysis dataset (IMDAA) overestimates the precipitation intensity as compared to the daily gridded precipitation (IMD), specifically over the Himalayan foothills. Although the

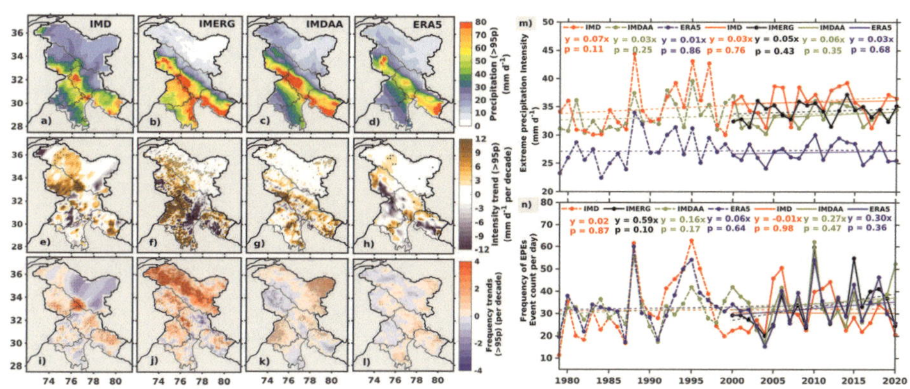

Figure 2.
Precipitation intensity (> 95th percentile) and trends in intensity and frequency of EPEs for the summer season. Set of plots (a-d) represents intensity climatology, (e-h) shows intensity trend (per decade), and (i-l) represents frequency trend (events per decade) of EPEs for IMD, IMERG, IMDAA, and ERA5 respectively. Subplots (m, n) represent the area averaged time series in intensity and frequency of EPEs for entire duration (1979–2020) in dotted line and recent decades (2000–2020) in solid line.

datasets show heterogeneity in spatial distribution of long-term trends of extreme precipitation intensity, specific hotspots in the Himalayas show significant increasing trends (up to 12 mm day^{-1} per decade; **Figure 2e–h**). In addition to the increasing intensity of EPEs, these same hotspots have been observed for the increasing frequency of EPEs in Himachal Pradesh, Uttarakhand, and Jammu and Kashmir (**Figure 2i–l**). However, relatively mixed trends for frequency of EPEs over different regions of WH is clearly visible. The northern part of the study region is an exception in terms of opposite grid wise trends of frequency and intensity of extremes. To have a more elaborate understanding, we have further investigated the temporal trends of precipitation extremes (both intensity and frequency) through an area-averaged time-series of EPEs at seasonal scale in different gridded datasets (**Figure 2m–n**). The time-series of intensity of daily precipitation extremes shows an increasing trend for all datasets in the long-term as well as recent decades (**Figure 2m**). In addition, all datasets agree on the long-term rise in frequency of EPEs over the region as well as during recent decades, exception being only IMD which shows negative trend in recent decades (**Figure 2n**).

3.2.2 Winter season

The geographical distribution of multi-year seasonal winter precipitation extremes' intensity and decadal trends in each dataset (after re-gridding) during 1979–2020 as per the duration of data availability for each dataset have been presented in **Figure 3**. Looking into the climatology of extreme precipitation intensity (> 95th percentile), considerable heterogeneity in extreme precipitation amounts as well as patterns are observed among different datasets, highlighting the role played by complex regional topographical variations. However, the highest precipitation amounts are observed along the western Himalayan foothills in all datasets (**Figure 3a–d**). The spatial extent of precipitation is maximum with high intensity over J&K, followed by HP and UK, respectively. This is due to the fact that vigor of WDs decreases as they move from J&K along WH towards central Himalayas [60]. Daily extreme

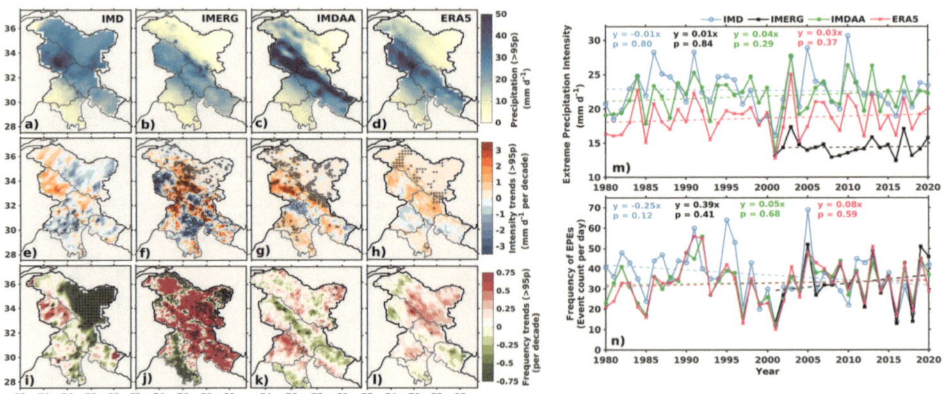

Figure 3.
Precipitation intensity (> 95th percentile) and trends in intensity and frequency of EPEs for the winter season. Set of plots (a-d) represents intensity climatology, (e-h) shows intensity trend (per decade), and (i-l) represents frequency trend (events per decade) of EPEs for IMD, IMERG, IMDAA, and ERA5 respectively. Subplots (m, n) represent the area averaged time series in intensity and frequency of EPEs for entire duration (1979–2020) in dotted line and recent decades (2000–2020) in solid line.

precipitation amounts are found to be reaching beyond 50 mm day^{-1} in IMDAA. Although it becomes important to note here that, like any reanalysis product IMDAA too, has been found to overestimate regional precipitation amounts [1].

The spatial variations of long-term trends per in extreme precipitation intensity although exhibit considerable heterogeneity, however, significant (confidence level = 0.95) increasing trends (up to 3.5 mm day^{-1} per decade) over some parts of HP, UK, JK and Ladakh can be observed (**Figure 3e–h**). Moreover, the trends in IMERG are very intense and highly significant, implying the strengthening of extreme precipitation intensities in the recent decades. Along with extreme precipitation intensity, the frequency of occurrence of such extreme events at different grid locations also seems to be on the rise over some parts of JK, northern part of HP and some parts of Punjab (**Figure 3i–l**). Again, IMERG shows highly significant rising trends of extreme precipitation intensity over most regions of WH.

Further, we have investigated the trends for area-averaged precipitation extremes at seasonal scale in different re-gridded datasets over the entire study domain (**Figure 3m–n**). All datasets including IMDAA clearly indicate that precipitation extremes are not only becoming frequent but are also intensifying over time, exception being IMD which shows a decreasing trend. Although it is important to note that none of the trends pass the significance test.

IMDAA and ERA5 show that precipitation intensity of extremes has increased by almost 16% and 23% respectively from 1979 to 2020 (**Figure 3m**). The magnitude of increase in intensity observed in IMERG from 2000 to 2020 is about 23.5%. In terms of rise in frequency of EPEs, IMDAA shows an increase of about 14 events (grid wise) per day seasonally over the domain since 1979 (**Figure 3n**). ERA5 agrees well on the enhancing frequency too but shows a slightly lesser increase of about 8 events per day. However, the rise is much sharper in the recent decades observed in IMERG satellite data which shows an increase over approximately 32 events per day in the region. The findings highlight the fact that extreme precipitation conditions are strengthening recently. The obtained results are in compliance with individual station-based trends over WH reported in some studies [40, 41] as well as other studies based on gridded observations and satellite datasets [14, 33], thus indicating a rise in precipitation extremes. Considering that the precipitation extremes are on a rise over the region, it becomes critical to understand the possible causes for these enhancements through trends for various dynamics and regional atmospheric conditions contributing to rise in precipitation extremes over WH.

3.3 Trends for different dynamic and thermodynamic controls related to EPEs

Any changes in weather and climate extremes are generally related to local exchanges in heat, moisture, and other thermodynamic quantities as well as dynamic changes. Although dynamic and thermodynamic processes in the atmosphere are interlinked, it is important to separately investigate their roles for variations in climate extremes (e.g. [61]). The thermodynamic controls of precipitation extremes are associated with an enhancement in the atmospheric moisture content, the most basic assumption being that precipitation extremes portray a tendency to rise in a warming climate, as per Clausius–Clapeyron relationship [62, 63]. Several studies propose the direct link for amplification of extremes over the WH region with increasing temperatures as well as atmospheric moisture content during both summer and winter seasons (e.g. [14, 64, 65]). Changes in the thermodynamic signatures significantly contribute to variations in precipitation patterns.

3.3.1 Summer season

In order to understand the variability of precipitation and the underlying factors might be contributing to the rise of extreme precipitation over the WH during summer monsoon season, seasonal trends of various atmospheric parameters, including hydrometeors, have been investigated. Interannual variations of tropospheric air temperatures at different levels and their trends for long-term as well as recent decades are shown in **Figure 4a–b**. The results show that the upper level (200 hPa) and mid-tropospheric (500 hPa) temperatures are significantly increasing, and in recent decades this increase is much sharper, which indicates the warming in the upper levels of troposphere over the WH (**Figure 4a**).

However, the seasonal trends at the lower level (850 hPa) and near-surface (1000 hPa) show a relative cooling trend in the long term (1979–2020), whereas the recent decades show warming at these near surface levels (**Figure 4b**). It is well known that global warming and related changes in the atmosphere above the WH generate EPEs, flash floods, cloud bursts, river flooding, landslides etc. [23, 66]. The overall result suggests that the increased precipitation intensity and frequency over the WH are directly associated with warming. Additionally, we have investigated the climatology of specific humidity at upper-level and low levels over the study region (**Figure 4c–d**). It has been observed that the specific humidity shows an increasing trend during the summer monsoon even though it is not significant at upper level but mid-tropospheric and lower-tropospheric specific humidity exhibits significantly increasing trends that is indicative of possibility of increased evaporation which can consequently contribute to increased precipitation, thus supporting the rise in precipitation intensity and frequency trends over the WH.

The interannual variability of cloud hydrometeors, total column of snow water, total column ice water, total column liquid water, and total column rain water in the long-term (1979–2020; dotted pink line) and recent decades (2000–2020; black line)

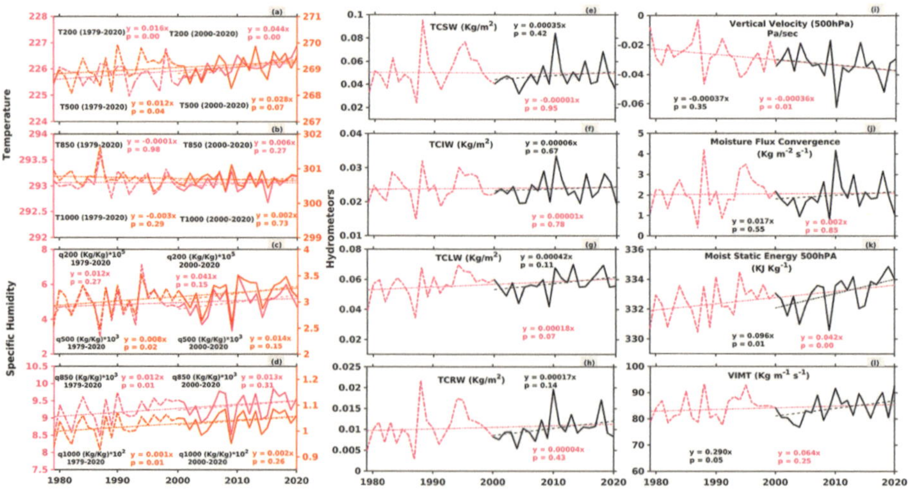

Figure 4.
Time series of the temperature (a-b) and specific humidity (c-d). Total column variables (e) snow water, (f) ice water, (g) liquid water, and (h) rainwater. Subplot (i) represents vertical velocity at 500 hPa, and (j) represents moisture flux convergence, (k) represents moist static energy, and (l) represents vertically integrated moisture transport using ERA5 reanalysis data during 1979–2020 for summer (JJAS) season. Dotted line represents the trend in entire duration and solid line indicate recent decades.

are shown in **Figure 4e–h**. Several studies have found that changes in cloud microphysical properties can have an effect on the simulated mesoscale dynamics of extreme events (e.g. [31, 67]). All the hydrometeors exhibit increasing yet insignificant long-term trends except for total column liquid water, which shows a significantly increasing trend (**Figure 4g**). It is worth noting that all four hydrometeors reveal comparatively sharper increasing trends in recent decades when compared to the entire study period. Studies report that the presence of atmospheric aerosols potentially assists an alteration of these cloud properties leading to more precipitation under favorable atmospheric conditions (e.g. [68]). This could be a possible explanation for the role of natural as well as anthropogenic forcings to increasing levels of precipitation extremes. Although a clear justification of these possibilities is beyond the scope of this study.

We further investigate extreme precipitation events by examining the underlying changes in other related dynamical and thermodynamic parameters whose characteristics provide crucial information about atmospheric conditions, which is important in the case of extreme precipitation events. Generally, the variations in dynamical components are caused by changes in vertical motion, whereas variations in atmospheric water vapor lead to changes in thermodynamic components [69]. **Figure 4i** displays a decreasing trend of 500 hPa vertical velocity (Pa s^{-1}) over the study domain, indicating a rise in convection over the WH which can favor cloud formation. This rising motion causes supersaturation, which is the primary cause of cloud droplet nucleation, condensation of water vapor into liquid water droplets and eventually precipitation [70, 71]. Therefore, increasing strength of vertical velocity over time is closely related and has far-reaching implications for vertical water, mass transport, and extreme precipitation.

Our study further investigates the roles of variability in atmospheric moisture transport over the topographic regimes of WH in case they show any contribution towards the rise of summer extreme precipitation. To accomplish this, we have investigated the trends for vertically integrated moisture flux convergence (VIMFC) and vertically integrated moisture transport (VIMT) over WH during the ISM which are given by,

$$VIMFC = -\frac{1}{g}\int_{psurf}^{ptop}\left(\frac{duq}{dx}+\frac{dvq}{dy}\right)dP \quad (1)$$

$$VIMT = \frac{1}{g}\int_{300\ hPa}^{1000\ hPa} qV\,dP \quad (2)$$

where, q is specific humidity, V is the horizontal velocity, and dP denotes the vertical incremental change in pressure.

Observed increasing seasonal trends in VIMFC (**Figure 4j**) directly characterize the behavior of EPEs and provide a favorable condition for increasing trends of EPEs in the WH. The region receives moisture directly from the Arabian Sea through southwesterlies and Bay of Bengal from north-easterlies [11] during southwest monsoon. Increasing levels of VIMT observed in **Figure 4l** explains that more moisture is getting transported to the region over the time. Further, our study notes that VIMT shows a significantly increasing and comparatively sharper trend over the WH during the recent decades, which constitutes enhanced seasonal moisture transport and, thus more precipitation. MSE is the one of the most important thermodynamic parameters, defined as the total sum of an air parcel's internal and gravitational potential energy.

$$MSE = C_P T + gz + L_v q \tag{3}$$

where, C_P is the specific heat capacity at constant pressure, T is the air temperature, g is the gravitational acceleration, z is the geopotential height, L_v is the latent heat of vaporization, and q is the specific humidity. The MSE at 500 hPa shows a significantly increasing trend, which reveals a higher atmospheric instability during the summer monsoon season **Figure 4k**. When the MSE is imported from the surrounding environment, it destabilizes the atmosphere by heating and humidifying it, resulting in deep convective precipitation [72, 73].

3.3.2 Winter season

The examination of area-averaged wintertime trends for mean air temperatures and specific humidity at different tropospheric levels (200 hPa, 500 hPa, 850 hPa and 1000 hPa) over the study region in ERA5 reanalysis are presented in **Figure 5a–d**. The trends indicate an increase in the air temperatures at all considered levels in the troposphere indicating that both lower and upper atmospheric temperatures are on a rise in western Himalayas, a clear indication of global warming effects. At the same time, atmospheric water vapor concentrations seem to be on a significant rise specifically in the upper and lower tropospheric levels. Conclusively, we can infer that changes observed in these regional thermodynamic variables are crucial and contributing to rising trends for precipitation extremes over the region. However, no discernible trend in middle troposphere vertical velocity has been observed over the region.

Further, we aimed to understand if there is any role of dynamical signatures of the atmosphere in influencing the rise in precipitation extremes over WH. We have focused on the variability associated with transient activity of westerly troughs

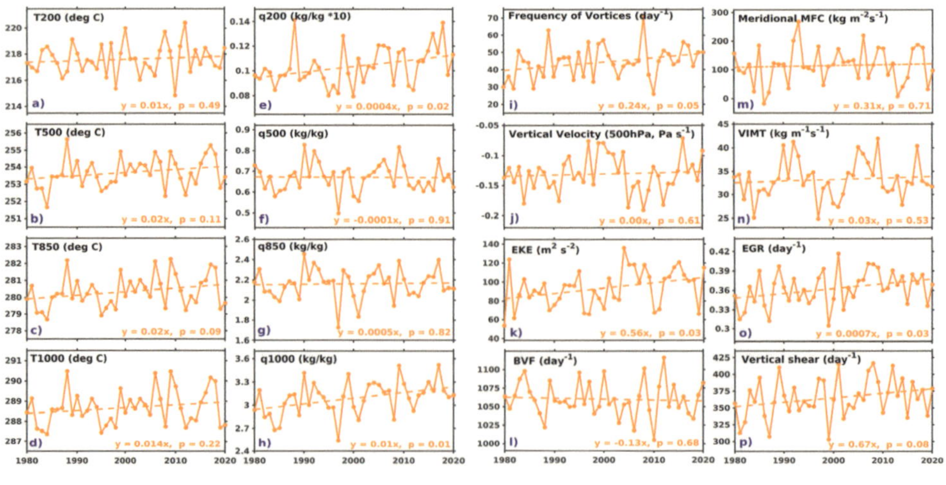

Figure 5.

Time series of the temperature (a-d), specific humidity (e-h), frequency of vortices (i), vertical velocity (500 hPa, j), eddy kinetic energy (k), brunt–Väisälä frequency (l), meridional moisture flux convergence (m), vertically integrated moisture transport (n), eady growth rate (o), and vertical shear (p) during 1979–2020 for winter (DJF) season in ERA5 reanalysis.

(vortices) over WH. Our analysis considers total counts of cyclonic (relative) vorticity in each winter season to measure the activity of troughs over WH. The occurrences of vorticity (500 hPa) exceeding 1.5 standard deviation at individual grid points over WH for each winter season has been counted and the trends have been observed (**Figure 5i**). An increase in the frequency count of vortices over the region is found, which indicates that the formation of these troughs is becoming more frequent in the recent decades. Such conditions can lead to development of deep convection in the presence of enough moisture and thus can create favorable conditions for heavy precipitation. Moreover, moisture flux convergence plays an important role in inducing heavy precipitation through the deepening of such vortices, thus, we also investigated the trends for vertically integrated meridional moisture flux convergence over the region. **Figure 5m** shows the time-series for moisture convergence associated with meridional $\left(-\frac{\partial vq}{\partial y}\right)$ winds, area-weighted over WH.

The results reveal a slight increasing trend, though insignificant, in moisture flux convergence associated with meridional winds. Moisture transport from the Arabian Sea into WH during winter has been designated as a crucial moisture input source during extremes (see [34]). As our findings suggest an enhancement in both transient activity of westerly troughs and moisture flux convergence, it can be concluded that variability in dynamic responses of the atmosphere in conjunction with thermodynamic variations might be a major contributing factor for increasing trends of extreme precipitation over the region. Mediterranean, Caspian, Red and Arabian seas are primary contributors for eastward moisture advection towards WH leading to moisture availability for precipitation [74, 75]. Here, we have tried to investigate the trends for winter seasonal moisture supply over WH (**Figure 5n**) through time series for VIMT which reveals a clear increasing trend in the moisture transport over WH through the years. This implies that with more moisture available over the region, conducive conditions for development and sustenance of heavy precipitation events can be created.

WDs, termed as immature baroclinic waves [76], develop and intensify primarily through atmospheric baroclinic instability, known to be generated by the meridional gradient of temperature and vertical shear of the background subtropical westerly flow [14]. Strong upper-atmospheric baroclinicity is generally observed in the locality of the subtropical jet during winter season [77]. The baroclinic processes partially influence the vertical velocities in the region majorly through a coupling between the background westerly flow (jet), WDs, and the orography and, thus leads to precipitation over WH [56]. The baroclinic instability in the atmosphere can be measured through maximum Eady growth rate (EGR) which follows the maximum growth rates to configure the Eady problem [78, 79] and is given as:

$$\sigma_E = 0.3098\frac{|f|\left|\frac{\partial U(z)}{\partial z}\right|}{N} \tag{4}$$

where, f is the Coriolis parameter, $U(z)$ is the vertical profile for zonal component of wind, z is the vertical coordinate and, N refers to the Brunt–Väisälä/buoyancy frequency defined by,

$$N^2 = \frac{g}{\theta}\frac{\partial\theta}{\partial z} \tag{5}$$

where, g is the acceleration due to gravity, and θ is the potential temperature. The buoyancy frequency represents atmospheric static stability. We have calculated EGR between two levels (200 minus 850 hPa) and further looked into its long-term trends (**Figure 5o**).

It is evident that the trends for EGR are increasing with high statistical significance. This highlights the fact that baroclinic instability is on a rise in the atmosphere over WH. A more baroclinically unstable atmosphere favors an intensification of WDs and can lead to heavy precipitation [13, 14, 41, 76, 80] have reported enhanced baroclinicity over WH during the recent decades and suggested the potential role of zonally asymmetric changes in the wintertime circulation caused by the elevation dependent climate warming signal. Further, we tried to investigate whether this enhancement in the regional baroclinicity, and consequently EPEs, is contributed through changes in the vertical shear of zonal wind (**Figure 5p**) or static stability (buoyancy frequency; **Figure 5l**). The results revealed that there has been an increase in the vertical shear over the period in WH and at the same time static stability of the region is observing a decreasing trend, meaning a more unstable atmosphere to intensify WDs. Further, it is clear that changes in both vertical shear and static stability are responsible for enhancing the regional baroclinicity and thus increase precipitation extremes over the region.

Various studies suggest that higher kinetic energy in the atmosphere in response to jet helps in the growth and intensification of WDs due to their baroclinic nature [13, 81]. The dynamical variations over WH are largely characterized by high frequency transient eddies in the atmosphere, which result from the conversion of the available potential energy into kinetic energy through baroclinic instability [82–84]. Therefore, we have investigated the trends for upper tropospheric (200 hPa) eddy kinetic energy (EKE) in ERA5 to understand the role of localized impacts of westerly flow and consequent energy exchange processes in the atmosphere for fueling the rise of precipitation extremes. EKE is defined as the kinetic energy associated with the time-varying component of the horizontal velocity field.

$$EKE = \frac{1}{2}\left(u'^2 + v'^2\right) \tag{6}$$

$$u = \overline{u} + u' \tag{7}$$

$$v = \overline{v} + v' \tag{8}$$

here, u and v are horizontal velocity components, u' and v' denote time-varying velocity components whereas, $\overline{u}$ and $\overline{v}$ represent the time mean velocity components. A bandpass filter of 2–10 days has been applied to the anomalies of horizontal velocity fields to filter out the mesoscale transient eddies and their trends have been looked into. The findings reveal that EKE over the region has been increasing significantly (**Figure 5k**) which provides preferable conditions for intensifying WDs over WH and thus can contribute to heavy precipitation over the region.

4. Summary and conclusion

The projected rise in precipitation extremes under the warming climate is a key matter of concern. This study investigates the spatiotemporal variations and trends for intensity and frequency of precipitation extremes over the western Himalayan region during summer and winter monsoon seasons. Furthermore, the potential dynamical and thermodynamic controls of precipitation have been explored to understand their possible roles in the increase of such extreme precipitation events over the region. Since the drivers and contributing atmospheric factors for summer and winter monsoons over WH are completely different, an effort has been made to explain the possible causes for increasing extremes for individual seasons separately. Based on the key findings from our study, the following conclusions can be drawn.

- WH receives precipitation mostly in liquid form during summer monsoon whereas solid precipitation is found to be the dominant form of precipitation in winter season.

- The trends for intensity and frequency of precipitation extremes over the region indicates that extremes are intensifying and becoming more frequent over the years in the western Himalayas during both seasons, specifically during recent decades. Although the trends show a somewhat heterogeneous and mixed pattern, intriguingly several pockets in Himachal Pradesh, Uttarakhand, and Jammu & Kashmir are exhibiting an increase in the frequency as well as the intensity.

- This increase seems to be directly associated with rising trends for air temperatures and atmospheric water content during both seasons, which signifies the role played by warming climate, providing the feedback of moisture for evaporation and enhanced cloud formation.

- During the summer season, increasing trends for most of the total column cloud hydrometeors, with significance in case of liquid water signals to the influence of cloud microphysical properties in the rise of EPEs.

- Moreover, the increase in the strength of vertical velocity over time during summer monsoon combined with an upward trend in 500 hPa MSE can make the overall atmosphere unstable and prone to more precipitation. What is more important to note is that this instability is further extended in the moisture transport where we observe increasing trends in VIMT and the VIMFC. This ever-growing availability of moisture can have serious impacts on the frequency and the intensity of EPEs over the Himalayas specifically along the Himalayan foothills.

- The possible dynamics related to increase of EPEs in winter season relates to a rise in formation of transient westerly troughs/vortices over the study region combined with an enhancement of regional moisture flux convergence associated with meridional winds as well as moisture transport. This explains the contribution of increased moisture supply possibly from Arabian sea in intensifying the vortices and deepening the convection.

- The enhancing baroclinicity and zonal vertical shear and reduced levels of static stability in the atmosphere is highly supportive of rising EPEs in the WH. Additionally, the winter atmosphere over WH is observing an increase in the amount of kinetic energy over the years which implies the availability of higher energy to feed and intensify WDs over the region.

- Lastly, the newly developed high-resolution reanalysis, IMDAA, is realistic in representing the regional extreme precipitation distribution and trends for intensity and frequency, in comparison with other utilized datasets.

In summary, we would like to acknowledge that the substantial variations and the spatial heterogeneity in magnitude exist between different data products over WH mainly due to lack of in-situ data and complex orography. Such uncertainties can be reduced by increasing the coverage of ground-based in-situ observational networks over the region. The climate warming signal with its positive feedback mechanism in terms of enhancing the available moisture content in the atmosphere is a crucial factor influencing the rise of precipitation extremes over WH. Significant variations of various dynamics supported by local thermodynamic processes have also been observed, helping the growth and intensification of EPEs. However, future work is required to understand the underlying physical processes and their interactions with regional orography during such events to have a more elaborate understanding of the feeding mechanisms.

Acknowledgements

This research work was supported by the Science and Engineering Research Board, Department of Science and Technology, Government of India under the "Start-up Research Grant (SRG) scheme" (Grant SRG/2020/001857) and Prime Minister's Research Fellowship (PMRF), Ministry of Education, Government of India. Authors Nischal and Rohtash gratefully acknowledge the financial assistance from PMRF.

Conflict of interest

The authors declare no conflict of interest.

Nomenclature

WH	Western Himalayas
ISM	Indian summer monsoon
EPEs	Extreme precipitation events
WD	Western disturbances
ARs	Atmospheric rivers
IPCC	Intergovernmental Panel on Climate Change
IMDAA	Indian Monsoon Data Assimilation and Analysis
IMD	India Meteorological Department
GPM-IMERG	Integrated MultisatellitE Retrievals for Global Precipitation Measurement
MSE	Moist static energy
EGR	Eady growth rate
EKE	Eddy kinetic energy
J&K	Jammu and Kashmir
HP	Himachal Pradesh
UK	Uttarakhand
VIMFC	Vertically integrated moisture flux convergence
VIMT	Vertically integrated moisture transport

Author details

Nischal Sharma[1], Rohtash Saini[1], Sreehari K[1], Akash Pathaikara[1,2], Pravin Punde[1,3] and Raju Attada[1*]

1 Department of Earth and Environmental Sciences, Indian Institute of Science Education and Research Mohali, Punjab, India

2 Division of Environmental Science and Technology, Pohang University of Science and Technology, South Korea

3 Department of Physics and Technology, UiT The Arctic University of Norway Tromsø, Norway

*Address all correspondence to: rajuattada@iisermohali.ac.in

References

[1] Nischal, Attada R, Hunt KM. Evaluating winter precipitation over the western Himalayas in a high-resolution Indian regional reanalysis using multi-source climate datasets. Journal of Applied Meteorology and Climatology. 2022;**61**:1607-1627. DOI: 10.1175/JAMC-D-21-0172.1

[2] Rohtash, Attada R. Analysis of Himalayan summer monsoon rainfall characteristics using Indian high-resolution regional reanalysis. International Journal of Climatology. Revision submitted

[3] Singh P, Ramasastri KS, Kumar N. Topographical influence on precipitation distribution in different ranges of western Himalayas. Hydrology Research. 1995;**26**:259-284

[4] Dimri AP. Wintertime land surface characteristics in climatic simulations over the western Himalayas. Journal of Earth System Science. 2012;**121**(2):329-344

[5] Beniston M, Keller F, Koffi B, Goyette S. Estimates of snow accumulation and volume in the Swiss Alps under changing climatic conditions. Theoretical Applied Climatology. 2003; **76**:125-140

[6] Anders AM, Roe GH, Hallet B, Montgomery DR, Finnegan NJ, Putkonen J. Spatial patterns of precipitation and topography in the Himalaya. Special Papers-Geological Society of America. 2006;**398**:39

[7] Shekhar MS, Chand H, Kumar S, Srinivasan K, Ganju A. Climate-change studies in the western Himalaya. Annals of Glaciology. 2010;**51**(54):105-112

[8] Kulkarni A, Patwardhan S, Kumar KK, Ashok K, Krishnan R. Projected climate change in the Hindu Kush–Himalayan region by using the high-resolution regional climate model PRECIS. Mountain Research and Development. 2013;**33**(2):142-151

[9] Shrestha D, Singh P, Nakamura K. Spatiotemporal variation of rainfall over the central Himalayan region revealed by TRMM Precipitation radar. Journal of Geophysical Research: Atmospheres. 2012;**117**(D22106). DOI: 10.1029/2012JD018140

[10] Bharti V, Singh C, Ettema J, Turkington TA. Spatiotemporal characteristics of extreme rainfall events over the northwest Himalaya using satellite data. International Journal of Climatology. 2016;**36**(12):3949-3962

[11] Aggarwal D, Attada R, Shukla KK, Chakraborty R, Kunchala RK. Monsoon precipitation characteristics and extreme precipitation events over Northwest India using Indian high resolution regional reanalysis. Atmospheric Research. 2022;**267**:105993

[12] Vellore RK, Kaplan ML, Krishnan R, Lewis JM, Sabade S, Deshpande N, et al. Monsoon-extratropical circulation interactions in Himalayan extreme rainfall. Climate Dynamics. 2016;**46**(11): 3517-3546

[13] Hunt KM, Turner AG, Shaffrey LC. The evolution, seasonality and impacts of western disturbances. Quarterly Journal of the Royal Meteorological Society. 2018;**144**(710):278-290

[14] Madhura RK, Krishnan R, Revadekar JV, Mujumdar M, Goswami BN. Changes in Western disturbances over the western Himalayas in a warming environment.

Climate Dynamics. 2015;**44**(3):1157-1168

[15] Cannon F, Carvalho L, Jones C, Bookhagen B. Multi-annual variations in winter westerly disturbance activity affecting the Himalaya. Climate dynamics. 2015;**44**(1):441-455

[16] Ramaswamy C. On the sub-tropical jet stream and its role in the development of large-scale convection. Tellus. 1956;**8**(1):26-60

[17] Hasson S, Lucarini V, Khan MR, Petitta M, Bolch T, Gioli G. Early 21st century snow cover state over the western river basins of the Indus River system. Hydrology and Earth System Sciences. 2014;**18**(10):4077-4100

[18] Messerli B, Viviroli D, Weingartner R. Mountains of the world: Vulnerable water towers for the 21st century. AMBIO: A journal of the human. Environment. 2004;**33**(sp13):29-34

[19] Nandargi S, Dhar ON. Extreme rainstorm events over the Northwest Himalayas during 1875–2010. Journal of Hydrometeorology. 2012;**13**(4):1383-1388

[20] Houze RA Jr. Orographic effects on precipitating clouds. Reviews of Geophysics. 2012;**50**(1):RG1001

[21] Bookhagen B, Burbank DW. Toward a complete Himalayan hydrological budget: Spatiotemporal distribution of snowmelt and rainfall and their impact on river discharge. Journal of Geophysical Research: Earth Surface. 2010;**115**(F3):F03019

[22] Priya P, Mujumdar M, Sabin TP, Terray P, Krishnan R. Impacts of indo-Pacific Sea surface temperature anomalies on the summer monsoon circulation and heavy precipitation over Northwest India–Pakistan region during 2010. Journal of Climate. 2015;**28**(9):3714-3730

[23] Dimri AP, Chevuturi A, Niyogi D, Thayyen RJ, Ray K, Tripathi SN, et al. Cloudbursts in Indian Himalayas: A review. Earth-Science Reviews. 2017;**168**:1-23

[24] Houze RA, Rasmussen KL, McMurdie L, Chaplin MM, Kumar A. Synoptic and Mesoscale Aspects of the June 2013 Flooding in Uttarakhand. India: Mon. Weather Rev., submitted; 2017

[25] Hock R, Rasul G, Adler C, Cáceres B, Gruber S, Hirabayashi Y, et al. High mountain areas. In: IPCC Special Report on the Ocean and Cryosphere in a Changing Climate. Cambridge, UK and New York, NY, USA: Cambridge University Press; 2019. pp. 131-202

[26] Houze RA, Rasmussen KL, Medina S, Brodzik SR, Romatschke U. Anomalous atmospheric events leading to the summer 2010 floods in Pakistan. Bulletin of the American Meteorological Society. 2011;**92**(3):291-298

[27] Toffoli A, Bitner-Gregersen EM, Osborne AR, Serio M, Monbaliu J, Onorato M. Extreme waves in random crossing seas: Laboratory experiments and numerical simulations. Geophysical Research Letters. 2011;**38**(6):L06605

[28] Pattanaik DR, Rajeevan M. Variability of extreme rainfall events over India during southwest monsoon season. Meteorological Applications: A Journal of Forecasting, Practical Applications, Training Techniques and Modelling. 2010;**17**(1):88-104

[29] Rasmussen KL, Houze RA. A flash-flooding storm at the steep edge of high

terrain: Disaster in the Himalayas. Bulletin of the American Meteorological Society. 2012;**93**(11):1713-1724

[30] Guhathakurta P, Sreejith OP, Menon PA. Impact of climate change on extreme rainfall events and flood risk in India. Journal of Earth System Science. 2011;**120**(3):359-373

[31] Rasmussen KL, Hill AJ, Toma VE, Zuluaga MD, Webster PJ, Houze RA Jr. Multiscale analysis of three consecutive years of anomalous flooding in Pakistan. Quarterly Journal of the Royal Meteorological Society. 2015;**141**(689): 1259-1276

[32] Singh D, Tsiang M, Rajaratnam B, Diffenbaugh NS. Observed changes in extreme wet and dry spells during the south Asian summer monsoon season. Nature Climate Change. 2014;**4**(6): 456-461

[33] Krishnan R, Sabin TP, Madhura RK, Vellore RK, Mujumdar M, Sanjay J, et al. Non-monsoonal precipitation response over the Western Himalayas to climate change. Climate Dynamics. 2019;**52**(7): 4091-4109

[34] Hunt KM, Turner AG, Shaffrey LC. The impacts of climate change on the winter water cycle of the western Himalaya. Climate Dynamics. 2020 Oct; **55**(7):2287-2307

[35] Ballesteros-Cánovas JA, Trappmann D, Madrigal-González J, Eckert N, Stoffel M. Climate warming enhances snow avalanche risk in the Western Himalayas. Proceedings of the National Academy of Sciences. 2018; **115**(13):3410-3415

[36] Wang SY, Davies RE, Huang WR, Gillies RR. Pakistan's two-stage monsoon and links with the recent climate change. Journal of Geophysical Research: Atmospheres. 2011;**116**(D16):D16114

[37] Wu J, Xu Y, Gao XJ. Projected changes in mean and extreme climates over Hindu Kush Himalayan region by 21 CMIP5 models. Advances in Climate Change Research. 2017;**8**(3):176-184

[38] Na Y, Lu R, Fu Q, Kodama C. Precipitation characteristics and future changes over the southern slope of Tibetan plateau simulated by a high-resolution global nonhydrostatic model. Journal of Geophysical Research: Atmospheres. 2021; **126**(3):e2020JD033630

[39] Fu Y, Ma Y, Zhong L, Yang Y, Guo X, Wang C, et al. Land-surface processes and summer-cloud-precipitation characteristics in the Tibetan plateau and their effects on downstream weather: A review and perspective. National Science Review. 2020;**7**(3):500-515

[40] Shekhar MS, Rao NN, Paul S, Bhan SC, Singh GP, Singh A. Winter precipitation climatology over Western Himalaya: Altitude and range wise study. Journal of Indian Geophysics Union (March 2017). 2017; **21**(2):148-152

[41] Narasimha Rao N, Devi U, Shekhar MS, Singh GP. Trends of winter precipitation extremes over northwest Himalaya. Hydrological Sciences Journal. 2021;**66**(13):1882-1891

[42] Nayak MA, Azam MF, Lyngwa RV. ERA5-based database of atmospheric Rivers over Himalayas. Earth System Science Data Discussions. 2021:1-35 [Preprint]

[43] Seneviratne S, Nicholls N, Easterling D, Goodess C, Kanae S, Kossin J, et al. Changes in climate extremes and their impacts on the natural physical

environment. In: Managing the Risks of Extreme Events and Disasters to Advance Climate Change Adaptation, A Special Report of Working Groups I and II of the Intergovernmental Panel on Climate Change (IPCC). Cambridge, UK and New York, NY, USA: Cambridge University Press; 2012. pp. 109-230

[44] Mishra V, Ganguly AR, Nijssen B, Lettenmaier DP. Changes in observed climate extremes in global urban areas. Environmental Research Letters. 2015; **10**(2):024005

[45] Mukherjee MM. Global warming and climate change in India: A social work perspective Whanake. The Pacific Journal of Community Development. 2017;**3**(1):28-36

[46] Min SK, Zhang X, Zwiers FW, Hegerl GC. Human contribution to more-intense precipitation extremes. Nature. 2011;**470**(7334):378-381

[47] Vellore RK, Bisht JS, Krishnan R, Uppara U, Di Capua G, Coumou D. Sub-synoptic circulation variability in the Himalayan extreme precipitation event during June 2013. Meteorology and Atmospheric Physics. 2020;**132**(5): 631-665

[48] Palazzi E, Von Hardenberg J, Provenzale A. Precipitation in the Hindu-Kush Karakoram Himalaya: Observations and future scenarios. Journal of Geophysical Research: Atmospheres. 2013;**118**(1):85-100

[49] Norris J, Carvalho LM, Jones C, Cannon F. WRF simulations of two extreme snowfall events associated with contrasting extratropical cyclones over the western and central Himalaya. Journal of Geophysical Research: Atmospheres. 2015;**120**(8):3114-3138

[50] Raju A, Parekh A, Chowdary JS, Gnanaseelan C. Assessment of the Indian summer monsoon in the WRF regional climate model. Climate Dynamics. 2015; **44**(11):3077-3100

[51] Pai DS, Rajeevan M, Sreejith OP, Mukhopadhyay B, Satbha NS. Development of a new high spatial resolution (0.25× 0.25) long period (1901-2010) daily gridded rainfall data set over India and its comparison with existing data sets over the region. Mausam. 2014;**65**(1):1-8

[52] Huffman GJ, Bolvin DT, Braithwaite D, Hsu K, Joyce R, Xie P, et al. NASA global precipitation measurement (GPM) integrated multi-satellite retrievals for GPM (IMERG). Algorithm Theoretical Basis Document (ATBD) Version. 2015;**4**(26)

[53] Rani SI, Arulalan T, George JP, Rajagopal EN, Renshaw R, Maycock A, et al. IMDAA: High-resolution satellite-era reanalysis for the Indian monsoon region. Journal of Climate. 2021;**34**(12): 5109-5133

[54] Punde P, Nischal AR, Aggarwal D, Radhakrishnan C. Numerical simulation of winter precipitation over the Western Himalayas using a weather research and forecasting model during 2001–2016. Climate. 2022;**10**(11):160

[55] Hersbach H, Bell B, Berrisford P, Hirahara S, Horányi A, Muñoz-Sabater J, et al. The ERA5 global reanalysis. Quarterly Journal of the Royal Meteorological Society. 2020;**146**(730):1999-2049

[56] Baudouin JP, Herzog M, Petrie CA. Cross-validating precipitation datasets in the Indus River basin. Hydrology and Earth System Sciences. 2020;**24**(1): 427-450

[57] Mann HB. Nonparametric tests against trend. Econometrica: Journal of

the Econometric Society. 1945;**13**(3):245-259

[58] Kendall MG. Rank Correlation Methods. 4th ed. Charles Griffin: London, UK; 1975

[59] Wang F, Shao W, Yu H, Kan G, He X, Zhang D, et al. Re-evaluation of the power of the mann-kendall test for detecting monotonic trends in hydrometeorological time series. Frontiers in Earth Science. 2020;**8**:14

[60] Meher JK, Das L. Gridded data as a source of missing data replacement in station records. Journal of Earth System Science. 2019;**128**(3):1-4

[61] Shepherd TG. Atmospheric circulation as a source of uncertainty in climate change projections. Nature Geoscience. 2014;**7**(10):703-708

[62] Allen MR, Ingram WJ. Constraints on future changes in climate and the hydrological cycle. Nature. 2002;**419**:224-232

[63] Trenberth KE, Dai A, Rasmussen RM, Parsons DB. The changing character of precipitation. Bulletin of the American Meteorological Society. 2003;**84**(9):1205-1218

[64] Nandargi S, Dhar ON. Extreme rainfall events over the Himalayas between 1871 and 2007. Hydrological Sciences Journal. 2011;**56**(6):930-945

[65] Singh V, Goyal MK. Spatio-temporal heterogeneity and changes in extreme precipitation over eastern Himalayan catchments India. Stochastic Environmental Research and Risk Assessment. 2017;**31**(10):2527-2546

[66] Raj S, Shukla R, Trigo RM, Merz B, Rathinasamy M, Ramos AM, et al. Ranking and characterization of precipitation extremes for the past 113 years for Indian western Himalayas. International Journal of Climatology. 2021;**41**(15):6602-6615

[67] Hazra A, Chaudhari HS, Ranalkar M, Chen JP. Role of interactions between cloud microphysics, dynamics and aerosol in the heavy rainfall event of June 2013 over Uttarakhand, India. Quarterly Journal of the Royal Meteorological Society. 2017;**143**(703):986-998

[68] Choudhury G, Tyagi B, Vissa NK, Singh J, Sarangi C, Tripathi SN, et al. Aerosol-enhanced high precipitation events near the Himalayan foothills. Atmospheric Chemistry and Physics. 2020;**20**(23):15389-15399

[69] Chou C, Chen CA, Tan PH, Chen KT. Mechanisms for global warming impacts on precipitation frequency and intensity. Journal of Climate. 2012;**25**(9):3291-3306

[70] Mukhopadhyay P, Taraphdar S, Goswami BN, Krishnakumar K. Indian summer monsoon precipitation climatology in a high-resolution regional climate model: Impacts of convective parameterization on systematic biases. Weather and Forecasting. 2010;**25**(2):369-387

[71] Varikoden H, Revadekar JV. On the extreme rainfall events during the southwest monsoon season in northeast regions of the Indian subcontinent. Meteorological Applications. 2020;**27**(1):e1822

[72] Pillai PA, Sahai AK. Moist dynamics of active/break cycle of Indian summer monsoon rainfall from NCEPR2 and MERRA reanalysis. International Journal of Climatology. 2014;**34**(5):1429-1444

[73] Adames ÁF, Ming Y. Moisture and moist static energy budgets of south Asian monsoon low pressure systems in GFDL AM4. 0. Journal of the Atmospheric Sciences. 2018;**75**(6): 2107-2123

[74] Dimri AP, Niyogi D, Barros AP, Ridley J, Mohanty UC, Yasunari T, et al. Western disturbances: A review. Reviews of Geophysics. 2015;**53**(2):225-246

[75] Baudouin JP, Herzog M, Petrie CA. Synoptic processes of winter precipitation in the upper Indus Basin. Weather and Climate Dynamics. 2021; **2**(4):1187-1207

[76] Hunt KM, Turner AG, Shaffrey LC. Extreme daily rainfall in Pakistan and North India: Scale interactions, mechanisms, and precursors. Monthly Weather Review. 2018;**146**(4): 1005-1022

[77] Singh MS, Agnihotri CL. Baroclinity over India in winter and its relation to western disturbances and jet streams-part I. Mausam. 1977;**28**(3):303-310

[78] Eady ET. Long waves and cyclone waves. Tellus. 1949;**1**(3):33-52

[79] Simmonds I, Lim EP. Biases in the calculation of southern hemisphere mean baroclinic eddy growth rate. Geophysical Research Letters. 2009; **36**(1):L01707

[80] Sankar NV, Babu CA. Role of vorticity advection and thermal advection in the development of western disturbance during north Indian winter. Meteorology and Atmospheric Physics. 2020;**132**(4):515-529

[81] Pujol MI, Larnicol G. Mediterranean Sea eddy kinetic energy variability from 11 years of altimetric data. Journal of Marine Systems. 2005;**58**(3–4):121-142

[82] Pedlosky J. Instability theory. In: Geophysical Fluid Dynamics. New York, NY: Springer; 1987. pp. 490-623

[83] Wirth V, Riemer M, Chang EK, Martius O. Rossby wave packets on the midlatitude waveguide—A review. Monthly Weather Review. 2018;**146**(7): 1965-2001

[84] Orlanski I, Chang EK. Ageostrophic geopotential fluxes in downstream and upstream development of baroclinic waves. Journal of Atmospheric Sciences. 1993;**50**(2):212-225

Chapter 11

Integrating Google Earth Engine and Decametric Sentinel 2 Images for Analysis of Vegetation Pre and Post the Disaster at Brumadinho, Brazil

Rodrigo Martins Moreira and Maria Paula Cardoso Yoshii

Abstract

This paper presents the application of the normalized difference vegetation index to assess the vegetation dynamics for the period between years 2017 and 2021 at Brumadinho, MG, Brazil. The normalized difference vegetation index was calculated using a Google Earth Engine script applying Sentinel 2 data with a spatial resolution of 10 meters, to quantify the extent of the affected area and assess the vegetation dynamic after the disaster. The Dwass-Steel-Crichlow-Fligner test for nonparametric data was used for a pairwise comparison between years and the confidence interval was calculated using bootstrap with 9999 repetitions. The total area affected by the dam brake was 2662 ha. The NDVI values presented a statistically significant decrease from 2017 to 2019, with little increase until 2021. Mean NDVI values were 0.314003 [0.31028; 0.317564], 0.339887 [0.336591; 0.343231], 0.145814 [0.144004; 0.1476], 0.1495 [0.147676; 0.15128], and 0.15572 [0.153727; 0.15774] for 2017–2021, respectively. According to the results, we conclude that the vegetation in the affected area did not fully recover.

Keywords: dam-break, vegetation index, disaster, remote sensing, cloud computing

1. Introduction

In recent decades, there has been an increase in concern about environmental disasters. This concern fosters the need for emergency management to mitigate socio-economic consequences [1]. Disaster management is defined as the field of science that develops and applies technologies, planning, and management to deal with extreme events. Whether of natural or anthropic origin, events are managed that can kill or injure people and animals, as well as cause extensive damage to properties and communities [2].

Occurred in 25-01-2019, the catastrophe of Brumadinho released 43 million m^3 of iron ore tailings enters the list of the biggest mining disasters of history, being classified by number of deaths: Bulgaria, 1966, lead-zinc tailings (488 deaths); Brazil (Brumadinho), 2019, iron ore tailings (363), Chile, 1965, copper tailings (300) and China, 2008, iron tailings (277) [3].

Several challenges arise when dealing with disasters, the main ones being the need to analyze large spatial extensions in a short time [4]. Both in the drafting of evacuation routes, or in the analysis of potential risks, managers deal with aspects related to space. Still, in the context of developing countries, they face challenges related to a lack of financial resources and trained analysts [5]. In this context, geographic information systems (GIS) and remote sensing translate as tools to support geographic analysis, through storage, processing, and access to spatialized information [6]. Several studies assessing the social, economic, and ecologic impacts over flood areas using GIS and orbital remotely sensed data have been deployed [7–9].

In this context, the normalized difference vegetation Index (NDVI) is a ratio index calculated using the difference between near-infrared reflectance and red reflectance to their sum, being widely used to assess vegetation dynamics before and after disasters [10–12]. Rotta et al. [13] assessed pre-disaster scenarios and the causes to the dam collapse using satellite-driven soil moisture index, multispectral high-resolution imagery, and Interferometric Synthetic Aperture Radar (InSAR) products to assess pre-disaster scenarios and the direct causes of the tailings dam collapse. Cheng et al. [14] used Landsat 8 operational land imager products to assess sediment concentration and found an increase of sediments in the Paraopeba River due to the mudflow. Gama et al. [15] using Sentinel −1 InSAR data were able to observe persistent trends of deformation on the crest, middle and bottom sectors of the dam, which may have caused the collapse.

Nonetheless, to the researchers' knowledge, there is a gap regarding the assessment of vegetation recovery for the impacted area for subsequent years and compared to the previous dynamic. This is due to the large quantity of images that would require a large amount of time and processing capacity. For example, for an entire year, there would be 73 Sentinel 2 images to process. In this chapter, we assessed the NDVI reflectance for 2018, 2019, and 2020, a total of 217 images with a spatial resolution of 10 meters. To tackle this problem, we used the Google Earth Engine cloud computing environment.

Therefore, the research question that led the discussion is "how has the vegetation recovered after two years of the disaster?" Therefore, the aim of this work is to assess vegetation dynamics by analyzing the reflectance values for the NDVI index for 2018, 2019, and 2020.

1.1 Remote sensing applied to analyzes of environmental impacts of disasters

In a post-disaster crisis situation, environmental impact managers and analysts need accurate information in the short term. The information needs to show the spatial and temporal scale of what has happened and what can happen. Based on this information, the necessary resources will be allocated to leverage immediate responses to contain the damage. Access to spatialized information allows the preparation of plans to anticipate contingencies; evaluation of possible scenarios; efficiently and effectively; and actions for recovery and reconstruction of damages [16].

Thus, GIS and remote sensing translate into a key tools in impact analysis and disaster management. This technology has the capacity to aggregate socioeconomic information; images of remote sensing; storage, manipulation, queries, and data analysis; and, more importantly, the visualization of the data [17].

The use of GIS for disaster management can support several aspects of decision-making, such as:

- Disaster prediction: disaster impacts extent and possible impacts to high-importance areas [18];
- Vulnerability analysis: database spatialization regard services such as hospitals, police stations, fire brigades, and shelters, which can be used to mitigate the situation; and other structures with the potential to aggravate the situation, such as dams and effluent treatment plants [19];
- Evaluation of the magnitude of the damages: analyzes of the geographic extent of the impacts caused by the disaster, it assists decision-makers in the elaboration of containment measures for priority areas, such as nurseries, hospitals, schools, and nursing homes [20];
- Identification of hazardous materials: identifies locations that contain materials with potential for contamination, bringing the type of material, quantity, and reactive aspects, such as gas stations and pharmaceutical laboratories [21];
- Human resources: identifies the location and personal information of individuals with skills or experience useful in emergency response situations [22];
- Resources inventory: aggregates data regarding shelters and provision of resources such as food and water. Equipment is also accounted, such as boats, motor vehicles, and kite trucks; brings resources that can be made available by neighboring cities [23];
- Infrastructure: aggregates information regarding roads, airports, bridges, and evacuation routes; in addition to structures such as electrical networks, water supply networks, and sewage collection networks [24].

2. Materials and methods

The disaster occurred in the Brumadinho Municipality, at Minas Gerais State, Brazil. The Córrego do Feijão Dam, the Paraopeba river, and the affected area by the dam break are presented in **Figure 1**. The impacted area was obtained by the researchers by setting a threshold of values less than 0, converting it in a Boolean image, converting it to a vectoral file, and by overlaying selecting the polygon that matched the crisis image.

2.1 Socioeconomic and environmental description

With an estimated population of 39,520 [25], a Municipal Human Development Index of 0.747 and GDP per capita of more than R $ 40 thousand, the municipality of Brumadinho belongs to the Metropolitan Region of Belo Horizonte and is inserted in the Quadrilátero Ferrífero where the main economic activity is iron mining. It is crossed by the Paraopeba River and a member of the Aguas Claras and Rio Manso River Basin, a source of supply for about 28% of the population.

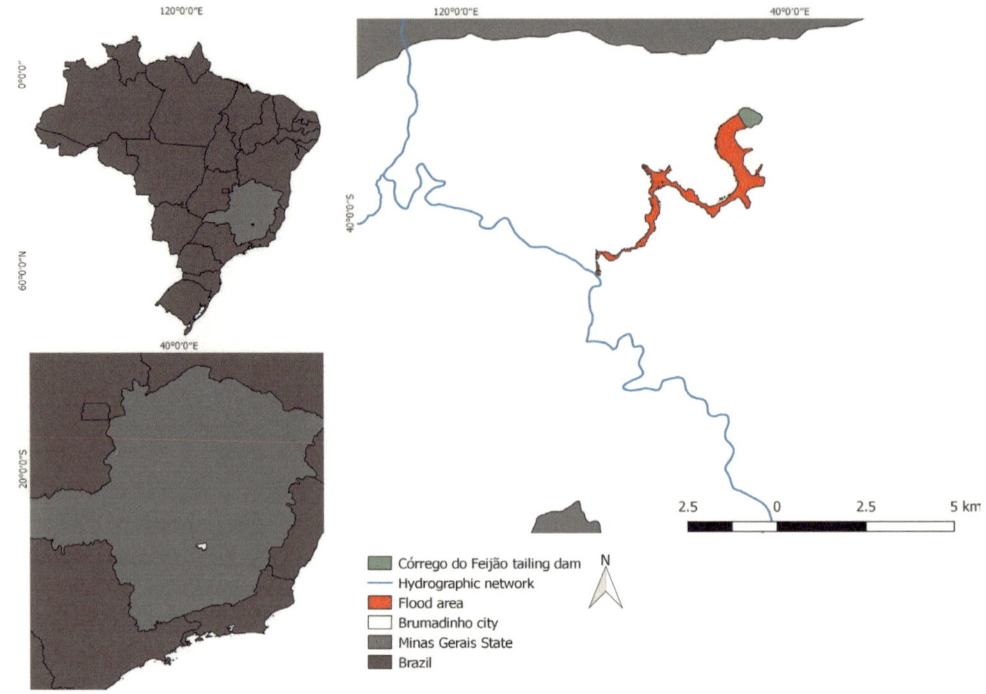

Figure 1.
Study area location.

The municipality has as a predominant biome the Atlantic Forest and some remnants of Cerrado, protected by the Special Protection Area (APE) of the Rio Manso; Environmental Protection Area (APA) Sul—RMBH; APA Inhotim; APE Catarina; Serra do Rola-Moça State Park, Serra da Moeda Natural Monument, Serra do Rola-Moça Natural Monument and Private Reserves of Natural Heritage—RPPN. About its climate, Brumadinho is located in the zone of influence of the Climate Cwa—Tropical Altitude, with summer rains, hot summers, and dry winters.

2.2 Flowchart with research steps

The steps for the work deployment are presented in **Figure 2**. The description of each step is presented below. All data were managed with QGis version 3.10 [26].

2.3 Acquisition of geographic data

The georeferenced images were retrieved from the United States Geological Survey. Multispectral imager sensor images from Sentinel-2A satellite with 10 meters resolution were used. Sentinel-2 images used to deploy the real color composite (RGB432) details are described in **Table 1**.

2.4 3D modeling

For hypsometry mapping and 3D modeling, the Shuttle Radar Topography Mission (SRTM) data were used. The SRTM data were acquired from the Topodata [27] web platform, a Brazilian program that resampled the SRTM data to 30 meters resolution.

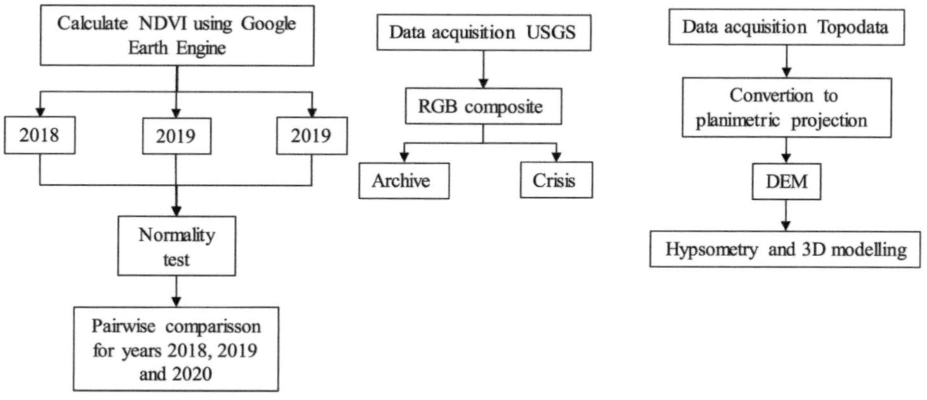

Figure 2.
Research procedures flowchart.

Satellite/sensor	Wavelength (nm)	Period	Acquisition date	Track
Sentinel 2A—multispectral instrument (MSI)	Blue—458–523	Archive	01/22/2019	131/241
	Green —543–578	Crisis	02/01/2019	131/241
	Red—650–680			
	Red edge—698–713			
	Red edge—733–748			
	NIR—785–899			
	NIR narrow—855–875			
	SWIR—1565–1655			
	SWIR—2100–2280			

Table 1.
Sentinel-2 images characteristics.

The SRTM provided by the Topodata program comes in WGS1984 datum. For analytic and standardization purposes, all data were converted to SIRGAS2000 Universal Transverse Mercator (UTM) zone 21S datum.

2.5 NDVI calculation using google earth engine

For this study, researchers deployed a Google Earth Engine script to calculate normalized difference vegetation index (NDVI) values for the periods of January 1th, 2018 to December 31th, 2018; January 1th, 2019 to December 31th, 2019; and January 1th, 2020 to December 31th, 2020, according to the sentinel 2 constellation revisit period of 5 days. Level 2 (L2) orbital remote sensing products from the Sentinel-2 satellite were used, the L2 algorithm generates ortho corrected, atmospherically corrected, and with bottom-of-atmosphere reflectance. The constellation of two satellites—Sentinel-2A and Sentinel-2B—orbit the Earth at an altitude of 786 km but are separated by 180° to optimize global coverage and revisit times.

The NDVI, proposed by Rouse et al. [28], is the most common ratio index for remote sensing environmental analysis. It allows us to identify areas with dense vegetation and areas with no vegetation. It is calculated using the near infrared (NIR) and red bands, applying the following equation:

$$NDVI = \frac{(\rho NIR - \rho Red)}{(\rho NIR + \rho Red)} \quad (1)$$

The NDVI presents values that vary from −1 to 1, where values below 0 are considered non-vegetated areas, such as bare soil, mud, or water, and as closer to 1 get, the healthier the vegetation.

This index presents high accuracy for comparisons in time and spatial scales, considered suitable for analyses of areas affected by floods. Flooded areas generally present the characteristics of water or mud presence. These surfaces present low reflectance patterns in the red and NIR spectrum, while vegetated areas present higher reflectance patterns [7].

2.6 Statistical analysis

The descriptive statistics were calculated and confidence interval was obtained using the Bootstrap method with 9999 repetitions. The Shapiro–Wilk test was used to verify normality, results ($p < 0.001$) show no adherence to normality. The Dwass-Steel-Crichlow-Fligner test, described by Dwass [29], Steel [30, 31], Douglas and Michael [32] for independent nonparametric samples, was used to test if there was a significant difference between years. The pairwise comparison was deployed using NDVI values for 2018, 2019, and 2020. The test assumed alpha equals 0.05.

3. Results and discussion

3.1 Hypsometry of the study area

The Córrego do Feijão dam was located at the right top corner of **Figure 3**. As it can be noticed, the dam was located at a higher altitude than the Paraopeba river. This river provides drinking water for several municipalities along its course, among other types of usage such as industries, irrigation, and for livestock. The presented data are key for the study of tailing dam location, since it provides information for decision-makers to prevent the installation of these dams near human settlements. The SRTM data can be used to model the impacts of dam breaks and the main areas that will be affected. This method can be used in the environmental licensing of tailing dams.

The digital elevation model (DEM) deployed using SRTM data is key for flow accumulation analysis [33]. Mubareka et al. [34] present a study, using SRTM data to predict settlement location and density at a 90 m resolution. The terrain representation in 3D allows decision-makers to tackle priority areas to focus efforts.

3.2 Environmental impacts in Brumadinho, MG, Brazil

In addition to the 363 confirmed deaths, the mud devastated homes, buildings, businesses, and families, leaving only the trail of destruction and uncertainties. Some of the families that lived there had small land where they developed subsistence

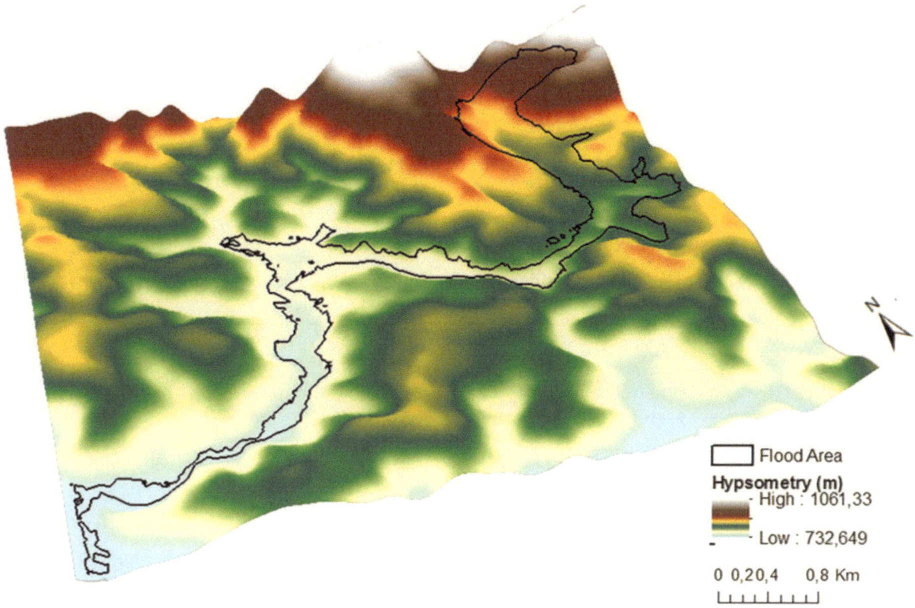

Figure 3.
3D digital elevation model representation of the study area.

agriculture as their main source of income, just as fishermen and rural communities lost all their material possessions.

Following the catastrophe, the National Human Rights Commission carried out a mission aimed at promoting qualified listening and proposing emergency actions for the affected populations. During the mission, the possible impacts resulting from the rupture of Córrego do Fundão tailings dam were raised, among them the mortality of specimens in all trophic chains; harming the conservation status of species already listed as endangered and the entry of new species into the list of threatened species, and undermining the structure and function of associated aquatic and terrestrial ecosystems [35].

Figure 4 presents the archive image acquired 01/22/2019 from Sentinel MSI (left) and the crisis image from 02/29/2019 before the Córrego do Fundão tailing dam failure (right).

3.3 Vegetation loss due to Córrego do Fundão tailings dam failure

The Atlantic Forest biome prevailed in the area affected by the tailings Dam I failure. This biome holds up to 8% of the world's species and is recognized as a hotspot for biodiversity conservation. As a result of anthropic activities, in recent years, its occupied territory has plummeted to less than 15% [36], thus being a global conservation priority [37].

Although mining activity is predominant in the area, besides the Atlantic Forest parks protection, activities, such as agriculture and livestock, both for subsistence, were responsible for a small portion of land use and occupation [38]. Of the affected area, the major affected area corresponds to Atlantic Forest vegetation, and the rest are activities carried out by the population, such as housing and agriculture. It should be noted that food production, food security, and community health were strongly

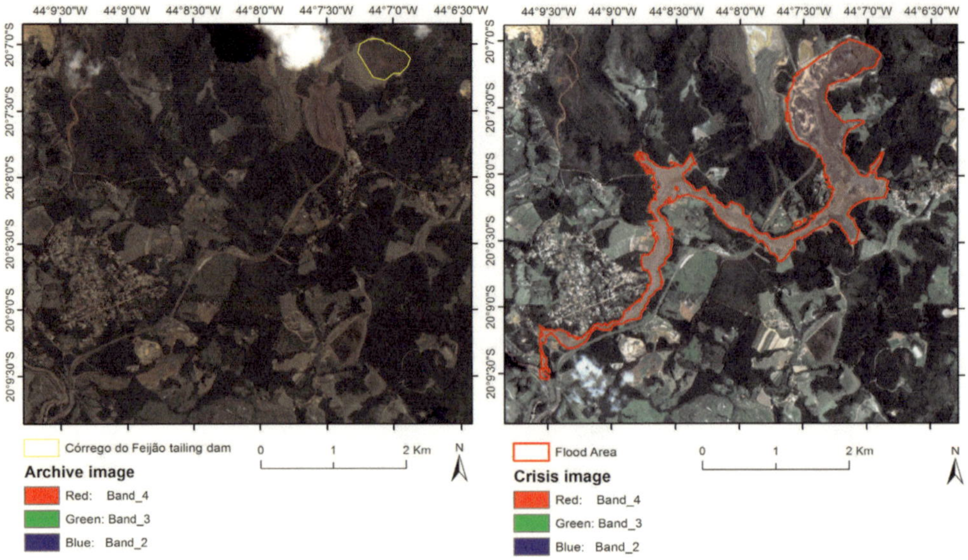

Figure 4.
Archive image from 22 to 01-2019 on the left panel and crisis image from 01 to 02-2019 on the right panel of the study area.

and directly affected since natural resources such as soil, water, and ecosystem interactions were compromised.

With the Vale S/A tailings dam failure in Brumadinho (MG), 11.7 million cubic meters of high silica and iron sludge were dumped under an area of Atlantic Forest native vegetation.

Figure 5 presents the area affected by the dam break, representing a vegetation loss of 2662 ha. The difference can be identified due to high reflectance values in the NIR band by mud-covered areas. Vegetated areas have high absorbance of the green spectrum. As seen in the NDVI Archive histogram, peaks of reflectance can be noticed in 0.8 μm, which shows a high quantity of green vegetation in the area, after the dam break, it can be noticed a significant decrease in peak values, now in 0.1 μm.

The NDVI is the most used remote sensing index for vegetation loss studies [39], for example, in the studies of Mariana environmental disaster and analysis of native Atlantic Forest loss [13]. We used Google Earth Engine to calculate the NDVI, using the affected area polygon, for years 2018, 2019, and 2020. It can be noticed, in **Figure 6**, lower values for NDVI in the crisis year.

The mean values, with lower and upper bootstrap confidence intervals, for NDVI for year 2017 were 0.314003 [0.31028; 0.317564], for NDVI in year 2018 values were 0.339887 [0.336591; 0.343231], for year 2019 the values were 0.145814 [0.144004; 0.1476], for year 2020 values were 0.1495 [0.147676; 0.15128], and for year 2021 the values were 0.15572 [0.153727; 0.15774]. The Dwass–Steel–Crichlow–Fligner test for nonparametric data presented a statistically significant ($p < 0.001$) decrease in vegetation when comparing 2018–2019 and 2020 values. The descriptive and inferential statistics are presented in **Tables 2** and **3**.

Table 3 displays the inferential statistics for the NDVI values for the period between 2018 and 2021. The results clearly show that the vegetation did not recover from the disaster in January 2021, with significant difference between the years pre and post the disaster, with exception of the year 2018.

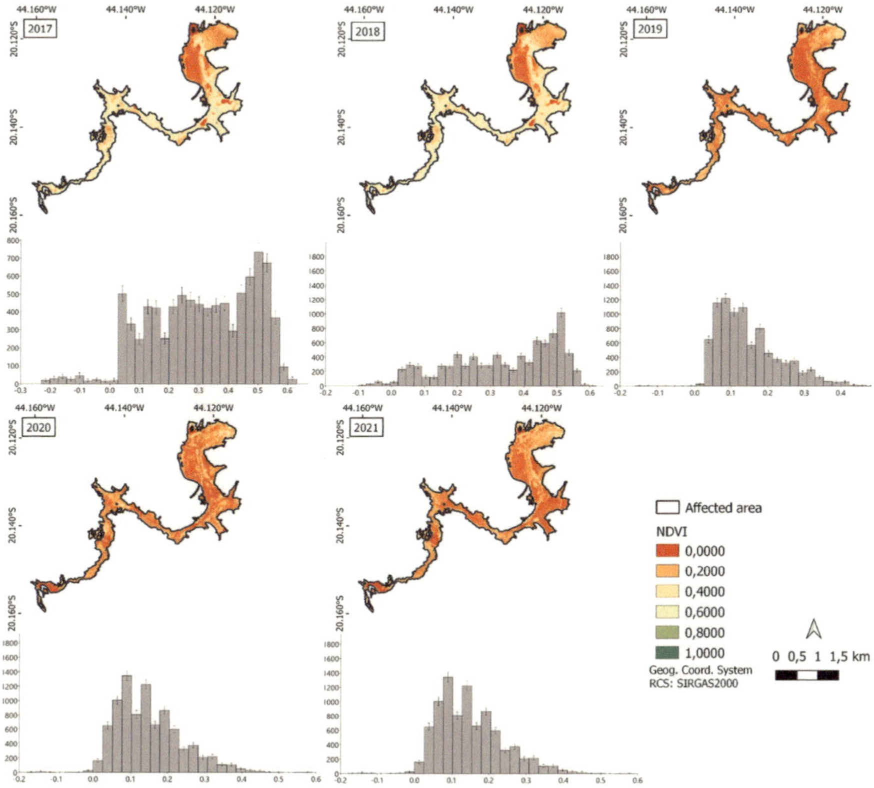

Figure 5.
Vegetation dynamics and histograms of NDVI values for years 2017–2021.

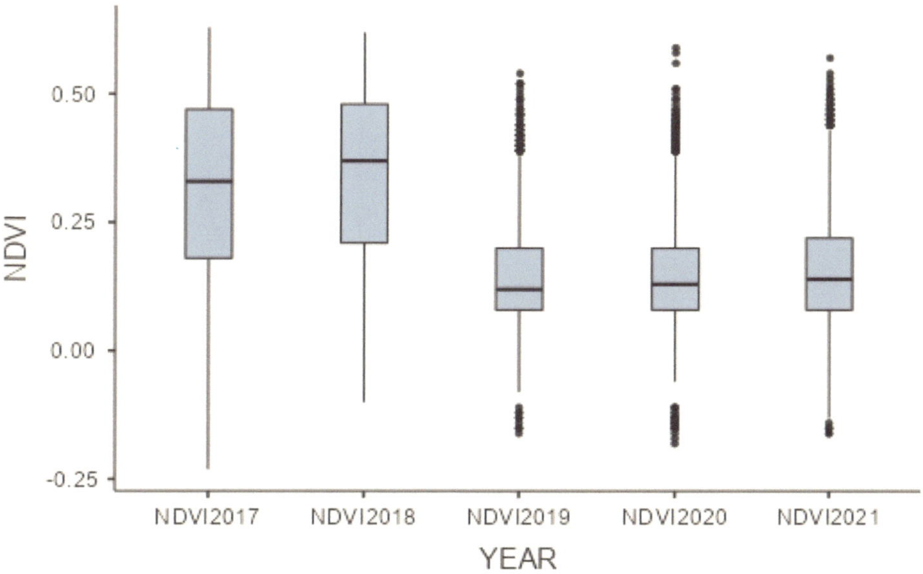

Figure 6.
NDVI values for years 2017–2021.

	2017	Bootstrap	2018		Bootstrap		2019	Bootstrap		2020	Bootstrap		2021	Bootstrap	
		Lower conf.	Upper conf.		Lower conf.	Upper conf.		Lower conf.	Upper conf.		Lower conf.	Upper conf.		Lower conf.	Upper conf.
N	8817	8817	8817	8817	8817	8817	8817	8817	8817	8817	8817	8817	8817	8817	8817
Min	−0.23			−0.1			−0.16			−0.18			−0.16		
Max	0.63			0.62			0.54			0.59			0.57		
Sum	2768.56	2735.74	2799.96	2996.78	2967.72	3026.27	1285.64	1269.68	1301.56	1318.19	1302.06	1333.9	1373.03	1355.41	1390.86
Mean	0.314003	0.31028	0.317564	0.339887	0.336591	0.343231	0.145814	0.144004	0.1476	0.1495	0.147676	0.15128	0.15572	0.153727	0.15774
Std. error	0.001848	0.001825	0.00187	0.00171	0.001691	0.001729	0.000945	0.000929	0.000961	0.000935	0.000918	0.000951	0.00105	0.001033	0.001067
Shapiro–Wilk *p*	<.0001			<.0001			<.0001			<.0001			<.0001		

Table 2.
Descriptive statistics for NDVI values according to the period of 01-01-2018 *to* 31-12-2018, 01-01-2019 *to* 31-12-2019, *and* 01-01-2020 *to* 31-12-2020.

Comparisons		p	Comparisons		p
2017	**2018**	**1**	**2018**	**2020**	**<0.001***
2017	2019	0.082*	2018	2021	0.004*
2017	2020	0.006*	2019	2020	0.808
2017	2021	0.019*	2019	2021	0.994
2018	2019	0.027*	2020	2021	0.969

**Values which reject the null hypothesis with alpha equals to 0.05.*

Table 3.
Inferential statistics for NDVI values pairwise comparison to the period of 2018–2021.

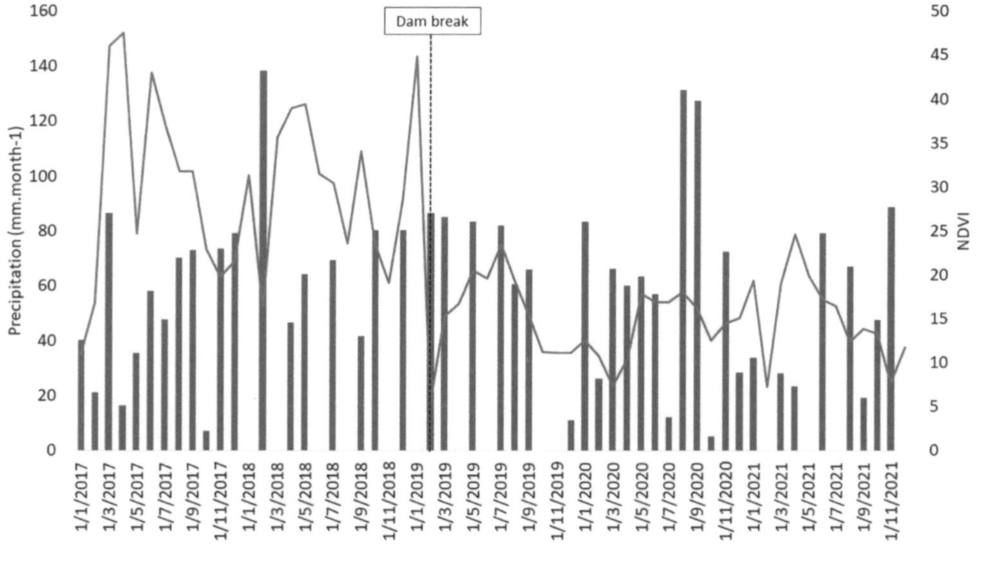

Figure 7.
NDVI and precipitation (mm.Month^{-1}) for the period between 2017 and 2021.

Silveira et al. [40] used Landsat 8 images of before and after the Mariana, MG disaster to detect vegetation loss in the affected area. Their conclusions were that the index produced “highly accurate maps of areas affected by post-dam-failure flooding in the region. This approach can be used in many other contexts for rapid and accurate assessment of such land-cover change.” (**Figure 7**).

3.4 Disaster management

Disasters are events that escape to normality, involving large negative environmental, economic, and social impacts. Their environmental and socio-environmental consequences can be reversible or not [41]. Its origin can be natural or anthropogenic, it is currently considered that they are, in general, products of interrelation between human activities and natural phenomena [42–44] as an example of the disaster that occurred in the municipality of Brumadinho, MG in February 2019.

Disaster management presents itself in a cycle divided into mitigation, preparation, response, and recovery. Mitigation implements measures that can eliminate or reduce the degree of risks and hazards. Preparation is where actions are taken in advance in order to develop effective mechanisms to respond to the event. It is followed by the response that is the actions to be implemented as soon as the event occurs minimizing the damage and the recovery phase, that is, the reestablishment of the area and its actions to reach normality [45–48].

In the last 10 years, according to the Université Catholique de Louvain's EM-DAT [49], 10 disasters were registered in Brazil between 2010 and 2019 including industrial, transportation, and assorted. Among them is the disaster in the municipality of Mariana (MG) with a total of 25 deaths and the Brumadinho disaster in early 2019 with 363 deaths. Furthermore, it must be accounted for the loss of Atlantic Forest native vegetation, water quality degradation, and livestock deaths as environmental developments to this catastrophe.

In response to these events, Brazil has a specific legislation for disasters, where the Federation, states, municipalities, and organized society have defined roles. The Law 12,608—2012 [50] establishes the responsibilities, goals, and directives of planning with the objective of reducing disasters. This law foresaw the zoning for land use and occupation, creation of a database for risk areas monitoring, prevention, response, and mitigation plans. Nevertheless, its efficiency was absent in disasters such as those

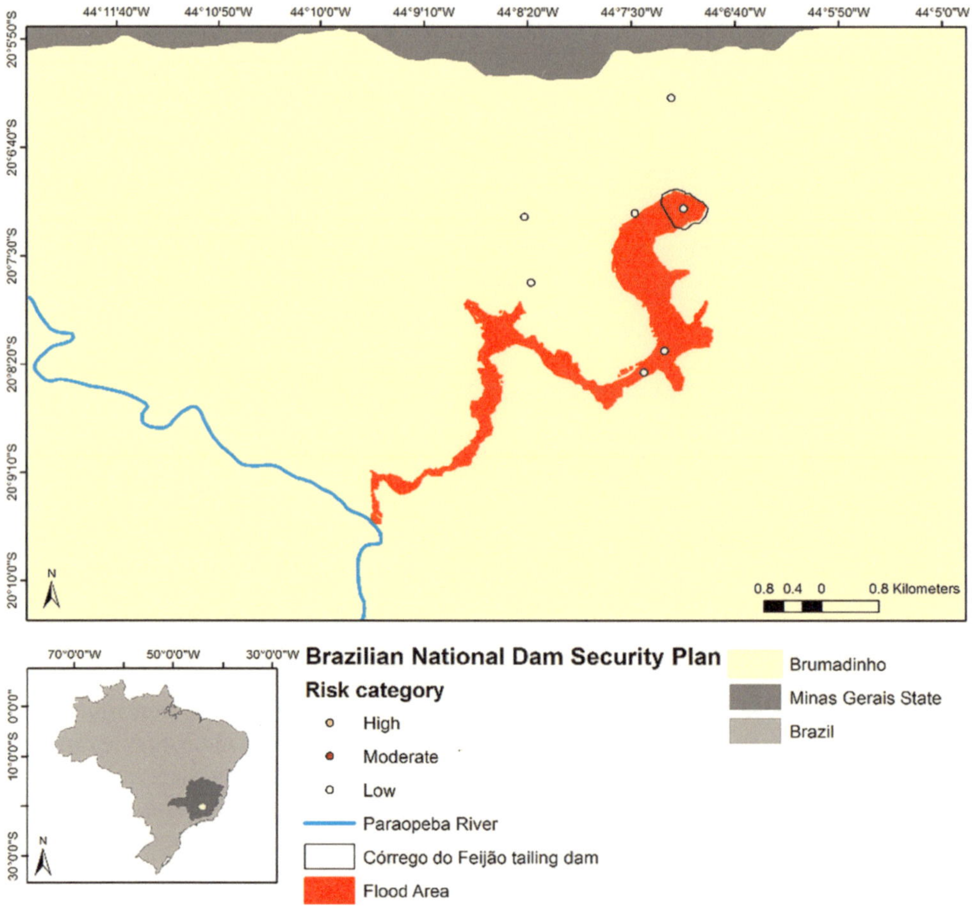

Figure 8.
Tailing dams risk categories according to the Brazilian National dam Security Plan.

mentioned above, it is exposed that there is a need for structuring and implementing disaster management plans.

3.5 Environmental regulation regarding the tailings dam I

The Tailings Dam I, broken in January 2019, belonged to the Córrego do Feijão Dam Complex, it was 87 meters high, medium size, and stored iron ore. According to the Brazilian National Dam Security Plan (PNSB), the dam was categorized as low risk and had high potential damage. This complex belongs to Vale, the largest Brazilian mining company. In this complex, there are still five small dams classified as low risk, as seen in **Figure 8**, with potential damage between medium and high, **Figure 9** [51]. The dam's potential damage is a function of the human life's potential loss and economic, social, and environmental impacts. In addition to these, the municipality of Brumadinho has another 20 small and medium-sized dams, between low and medium risk [52].

In 2017, the Environmental Impact Report (EIA/RIMA) was registered at the State Environmental Foundation of Minas Gerais.

The EIA/RIMA is a direct product of the Environmental Licensing process. It is a regulatory instrument in which the government, represented by environmental

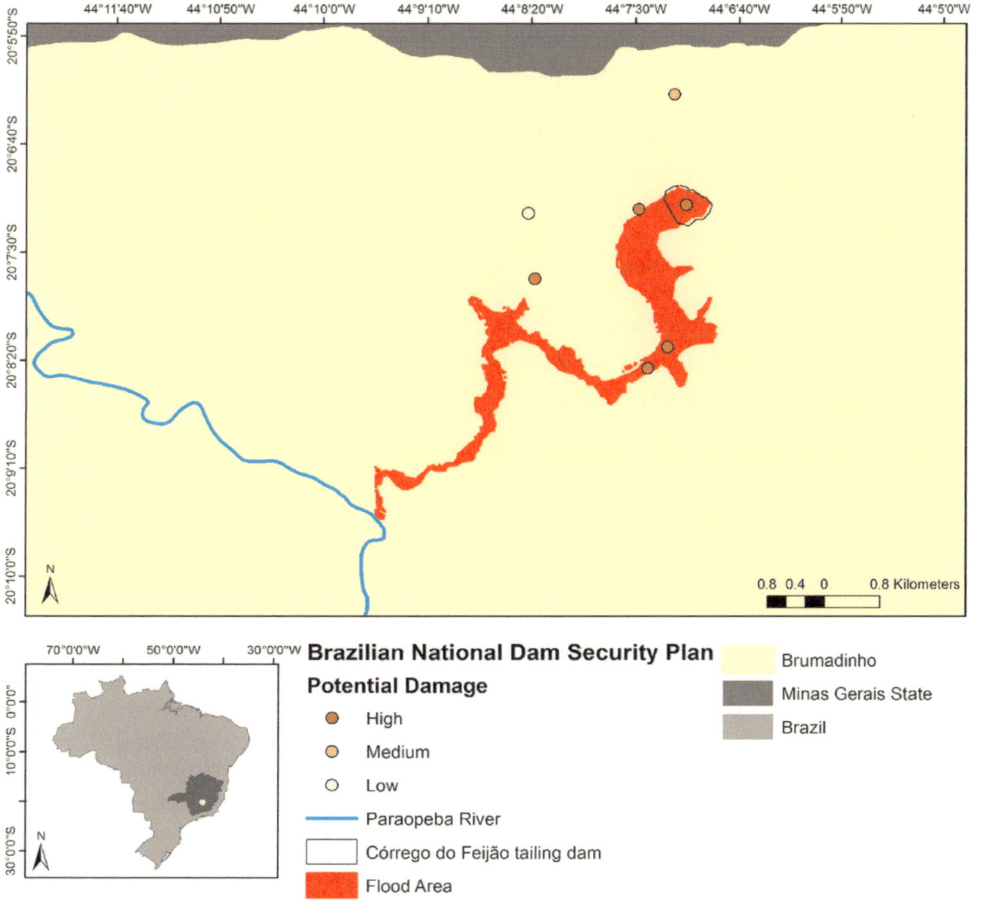

Figure 9.
Tailing dams damage potential according to the Brazilian National dam Security Plan.

agencies, authorizes and monitors the installation and operation of activities that appropriate from natural resources or that are considered effective or potentially polluting. It is mandatory for entrepreneurship, to seek environmental licensing from the competent agency, from the initial stages of its planning and installation until its effective operation.

Environmental impact study and the environmental impact report are included in this context as a regulatory requirement, established by the National Environmental Council (CONAMA) Resolution 001/86. It consists of in situ studies in soil, water, and air to verify if the prospective area contains environmental liabilities. Furthermore, present studies regarding the socio-economic-environmental relationship will be affected by the implementation of the enterprise.

The EIA/RIMA proposed licensing the operational continuity of the Córrego do Feijão dams until the year 2029. It was approved in November 2018 by the Superintendency of Priority Projects (SUPPRI) linked to the State Secretariat of Environment and Sustainable Development of the State of Minas Gerais. Palagi and Javernick-Will [53] states that a major constraint for policy-making and application is "institutional norms and cultural beliefs considerably narrowed the range of options post-disaster decision-makers perceived as viable, appropriate, or compassionate."

Vale has expressed in an official note regarding the risk management actions that had been implemented before and during the event, such as alarms, leakage routes, and maintenance and monitoring of dams. Regardless of these actions, it is not yet known exactly what triggered the rupture of the dam. As in the Mariana disaster that occurred in 2015, some hypotheses were raised about the disaster in Brumadinho. To solve any doubts regarding this and other possible dam breaks, not only of Vale but other mining companies, the Public Prosecutor's Office continues to follow the developments, as well as the documents issued where possible problems regarding the dam structure were confirmed.

As a consequence of this act, a stronger posture is expected by the government agencies. On the contrary, the rural and economic development-focused parties in the Brazilian senate propose the flexibilization of the environmental laws. For example, reducing the bureaucratic criteria for the environmental licensing, main regulation instrument in Brazil. In this sense, integrating technology and decision-making is key for disaster management.

4. Conclusions

The vegetation in the area affected by the dam break in Brumadinho did not present a statistically significant recovery according to NDVI values. The total area affected by the dam brake was 2662 ha. The NDVI values presented a statistically significant decrease from 2017 to 2019, with little increase until 2021, NDVI values for year 2017 were 0.314003 [0.31028; 0.317564], for year 2018 the NDVI values were 0.339887 [0.336591; 0.343231], for year 2019 the NDVI values were 0.145814 [0.144004; 0.1476], for year 2020 the NDVI values were 0.1495 [0.147676; 0.15128], and for year 2021 the NDVI values were 0.15572 [0.153727; 0.15774]. The study shows that GIS technology is key for post disaster impacts assessment and monitoring of vegetation recovery and aiding strategic planning of decision-makers. GIS presents low cost and rapid response tools for disaster management, allowing it to focus efforts in priority areas. Furthermore, GIS must be integrated in disaster prevention policies, integrating environmental licensing development processes.

Regarding the information gathered during the study, it is concluded that although Brazil has regulatory instruments, they must be revised for efficient application.

Once the company responsible had employed the disaster management tools, such as GIS-based decision-making, with prevention actions, rapid response to events, and recovery, much of the damage would be minimized and even avoided. Still on disaster management, this is also government responsibility. Decisions must be community driven, where agile responses to events and community well-being are guaranteed, regardless of responsibility.

Approaching all environmental impacts, besides the loss of Atlantic Forest vegetation, directed studies on aquatic and terrestrial ecosystems are of great relevance to explain and elucidate all the damage caused, supporting future discussions about legislation of dam failures and other projects that endanger environmental health. These discussions must focus on prevention other than remediation.

Author details

Rodrigo Martins Moreira[1]* and Maria Paula Cardoso Yoshii[2]

1 Geomatics and Statistics Laboratory, Department of Environmental Engineering, Federal University of Rondõnia, Ji-Paraná, Rondônia, Brazil

2 University of São Paulo, São Carlos, São Paulo, Brazil

*Address all correspondence to: rodrigo.moreira@unir.br

References

[1] Kumari KP, Srinivas K. Applications of RS and GIS techniques for disaster studies in east Godavari District, Andhra Pradesh, India. In: Proceedings of International Conference on Remote Sensing for Disaster Management. Cham: Springer; 2019. pp. 199-214

[2] Jacobi P, Roberto; Aledo, Antonio; Warner, Jeroen. Sobre a necessidade de tratar dos desastres no contexto da sociedade de risco. Ambiente & Sociedade. 2014;**17**(4):1-4

[3] Carmo FFD et al. Fundão tailings dam failures: The environment tragedy of the largest technological disaster of Brazilian mining in global context. Perspectives in Ecology and Conservation. 2017;**15**(3): 145-151

[4] Schipper L, Pelling M. Disaster risk, climate change and international development: Scope for, and challenges to, integration. Disasters. 2006;**30**(1):19-38

[5] Samela C, Albano R, Sole A, Manfreda S. A GIS tool for cost-effective delineation of flood-prone areas. Computers, Environment and Urban Systems. 2018;**70**:43-52

[6] Boonmee C, Arimura M, Asada T. Facility location optimization model for emergency humanitarian logistics. International Journal of Disaster Risk Reduction. 2017;**24**:485-498

[7] Arabameri A et al. Flash flood susceptibility modelling using functional tree and hybrid ensemble techniques. Journal of Hydrology. 2020;**587**:125007

[8] Felix AY, Sasipraba T. Spatial and temporal analysis of flood hazard assessment of Cuddalore District, Tamil Nadu, India. Using geospatial techniques. Journal of Ambient Intelligence and Humanized Computing. 2021;**12**(2):2573-2584

[9] Soltani K et al. Mapping the spatial and temporal variability of flood susceptibility using remotely sensed normalized difference vegetation index and the forecasted changes in the future. Science of the Total Environment. 2021; **770**:145288

[10] Buma B. Evaluating the utility and seasonality of NDVI values for assessing post-disturbance recovery in a subalpine forest. Environmental Monitoring and Assessment. 2012;**184**(6):3849-3860

[11] Dutra DJ, Elmiro MAT. Avaliação de índices espectrais obtidos com imagens sentinel-2 e landsat 8 antes e após rompimento da barragem da mina do feijão, brumadinho–mg. Geociências. 2020;**39**(2):517-523

[12] de Souza Diniz F, Maria J, et al. Avaliação do Sentinel-2, NDVI e MLME para Mapeamento do Uso e Cobertura da Terra. Anuario do Instituto de Geociencias. 2020;**43**(2)

[13] Rotta LHS et al. The 2019 Brumadinho tailings dam collapse: Possible cause and impacts of the worst human and environmental disaster in Brazil. International Journal of Applied Earth Observation and Geoinformation. 2020;**90**:102119

[14] Cheng D et al. Watch out for the tailings pond, a sharp edge hanging over our heads: Lessons learned and perceptions from the Brumadinho tailings dam failure disaster. Remote Sensing. 2021;**13**(9):1775

[15] F Gama F et al. Deformations prior to the Brumadinho dam collapse

revealed by Sentinel-1 InSAR data using SBAS and PSI techniques. Remote Sensing. 2020;**12**(21):3664

[16] Poursanidis D, Chrysoulakis N. Remote sensing, natural hazards and the contribution of ESA sentinels missions. Remote Sensing Applications: Society and Environment. 2017;**6**:25-38

[17] Sharma T, Prasad P, et al. Review of flood disaster studies in Nepal: A remote sensing perspective. International journal of disaster risk reduction. 2019; **34**:18-27

[18] Tabata T, et al. Earthquake disaster waste management reviews: Prediction, treatment, recycling, and prevention. International Journal of Disaster Risk Reduction. 2019;**36**:101119

[19] Canevari-Luzardo L, et al. Using partial participatory GIS in vulnerability and disaster risk reduction in Grenada. Climate and Development. 2017;**9**(2): 95-109

[20] Olyazadeh R, Aye ZC, Jaboyedoff M, Derron MH. Prototype of an open-source web-GIS platform for rapid disaster impact assessment. Spatial Information Research. 2016;**24**(3):203-210

[21] Nascimento VF, Sobral AC, Andrade PR, Yesiller N, Ometto JPHB. Natural disaster risk in municipal solid waste disposal sites using GIS: A case study in São Paulo state, Brazil. Journal of Water Resource and Protection. 2017;**9**(11):1213

[22] Mccormick S. New tools for emergency managers: An assessment of obstacles to use and implementation. Disasters. 2016;**40**(2):207-225

[23] Rodríguez-Espíndola O, Albores P, Brewster C. Disaster preparedness in humanitarian logistics: A collaborative approach for resource management in floods. European Journal of Operational Research. 2018;**264**(3):978-993

[24] Abdalla RM, Alharbi SM. The role of GIS technology in assessing critical infrastructure resiliency during emergencies in the City of Jeddah, Saudi Arabia. International Journal of Emergency Management. 2017;**13**(2): 183-191

[25] IBGE. Panorama da cidade de Brumadinho. 2019. Available from: https://cidades.ibge.gov.br/brasil/pr/brumainho/panorama. [Accessed: 2019]

[26] QGIS Development Team. QGIS geographic information system. Open Source Geospatial Foundation Project. 2021. Available from: http://qgis.osgeo.org

[27] Topodata: banco de dados geomorfométricos do Brasil. Variáveis geomorfométricas locais. São José dos Campos; 2011. Available from: http://www.dsr.inpe.br/topodata/

[28] Rouse JW et al. Monitoring the Vernal Advancement and Retrogradation (Green Wave Effect) of Natural Vegetation. Vol. 371. Greenbelt, MD: NASA/GSFC Type III Final Report; 1974

[29] Dwass M. Some k-sample rank-order tests. Contributions to probability and statistics. 1960

[30] Steel RGD. A rank sum test for comparing all pairs of treatments. Technometrics. 1960;**2**(2):197-207

[31] Steel RGD. Some rank sum multiple comparison tests. Biometrics. 1961;**17**: 539-552

[32] Douglas CE, Michael FA. On distribution-free multiple comparisons in the one-way analysis of variance.

Communications in Statistics-Theory and Methods. 1991;**20**(1):127-139

[33] Gaurav K, Sinha R, Panda PK. The Indus flood of 2010 in Pakistan: A perspective analysis using remote sensing data. Natural Hazards. 2011;**59**(3):1815

[34] Mubareka S, Ehrlich D, Bonn F, Kayitakire F. Settlement location and population density estimation in rugged terrain using information derived from Landsat ETM and SRTM data. International Journal of Remote Sensing. 2008;**29**(8):2339-2357

[35] Cndh CN, Dos DH. Relatório Da Missão Emergencial A Brumadinho/Mg Após Rompimento Da Barragem Da Vale S/A. Brasília, DF: Conselho Nacional dos Direitos Humanos; 2019

[36] De Lima RAF et al. How much do we know about the endangered Atlantic Forest? Reviewing nearly 70 years of information on tree community surveys. Biodiversity and Conservation. 2015;**24** (9):2135-2148

[37] Omachi CY et al. Atlantic Forest loss caused by the world's largest tailing dam collapse (Fundão dam, Mariana, Brazil). Remote Sensing Applications: Society and Environment. 2018;**12**:30-34

[38] Pereira LF, Cruz GDB, Guimarães RMF. Impactos do rompimento da barragem de rejeitos de Brumadinho, Brasil: uma análise baseada nas mudanças de cobertura da terra. Journal of Environmental Analysis and Progress. 2019;**4**(2):122-129

[39] Evangelides C, Nobajas A. Red-edge normalised difference vegetation index (NDVI705) from Sentinel-2 imagery to assess post-fire regeneration. Remote Sensing Applications: Society and Environment. 2020;**17**:100283

[40] Silveira EMDO, Acerbi Júnior FW, Mello JMD, Bueno IT. Object-based change detection using semivariogram indices derived from NDVI images: The environmental disaster in Mariana, Brazil. Ciência e Agrotecnologia. 2017;**41** (5):554-564

[41] Ganem RS.. Gestão De Desastres No Brasil. 2012

[42] Comfort LK. Fragility in disaster response: Hurricane Katrina. The Forum. 2005;**3**(3):0000102202154088841090

[43] Cunha GR, Pires JLF, Pasinato A. Uma discussão sobre o conceito de hazards e o caso do furacão/ciclone Catarina. v. 36. Brasília/DF: Ministério da Agricultura, Pecuária e Abastecimento; 2004. p. 21

[44] Glickman TS, Golding D, Silverman ED. Acts of god and acts of man: Recent trends in natural disasters and major industrial accidents. In: Washington, D. C: Resources for the Future, Discussion Paper CRM. 1992. pp. 92-02

[45] Bertazzo TR et al. Revisão da literatura acadêmica brasileira sobre gestão de operações em desastres naturais com ênfase em logística humanitária. Transport. 2013;**21**(3):31

[46] Cova TJ. GIS in Emergency Management. 1999. p. 19

[47] Leiras A et al. Literature review of humanitarian logistics research: Trends and challenges. Journal of Humanitarian Logistics and Supply Chain Management. 2014;**4**(1):95-130

[48] Valencio NFLS, et al. Chuvas no Brasil: representações e práticas sociais. Revista Política e Sociedade. Florianópolis. 2005;**07**:163-183

[49] EM-DAT. The Emergency Events Database. Brussels, Belgium: Universite

catholique de Louvain (UCL)—CRED, D. Guha-Sapir; 2019

[50] Brasil, Law 12.608/2012. 12.608. 2012

[51] Brasil. Brazilian National Dam Security Plan. 2019

[52] Dnpm DN, De PM. Cadastro Nacional de Barragens. Brasília, DF: [s. n.]; 2016

[53] Palagi S, Javernick-Will A. Institutional constraints influencing relocation decision making and implementation. International Journal of Disaster Risk Reduction. 2019;**33**:310-320

Chapter 12

Interconnection among River Flow Levels, Sediments Loads and Tides Conditions and Its Effect on the Coastal Wetlands Reduction

Jesús Gracia-Sánchez, Judith G. Ramos, Liliana Marrufo Vázquez, Javier Osnaya Romero and Víctor M. Ortíz-Martínez

Abstract

When the river supplies water to vulnerable environments, such as marshlands, it is vital to establish the expected impact mostly under a changing climate, and moreover, if a dam is being projected to solve energy demands. Soil characteristics, specifically sediment composition, are exposed to changes that modify this type of ecosystem and are rarely investigated. For this, a discharge period for an average historical year was analyzed to evaluate the magnitudes of the flows, with or without a dam. Also, it helped to identify the modification of the hydrodynamic regime between the sea and the lagoon system, particularly during the dry season but also checking the behavior in the rainy season. Results showed that the main problem with the construction of the dam on the San Pedro-Mezquital river would be the effect of a controlled flow that reaches the wetlands of the alluvial plains, affecting the sediment load in the estuarine and coastal ecology. However, after a readjustment period, the dam neither significantly changes the previous flood conditions of the coastal plain nor the sediment load will be a problem. However, if an additional sediment load is required to maintain the coastal microhabitats, there are different ways to provide it.

Keywords: marshlands, river connectivity, wetlands, Marismas Nacionales, dam

1. Introduction

A river needs to be understood as part of a changing system, so it cannot be studied in isolation. In this way, the benefits throughout the basin at different levels (high, middle, and low) and scales can be observed, which will favor the development of social and ecosystem services. However, to achieve comprehensive management of a basin, it is important to understand some of the main obstacles to river management, such as the changes that have originated through several actions, such as land use management (changes of coverage and use), climate change, development of aquatic resources and expansion of the industrial sector, dams, among others. These actions have considered changes in the runoff pattern, the quality of the discharges to the river, the size distribution, and the load of transported sediments, as well as changes in the structure of fluvial and riparian micro-habitats.

Hydrological processes are governed by a large number of biophysical and climatic variables. No matter how small, any change made to sensitive and highly vulnerable ecosystems generates large changes. Thus, when the change is not so small and considers the installation of a dam, there is a modification that alters the energy flows, such as the precipitation and temperature, and, in consequence, the evapotranspiration (ET), groundwater recharge, runoff, and water bodies storage [1, 2]. This implies that the cause/effect relationship of the realized changes will have significant environmental implications, such as the loss of biodiversity, alterations in hydrological processes, and land degradation. When it comes to natural or artificial reservoirs, runoff can be attenuated or restricted, although secondary runoff associated with overflows can even occur, changing the ecosystem upstream and downstream in the river [3]. The main responsible for this type of change is associated with humans and their productive activities [2], including river regulation and global warming change, which are the main drivers of changes in the flow regime of rivers identified by [4]. In particular, a dam that, on one side, it will provide several benefits to people, such as diversion of water to agricultural or urban areas, electricity production, aid in navigation, and flood control [5, 6]. However, on the other side, there are also negative changes affecting the estuarine and coastal ecology, which has resulted, in many places, in the loss of habitats such as wetlands. In particular, vegetative populations, such as riparian or coastal vegetation (e.g., mangroves), show a structure defined by the availability of freshwater, nutrient recycling, tidal flows, frequency of flood periods, physical characteristics of sediments, and water chemistry [7–9]. As these ecosystems could be highly affected by the changes in the rivers and their plains, a geomorphological analysis is necessary, and it includes the vegetation and environment, as formations of geoforms making the system highly unstable). In general, the installation of dams is associated with the loss of natural structures, changes in the grain size distribution (structure and substrate of the river bed), and interruption of the sediment transport of the river bed (continuity of the river). Thus, it is important to consider that inducing a discontinuity in longitudinal migration will likely lead to fragmentation of the aquatic fauna. Also, special attention needs to be made to the hydrology alterations caused by river regulation and climate change. The continued temperature increase, as well as the reduction of water supply (precipitation), have the potential to modify the timing and magnitude of a river flow [10, 11]. Duan and Cai [12] noticed that global warming had affected the snow accumulation in winter; and its melting in spring, resulting in short timing with peak runoff and the increment of discharges earlier in the water year; thus, there are changes in the annual flow [13]. For instance, Duan and Cai [12] observed that flood risk was reduced in cold watersheds corresponding to high latitudes and altitudes. In contrast, an increase in flood risk was registered by Allamano et al. [14] due to the increase in temperature and precipitation intensity. It will be common for rain-dominant basins to present floods related to storms in the wet season [15]. Precipitation is characterized by frequency and duration, which defines its magnitude, as well as soil moisture. Thus, a reduction/increment of high flow magnitude is significantly connected to changes in precipitation characteristics, climate warming, and the spatial heterogeneity of the terrain [12]. Maskey et al. [4] pointed out that there is no single climate model that explains the behavior of the hydrological processes. Still, it is necessary to understand both impacts and effects caused by the water bodies to set realistic managed environmental flows that cope with their ecological goals. Although, Maskey et al. [4] found that "the flow regimes downstream of a dam are likely more altered by reservoir operations than by climate change." Also, an intermittent flow created, as a result of the reservoir operations, affects the static instream flow requirements that might impact the movement, establishment, and environmental signals for aquatic life.

1.1 Tidally-driven connectivity

Streams, rivers, lakes, wetlands, and marshes are aquatic ecosystems that interact because of their ability to import and export material and energy altering the fluxes of these materials. These interactions require connectivity, where various transport mechanisms in a heterogeneous landscape define the degree to which components of a river system are joined or connected to other water bodies [16, 17]. Connectivity between freshwater and marine aquatic habitats offers benefits, such as migration of coastal habitats and species, protection of inland habitats against storms, waves, and seawater, food production, biodiversity protection, flood protection, and erosion control. In general, connectivity in these ecosystems is defined by characteristics of the physical landscape, climate, biota, and human impacts. The last one has partially restricted or completely obstructed tidally-driven aquatic habitats in coastal areas due to human developments (reclamation, species invasion, and environmental pollution), and infrastructure [18]. Thus, many coastal habitats have been disappearing because there is not enough space for inland movements. It is not only a longitudinal consideration, but it also requires studying the lateral and vertical movement since the first links ditches and floodplains, and the second the surface and subsurface [18]. It is important to contemplate the three of them in order to achieve good projects of maintenance or adaptation of these areas. In addition, it is necessary to consider that wetlands in riparian or floodplain areas can have bidirectional lateral hydrologic flows, whereas the type of connection between them only influences wetlands in non-floodplain areas. Also, it is necessary to consider whether vertical movement corresponds to an expansion or contraction of the river network since each one can affect the duration and timing of flow in this network [16].

An interesting aspect related to rivers is the transport of nutrients, sediments, chemicals, organic particles, microbes, detritus of various size classes, and living organisms downstream through wetlands, deltas, estuaries, and other downstream systems. Leibowitz et al. [16] identified the proportion of the material from (or reduced by) streams and wetlands, the residence time of the material in the downstream water; and the relative importance of the material to river function or ecosystem services as the main factor that influences at the material and energy fluxes from streams, and wetlands downstream. As connectivity can be defined as the degree to which components of a system are connected and interact through various transport mechanisms, it varies over time and space, and there are several methods to characterize or quantify it at the watershed scale, such as field hydrological monitoring, hydrological models, connectivity index, remote sensing approach (e.g., LIDAR data or aerial imagery) and graph theory [18]. Park and Latrubesse [19] used remote sensing to map the seasonal water extent and quality variabilities. Also, they identified channel-floodplain connection thresholds and validated results using field measurements. They confirmed that hydrological connectivity processes happen mainly through floodplain channels having specific river level thresholds over space and time characterized by different recharging conditions (through sequential river pulses), water residence, and recessional periods. In addition, Freeman et al. [20] identified that changes upstream of a river affect directly downstream ecosystems, which are also associated with sea conditions, such as flooding, frequency by the tidal and cumulative flooding time. The last two are also responsible for the ecological structure and functions of coastal ecosystems, such as estuaries, mangroves, and tidal flats [21]. Thus, the main goal of this research was to quantify the impact of a dam operation in both river and marshland systems as a function of their connectivity by looking at its spatial (part of the basin) and temporal (dry and wet seasons) variations. For this, the hydrology was modeled for a period of discharges to evaluate the magnitudes of the flows, with or without the dam. This allowed identifying the modification of the hydrodynamic regime between the sea and the lagoon system during the dry and wet seasons.

2. Study area

The study area is in the state of Nayarit, which has the largest number of mangroves together with the state of Sinaloa, sharing the Teacapán-Agua Brava-Marismas Nacionales lagoon system [22]. The rivers that drain the system are Cañas, Acaponeta, Rosamorada, Bejuco, San Pedro, and Santiago, with 161,515 km^2. This study area is located in the physiographic province of the coastal plain of the Pacific, in the sub-province of the delta of Santiago. Along the river, lands of the deltaic landscape system characterized by marshes with coastal lagoons and parallel bars of old coastlines can be found (**Figure 1**). The soil types correspond to fluvisols, cambrisols, and feozem with some gleysols. The feozem is well-drained, fertile, and productive soils. Thus, these plains are excellent for agriculture. In general, more of the landscapes are converted into agricultural lands.

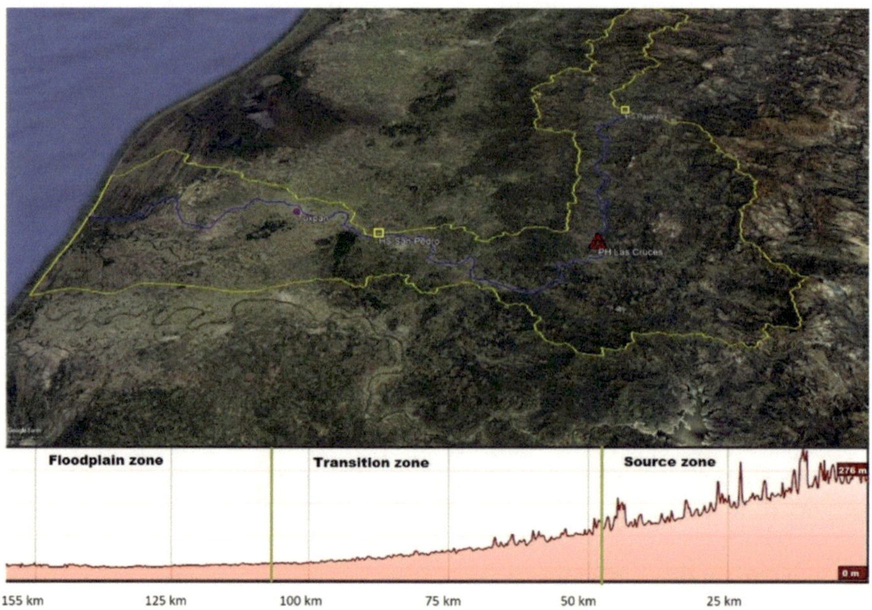

Figure 1.
General scheme of the San Pedro river from the Las Cruces Hydroelectric Project (PHLC) to the Marismas Nacionales zone.

Climate is sub-humid and warm with summer rainfall and semi-warm sub-humid. The average annual precipitation is between 1000 and 1500 mm, reaching more than 2000 mm in September and October. However, rainfall is also associated with the presence of the Mexican monsoon and the occurrence of hurricanes and tropical storms, which may appear between June to October each year. Vegetation includes agriculture occupying 10.67% of the alluvial plains with a low slope (less than 1 m) of the high plateau with deep and moderately deep soils. On both river banks, two irrigation districts (DR) are developed: 052 and 043 are at the present.

Bojórquez et al. [23] identified that the deltaic plains are made up of alluvial material. The presence of dynamic processes is manifested in erosion (in the channel) and accumulation along the flood plain (terraces and dikes). This type of geomorphological landscape in the lower part of the San Pedro-Mezquital river basin, dividing it into (a) intermediate fluvial plain (second level of a fluvial terrace at heights of 5–10 m, from which the rivers overflow in extraordinary flows), (b) low plain with fluvio-marine influence, and (c) fluvial flood plain. In the latter, it is found in the current riverbeds with fluvisols subjected to the action of the systematic flooding of the Santiago, San Pedro, and Acaponeta rivers, where a rejuvenation process is active, which is manifested by the presence of fluvic material in the first 50 cm at the thick soil. In the Marismas Nacionales (MaNas, National Marshes) region, there are few evolved mineral soils of unconsolidated colluvial-marine contribution with moderate erodibility and salinization or sodification [23].

The MaNas is estuary's most important wetlands system, where the mangrove is the dominant vegetation (15–20% of the total of this valuable ecosystem) with species reaching up to 20 or 30 m tall and occupying almost 175,000 hectares [24]. Other vegetation types are halophytic and tular vegetation; the distributions of mangroves and halophytic vegetation clearly respond to a hydraulic regime of flooding. However, these areas have been affected by human beings who have modified the riverbed not only in the mangrove area but also upstream with deforestation and the creation of dams or canals to guarantee agricultural activities. Blanco y Correa et al. [24] noted that from the total area of the estuary, around 135,000 ha (77%) suffered from tidal disturbance; between 5000 and 10,000 ha (from 2.8 to 5.5%) have disappeared under the sea due to coastal erosion, and 5000 ha (2.8%) represented effects from the hydro-sedimentary imbalance of its river. Thus, more than 15,000 ha (8.86%) have an accumulated and synergistic environmental deterioration of at least four to six decades of interaction with different factors, such as poor water, sediment quantity, and quality.

As the MaNas area is important for its high biodiversity and ecosystem services, it is required to take care of its mangroves and, in general, of the entire wetland system. In this way, the real impact that river systems have in the face of natural and anthropic changes, such as altering rivers' capacity to recover or presenting different losses in the hydrological basin needs to be considered. Some actions to protect these ecosystems are the reduction of the possible erosion of beaches, eutrophication of bodies of water, loss of habitat, loss of soil fertility, and reducing the vulnerability to storms or tidal waves, among others.

It is important to recognize that in the MaNas area, the contribution of water is associated with the rain and runoff from the San Pedro river and the effect of the underground movement of water, which means that it can move from neighboring lateral areas into and out of the areas bounded by the watershed. In other words, the hydraulic behavior of the MaNas zone owns complex hydrological connectivity due to the existing interaction between freshwater and marine water. There are associations between the periods of sedimentation and marine regression, which largely affects the sediments' character (differentiation of mechanical particles and/or carbon content in the thickness of the soil profile), the spatial distribution of salinity, and the predominant type of vegetation. This makes MaNas different from the rest of the basin, being the movement of subsurface water decisive. In fact, the drainage patterns of the lower courses of the San Pedro-Mezquital river changed due to the deposition of sediments and the construction of protection levees on the riverbanks. Although the dam will signify greater flood control, at the same time, it will introduce changes in the coastline, intensifying the erosive processes of channels, and the accumulation of sediments in the coastal lagoons.

2.1 PHLC: reservoir characteristics

Las Cruces Hydroelectric Project (PHLC) curtain is located in the state of Nayarit on the San Pedro river, approximately 80 km from the river's discharge to its floodplain to later drain into the MaNas lagoon area and, finally, to the Pacific Ocean. **Figure 2** shows a general scheme of the PHLC contribution basin, having considered a tributary area of 24,879 km^2, as well as the general plan of the reservoir, which has a maximum flood length of approximately 60 km on the San Pedro river considering the elevation of the OMWL (ordinary maximum water level) at 238 masl.

The most important data of the reservoir concerning this study are: (a) OMWL capacity of 2267 × 10^6 m^3 and (b) extraordinary maximum water level (EMWL) capacity of 2485 × 10^6 m^3. **Figure 3** shows the runoff and sediment volumes at the inlet and outlet of the reservoir, as well as the hydrometric stations (EH) of Pajaritos

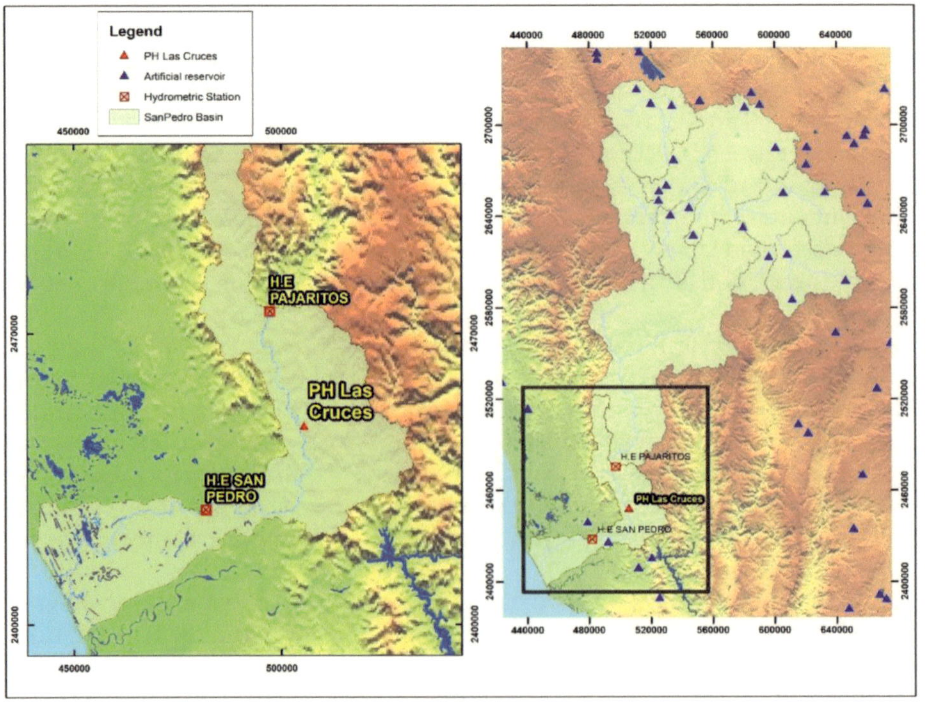

Figure 2.
General plan of the reservoir and the contribution basin to the PHLC. Source: Ref. [25].

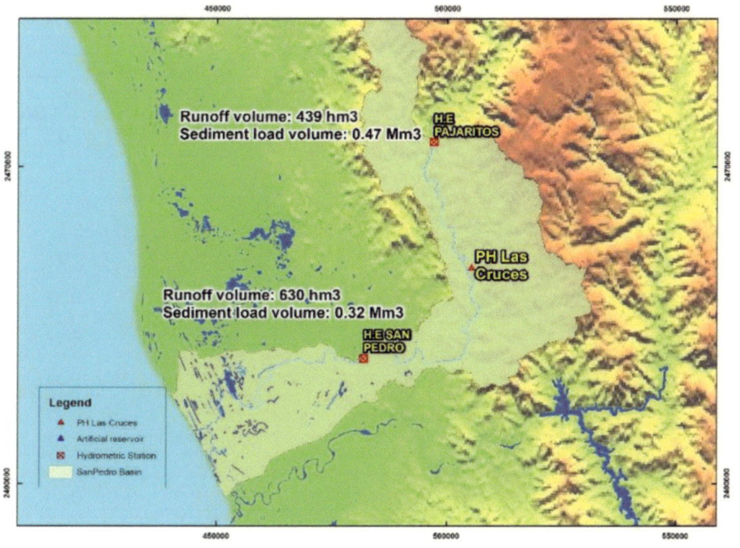

Figure 3.
General plan of the reservoir: runoff and sediment load. Source: Ref. [25].

and San Pedro. The bathymetry of the reservoir was obtained from the LIDAR information provided by [25], and the hydrometric information used to obtain the average monthly discharges was obtained from Ref. [26].

Table 1 shows the monthly average discharge values and the monthly mean concentrations of transported fine solids obtained from the monthly sediment averages. The monthly mean concentrations of fine solids were obtained from the quotient of the monthly sediment volumes divided by the runoff volume.

Annual precipitation throughout the San Pedro river basin shows high variability. However, there is a clear trend, with precipitation being greater as one move toward the outlet of the basin. In fact, the highest rainfall occurs in the MaNas area, where it reaches more than 1.5 m per year. The average annual discharge calculated at the EH San Pedro station was 87.2 $m^3 s^{-1}$, the maximum 204.7 $m^3 s^{-1}$, and the annual minimum 24.6 $m^3 s^{-1}$. According to the records, the maximum annual average discharge was presented in 2008, whereas the minimum was in 2005. The San Pedro river basin recorded 53 floods from 1944 to 2004, causing significant socioeconomic damage [26]. It was determined that the flood season occurs mainly from July to November. These floods are associated with the fact that the San Pedro river basin is subject to extraordinary meteorological events. In particular, the rainfall in this area is due to the presence of the Mexican monsoon, which causes above-average rainfall, although not always with the same intensity during the months of June, July, and August. The monsoon is produced by a break in the atmospheric circulation in the northwest that generates the release of latent heat and a rise in temperature and wind and, consequently, an increase in humidity. The humidity flow begins in the so-called intertropical zone of convergence and is accentuated by the tropical waves caused by the trade winds, which come from the central Atlantic, generating a displacement of humidity toward the Pacific coastal zone. But, also, droughts occurrence was identified for two historical periods, from 1976 to 1980 and 1995 to 2001, with 5–7 successive years of null flow, respectively. Regarding the river regime, flows between 1000 and 1400 $m^3 s^{-1}$ were observed in the case of very dry years, whereas in average years,

Month	Discharge ($m^3 s^{-1}$)	Concentration (kg monsoom^{-3})
January	4.76	0
February	2.29	0
March	0.22	0
April	0.02	0
May	0.00	0
June	8.74	0
July	78.53	0.89
August	40.91	0.47
September	44.83	0.53
October	40.478	0.27
November	67.03	1.67
December	111.34	0.44

Table 1.
Average monthly discharges and concentration data.

the accumulated flows are around 2200 $m^3 s^{-1}$, and in very wet years of a more than 4000 $m^3 s^{-1}$. If one considers a design flood with a return period (Tr) of 10,000 years, peak flows of 12,237.8 $m^3 s^{-1}$ and 12,867.4 $m^3 s^{-1}$ were obtained. When this design avenue passed through the spillway, a maximum discharge rate of 9048 $m^3 s^{-1}$ was obtained, reaching a maximum level of 2.3 m, which is less than the EMWL.

3. Method

To evaluate the impact of the operation of the PHLC in the lagoon and MaNas zones, a flash flood period for an average historical year from September 19th to October 6th, 2010 was analyzed. Two conditions were applied: without the dam using the hydrograph measured during the mentioned period, and with the dam using the turbine hydrograph plus the contributions of its own basin. **Figure 4** shows the two hydrographs considering these scenarios.

The operation of the PHLC in the lagoon and MaNas zones shows that the two hydrographs, with and without the dam are practically the same, so one can conclude that the alteration of the flow in flash floods due to the presence of the PHLC would be minimal, without any important consequence for the downstream areas. Another fundamental aspect to consider is the sedimentation sites and granulometric characteristics of the material. For that, Gracia-Sánchez et al. [27] applied physical models looking for a reliable representation of the reservoir entrance. Thus, sites with materials between 1.0 and 0.1 mm were identified for their possible extraction and, consequently, to refine or adjust the model results. With this information a numerical model was developed founding that only four sites required an adjustment (**Figure 5**), these sites are the mouths of Camichín (SEC1) and Palapares (SEC11), the Toluca Lagoon (SEC25), and the Mexcaltitlán lagoon (SEC85). However, even though the numerical model represents well the deposit of sediment at the entrance of the reservoir, there is no interpretation of the granulometric distribution that occurs in the deposit, particularly in the case of the PHLC, which is crucial to define the future transport downstream the curtain.

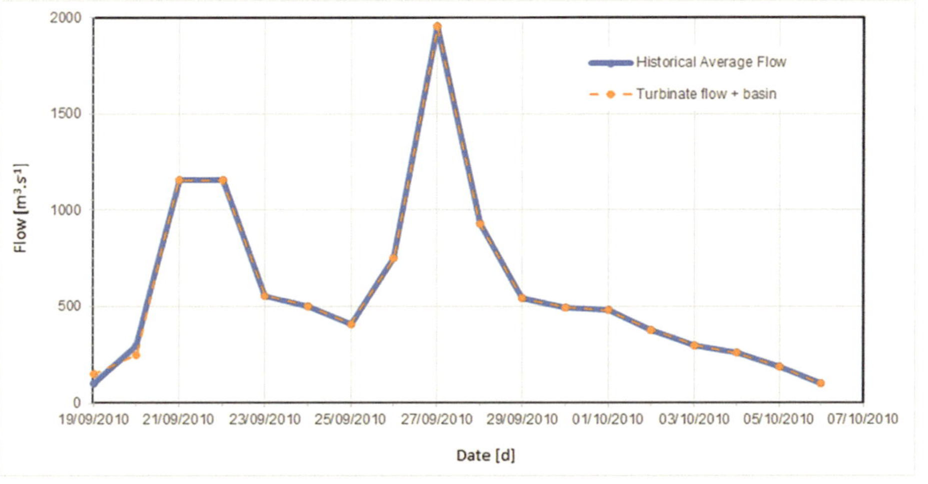

Figure 4.
Hydrographs for the simulation scenarios in the flash flood season.

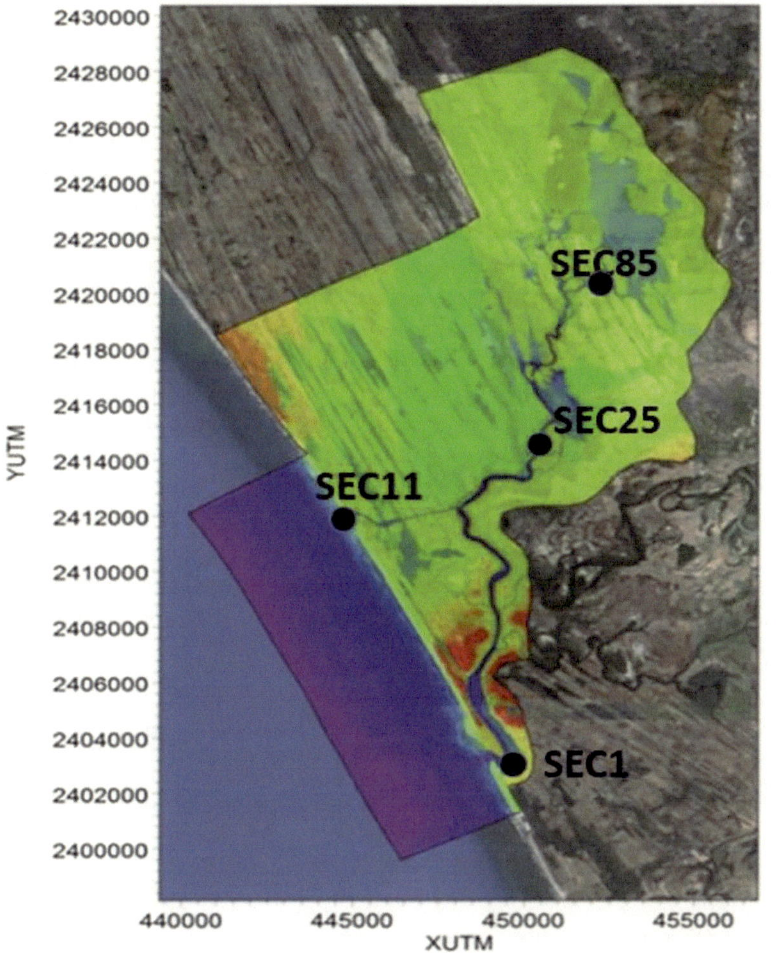

Figure 5.
Flash flood analysis sites.

4. Results and discussions

To understand the effect that the PHLC would have on the flow downstream, a simulation was carried out considering two scenarios: a historical condition and the PHLC for dry and rainy seasons.

4.1 PHLC functioning during the dry season

Table 2 presents the discharges observed in the lagoon system without and with dam during the dry season in the sections identified along the river. The negative signs in the average discharge indicate that the direction of the flow is toward the sea.

When the average flow rate is 0 $m^3 s^{-1}$, it indicates that during the analyzed period, the volume of water that entered the system is the same as the one that left. If this is different from 0 with a positive sign, it indicates that, during the studied period, the volume of water that enters is greater than the one that leaves and, on the contrary, if the sign is negative, it indicates that more water leaves than the one that enters the system. In this way, it is observed that with the operation of the PHLC, when the discharge in the San Pedro river increases, the volume of water that leaves the system toward the sea increases (**Table 3**).

Also, it was observed that when the levels in the lagoon system increase due to the operation of the PHLC during the dry season, the difference in levels between the system and the sea is modified by: (a) in high tide conditions the gradient is lower; therefore, the flow of water from the sea to the system decreases, a situation that is reflected in the reduction of the discharges that enter through the sea toward the lagoons; and (b) low tide condition, the gradient is greater; therefore, the flow of water toward the sea increases as well as in the flows that leave the system. In particular, in the Camichín mouth (SEC25), there is a major exchange of water between the sea and the lagoon system, leaving the Palapares mouth with a reduced exchange (SEC11). This water exchange defines the maximum, average, and minimum water levels at each site with respect to the two established conditions (without and with the PHLC) as shown in **Table 4**.

As highlighted in **Table 4**, in the four sites, the water level impact due to the operation of the dam is not significant. This is confirmed in **Table 5**, which shows the maximum, average, and minimum variations reached for each site.

The water level variation associated with the operation of the PHLC during the dry season showed that at the Camichín mouth there are no differences, but in the Palapares mouth and in both lagoons (Mexcaltitán and Toluca), there is an increase in the average level of the order of 1 cm. Thus, the average water level for the four sites is 0.18 m, 0.14 m, 1.0 m, and 1.6 m, respectively. In addition to the discharge and water

Discharge ($m^3 s^{-1}$)	SEC1		SEC11		SEC25		SEC85	
	Without	With	Without	With	Without	With	Without	With
Maximum	426.5	423.0	157.3	156.1	231.3	227.5	79.4	78.9
Reflux	423.0	425.4	115.4	116.5	199.6	205.1	65.1	69.8
Average	−5.1	−9.3	0.1	−1.8	−7.0	−12.2	−4.2	−8.7

Table 2.
Maximum (toward the system), reflux (toward the sea), and average discharges in the lagoon system, without and with dam during the dry season.

Discharges variation ($m^3 s^{-1}$)	SEC1	SEC11	SEC25	SEC85
Maximum flow	−3.5	−1.2	−3.8	−0.5
Reflux	2.4	1.1	5.5	4.7
Average	−4.2	−1.9	−5.2	−4.5

Table 3.
Discharges variation for the operation of the PHLC in the dry season.

Water level (m)	Tidal wave	SEC1		SEC11		SEC25		SEC85	
		Without	With	Without	With	Without	With	Without	With
Maximum	0.73	1.01	1.01	0.88	0.88	2.53	2.53	2.79	2.79
Average	−0.01	0.17	0.17	0.13	0.14	1.03	1.04	1.57	1.58
Minimum	−0.74	−0.67	−0.67	−0.63	−0.63	−0.49	−0.49	−0.10	−0.10

Table 4.
Maximum, average, and minimum water levels without and with dam during the dry season.

Water level variation (m)	Camichín Mouth	Palapares Mouth	Toluca lagoon	Mexcaltitán lagoon
Maximum	0.00	0.00	0.00	0.00
Average	0.00	0.01	0.01	0.01
Minimum	0.00	0.00	0.00	0.00

Table 5.
Variation of levels due to the operation of the P. H. Las Cruces (PHLC) during the flash flood season.

levels data, other aspects to be considered in the operation of the lagoon system are velocity and salinity.

Reviewing velocities for the four sites, one can find that the main changes were observed in both mouths of Camichín and Palapares. In the first, the average flow velocity for both scenarios (without and with a dam) is around 0.933 m s^{-1}. However, the maximum velocity was 2.097 m s^{-1} for current conditions and 2.099 m s^{-1} for the project conditions, so there would be an increase of the order of 0.002 m s^{-1}, which is not significant. At the Palapares mouth, an average velocity of 0.840 m s^{-1} for the current condition, and 0.842 m s^{-1} for the project condition were observed, which shows an increase of the order of 0.002 m s^{-1}, which is also not significant. The maximum velocity registered was 1.807 m s^{-1} for the current conditions and 1.808 m s^{-1} with a dam; thus, there is an increment due to the operation of the project of 0.001 m s^{-1}.

In 2016, the Institute of Marine Sciences and Limnology (ICMyL) of the UNAM established an initial salinity of the system and concluded that there is an important change from almost 0 in the river to close to 30 PSU in the lagoons, and then a gradual increase from the lagoon to the sea, as shown in **Figure 6**. Salinity is measured in practical salinity units (PSU) or ppt (parts per thousand).

To estimate the salinity of the system under PHLC operation, simulations were carried out considering that the San Pedro river has fresh water, while the lagoon

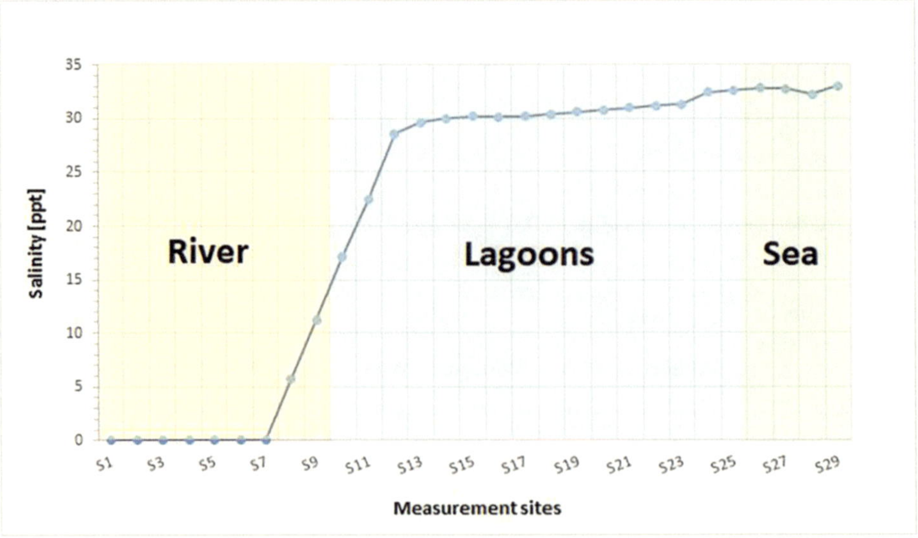

Figure 6.
Salinity measured in the system in May 2014.

system presents high salinity concentrations (29–32 ppt), and close to the sea the salinity is 33 ppt. Thus, for the dry season, the average salinity of 32.9 ppt was obtained in the mouths of Camichín and Palapares, as well as in the lagoons of Toluca and Mexcaltitán in current conditions, and with the dam, the salinity decreased by only 0.1 ppt. The latter is due to the PHLC operation, where it is expected a greater contribution of freshwater coming from the San Pedro river; thus, the salinity concentration decreases mainly in the Mexcaltitán and Toluca Lagoons, with average values of 11 and 3 ppt, respectively. Similarly, in the mouths of Camichín and Palapares, the salinity decreases on average by 0.1 and 0.8 ppt, respectively.

4.2 Lagoon behavior in the dry season and its relationship with the tides

The water level behavior within the lagoon system is mainly determined by the tidal condition and its influence reaches the Mexcaltitán lagoon, at a distance of around 25 km from the Camichín mouth. With the operation of the PHLC during the dry season, there is an increase in the level of the lagoon system, in the Mexcaltitán lagoon, the average increment is around 0.06 m, while in the Toluca lagoon is around 0.02 m. Thus, when comparing and calculating the maximum amplitude of the water levels in the sea and in each of the lagoons, it was found that in the sea the amplitude is 1.85 m, while in the Toluca lagoon is 0.95 m without and with the dam, and in the Mexcaltitán lagoon is 0.3 m under current conditions with an increment of 0.33 m with the dam. As one can observe, the amplitude of the water levels in the Toluca lagoon does not have a significant impact due to the operation of the hydroelectric project. However, in the Mexcaltitán lagoon, the amplitude of the water levels increases by around 3 cm due to the contributions of the San Pedro river during the dry season. Thus, in the Toluca lagoon, the percentage of tidal damping is around 48% for both scenarios, while in the Mexcaltitán lagoon, it is around 84% under current conditions and around 82% under hydroelectric project conditions.

In addition, velocities values do not present significant differences in the dry season considering the PHLC operation as increments of just 0.01 and 0.02 m s^{-1} were observed. Therefore, the sediment transport conditions are not substantially modified in the lagoon system. This was confirmed since the velocities are determined by the tide condition, when the maximums are reached, close to the high and low tides, the system has the capacity to put the sediment in suspension.

As the largest discharges in the entire system were at the mouth of Camichín, it is the site where the main exchange of fresh water with the sea occurred. The maximum flow entering the system under current conditions was calculated at around 427 m^3 s^{-1}, whereas the one leaving the system was 423 m^3 s^{-1}. With the operation of the hydroelectric project, a reduction of the maximum flow that enters the system through the sea of the order of 4 m^3 s^{-1} was observed; however, the flow that leaves increased by the order of 2 m^3 s^{-1}. On the contrary, at the mouth of Palapares, under current conditions, the maximum flow that entered the system through the sea was 157 m^3 s^{-1} and the one leaving was 115 m^3 s^{-1}. With the operation of the project, there was a decrease in the flow that enters of around 1 m^3 s^{-1}, while the flow that leaves increased in the order of 1 m^3 s^{-1}.

4.3 PHLC operation during the rainy season

Applying numerical modeling during the period of flash floods, the PHLC operation does not have significant modifications in the hydrodynamics and salinity of the lagoon system. Thus, the water level behavior in the Mexcaltitán and Toluca lagoons, in both conditions (without and with the project), is mainly determined by the contributions of the San Pedro river. Unlike the mouths of Camichín and Palapares, where their behavior is mainly determined by the influence of the tide, with an increase in its average level due to the contributions of the river. In general, the water level in the lagoon system for the flash flood season, is above the mean sea level, of the order of 18 cm at the Camichín mouth, 14 cm at the Palapares mouth, and 1 and 1.6 m at the Toluca and Mexcaltitán lagoons, respectively. Thus, if the project comes into operation, it would be expected an average increase in the water level of the order of 1 cm in the Camichín and Palapares mouths. Whereas, the maximum water levels in the Mexcaltitán and Toluca would reach around 2.8 and 2.5 m, respectively.

Regarding the flow velocity, this increase considerably throughout the lagoon system in the rainy season. In the dry season, the velocities analyzed never were above 1 m s^{-1}, while in the flood season, the speeds were close to 2 m s^{-1} at the mouths of Camichín and Palapares. On the contrary, the simulation with the operation of the PHLC showed that the speed does not present significant differences with respect to the current conditions calculating increments between 0.001 and 0.002 m s^{-1}.

Looking at the discharge values during the rainy season, and for most of the simulation time, the discharge moves out through the mouths toward the sea and exceeds 1000 m^3 s^{-1}, which corresponds to the peak of the flood. With the operation of the project, there is an average increment in the system discharges of 1 m^3 s^{-1}. In this way, it is considered that during flash floods, practically a constant gradient of the lagoon system toward the sea is established.

Undoubtedly, the greatest change observed in the rainy season is in the salinity concentration of the entire system. This is due to the contribution of freshwater from the San Pedro river at that time that reaches 0 PSU in a short period (2 days) and this behavior is not modified with the operation of the PHLC.

Also, the natural conditions of flooding analysis were assessed (without ribs), and the results showed that there is no significant change because the disturbance in the average annual maximum discharge is not representative. In fact, it is expected a possible attenuation of floods during the rainy season, which is more related to the existence of the reservoir than to its operating policy. This is because, unlike the minimum annual discharges, which are very low in the EH San Pedro, any change in the operation of the reservoir would alter the flooding conditions. This means that only by reducing the operating volume to the desired minimum, the minimum average annual discharge frequency curves (ecological expenditure) could be preserved. Thus, the most important aspect is to establish operating policies that can meet the generation needs and that these policies are viable for the San Pedro river ecosystem conservation. Another determining factor observed was that flood volumes and times are related to sediment dynamics. Thus, larger floods stir up sediment deposition, while shorter flood times alter the rate of sediment deposition.

4.4 Coastal wetlands reduction

The coastal marshes of the San Pedro river own a complex hydrological connectivity due to the existing interaction between freshwater and marine water, which affects largely the spatial distribution of salinity and vegetation in this region. The greatest changes observed in the Río San Pedro-MaNas ecosystem occurred in the alteration of the land cover and use, generating modifications of the hydrology and the local landscape, as well as altering the flood patterns responsible for the functional dynamics of the system, the maintenance, productivity and services that these ecosystems naturally provide to societies. Currently, the construction of any infrastructure can be seen as disastrous as it would profoundly modify the physiography of the entire basin, but at the same time dams, reservoirs and canals are an aid that would help to distribute hydrological resources in a constant manner and control possible flood events. Although the presence of a dam in the middle basin of the San Pedro-Mezquital river would generate initial changes, after the adjustment period, the system would be expected to work supportive of the MaNas conservation by reinforcing the marine influence in the area. This, in turn, would lead to a re-elaboration of the sediments of the coast, which may favor an effective wave that transports sand to the most deprived areas.

The fine sediment constitutes one of the most important aspects in the continuity of the sediment as it plays a determining role in the zone of the lagoons and mangroves for its sustainability. So far, what can be obtained from the work carried out is that the fine sediment will not be significantly retained in the PHLC reservoir. Rather it will re-distribute over time, that is, it will continue passing the same fine sediment but distributed differently over time. In fact, the first results indicate that the amount of fine sediment that would pass through the PHLC reservoir is not significant with respect to the total that reaches the mangrove area since the contribution of fine sediment from the San Pedro river is 0.28×10^6 m^3, while the amount contributed downstream of the EH San Pedro is 1.62×10^6 m^3 and this reaches 2.19×10^6 m^3 at the outlet to the sea. Therefore, it can be stated that in the hypothetical case, if all the fine material is retained in the PHLC reservoir, only 10% of the material that reaches the sea would be modified, but the reservoir, as mentioned, cannot retain this type of sediment, even if the dam acts as a blocking of the transported sediments to avoid that a lack of sediments downstream could become critical, maintaining the floodplain ecosystem. Auel et al. [28] indicated that it is possible to achieve sediment management to maintain the sediments downstream. Some techniques used for this purpose are flushing, sluicing, and bypassing [29, 30]. However, in the simulations carried out by [27], the main deposit of the thick material occurs in the tail of the vessel, for which its extraction must be done through a mechanical mechanism. In theory, the sediment can be moved to the curtain by emptying the reservoir several times, since it is a cannoned channel, that is, it is narrow and the water can reach high speeds for sediment transport. Therefore, if the coarse sediment is really needed downstream of the curtain, an option would be to use some mechanical methods such as dredging and transporting the material to the sites where the coarse sediment is needed.

The operation of the PHLC in the lagoon and MaNas zones shows that flash floods would be minimal, without any important consequence for the downstream areas. Even looking at the velocity results, the operation of the PHLC would not significantly modify the exchange conditions and water contributions to the lagoon system during a flash flood event. However, it is important to mention that with the operation of the PHLC, discharges on the San Pedro river would increase the volume of water that leaves the system toward the sea. Another important result of the operation of the PHLC is that the concentration of salinity would decrease due to the increase in the contributions of freshwater (constant flow) from the San Pedro river during the dry season.

It is during the dry season, when the water levels in the lagoon system increase as a result of the PHLC operation, the difference in levels between the lagoon system and the sea is modified according to:

- In high tide conditions, the gradient is lower; therefore, the flow of seawater into the system decreases. This situation is reflected in the reduction of the flow that enters through the sea toward the lagoons.

- In low tide conditions, the gradient is greater; therefore, the flow of freshwater toward the sea increases, and it is reflected in the increase in the flow that leaves the system.

Twilley and Brinson [31] observed that rising sea level is the main force affecting coastal wetlands, basically because there is no possible migration inshore of them. Migration takes place when the reduction of freshwater flow upstream accelerates salinity intrusion inland but also if the landscape has the possibility to allow mangroves to colonize inland marsh habitats. However, this kind of mangrove migration is almost impractical because of urbanization, tourism, agriculture, and aquaculture development in the zone. Thus, there is a significant reduction of land for this kind of refuge area in many coastal zones, such as the MaNas case.

5. Conclusions

The main problem with the construction of the dam on the San Pedro-Mezquital river would be the effect of a controlled flow that reaches the wetlands of the alluvial plains, affecting the estuarine and coastal ecology. However, it was observed that the behavior of the flood is the same; thus, the PHLC project does not change the flood conditions of the plain.

Also, the fine material will not represent any continuity problem and will only travel downstream, with a somewhat different distribution in time than what currently occurs, but it will be able to reach the mangrove area without any problem. Opposite to the coarse material that will effectively be retained in the reservoir and, specifically, at the reservoir entrance 60 km from the curtain. However, the amounts of sediment that are calculated in the range from 1.0 to 0.1 mm are of the order of 100,000 m^3 per year on average and this is only a small part of all the sediment that will be stored at the reservoir entrance.

Finally, an ecological cost should be assessed considering the contributions of the base flow (underground) since this can substantially alter the ecosystem. Especially, during the dry season, when it would be a higher discharge for these ecosystems. However, the ecosystems from these regions have already evolved to adapt to the dry season, which is necessary for their survival. Thus, the most probable loss of these marshlands is associated mainly with human activities rather than the hydrological behavior of the system.

Acknowledgements

This work is financially supported by the "Programa de Apoyo a Proyectos de Investigación e Innovación Tecnológica (PAPIIT)," UNAM. Project grant num. IN106818 (2018-2019).

Acronyms and abbreviations

CFE	Federal Electricity Commission
DR	irrigation districts
EH	hydrometric stations
EMWL	extraordinary maximum water level
ET	evapotranspiration
ICMyL	Institute of Marine Sciences and Limnology
LIDAR	light detection and ranging
MaNas	Marismas Nacionales
OMWL	ordinary maximum water level
PHLC	Las Cruces Hydroelectric Project
ppt	parts per thousand
PSU	practical salinity unit
SEC	bathymetry sections
U.S. EPA	U.S. Environmental Protection Agency

Author details

Jesús Gracia-Sánchez, Judith G. Ramos*, Liliana Marrufo Vázquez,
Javier Osnaya Romero and Víctor M. Ortíz-Martínez
Hydraulic Coordination, Instituto de Ingeniería, Universidad Nacional Autónoma de México, Ciudad de México, Mexico

*Address all correspondence to: jramosh@iingen.unam.mx

References

[1] Chase TN, Pielke RA Sr, Kittel TGF, Nemani RR, Running SW. Simulated impacts of historical land cover changes on global climate in northern winter. Climate Dynamics. 2000;**6**:93-105. DOI: https://link.springer.com/article/10.1007/s003820050007

[2] Garg V, Nikama BR, Thakur PK, Aggarwal SP, Gupta PK, Srivastav SK. Human-induced land use land cover change and its impact on hydrology. HydroResearch. 2019;**1**:48-56. DOI: 10.1016/j.hydres.2019.06.001

[3] Kusumastuti DI. The effects of threshold nonlinearities on the transformation of rainfall to runoff to flood in a lake dominated catchment system [thesis]. Perth: University of Western Australia; 2006

[4] Maskey ML, Facincani Dourado G, Rallings AM, Rheinheimer DE, Medellín-Azuara J, Viers JH. Assessing hydrological alteration caused by climate change and reservoir operations in the San Joaquin River basin, California. Frontiers in Environmental Science. 2022;**10**:765426. DOI: 10.3389/fenvs.2022.765426

[5] Grantham TE, Viers JH, Moyle PB. Systematic screening of dams for environmental flow assessment and implementation. Bioscience. 2014;**64**:1006-1018. DOI: 10.1093/biosci/biu159

[6] Kingsford RT. Ecological impacts of dams, water diversions and river management on floodplain wetlands in Australia. Austral Ecology. 2000;**25**:109-127. DOI: http://citeseerx.ist.psu.edu/viewdoc/download?DOI=10.1.1.460.5957&rep=rep1&type =pdf

[7] Flores-Verdugo FJ, Agraz-Hernández CM, Benitez-Pardo D. Ecosistemas acuáticos costeros: importancia, retos y prioridades para su conservación. In: Sánchez Ó, Herzig M, Peters E, Márquez-Huitzil R, Zambrano L, editors. Perspectivas sobre conservación de ecosistemas acuáticos en México. Instituto Nacional de Ecología. Mexico: SEMARNAT; 2007. p. 97. ISBN: 978-968-817-856-0

[8] Lugo AE, Snedaker SC. The ecology of mangroves. Annual Review of Ecology and Systematics. 1974;**5**:39-64. DOI: 10.1146/annurev.es.05.110174.000351

[9] Stoddard J, Larsen DP, Hawkins CP, Johnson RK, Norris RH. Setting expectations for the ecological condition of streams: The concept of reference condition. Ecological Applications. 2006;**16**(4):1267-1276. DOI: 10.1890/1051-0761(2006)016[1267:SEFTEC]2.0.CO;2

[10] Pachauri RK, Reisinger A. Climate Change 2007. Synthesis Report. Contribution of Working Groups I, II and III to the Fourth Assessment Report. Switzerland; 2018. 104 p. Available from: http://www.ipcc.ch/ipccreports/ar4-syr.htm. [Accessed: 05 May 2020]

[11] Poff NL, Zimmerman JKH. Ecological responses to altered flow regimes: A literature review to inform the science and management of environmental flows. Freshwater Biology. 2010;**55**:194-205. DOI: 10.1111/j.1365-2427.2009.02272.x

[12] Duan L, Cai T. Changes in magnitude and timing of high flows in large rain-dominated watersheds in the cold region of north-eastern China.

Water. 2018;**10**(11):1658. DOI: 10.3390/w10111658

[13] Hidalgo HG, Das T, Dettinger MD, Cayan DR, Pierce DW, Barnett TP, et al. Detection and attribution of streamflow timing changes to climate change in the Western United States. Journal of Climate. 2009;**22**:3838-3855. DOI: 10.1175/2009jcli2470.1

[14] Allamano P, Claps P, Laio F. Global warming increases flood risk in mountainous areas. Geophysical Research Letters. 2009;**36**:392-395. DOI: 10.1029/2009GL041395

[15] Hamlet AF, Lettenmaier DP. Effects of 20th century warming and climate variability on flood risk in the Western U.S. Water Resources Research. 2007;**43**:W06427. DOI: 10.1029/2006WR005099

[16] Leibowitz SG, Wigington PJ Jr, Schofield KA, Alexander LC, Vanderhoof MK, Golden HE. Connectivity of streams and wetlands to downstream waters: An integrated systems framework. American Water Resources Association. 2018;**54**(2):298-322. DOI: 10.1111/1752-1688.12631

[17] US EPA. Connectivity of Streams and Wetlands to Downstream Waters: A Review and Synthesis of the Scientific Evidence (Final Report). Washington, DC, EPA/600/R-14/475F: U.S. Environmental Protection Agency; 2015. DOI: https://cfpub.epa.gov/ncea/risk/recordisplay.cfm?deid=296414

[18] Li Y, Xu J, Wright A, Qiu C, Wang C, Liu H. Integrating two aspects analysis of hydrological connectivity based on structure and process to support muddy coastal restoration. Ecological Indicators. 2021;**133**:108416. DOI: 10.1016/j.ecolind.2021.108416

[19] Park E, Latrubesse EM. The hydro-geomorphologic complexity of the lower Amazon River floodplain and hydrological connectivity assessed by remote sensing and field control. Remote Sensing of Environment. 2017;**198**:321-332. DOI: 10.1016/j.rse.2017.06.021

[20] Freeman MC, Pringle CM. An Jackson CR: Hydrologic connectivity and the contribution of stream headwaters to ecological integrity at regional scales. JAWRA. 2007;**43**(1):5-14. DOI: 10.1111/j.1752-1688.2007.00002.x

[21] Luo M, Wang Q, Qiu D, Shi W, Ning Z, Cai Y, et al. Characteristics of hydrological connectivity and ecological effects of a typical tidal channel system in the Yellow River Delta. Journal of Beijing Normal University (Natural Science). 2018;**54**(1):17-24. DOI: 10.16360/j.cnki.jbnuns.2018.01.003

[22] Valdez-Hernández JI, Acosta-Velázquez J, Ruiz-Luna A, Solís-Venegas JA, Rocha-González V. Criterios para la selección del sitio de manglar Teacapán - Agua Brava – Marismas Nacionales [Internet]. Comisión Nacional para el Conocimiento y Uso de la Biodiversidad (CONABIO). Sitios de manglar con relevancia biológica y con necesidades de rehabilitación ecológica. México: CONABIO; 2009. Available from: http://www.conabio.gob.mx/conocimiento/manglares/doctos/caracterizacion/PN10_Teacapan_Agua_Brava_Marismas_Nacionales_caracterizacion.pdf. [Accessed: 23 March 2018]

[23] Bojórquez I, Nájea O, Hernández A, Flores F, González A, García D, et al. Particularidades de formación y principales suelos de la llanura costera norte del estado de Nayarit, México. Cultivos Tropicales. 2006;**27**(4):19-26. DOI: https://www.redalyc.org/pdf/1932/193215912003.pdf

[24] Blanco y Correa M, Flores Verdugo F, Ortiz Pérez MA, De la Lanza G, López Portillo J, et al. Diagnóstico Funcional de Marismas Nacionales. Final Report of the Coordination Agreements between the Autonomous University of Nayarit and the National Forestry Commission sponsored by the Government of the United Kingdom, Tepic, Nayarit. 2011190 p, 84 maps p 1 DVD. ISBN 978-607-7868-35-4

[25] CFE: PROYECTO HIDROELÉCTRICO LAS CRUCES, Capítulo II, Descripción de las obras, Manifestación de impacto ambiental, CFE. Ciudad de México: Federal Electricity Commission (CFE); 2014

[26] Domínguez R: Complementación Hidrológica, de Sedimentos en Suspensión y de Políticas de Operación para Extracción de Sedimentos del Embalse del P H Las Cruces, Nay, Prepared by the Instituto de Ingeniería, UNAM, to the Federal Electricity Commission (CFE). Ciudad de México: 2015

[27] Gracia-Sánchez J, Luna-B JC, Osnaya-R J, Ortíz-M V, Carrizosa-E E, Franco V. Estudios complementarios en materia de políticas de operación, diagnóstico de zonas costeras, transporte de sedimentos y manejo de cuenca, del proyecto hidroeléctrico Las Cruces, Nayarit (2da etapa), (Informe Final Modelos Físicos). Ciudad de México, México: Instituto de Ingeniería, UNAM to the Federal Electricity Commission (CFE); 2016

[28] Auel C, Kobayashi S, Sumi T, Takemon Y. Effects of sediment bypass tunnels on sediment grain size distribution and benthic habitats. In: Wieprecht S, Haun S, Weber K, Noack M, Terheiden K, editors. Proceedings of the 13th International Symposium on River Sedimentation; Stuttgart, Germany. 1st ed. London: CRC Press; 19-22 September, 2016. pp. 825-832. DOI: 10.1201/9781315623207

[29] Kondolf GM, Rubin ZK, Minear JT. Dams on the Mekong: Cumulative sediment starvation. Water Resources Research. 2014;**50**:5158-5169. DOI: 10.1002/2013WR01465

[30] Morris GL, Fan J. Reservoir Sedimentation Handbook. New York: McGraw-Hill Book Co.; 1998. 13.1-13.25 p. ISBN: 0-07-043302-X

[31] Twilley RR, Brinson MM. Consequences for wetlands of a changing global environment. In: Batzer DP, Sharitz RR, editors. Ecology of Freshwater and Estuarine Wetlands. 2nd ed. Oakland, California: University of California Express; 2014. pp. 261-286. ISBN: 978-0-520-27858-5

Chapter 13

Risk Reduction

Roberto Richardson Varas

Abstract

The existing high alluvial risk Basin Lluta "Chile," due to the presence of large mountains subject to inclement weather, coupled with the growing urban pressure to use sectors downstream increasingly exposed to this type of event, justify studies of the behavior of debris flows and the ability to mitigate its effects by the construction of protection and control. For this purpose, made the hidrogeomorfológica map for this basin. The surface of this basin is 3447 km^2 and is in a desert area. It is characterized by low rainfall and therefore all its soils are devoid of vegetation except for the low part of the valley where the land use is agricultural. The natural regime has permanent basin runoff to the sea all year round with an average flow of 0.45–1.44 m^3seg. The impact has increased the flow, but homogeneity is not presented with the full extent of the basin, showing notable spatial differences in the level of danger she originates. Different from what was described flood of 2001, which was reflected in the lower basin Lluta where road infrastructure was badly damaged by the effects of scouring and landslides, resulting in this reconstruction of road and railway bridges, implying a strong investment in works of river bank protection upstream of the bridges.

Keywords: alluvia risk, debris flow, runoff, water balance, hydrographic basin, infiltration, flooding, mitigation, runoff, discharge, suspended sediment, hydrogeomorphological chart

1. Introduction

The high alluvial risk existing in the Lluta basin due to the presence of large mountains subject to inclement weather, added to the growing urban pressure to use sectors of the lower basin that are increasingly exposed to this type of event, justify studies on the behavior of debris flows, and the possibility of mitigating their effects through the construction of protection and control works. For this purpose, the hydrogeomorphological chart for this basin has been prepared, being a management support tool, the main objective of this work.

The hydrographic basin of the river Lluta and its valley of the same name is located in the XV Region of Chile, comprised between parallels 18° and 18°30' South Latitude and meridians 70°20 and 69°22' West Longitude, in the Provinces of Arica and Parinacota, whose main river flows into the Pacific Ocean in the coastal sector called Chacalluta Valley.

This basin, in a natural regime, presents permanent surface runoff to the sea throughout the year with an average flow of 0.45m^3sec. at 1.44m^3sec., the Lluta river has carved out a quite narrow and deep valley, which is limited by quite steep and high slopes. The high alluvial risk existing in this basin due to the presence of large mountains subject to inclement weather, added to the growing urban pressure to use sectors of the lower basin, justify studies on the behavior of debris flows, and the possibility of mitigating their effects through the construction of protection and control works. The impact of the increase in flow does not occur homogeneously throughout the basin, showing notorious spatial differences in terms of the level of danger that it causes. Totally different from what was described as the 2001 flood, which manifested itself in the lower Lluta basin, where the road infrastructure suffered great damage due to undermining and landslides, resulting in the reconstruction of road and railway bridges, implying a strong investment in fluvial protection works on the banks upstream of the bridges, that is, fluvial defense works were carried out in specific areas.

2. Application

There are numerous works that show the utility or value of the hydrogeomorphological chart in the treatment or study of water basins and the planning of defenses against unexpected floods. In Ref. [1, 2], in their document, "An index for the hydrogeomorphological assessment of fluvial systems" points out that fluvial dynamics is the key not only to the functioning but also to the ecological, landscape, and environmental value of fluvial systems. If you want to conserve a river as an ecosystem and as an environmental corridor in the territory, first of all, its hydrogeomorphological dynamics must be protected, because this is what will guarantee the protection of each and every one of the elements of the system and its relationships. For this reason, the evaluation of the hydrogeomorphological functioning of river systems is essential to determine their ecological status, as well as their foreseeable trends, for which three evaluation parameters are considered:

- functional quality of the river system, including
    i. naturalness of the flow regime,
    ii. availability and mobility of sediments, and
    iii. functionality of the floodplain;
- channel quality, including
    i. naturalness of the layout and morphology in the plan,
    ii. continuity and naturalness of the bed and longitudinal and vertical processes, and
    iii. naturalness of the margins and lateral mobility;
- riverbank quality, including
    i. longitudinal continuity,
    ii. width, structure, naturalness, and
    iii. transversal interconnectivity.

DOI: http://dx.doi.org/10.5772/intechopen.109404

Hydrogeomorphological cartography, as shown in **Figure 1**, is a document with particular characteristics aimed at being a tool that serves to mitigate water risks with the purpose of supporting the results of the application of a set of measures aimed at reducing risk and eliminating vulnerability, that is, physical, social, and economic. Thus, mitigation also constitutes one of the most important activities, since it allows actions to be carried out in advance, with the purpose of significantly reducing the expected consequences of an event. This stage is the most efficient and economical in terms of investment of resources and social cost, and of infrastructure and the environment.

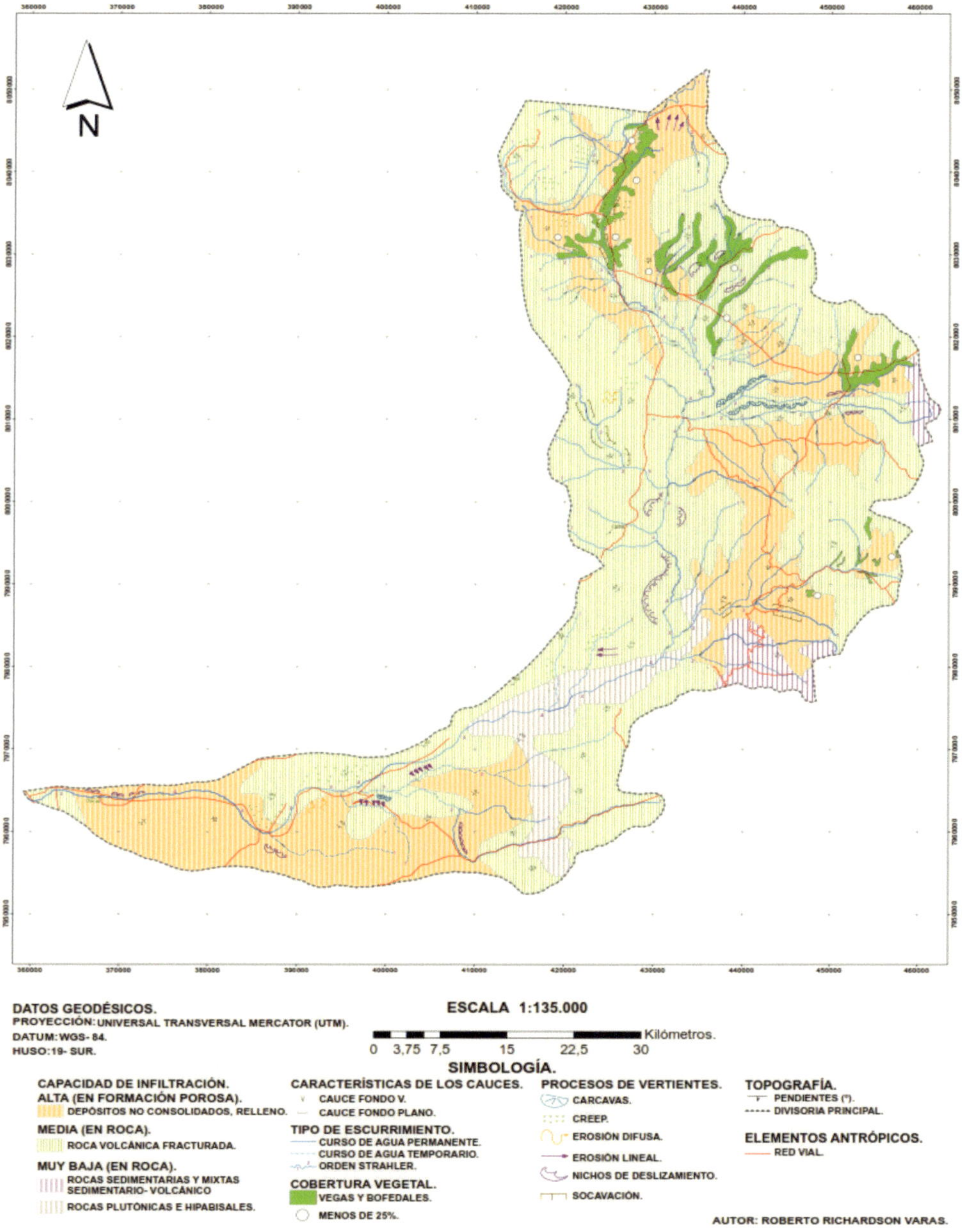

Figure 1.
Hydrogeomorphological chart of the Lluta river, Arica and Parinacota region, Lluta river basin slope map.

3. Description of the upper section of the hydro geomorphological chart of the Lluta River

Each number of this section described in **Figure 2** corresponds to the following:

1. Linear erosion

2. Creep

3. Medium infiltration capacity (fractured volcanic rock)

4. Permanent watercourse

5. Vegetables and bofedales

6. High infiltration ability (unconsolidated deposits-fill)

7. Order of strahler

8. Undercutting

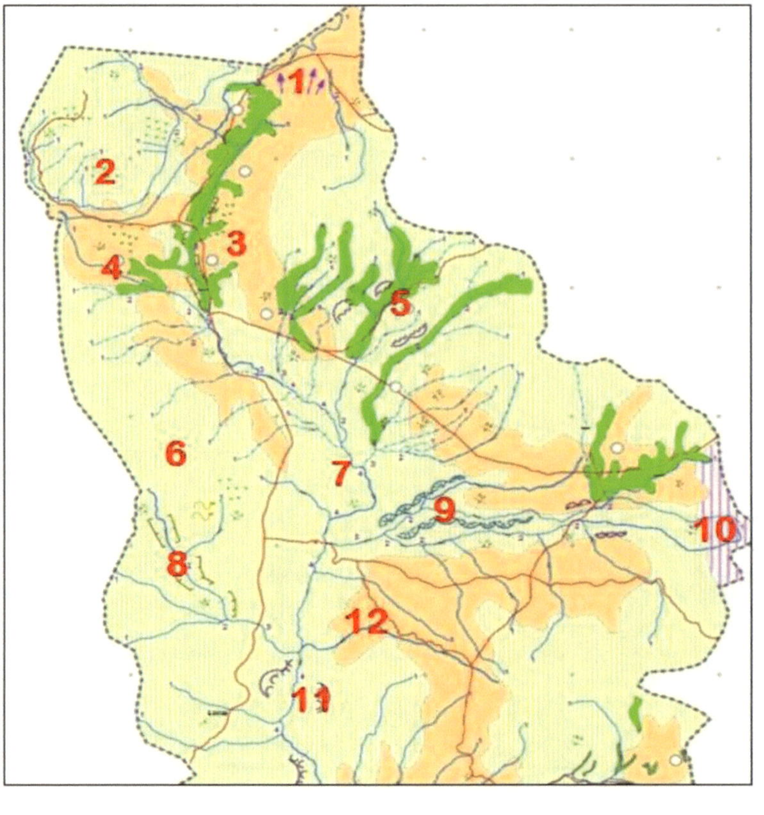

Figure 2.
Representation of the hydrogeomorphological chart of the Lluta River (high section).

9. Gullies

10. Very low infiltration capacity (sedimentary and mixed rocks)

11. Landslide niches

12. Slop

In the upper section, there are vegetation and bofedales, as shown in **Figure 3**, which in the same way are not enough so that the erosion processes are not activated. As shown in **Figures 2** and **4**, there is a permanent water course, which erodes material in this section, which is partly a consequence of its slope and there are arms around the permanent watercourse that reach the fourth order of Strahler in Ref. [3].

Figure 3.
Boreal. Vegetal behavior near the Lluta river, which is partly the one found mainly in the upper reaches of the Lluta basin, as shown by the red circle.

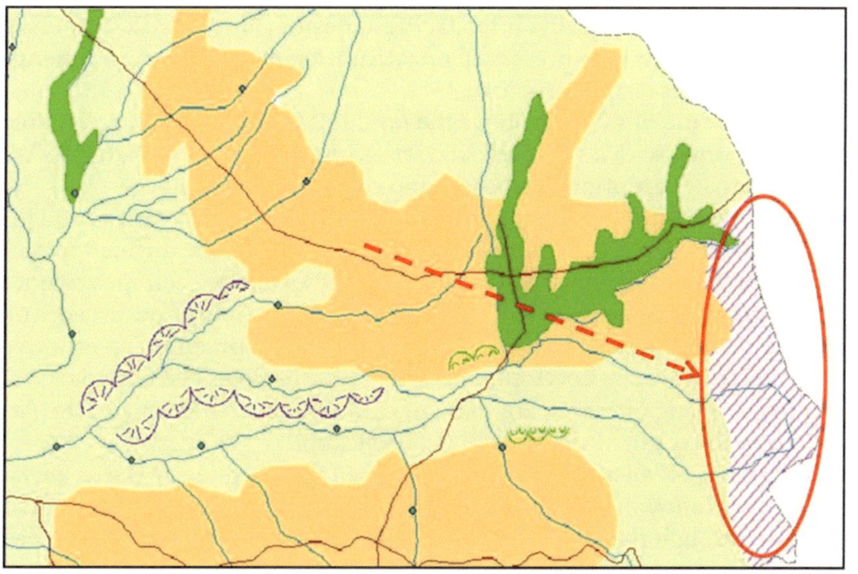

Figure 4.
Representación de arrastre de rocas sedimentarias y mixtas en el flanco SE respecto al cauce principal del Río Lluta.

The river has a high infiltration ability (unconsolidated deposits-fill) far from the river's flood bed. On the right flank of the Lluta river in this section, there is a part of land with a very low infiltration capacity (sedimentary and mixed rocks), since it corresponds to the deposit of material from the arms that reach up to the fourth order of Strahler. The river in its highest section presents in its flood bed a medium infiltration ability (fractured volcanic rock).

The upper basin is a complex climatic zone, characterized by intense cold and strong winds; which is added to the considerable height of the location and the great geographical distances that separate it from the main urban centers; Likewise, the geology of this study area has been determined in recent times, mainly by the predominant presence of units associated with the Tarapacá Volcanic Complex, whose products have their genesis in the volcanic events that occurred from the Pliocene to the Holocene, presented as stages evolutionary of such a complex. The eruptive events included processes of successive collapse of the volcanic building, pyroclastic flows, and lahars, among others in Ref. [4].

The type of permeability is primary in porous formation, and there are unconsolidated deposits, including fluvial, glacial, alluvial, lacustrine, alluvial, eolian sediments, aquifers of variable extension, generally stratified, free or semi-confined aquifers, variable permeability, and variable chemical quality.... Likewise, combined deposits or sedimentary rocks, such as sandstones, lutites, siltstones, and claystones from lacustrine or marine deposits, aquifers of variable extension.

Secondary permeability is also present in fractured volcanic rocks, consisting of lava flows, tuffs, and andesite breccias with intercalations of continental clastic sediment, rhyolitic and dacitic ignimbrite, poorly explored aquifers, of little-known extent and importance, attributable to aquifer characteristics of these formations, volcanoes of the Altiplano.

In the same way, permeability can be seen in an extremely low to absent condition in sedimentary and mixed rocks, sedimentary-volcanic rocks, lava flows, tuffs, gaps, ignimbrite with intercalations, shale, limestone, sandstone, and conglomerates; in general impermeable, they are considered basement of the aquifer fillings. Also, in plutonic rocks, such as granitic intrusive, impermeable basements, likewise, in volcanic rocks, lava flows and rhyolitic, dacitic andesitic, and basaltic pyroclastic deposits, associated with well-preserved or active ancient volcanoes, in general, they do not present aquifers characteristics.

The Tacora volcano of 6000 m above the nm., it is cataloged as an active volcano of the fumarolic type, with a stratovolcano structure, located 17°43'S; 69°46'W. As shown in **Figure 5**, an eruption of this volcano could generate lahars.

Lahar phenomena stand for great danger mainly because they are natural phenomena of high energy, considerable destructive power, and are difficult to control. However, they are relatively easy phenomena to predict due to their movement in the direction that the valleys follow. The prediction of the time of occurrence of this type of phenomenon is not trivial, but normally the development of an eruptive event can be the antecedent to the development of this type of flow. The material could go to two main existing ones that are the Caracarani river to the east and the Azufre "Sulfer" river to the south of the slopes of the volcano.

Lahars are flows of volcanic materials generated when meteoric water, surface runoff, and partial snowmelt mixes with these materials and moves, transporting them en masse through the ravines and designated river channels that have their headwaters in the slopes of the volcano.

The Azufre "Sulfur" river found in the highlands, east, and south sector of the Tacora volcano, is close to the Peruvian border and its name is due to the high content

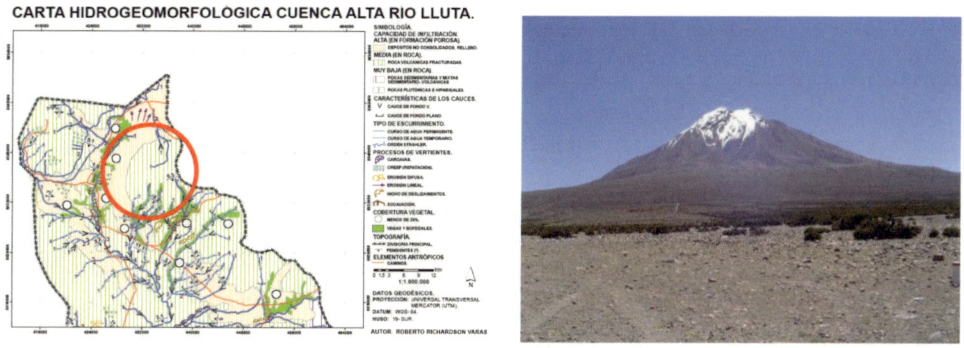

Figure 5.
Tacora volcano (source, photograph by Felipe Aguilera, University of Atacama, Chile). Main source of medium infiltration ability (fractured volcanic rock), as shown in red circle on the hydrogeomorphological chart.

Figure 6.
Photo of the slope of the Tacora volcano showing a fumarole. (Source, photo by Felipe Aguilera, University of Atacama, Chile). Likewise, the cause of unconsolidated deposits infill, which is dragged by the sulfur river, as reflected in the hydrogeomorphological chart.

of sulfur. It is a tributary of the Lluta river, as shown in **Figure 6**. In the upper areas of this river, there are volcanic rocks from the Miocene period.

The Azufre "Sulfur" river is formed from the meeting of several thermal springs, of which the Tacora stream is the main one, at the southwestern foot of the Tacora volcano. Its natural channel runs approximately 20 km in a SE direction.

The Asofre "Sulfur" River is currently diverted through an artificial canal to some evaporation ponds built on the Titire pampa, to avoid contamination of the Lluta. In the natural regime, the gauging practiced yield values of 30 l/s to 50 l/s as representative of the flow of the river as shown in **Figures 6** and 7.

For this area, it is proposed to install buried tubes in Ref. [5], (Japanese defense type slit dam, and others) in the area of greatest probability of runoff into these channels, thus the lahar loses speed, dissipates energy, and delays its arrival at the channel, although only due to a sudden change in slope when reaching the base level of the flat areas on which the volcanic buildings rise, it stops, as long as the slope is gentle.

Continuing with the work of the hydrogeomorphological chart, the Caracarani river is analyzed, it originates at the foot of the Laguna Blanca pass and receives contributions almost from the borderline with Peru, from the eastern slope of the Tacora volcano 5966 m, from the new snow-capped from Chupiquiña 5787 m, and from the western slope of Cerro Caracarani 5190 m.

Figure 7.
Inlet in the Azufre river, to divert the polluted waters toward the "parapets" Humapalca locality, commune of General Lagos, Parinacota Province, Chile. As the hydrogeomorphological chart shows, with the representation of a red arrow, the Asufre river was one of the main tributaries of the Lluta, and regardless of its diversion, the old materials deposited form part of the hydrogeomorphological chart of the Lluta.

Figure 8.
View of the "parapets" of the Azufre River, from the channel that diverts the waters of the river to these ponds or pools. Town of Humapalca, commune of General Lagos. (source, Elias Munoz) As the hydrogeomorphological chart shows, with the representation of a red arrow, the Sulfur River was one of the main tributaries of the Lluta, and regardless of its diversion, the old materials deposited form part of the hydrogeomorphological chart of the Lluta.

Due to the characteristics of the basin, the Caracarani River is considered the main tributary, from its source to Humapalca, as shown In **Figure 8**. It has a category 3 with low slope and on the sides to the course scour and creep processes can be seen, bearing in mind that pluviometric events, such as the one that occurred in the year 2001 (which have the probability of being generated possibly every 30 years), where the water fell during a period of 24 hours reached 20 mm. It may be of similar characteristics where the water fell on the entire surface of the basin in unison. However, for the conditions described and analyzed in the development of these points, it is pointed out that alluvial events can be generated by surface runoff. Likewise, it is necessary to increase the flow of the channels. For this, it is proposed to apply dams of gabion structures in this segment of the channel, whose purpose is to cushion the impact of the water flow in changes of slope and on the banks of watercourses. Thus, reducing its speed and filtering the sediments, that is, it allows regulating medium and large water flows, due to its great resistance in periods of flooding.

The consumption of the Caracarani is very variable and follows the changes in ambient temperature, with an average of 400 liters/sec. The variations in the capacities show flows from 255 to 640 liter/sec.

The proposed application for this section is a dike or a sink, for its application, it must be excavated until an impermeable layer is found and if this is not found, the soil must be leveled and firmly compacted. The first row of gabions must be buried in the bed, then a second row is arranged on top of the first, although only at the ends, which make up the spillway toward the center. This run should be placed 25 cm further upstream than the first to increase resistance to the impacts of the floodwater flow in Ref. [6]. With this, it is possible to regulate water flow due to its great resistance in periods of flooding.

Likewise, in this channel, a protection system must be provided in curved sections due to the fact that the undermining in curves causes greater depths and speeds in the inner part of the same and, as a consequence, that the fluvial courses move laterally. This form of undermining is of great importance due to the serious damage it causes. The most usual forms of direct protection of the outer edge of a curve are groynes and marginal revetments or protections. With both solutions, the high-speed current lines move away from the materials that form the shore and cannot be dragged.

Another option proposed for this stretch is to consider open dams (slit dams) for this river: permeable or semi-permeable dams are more recommended, since they do not hinder the ordinary dragging of the channel or the living organisms that travel along it long of him.

Advancing toward the south of the upper basin, applying the hydrogeomorphological chart, we find the Quebrada Allane: it is a sub-basin that drains an area of 219 km^2 and corresponds to the northern limit of the Pampa Cascachane, its bed is characterized by presenting a single channel with an average of slope of 5.8° and category 3, the bottom of the channel from its source to the first third downstream is flat, then it presents a V-shaped bottom, in most of its trajectory, it only presents a certain level of anastomosis in the last part of its course, just before delivering its waters to the Lluta river.

The Allane ravine falls into the Lluta from the east, its main tributary is the Colpitas river. The latter is born at the western foot of the pass of the seven turns, a site between the Luxone and Pacocagua hills. The Allane river contributes 50% to the Lluta in the year, that is, 0.53 m^3/sec., while the Lluta at that point for the same type of year, carries 0.86 m^3/sec. The bed of the Quebrada Allane is characterized by presenting a single channel in most of its trajectory, it only presents a certain level of anastomosis in the last part of its course, just before delivering its waters to the Lluta river, in Ref. [4].

In the vicinity of the Quebrada Allane, terrace escarpments are located between the glacis and the beds of the Lluta river and the Quebrada Allane; deep terrace escarpments are developed that reach an average height of approximately 100 m in the Lluta. Slide scars can be seen on the upper edge of the escarpment. At the base of these forms, numerous gravity cones develop. All these forms testify to the erosive processes that the escarpment is currently undergoing. The origin of this form is due to the dissection that the Lluta has carried out as a pre-existing river in the landscape in the face of the continuous uplift that the area has experienced since the middle tertiary in Ref. [7], giving rise to the encasing of the water course, in Ref. [8].

Therefore, as in the segment of the Caracarani river, as shown in **Figure 9**, the use of dikes with gabion structures is proposed, as well as the application of breakwaters, which are structures supported or embedded in the shore that are within the current, which divert the current lines away from the shore. In addition, it favors the depositing of materials carried by the river between them. Depending on the case, they can be simple construction solutions and, therefore, economical. The cost of its maintenance decreases over time. Even with the tip of a jetty eroded, the rest of the structure

Figure 9.
Fluviometric station of the general Directorate of Waters of Chile, on the Caracarani river, near the town of Humapalca, commune of General Lagos, Parinacota Province. The Caracarani river is one of the tributaries of the Azufre river and, consequently, part of the deposits that it drag downstream are part of the results of the Lluta basin that are reflected in the representation of the hydrogeomorphological chart.

continues to work, and the destruction of one of them does not seriously endanger the others. Notwithstanding the foregoing, periodic maintenance is required to ensure its permanence over time and enable simple repairs.

This course, which presents an area of gullies and slip niches, proposes the coatings or marginal protections that rest directly against the embankment of the shore and the bottom of the channel. These are parapets, and these are built with materials that cannot be carried away by the current. Between these materials and those of the edge, a filter is generally placed that prevents fine particles from escaping between the gaps in the marginal protection. Their main advantage is that they fix the shore definitively, without allowing any subsequent displacement following any curvature or configuration, although they require a more complicated and precise construction procedure than groynes; and, therefore, higher cost, in addition to requiring careful maintenance, since a failure, even of a small portion, endangers the entire protection structure.

For channeling work in the sectors with the highest speed (greater than 3 m/s) and in which stability and water tightness must be ensured, a reinforced parapet system is proposed based on a core of geotubes, in which the gravitational component is given by a geosynthetic tube filled with fine gravel or sand depending on availability.

From the data collected, it is pointed out that the river has very pronounced floods and that they develop rapidly, that is, the water level rises rapidly so that the runoff phenomenon is more assimilated to alluvial flow. This means that the work, in addition to resisting the erosive and dragging effect of the protection material, must consider the impact effect that a wave of these characteristics has on the work in Ref. [9].

Then, parallel to the Quebrada Allane, as shown in **Figures 10** and **11**, is the Quebrada de Colpitas. In this section, the slope processes are activated due to a steeper slope, these processes are creeping, undermining, gullies, slip niches, and linear erosion are located in the upper third of the development of the Lluta. The course of the Colpitas river, as shown in **Figure 10**, drains an area of 219 km^2 and has a 5° slope and is category 3. The bottom of this riverbed in its first third from its source downstream is flat and then its bottom is V-shaped. Like the Quebrada de Allane, the Quebrada de Colpitas, as shown in **Figure 9**, presents an area of gullies and landslide niches, for which the intervention in this course must be similar to the course of the Quebrada de Allane, as shown in **Figure 12**. Therefore, the application of defense is of the same characteristic as those indicated in the previous course, or else build alluvial control structures.

DOI: http://dx.doi.org/10.5772/intechopen.109404

Figure 10.
View of the Quebrada de Allane in the background of the Taapaca volcano. The road passes over the right flank of the volcano in the image, the highest point is 4802 m. the union of the Quebrada de Allane and Colpitas can also be seen, where rock formations with different colors and shapes can be seen, with heights over 200 meters and vegetation conditioned to the highland desert climate. It is located at 3900 m.a.s.l. and distance of 3 km from the hamlet of Alcérreca. The Colpitas river is one of the tributaries of the Azufre river and, consequently, part of the deposits that it drags downstream are part of the results of the Lluta basin that are reflected in the representation of the hydrogeomorphological chart.

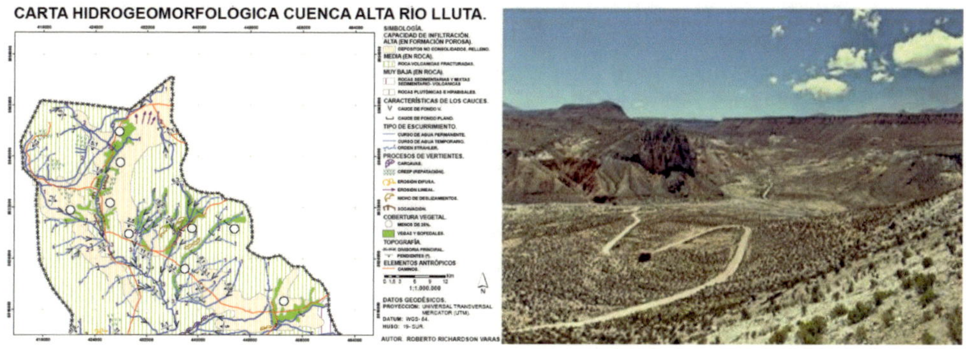

Figure 11.
The Allane ravine seen from the west. Contributions in material downstream, for the Lluta basin.

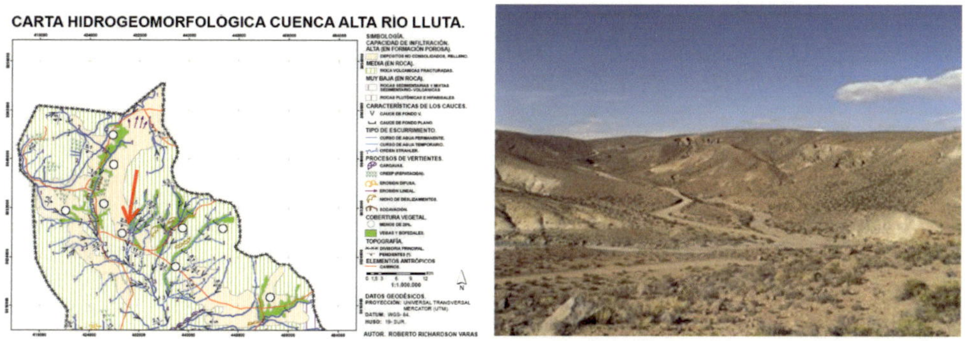

Figure 12.
Descent to the ford over the river Humapalca (Umapalca), on the road that joins the town of coronel Alcérreca (Alzérreca) and the town of Ancolacane. The Humapalca river as the cause of the current formation of the Lluta basin.

An important point to consider is through the general vision of the affected areas, where the spatial distribution of all the works is carried out, each one of them generates a protection zone on which its implementation will have a direct impact. Likewise, it is important to consider an interrelation of the works, which determines a correct functioning and fulfillment of the objectives. For example, in the control of gullies, the mere provision of dikes inside gullies without control of the runoff that causes this problem will be insufficient to stop its advance, so that at the head, at a distance greater than three times the depth of the gully in Ref. [10], a runoff control system must be available, by means of infiltration ditches or diversion channels among other possible alternatives to use.

It must be borne in mind that the gullies, sinkholes produced in rocks, and soils in sloping places due to the rainwater floods causing the so-called remounting erosion, contribute to the formation, normally, of ravines formed in the soft materials by the water of stream that, when there is a lack of sufficient plant cover, attacks the slopes digging long furrows with sharp edges.

The area's climate is characterized as a humid high-altitude steppe, with rainfall and temperatures highly influenced by geographic height. In the study area, annual rainfall reaches an average of approximately 300 mm, and these are concentrated mainly in the summer months, between December and March.

The base runoff of the Taapaca creek is produced mainly by the melting of the Nevados de Putre. Therefore, its higher average daily flows must occur in the summer months, and the gauged ones obtained during winter times are lower than those of summer. The maximum flows in the ravine are produced occasionally by the floods produced by the summer rains of the Altiplano winter (**Figure 13**).

The ravine is characterized by a well-formed channel with some very confined areas that show the sporadic occurrence of important flows, as shown in **Figure 14**. Its bed is, in general, stony with sectors with a high content of sand.

The flow measured at the confluence of its two headwater streams, in the Ruinas de Taapaca sector, averages 14 l/sec is increased to 30 l/sec. In the sector of the corrals downstream of Quebrada Milluni, due to the contributions coming from the affluent streams on the right side of the channel. The average flow measured is 30 l/sec. At the road crossing, it was estimated to be similar to that observed 1.3 km downstream.

Considering that the runoff of water in the Quebrada is produced by thaws, during the night there is no runoff, and it begins in the upper parts during the first hours

Figure 13.
Colpitas river during the highland winter rains of 2012. The Colpitas river is one of the tributaries of the Azufre river. So, part of the deposits that it carries downstream are part of the results of the Lluta basin that can be seen reflected in the representation of the hydrogeomorphological chart.

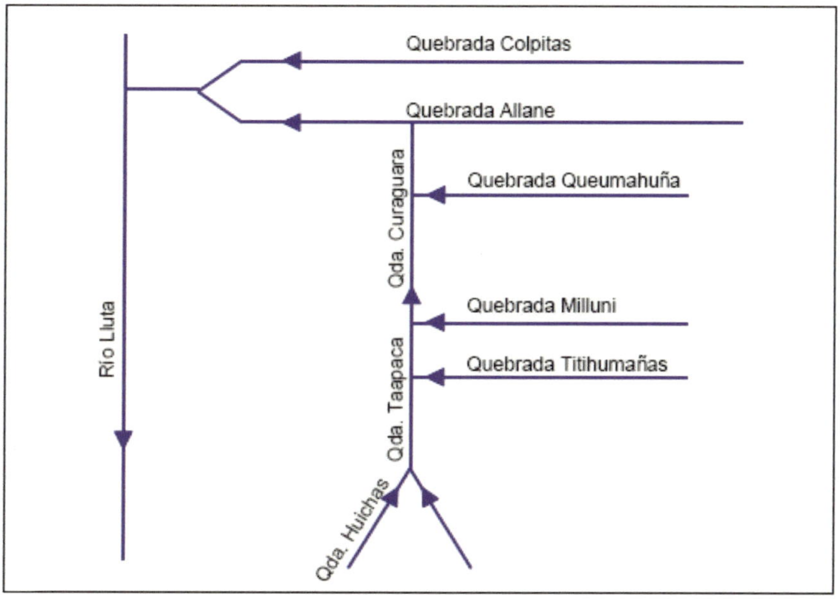

Figure 14.
Taapaca ravine water scheme as part of the Lluta river hydrographic basin.

of the day. In the corrals sector, a water front advance speed of 600 m/hr. has been measured, which decreases to 500 m/hr. downstream of this road, probably due to the decrease in the slope of the channel. Also bearing in mind that the Corrals sector is about 5–6 km from the confluence with the Allanes creek and, based on the data presented above, it is estimated that the waters would not be able to travel the distance to the Allanes creek before freezing or infiltrate, that is, the Taapaca creek in the area adjacent to the confluence with the Allanes creek.

Regarding eventual effects on the flora and vegetation of the sector, this is not sustained by the low flow that runs through the Taapaca creek, but, like all the vegetation of the Lauca national park, by the local rainfall regime.

Considering the characteristics of this ravine and its water contributions, background information obtained at the road crossing through the ravine approximately 1 km from its outlet to the Allane ravine, no water runs off and it is observed by dragging material and undermining, that the ravine has eventually runoff of high flows, which are attributed to floods resulting from the rains associated with the so-called Altiplanic winter.

The ravine is characterized by a well-formed channel with some very confined areas that show the sporadic occurrence of significant flows, much higher than what was observed. The characteristic of the flat channel bottom, its bed is, in general, stony with sectors with a high content of sand.

Therefore, only in the event of events, such as the one that occurred in 2001, this ravine would generate a contribution that would increase the bed of the Allane ravine. Therefore, to prevent an event of that magnitude or greater, it is proposed for this ravine to materialize the application of a control technique for floods, such as that affected by flood control works.

Regarding eventual effects on the flora and vegetation of the sector, this is not sustained by the low flow that runs through the Taapaca creek, but, like all the vegetation of the Lauca national park and surrounding area, by the local rainfall regime.

An important point of the analysis carried out in this area is to keep in mind the Tarapacá volcanic complex as defined by Clavero et al., in Ref. [11], this has historically been considered an inactive complex. However, the Spanish arrived in this part of the Andes, only about 450 years ago. In Ref. [8], the new data indicates that the Tarapacá volcanic complex is a dormant volcano, as shown the **Figure 15**, with strong potential for a future eruption. According to the spatial distribution of the domes and deposits in the Late Pleistocene-Holocene period, it is probable that, in the event of a renewal of activity, the products could be distributed on the S and SW flank of the complex. An injection of a new magmatic pulse could cause a deformation of the building, resulting in instability, and trigger a partial collapse of the upper flanks and domes, resulting in a debris avalanche toward the SSW. This sequence of events has been recurrent in stage IV (Late Pleistocene-Holocene, 0.45 Ma-Recent). In the case of an extrusion, explosion, and collapse of a dome, a pyroclastic flow can be caused, as occurred in the last stage (IV) of this volcanic complex.

In this regard, two possible scenarios could occur, the first of which could occur due to the development of an eruption between April and November, when there is a predominance of snow layers in the upper part of the complex as shown in **Figure 15**. Lahar flows may develop and probably be confined by the main valleys on the S and SW flanks of the complex, even affecting the international road that skirts Putre. Other flows could be caused in the direction of the western flank, affecting the main road that connects Putre with the towns that are further north in the Altiplano.

The second scenario, not associated with new volcanic activity, could develop during rainy seasons, from December to March. The rains induce the movement of small lahars each year, because of the remobilization of loose volcanoclastic material from the flanks of the complex. Although they rarely affect populated areas, they can often cut off roads. However, if an eruption were to occur, the availability of loose material could increase, and thus the volume of these flows could increase, favoring large discharges of material.

According to these scenarios shown through a statistical analysis extracted from the document in Ref. [8], on the Taapacá volcanic complex as shown in **Figure 15**, the volcanic eruptions in the last 30 ka, it is possible to say that the rate of eruptions is 1.6 per thousand years, with a recurrence of 450 years (95% confidence interval). The probability of occurrence of an eruption in the next 100 years can be estimated at 15%, and in the next 500 years, at about 56% in Ref. [8]. Considering the eruptive

Figure 15.
The Taapacá volcanic complex, stage II products (early late Pleistocene, 1.5–0.5 ma) and stage III (late Plesitocene, 0.5–0.47 ma) (source, manganese Los pumas environmental impact study, SGA 2011). Material from the Taapacá volcanic complex, is part of the Lluta basin deposit, as shown by the red circles.

history in the late quaternary and its composition, another eruption can be expected in the next few years (decades).

Continuing with the description of Taapaca, the two main formations are the Taipicagua and Chilcacagua streams. Further down, the Asiruni and Ancachi streams are tributary. In the surroundings, there are nine main springs whose waters irrigate different sectors of the Andean oasis.

Extending the application of hydro geomorphological mapping, from the point where the Azufre river and the Caracarani river meet and take the name Lluta river, it is seen in this sector of the Lluta riverbed and surroundings of the town of Putre in the direction of the Tacora volcano, outcrop mainly from stratified rocks and semi-joined deposits. Unconsolidated sedimentary deposits occur in a very minor proportion. More than 50% of the rocks and semi-merged deposits correspond to deposits of pyroclastic flows, lavas, and ignimbrites tuffs, mainly caused by Plio-Pleistocene volcanic activity. Older volcanic rocks, from the Miocene, outcrop 20 km to the west and about 30 km to the east of the Putre locality area.

Also, complementing what was shown above, a bibliographic compilation of theoretical-technical and contingency background was used, related to the type of study and area studied, supporteds by a visual interpretation of the surface processes, based on satellite images Landsat, SPOT, ASTER, and obtained from the Google Earth program. Plus, and topographical and geological charts, the result of which was to complement the identification of dynamic geoforms or quaternary deposits (ephemeral channels, gravitational forms, current, recent alluvial cones and fans, types of perennial channels, slopes, mass slides, etc.)

Bearing in mind that the Lluta has a category 4 in the initial stretch in the first 10 km approximately and the bottom of the river is flat and then it changes to a V-shape. At an approximate distance of 18 km to the north of the town of Putre, there are mass removal deposits, and about 30 km north-northeast of this locality, there are fluvioglacial and moraine deposits. In addition, this is an area that has medium infiltration ability, also, adding to this is the precipitation rate for topographic elevation ranges, which is what it shows, which could be the water supply.

In the Lluta study segment in the upper basin, monitoring data from two fluviometric stations have been shown on the Colpitas river in Alcerreca and the Lluta river in Alcerreca. For each of these stations, graphs of the monthly averages and the annual averages have been prepared. There is not a very marked seasonality; however, there is a tendency to register higher flows in the period from July to September.

For the design of the works to be considered, a rainfall analysis is previously needed to estimate magnitudes and intensities of rainfall in short periods of time (less than 24 hours, generally less than or equal to one hour), which directly affect the size of the works to be built. With this information and with the description of the characteristics of the land, the runoff that each conservation work must control (critical runoff) is determined, in ref [6], thus designing a diversion channel or infiltration ditch capable of transferring or retaining, respectively, in a safe way said runoff to receiving.

Although we understand that these data are only a conceptualization of reality, they show us the importance that this factor exerts on erosion, in the same way, the relevance of being able to reduce any of the characteristics of the slope, which will allow the deposition of the particles. Of soil displaced by the action of water.

From Humapalca to Allane, the route is about 26 km to the SE, and along this route, the Lluta receives from the northeast the Chuquiananta and Guancarane ravines that are born at the foot of the snow-capped mountains of those names and that are of little significance.

About 10 km upstream from point Allane where the ravine of that name, the river that flows into the ravine of that name, converges, the Lluta river begins to deepen its course, which came higher up at the level of the plateaus, to form a great canyon carved in the rhyolitic tuffs and in other little cohesive sediments that follow it downwards. Thus, in front of the Coronel Alcérreca station, a height of no less than 300 m can be seen between the box of the plateau and the bottom of the river.

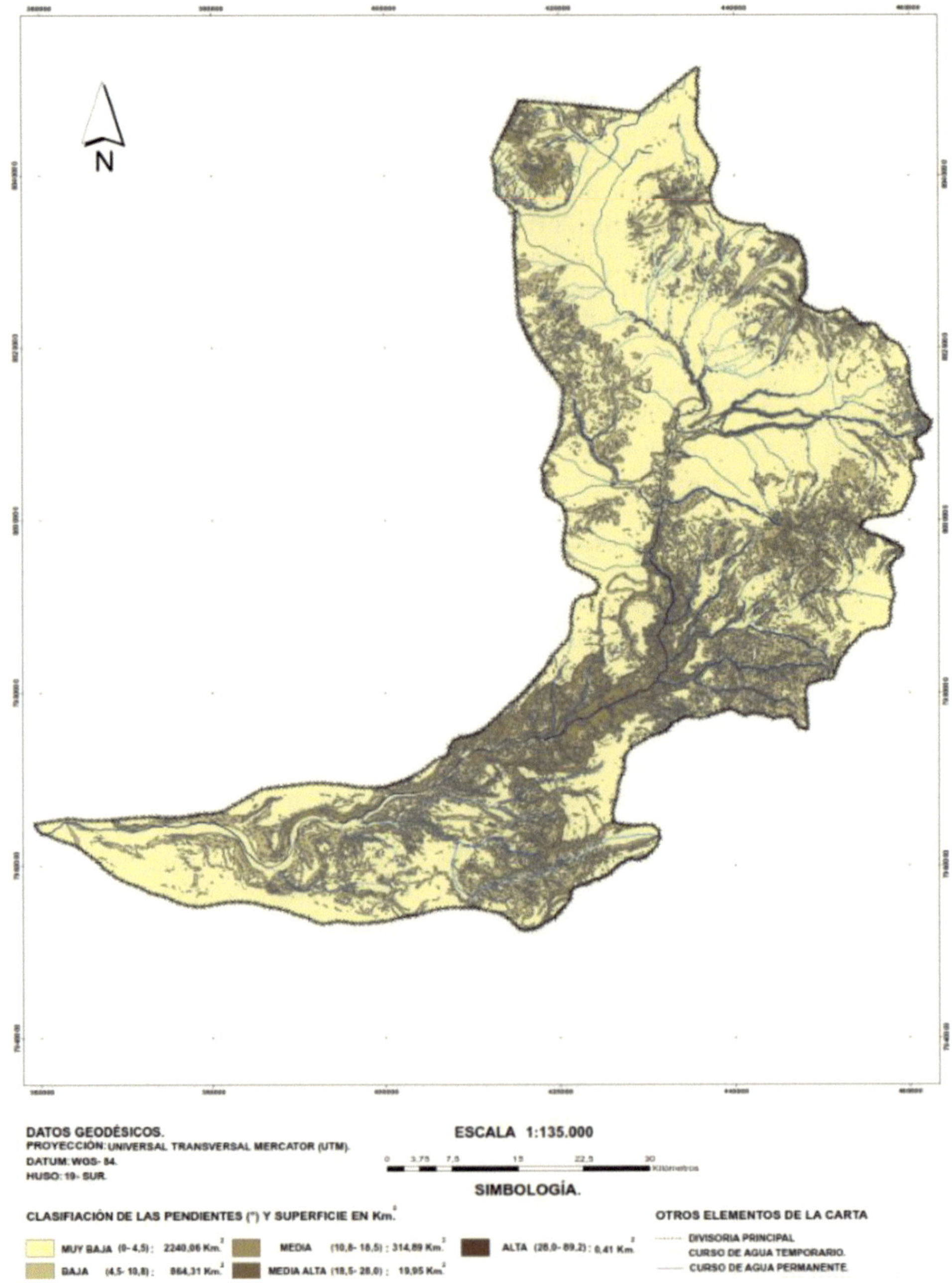

Figure 16.
Lluta river basin slopes map.

Continuing to the south, in the areas close to the Lluta riverbed on its sides, we find processes of slopes, such as landslide niches, an unstable zone, which slides with respect to a stable zone, through a surface or thin strip of land. Slippage occurs when the maximum tangential stress is reached at all points in the strip. For this type of slope, the blocks tend to roll and a vertical wall is required next to the trench so that the blocks do not try to get out.

The Lluta river continues in an increasingly narrow canyon heading north-south. In the next widening, called Jamiraya, the Putre ravine joins it on the left, which is born from the meeting of several watering holes originating in bofedales at the foot of the Nevados de Putre or Taapaca, as show the **Figure 15**. The gauging practiced in the Putre river in July 1968 showed a flow of 324 liters/sec for said cause in Ref. [12]. Shortly before the Putre junction, the Ancolacaya creek flows into the Lluta, parallel to it, with a very small contribution, estimated at less than 30 liters/sec.

The riverbed unfolds in a north-south direction, presenting an anastomosed pattern, with numerous diversions of channels between banks, mainly gravel together with accumulations of sand and gravel. Currently, the river has slightly affected its own bed in the sector, forming a small fluvial terrace, so the river box contains the bed itself, with a narrow terrace and a large number of gravity cones.

Analysis of the geological and geomorphological characteristics of the derived area in the analysis of average slopes, see slopes map, as shown in **Figure 16**, generating the identification of the potential activity of the geoforms, helping to determine the morphodynamic thresholds based on the slope ranges and their erosive potential. It can be considered that, according to the average slope values, all those slopes greater than 20° are recognized as potential generators of mass movements, and already over 30° of the average slope, as areas with high morphodynamics diversified in multiple processes, predominating the gravitational, in Ref. [8].

It is also due to considering what is related to the precipitations in the area that have an annual average value of 200 mm, while the evapotranspiration is 2000 mm. This makes the water balance in the soil very negative and infiltrations into the subsoil are minimal, limited only to precipitation events concentrated in very short periods of time.

4. Description of the middle section of the hydrogeomorphological chart of the Lluta river

The origins of the upper section of the Lluta river basin have been explained in detail, considering its foundations in terms of hydrographic and hydrological factors, which continue its journey downstream. Knowing the origins of its behavior and characteristics, the result of this is synthesized in the middle and low sections.

Each number in this section described in **Figure 15**, corresponds to the following:

1. Medium Infiltration Capacity (fractured volcanic rock)

2. Undercutting

3. Sliding Niches

4. High infiltration capacity (unconsolidated deposits-fill)

5. Linear Erosion

6. Very Low Infiltration Capacity (Sedimentary and Mixed Rocks)

7. Temporary watercourse

8. Vegetation less than 25%

In this section, it is joined, always on the left, by the Socoroma ravine, the last contribution of surface water that the Lluta receives. It is born in the snowless cordon that divides the Lluta basin from the Seco river basin to the south, a tributary of the San José river. In Coca, the last place of cultivation, the excess expense of spills and filters totals 25 liters/sec (Hydrographic Holes of Chile I Region of Tarapacá DGA).

Starting from Dos Hermanos, 77 km from the sea, the valley decreases its slope and widens. It begins in Chironte, 73 km from Chacalluta, the middle and lower course of the river with Agriculture Valley as shown in **Figure 17**.

Between Chironta and Boca Negra, several streams that are usually dry fall into the Lluta, but tend to grow with the summer highland rains. The main ones on the right are those of Chironte, in the sector of that name, and Palmani in the Vila Collo sector. Further down, on the south side, the Lluta receives the Chaquire ravines in Chaspisca and even further below Molinos, the Boca Negra ravine. Both are born in the Sierra de Huaylillas. From this point downwards, the Lluta increases with spring waters. From Boca Negra, the valley widens and the slope decreases to a value close to 2%.

In order to estimate the existing risk of flooding in the event of a major hydrometeorological event, the maximum flows have been obtained for return periods of 5, 10, and 20 years. For this, data from the fluviometric stations corresponding to Lluta in Alcérreca and Colpitas in Alcérreca have been used.

In the middle section of the Lluta river, as has already been mentioned, there is the consequence of the active processes in the upper section of the basin, large material that has been dragged, eroded, and undermined until reaching the middle section. It mainly constitutes an area for depositing sediments, which have been increasingly

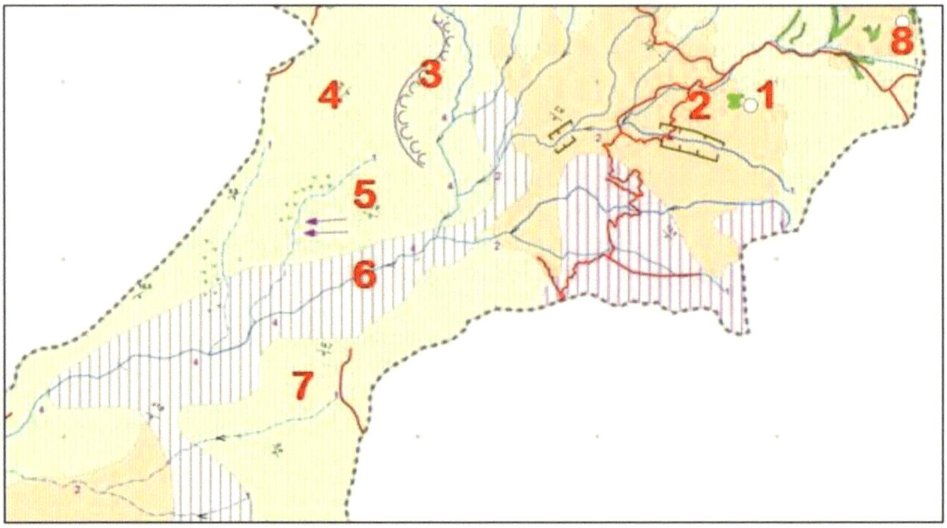

Figure 17.
Representation of the hydro geomorphological chart of the Lluta river (middle section) (source, from the author).

DOI: http://dx.doi.org/10.5772/intechopen.109404

physically, chemically, and biologically weathered upon reaching the middle section, so the material arranged in this section is mainly in boulders.

This is reflected in the hydro geomorphological chart with an extremely low infiltration ability (sedimentary and mixed rocks) in the riverbed as such. After the bed riverbed, we find medium infiltration ability (fractured volcanic rock). And a high infiltration ability the further away from the bed riverbed.

Vegetation of less than 25% is found, which decreases with respect to the high section. As there is less vegetation in this area, it prevents the normal flow of surface runoff from being lost, which is why the normal processes of slopes are triggered, such as linear erosion, undermining, slip niches, producing dragging of material toward the lower section of the river and, due to the force of scour and drag with which the surface runoff comes from the upper section, temporary watercourses are generated.

Already within the analytical description, in the area of influence, it was determined that the western sector of the course of the Lluta near Putre is the one with the greatest slopes. In general, the horizontal, gentle, and moderate slopes are associated with the bed of the Lluta river, the colluvial glacis, and the volcanic terrace, see the Map of slopes. In these last two, the sediment movements are conditioned to the dynamics coming from precipitation as it drains on the surface as sporadic laminar flows.

The area of acute danger is identified as all those forms that show current activity: the active valley slopes and the excavation slope, whose slopes are complex (great gradient variability, i.e., between 50 and 100%) and that are potentially exposed to events of colluvial origin, such as rock falls, debris flows, translational and rotational landslides, and complex movements, with great granulometric variability, from fine to coarse.

The area of intermediate danger, are all those that can border with areas of acute danger, being forms of ancient or recent origin and that can present slopes of 5 to 20°. Its risk is medium due to the potential danger that its immediate contiguity offers to slopes with acute risk, which through mass movements, such as debris flows, rock falls, rotational or translational landslides, and complex movements, feed the slopes making the escarpment It is considered that these phenomena will be less recurrent than the areas of acute danger.

5. Description of the lower section of the hydrogeomorphological chart of the Lluta river

The lower section of the Lluta river basin is a reflection of the normal process of the dynamics of a fluvial course since it is here where the processes activated with force and by the effect of gravity settle in the upper section, continuing in the middle section, and culminating in the lower or lower section of the river.

Each number of this section described in **Figure 18** corresponds to the following:

1. Linear Erosion

2. Sliding Niches

3. Temporary Watercourse

4. Creep

5. Permanent Watercourse

6. High Infiltration Capacity (Filled Unconsolidated Deposits)

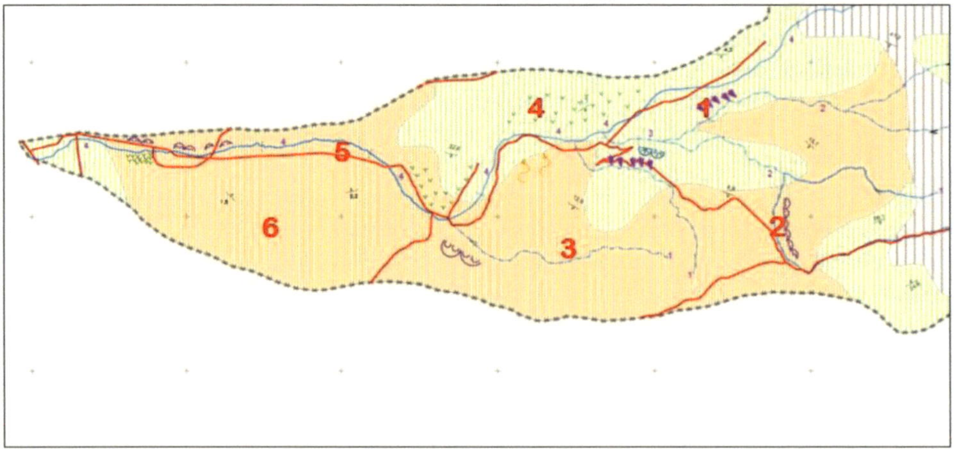

Figure 18.
Representation of the hydrogeomorphological chart of the Lluta river (lower section). (Source, Author's own).

In this section of the river, a high infiltration ability predominates (filling unconsolidated deposits), around which there is a permanent watercourse, as shown in **Figure 2** to next to the riverbed. In the lower section of the river and in the bed itself, there is a medium infiltration capacity ability.

Slope processes, such as linear erosion, slip niches, and creep, are activated to a lesser extent in the upper and middle reaches of the river, since here the materials that were dragged in the processes activated upstream have been deposited.

A lithological control is predominant toward the right bank of the river, where at the same time there is a depositional margin, and where at the same time the creep slope process is activated as shown in **Figure 19**.

The vegetation and bofedales in the lower section of the river are scarce, and the deposit of material dragged by gravity from the upper section predominates. In addition, facilities and equipment by human actions predominate in this area.

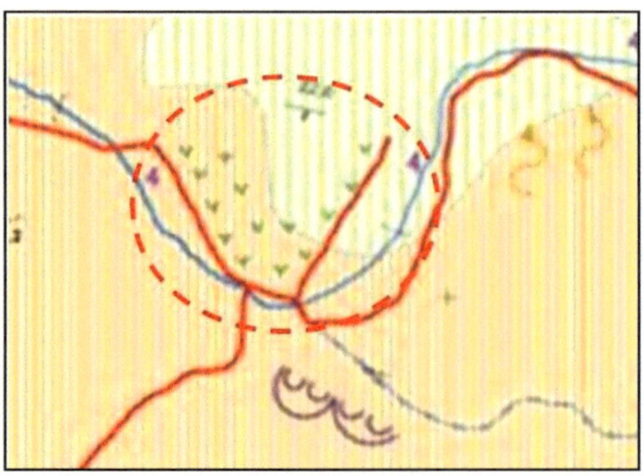

Figure 19.
Lithological control toward the left bank of the Lluta river, which shows that toward the right bank where a fluvial terrace has formed, there are creep processes. At the same time, on the right bank over said control, there are landslide niches.

6. Conclusion

Considering the hydrographic basin of the Lluta river as a case study, it is naturally characterized by a marked imbalance of the water-soil-vegetation system, which, associated with human activities and extreme hydrometeorological phenomena, gives it a torrential hydrological behavior that causes significant drag. With devastating effects of solids with alluvial load, which affect extreme damage to the lower part of the Lluta valley and the mouth of the beach area of the city of Arica.

Reflecting on the results of the analysis obtained for the entire basin in general, it can be concluded that it has a character of high torrentiality and susceptibility to erosion due to the basic conditions of the relief and the drainage system. The behavior regarding about the hydrological response of the basin is fundamentally determined found by the morphometric attributes of the catchment basin, in the upper course of the Lluta river. Thus, the middle and lower basin are very strongly influenced by the hydrological behavior of the upper basin.

In relation to the part and process elements studied, the hydro geomorphological chart is presented as a very complete tool, although it has deficiencies. The great dynamics offered by hydrographic systems require this type of cartography, which must be subject to a periodic updating system, updating its database, to like the process proven with regular digital cartography. Likewise, this product must be developed for all the basins susceptible to a process like the one presented, even more, considering the variables of climate change.

Author details

Roberto Richardson Varas
Geography, Faculty of Engineering, University Santo Tomas, Santiago, Chile

*Address all correspondence to: roberto.richardson@gmail.com

References

[1] Ollero A, Ballarín D, Díaz E, Mora D, Sánchez M, Acín V, et al. (IHG) Un índice para la valoración hidrogeomorfológica de sistemas fluviales. Madrid, Spain: Asociación Ibérica de Limnología; 2008. ISSN: 0213-8409

[2] Ollero A et al. Un índice hidrogeomorfológico (IHG) para la evaluación del estado ecológico de sistemas fluviales. Geographical. 2007;**52**:113-141

[3] Geografía Física de Arthur Strahler. Earth Science, Capitulo 13, Clasificación de los Climas y Regímenes Climáticos; 1989

[4] Ministerio de Obras Públicas (MOP), Dirección General de Aguas (DGA), Departamento de Estudio y Racionalización, INDERCO 1980. Compilación de los antecedentes del valle del Lluta

[5] Mikkan R. *Aguas Salvajes*. El Problema Aluvional de la Ciudad de Mendoza Universidad Nacional de Cuyo Facultad de Filosofía y Letras; 2007

[6] Francke Campaña S, Tokugawa K, Vargas Rona R. Manual de control de erosión. 3a. edición. Santiago, Chile: CONAF: JICA; 2009

[7] Seyfried H.; Wörner G.; Uhlig D.; Kohler I.; Calvo C. 1999. Introducción a la Geología y Morfología de los Andes en el norte de Chile. Changará volumen 30 N° 1, paginas 7-39 Universidad de Tarapacá Arica, Chile

[8] SGA. Soluciones en Gestión Ambiental. Estudio de Impacto Ambiental Manganeso los Pumas. 2011. Available from: https://www.sgasa.cl/sga/

[9] Arrau L. Plan Maestro de Aguas Lluvias de Arica, Manejo de los Cauces de los Ríos Lluta y San José y sus Desembocaduras, I Región. Consultoría PM-29, según Resolución D.O.H. N° 183, tramitada el 16 de enero de 2003. 2004

[10] Suárez de Castro. Conservación de suelo. San José, Costa Rica: Instituto Interamericano de Ciencias Agrícolas; 1979

[11] Clavero J, Sparks R, Pringle M, Valenzuela P, Edmundo & Gardeweg, M. Evolution and volcanic hazards of Taapaca volcanic complex. Central Andes of Northern—Journal of the Geological Society. 2004;**161**:603-618. DOI: 10.1144/0016-764902-065

[12] Niemeyer H. Hoyas hidrográficas de Chile: 1a. DGA: Región de Tarapacá; 1980. p. 1990

Chapter 14

Spatial-Temporal Relations of Flood Risk and Its Potential Dynamic Pressures in Myanmar

Hnin Wuit Yee Kyaw and Phone Pyai Tun

Abstract

Despite multiple challenges, floods remain the most frequently occurring hazard in Myanmar. Current developments of political instability, multidimensional insecurity, and associated economic crisis have burdened the existing vulnerabilities and inequalities of the Burmese people and their ecosystems. Diminishing adaptive capacities of degraded ecosystems, poor infrastructure, and extreme poverty, together with major livelihood dependency on climate-sensitive agriculture, will further increase flood risk. Moreover, other hazards such as COVID-19, heatwaves, and droughts may exacerbate flood impacts leading to compound disasters. Understanding how and which factors drive flood risk, and where they distribute are important to reduce flood risk, address its root causes, and prevent future flood damages by lessening exposures, vulnerabilities, and even hazards. We aim to compare the spatial-temporal distributions between dynamic pressures and flood risk, and identify the spatial relations on a national scale and within floodplains. We draw on socio-ecological risk assessment, systematic review, time-series analysis and modified t-test after testing spatial autocorrelations of dynamic pressures and flood risk. Our results show that many socio-ecological dynamic pressures driven by economic- and governance-related root causes had positive spatial relationships with flood risks. We recommend effective land use and environmental governance that consider compound and cascaded flood risk and investment in public services and infrastructure such as health and education to reduce vulnerabilities and increase resilience of Myanmar people.

Keywords: flood, risk, exposure, vulnerability, Myanmar, dynamic pressures, root causes, spatial-temporal relations, socio-ecological systems

1. Introduction

The disproportionate rise in climate-related impacts, and losses and damage in countries of the Global South may attribute to differences in vulnerability to those climate and disaster impacts. Yet, the issue is treating disasters separately from their underlying root-causes in their disaster- and climate-related management and adaptation practices, especially in those vulnerable countries. For example, climate change adaptation through nature-based solutions is not possible without addressing root causes of deforestation and forest degradation in those nations. There is a clear call for transformative change in dealing with vulnerabilities in those countries.

Disaster risk and impacts are increasingly becoming the products of more compound and complex causal mechanisms in today's interconnected world. Explanation of disaster risk needs to trace connections that join disaster impacts with a sequence of causal factors and processes. Cascading impacts are chains of disruptive disaster impacts on systems at different scales [1, 2]. This systemic nature of risk can be seen in the examples of the COVID-19 pandemic [3–6]. In other words, a disaster at the same time can be a driving factor to cause a larger and more calamitous compound disaster. This type of compound disaster can be especially worse in the incidence of social, economic, financial, health, and political crises, which are outside of the climate and disaster risk framing.

Here, root causes are defined as "an interrelated set of widespread and general processes within a society and the world economy" [7]. They are economic, demographic, and political processes that affect the distribution and allocation of resources among different social groups. Those causal mechanisms are spatially and temporally distant from those disastrous situations (unsafe conditions of vulnerability) to be clearly understandable at a glance [7–9]. They are usually "invisible" and "taken for granted" as they are embedded in socio-cultural beliefs and assumptions. For example, inadequate natural resource governance that leads to inequalities and deforestation, which in turn increases flood risk, cannot easily be mapped or easily visible as a causal factor. Therefore, they are difficult to express spatially and temporally. As a result, those root causes cannot be easily incorporated into the disaster risk management.

Therefore, it is essential to understand the spatial and temporal relations of risk drivers (root causes) and the places where risk occurs. Dynamic pressures are "processes and activities that translate the effects of root causes both temporally and spatially into unsafe conditions" [7]. Therefore, these pressures can be visualized over space and trends. In this chapter, we will explain spatial-temporal relations of dynamic pressures exercised by root causes and disaster risk using case study of Myanmar.

Myanmar provides a comprehensive case study to understand compound cascading risk driven by the interaction of deepened root causes. Myanmar has been experiencing multiple exogenous compound stressors and endogenous dynamic pressures driven by several root causes currently and dynamically over history. Myanmar is one of the countries, which is mostly affected by and is highly vulnerable to extreme climate events, according to GAR Report, Inform Risk Index 2017, and Global Climate Risk Index 2021 [10–12]. Given consistent inadequate governance, the dependency of Myanmar's economy on resource extraction and agriculture left Myanmar's forests, land, water, and environment in a seriously degraded condition [13–15]. This inadequate governance also deepens socioeconomic and class inequalities [13–15]. The fact that the majority of populations depend on their livelihoods in those degraded ecosystems and climate sensitive sectors results in limited capacities to cope and adapt to extreme climate events due to the degraded ecosystems. Long history of conflicts and the current development of political instabilities render already poor Myanmar population to face multidimensional insecurities and inequalities including food, job, and safety, to name a few [15, 16]. These recent developments disrupted humanitarian assistance, international development aid, food production, environmental management, and climate actions, resulting in conditions which can be extremely catastrophic in the face of extreme disasters [16]. Considering the increasing impacts of climate change in the near future, those extreme disasters are likely to occur. Public services and infrastructure, such as education and health services, have received little priority and attention for investment in their development, which is now seriously disrupted due to the recent changes [15]. Last COVID-19 incidences and associated vast mortalities due to shortages in oxygen supplies and health facilities prove weakening health infrastructure. The associated economic and financial crisis put additional millions of Burmese people into poverty. Social security systems and climate insurance development are yet in the very early stage of development to support recovery from disasters and climate impacts and for the risk transfer [17]. Official Development Assistance from outside Myanmar can be difficult due to the ethical challenges, safety, and entry issues [17]. Therefore, in case of any disasters, already vulnerable Myanmar people can be helpless, seeing the examples of 2008 Nargis Cyclone.

Despite those multiple challenges in Myanmar, flood remains the most frequent and has the highest contribution to average annual loss compared to all other hazards in the country [18]. Winseminus et al. (2016) estimated that 1.24 billion out of 1.81 billion population exposed to 1 in 100-year floods live-in South East Asia [19]. In 2015, a riverine flood resulted in over US$1 billion in damages, affected over 1.6 million people, and caused at least 149 deaths [20]. With climate change, Myanmar is likely to continue to experience river flood events at the scale of the 2015 floods or worse [21]. In Myanmar, some river flooding occurs almost every year during the monsoon season and communities now rely on these floods for nutrient deposition in their agricultural fields [22]. Communities have learnt to cope with and have adapted to these annual floods, harnessing their value [22]. However, as demonstrated by the impact of the 2015 flood, which was characterized to be of a 20–50-year return period [23], Myanmar has little capacity to cope and adapt with major flood events alone given the overlapping effects of different pressures.

Transformational risk reduction and adaptation require addressing root causes of flood risk to reduce its vulnerabilities and increase resilience. If only symptoms are addressed and root causes are not considered in the risk management, those root causes will be intensified in the future, enhancing compound pressures in a climate-changing world. The Sendai Framework for Disaster Risk Reduction calls for disaster risk management that considers all dimensions of disaster risk, including hazard, exposure, and vulnerability. To reduce flood risk, decision makers and stakeholders must have a holistic understanding of the underlying factors.

Studies have focused only on drivers that drive individual components of risk and any of the assessments have barely explored risk drivers that contribute all components of risk [24]. Unlike the Pressure and Release (PAR) model, this chapter will consider that root causes and dynamic pressures will not only lead to the vulnerabilities of the people, but they can also lead to all components of risks which are hazard, exposure, and vulnerability of socio-ecological systems.

Several studies attempted to understand drivers of flood risk. Flood risk was assumed to be caused by various factors: including physical factors such as river levees and river incision [25], climate-related factors such as Climate variations associated with Oscillations such as Pacific Decadal Oscillation (PDO) and El Niño Southern Oscillation (ENSO) [26], sea level rise [24, 27] and storm surges, typhoons and cyclones [24], social factors such as urbanization [24, 25, 28, 29], soil sealing [27], poverty, and limited education [28], environmental factors such as land use changes and deforestation [24], degraded wetlands and decreased drainage capacities [30], land subsidence [24, 27], governance-related factors such as failures of top-down approaches in flood risk management, lack of risk communications, lack of local information, limited citizen involvement in flood risk management [31], and subjective factors such as risk perceptions in shaping flood risk in terms of developing insurance-based flood management and risk transfer [32]. Abbas and Routray (2014) identified exposure to vectors, inaccessibility, damaged, and non-operated health care facilities as major factors of flood-induced health risks [33]. Assessments have likely prioritized physical and environmental risk drivers over social, economic, and governance-related drivers [24].

However, spatial-temporal relations of disaster risk and driving factors were rarely explored. A few studies investigated spatial-temporal relations of flood-related losses and damages and their influencing factors [34, 35]. However, those studies have a focus on the hazard and impacts part with limited vulnerability considerations in assessing the spatial-temporal distribution of disaster risk. We will aim to fill this gap by exploring spatial-temporal relations of dynamic pressures and flood risk that is considered as a function of hazards, exposure, and vulnerability, driven by dynamic pressures and root causes. There are three main objectives of the chapter. First, this chapter will investigate root causes and driving pressures of flood risk in Myanmar, considering potential compound and cascading risk, while the focus is on the flood risk. The spatial and temporal evolutions of dynamic pressures in relation to flood plain areas will be assessed. Second, this analysis will show the spatial distribution of flood risk based on the IPCC framing of risk [36]. Third, spatial relations of dynamic pressures and flood risk will be examined. Understanding the spatial and temporal relations of dynamic pressures (that translate from root causes to unsafe conditions of flood vulnerability) and flood risk can be used for transformative change to effective flood risk reduction.

2. Methodology

2.1 Case study

Myanmar is located in South East Asia (**Figure 1**). Its land size is approximately 678,500 km2 and it has an estimated population of 42.5 million people [37]. The country has a complex multi-level governance system with 330 townships at the lowest level of government [38]. Much of the population in Myanmar lives in poverty (MMPF, 2017) and the country was ranked 145 out of 189 on the Human Development Index in 2019 [39]. There are four major rivers in Myanmar with populations reliant on each [22]. The Ayeyarwady is the largest and the most important used in commerce and daily life [22]. Most of the country has a tropical monsoon climate, with river floods commonly occurring between May and October [22].

2.2 Conceptual framework

2.2.1 Understanding root causes and dynamic pressures

This chapter used the long-standing Pressure and Release Model (PAR) to understand how root causes were translated into unsafe conditions that are current vulnerabilities through dynamic pressures. As explained in the introduction, root causes are defined as "an interrelated set of widespread and general processes within a society and the world economy" [7]. Dynamic pressures are "processes and activities that

Figure 1.
Myanmar with labeled states. Townships are the next administrative level down.

translate the effects of root causes both temporally and spatially into unsafe conditions" [7]. Unsafe conditions are the specific conditions in which the vulnerability of the people and ecosystems is expressed in time and space when being exposed to a hazard. Unlike the original concept of the PAR model, this analysis will consider that root causes can lead to all components of risk: hazard, exposure, and vulnerability of socio-ecological systems.

2.2.2 Conceptual framework of flood risk

To define risk, this chapter built on the common risk framework from the IPCC 5th assessment report, where risk is the potential for adverse consequences and is a function of hazard, exposure, and vulnerability [36], and the PAR model where root causes and dynamic pressures are considered (**Figure 2**). In this analysis, we focused

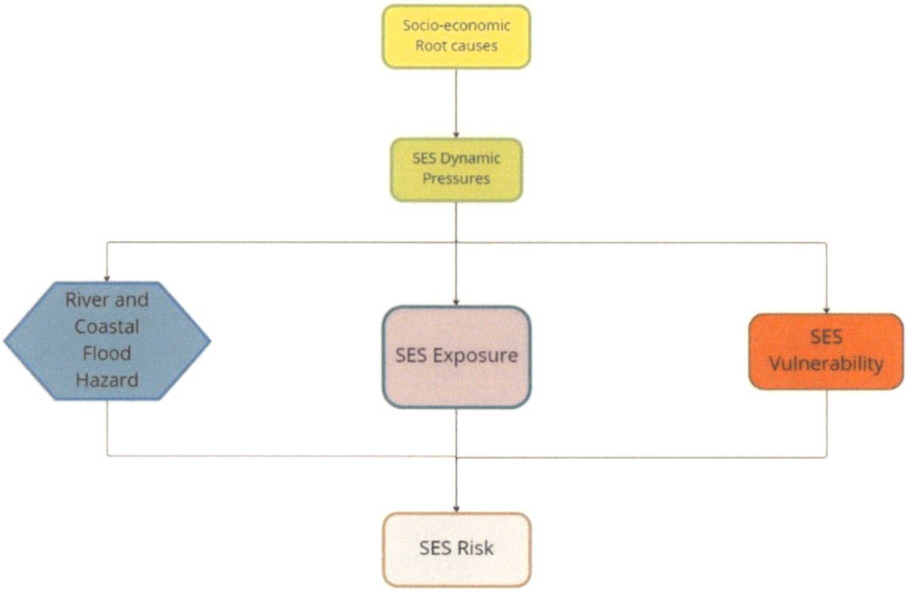

Figure 2.
Conceptual framework of socio-ecological system of flood risk.

on present-day flood risk to people, agricultural, and ecosystems. Exposure was defined as the presence of people and ecosystems within the extent of the hazard [40] —a 100-year river flood and coastal flood event. Vulnerability was defined as the propensity or predisposition of the Myanmar people to be adversely affected and encompassed the components of susceptibility and coping capacity [40]. Root causes and dynamic pressures have impacts on all elements of risk. As mentioned in the above section, it is the difference from the PAR model, where root causes have only drive vulnerability. In this framework, those underlying causes and dynamic pressures drive all elements of risk (**Figure 2**).

2.3 Workflow

Figure 3 outlines the methodological workflow of the quantitative index-based approach for the township level used in this analysis. All data were freely available, allowing for replication and validation of this study.

2.3.1 Identification of root causes and dynamic pressures of flood risk in Myanmar

To identify root causes and dynamic pressures, a systematic literature review was conducted to understand the main drivers and causes of flood risk in Myanmar. Two searches were conducted using Web of Science (WoS) and SCOPUS in November 2022. Search strings were constructed. The authors separately screened all titles and abstracts of the unique papers to determine relevance. Criteria included focus on Myanmar; risk and vulnerability to flood; and the exposure of people and ecosystems. Publications that met these criteria were selected. Papers that did not were excluded, including those that focused solely on flood hazard. Where one or both authors were

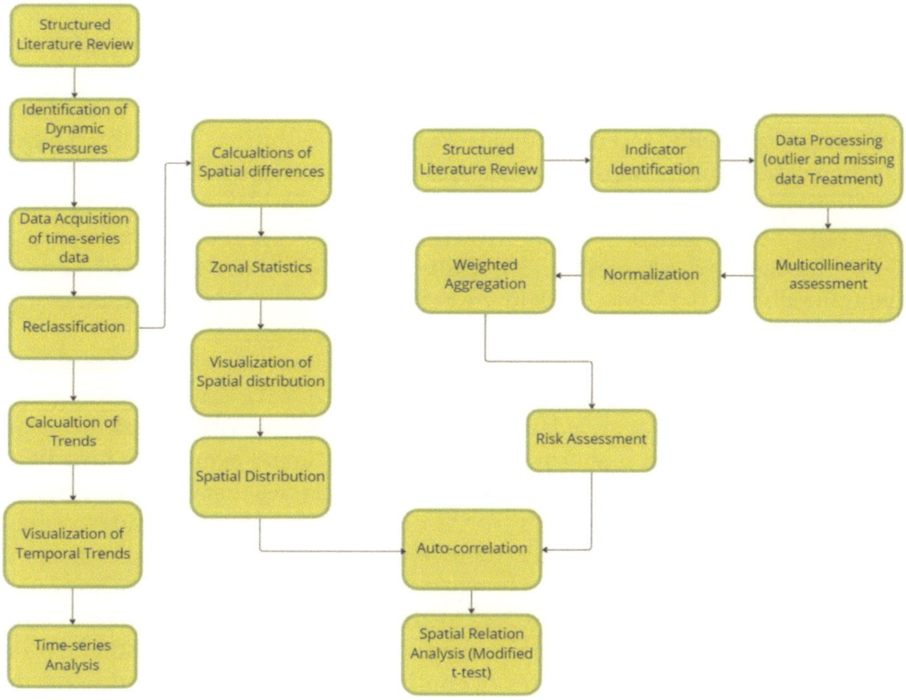

Figure 3.
Methodological workflow.

Searches	WoS	SCOPUS	Included
Search 1: (TITLE-ABS-KEY (flood) AND TITLE-ABS-KEY (risk OR vulner* OR resil* OR root AND cause* OR drive* OR pressur*) AND TITLE-ABS-KEY (eval* OR assess* OR profile OR index OR indic* OR multicriteria) AND TITLE-ABS-KEY (myanmar OR burma))	60	23	11
Search 2: (TITLE (myanmar OR burm*) AND TITLE-ABS-KEY (flood))	84	106	20
Google search “driving factors and root causes of flood risk in Myanmar” OR “Myanmar flood disaster risks and impacts”	—	—	3

Search strings use SCOPUS operators. For WoS, TOPIC replaced TITLE-ABS-KEY. TITLE remained the same.

Table 1.
Search terms used to capture papers to inform vulnerability indicator identification.

uncertain, the paper was read entirely by the authors to determine selection. To find additional gray literature a Google search was conducted. After screening the first 10 pages of results two reports were added to the literature review. At the end of the process, 34 papers were included for final review. A summary of searches is provided in **Table 1**.

Root causes and dynamic pressures were extracted from each relevant publication. After identifying relevant dynamic pressures in Myanmar, associated datasets were collected and performed following analysis to understand their spatial-temporal evolution.

2.3.2 Time series analysis

Time series analysis was performed to understand temporal aspect of dynamic pressures. It started with acquisition of dynamic datasets for available years. Most of datasets associated with land use changes such as deforestation, agricultural expansion, urbanization, loss of wetland areas, and mining are acquired from Regional Land Cover Monitoring System developed by SERVIR-Mekong [41] from 1987 to 2018. Conflict data were attained from the Armed Conflict Location & Event Data (ACLED) for the available period from 2019 to 2022 [42]. Land Surface Temperature was computed from the Terra Moderate Resolution Imaging Spectroradiometer (MODIS) dataset using Google Earth Engine (GEE).

After data acquisition of relevant datasets, each dataset was preprocessed for time series analysis. Datasets for forest and agriculture were reclassified according to given land cover classes in the regional land cover monitoring systems. The classes that defined "forests" included mangrove, forest, flooded forest, ever green broadleaf, mixed forest, and orchard and plantation forest. The classes that defined "agriculture" included rice, cropland, and aquaculture. After extracting targeting land cover classes, respective values per townships were calculated from each raster datasets using "zonal statistics as table" in ArcGIS for the time period from 1987 to 2018 to see trends of values per townships. For flood plain calculations, raster data of land-cover related dynamic pressures were clipped with flood plain boundaries [43]. Conflict data were preprocessed by counting the number of conflict events for each township for the period from 2019 to 2022. Types of conflict events included in the analysis were battles, explosions/remote violence, protests, riots, strategic developments, and violence against civilians according to the ACLED definition [42]. For land surface temperature data, average annual land surface temperature was calculated from daily land surface temperature and emissivity values from 2000 to 2022 (until 23rd November 2022). Flood data related to losses and damages in terms of mortalities and affected populations were acquired from EM-DAT dataset from 1965 to 2021 [20]. Visualization of time-series trends was made in Excel.

2.3.3 Flood risk assessment

2.3.3.1 Flood hazard/exposure

Extent of river flood and coastal flood for a 100-year return period was obtained from the Joint Research Centre (JRC) of European Commission [44] and the World Bank Data Catalog [45], respectively. ArcGIS was used to crop the global dataset to the national boundaries of Myanmar and limit the extent of the flooding to modeled depth above 20 centimeters.

To calculate hazard exposure, we determined the percentage of exposed population per township (e_soc_i) based on the modeled population distribution from WorldPop (WorldPop, 2020) (p_tot_i), and the population within the flood extent (p_exp_i) (Eq. 1. i refers to each township).

$$e\_soc_i = \frac{p\_exp_i}{p\_tot_i} \times 100 \tag{1}$$

The percentage of exposed population was normalized using linear min-max (Eq. 2) so that the range was reduced to between one (high exposure) and zero (no exposure) to create the exposure index (EI_i). This is a very common normalization method for indicator-based assessments [45].

$$EI_i = \frac{(E_i - E_{min})}{(E_{max} - E_{min})} \times 100 \quad (2)$$

In Eq. 2, E_i refers to the percentage of exposed population before transformation, E_{min} refers to the minimum value of exposure, and E_{max} refers to the maximum value of exposure.

2.3.3.2 Flood vulnerability

The same systematic review explained above was used to understand driving factors of flood vulnerability and to identify relevant indicators in Myanmar. These were grouped into susceptibility and coping and adaptive capacity indicators.

Next, data at the township level were collected from multiple sources including the 2014 census and other surveys. Based on data availability, 26 indicators were used in the final assessment (**Table 2**). Appendix A provides an overview of all desired indicators from the literature review, and the data sources of those used.

The indicators relating to access to telephone, internet, mobile, and radio were aggregated to one information indicator (c_ati), data for poorly constructed floors, and walls were aggregated to create one housing conditions indicator (s_wfl), and data for households with generators and solar energy were aggregated to an access to alternative electricity indicator (c_aes). Proxy variables were used for s_pov, s_vec, s_chr, s_con, c_ati, and s_wfl (for more information, see Appendix A).

As a third step, outliers were identified and treated following Damioli (2017) [50] using Microsoft Excel. Box plots based on the interquartile range and skewness and kurtosis were used to identify extreme values. Due to the paucity of data for triangulation to determine whether extreme values were errors, local expert knowledge of the authors was used. It was determined that only five indicators had outlying values that were errors (s_pov, s_vec, s_chr, s_fhh, and c_doc). These were treated using winzorization (see Appendix B).

Fourthly, the scale of missing data was assessed. Using findings from Downey and King (1998) on acceptable thresholds for missing data, no indicators were excluded as none had missing data above 20%. Ten townships[1] had missing data above 20% and were therefore marked as highly uncertain in the final risk and vulnerability assessments. Missing data were imputed using the IDW tool in ArcGIS and taking the mean from the output.

As a fifth step, multicollinearity analysis was conducted using Kendall's Tau and two-tailed approach for statistical significance in SPSS (IBM SPSS Statistics). This is a commonly used technique for non-normal data [51] with $r > 0.9$ indicating highly correlated datasets [52]. No issue of collinearity was detected (Appendix B).

Sixthly, the final set of indicators were normalized to a range between zero and one using the linear min-max approach. For indicators where high scores increase vulnerability (direction is positive), Eq. 3 was applied. For indicators where high scores lower vulnerability (direction is negative), values were inverted using Eq. 4. In Eqs. 3 and 4, X_i refers to the indicator value for a township (i) before transformation, X_{min}

[1] Mahaaungmyay, Pangsand, Narphan, Pangwuan, MongMao, Mongle, Cangaw, Ye, Latha, and Seikkan.

Indicator	Code	Data Source	Direction
Susceptibility			
Dependency ratio (people <15 and > 65 years / people between 16 and 64 years)	s_dep	[46]	+
Population below the poverty line (average annual income)	s_pov	[47]	—
Disability prevalence (population, %)	s_dis	[46]	+
Population with vector-borne diseases (malaria incidence per 10,000 people)	s_vec	[47]	+
Population with chronic illness (share of tuberculosis, dysentery, and hepatitis, %)	s_chr	[47]	+
Female-headed households (%)	s_fhh	[46]	+
Population affected by conflict in townships in the last year (number of conflict events per 10,000 people in the last year)	s_con	[42]	+
Households with access to safe sanitation (%)	s_san	[46]	—
Households with access to improved drinking water (%)	s_dri	[46]	—
Population living in poorly constructed housing (households with walls or floors constructed from leaf, bamboo, or earth, s_wal & s_flo average %)	s_wfl	[46]	+
Average travel time to the closest city (minutes)	s_ttc	[48]	+
Daily wage worker (%)	s_daw	[47]	+
Extent of Areas of Agricultural expansion in Flood plains (1987–2018) (%)	s_age	[41]	+
Extent of Areas of Urban extension in Flood plains (1987–2018) (%)	s_ure	[41]	+
Extent of Areas of Mining extension in Flood plains (1987–2018) (%)	s_mine	[47]	+
Extent of Areas of Forest Losses (1987–2018) (%)	c_ftl	[41]	+
Extent of Areas of Wetland Losses (1987–2018) (%)	c_wetl	[41]	+
Adaptive capacity			
Households with access to Information (telephone, mobile, internet, or radio, s tel., s mob, s int., & s_rad max %)	c_ati	[46]	—
Literacy rate, (%)	c_lit	[46]	—
Density of roads (road kernel density: km road/km2, search radius = 5 km)	c_dens	[49]	—
Households owning a boat (%)	c_boa	[46]	—
Number of doctors per 10,000 people	c_doc	[47]	—
Number of hospital beds per 10,000 people	c_bed	[47]	—
Extent of Forest Areas (%)	c_ft	[41]	—
Extent of Wetland Areas (%)	c_wet	[41]	—
Households with access to alternative electricity source (households using solar energy or a generator as the main source of lighting, c_sol & c_gen total %)	c_aes	[46]	—

Table 2.
Final list of indicators including code, data source, and direction.

refers to the minimum value of the indicator, X_{max} refers to the maximum value of the indicator, and X_i' refers to the indicator value after transformation.

$$X_i' = \frac{(X_i - X_{min})}{(X_{max} - X_{min})} \tag{3}$$

$$X_i' = \frac{(X_i - X_{max})}{(X_{min} - X_{max})} \tag{4}$$

Finally, indicators were given equal weights due to the absence of knowledge on their relative importance and aggregated using additive arithmetic aggregation into the vulnerability index VI_i (Eq. 5).

$$VI_i = \frac{\sum X_i')}{N} \tag{5}$$

In Eq. 5, X_i' refers to the normalized indicator values for the township and N refers to the number of indicators.

2.3.3.3 Flood risk

The vulnerability and exposure indexes were aggregated using two methods. Arithmetic mean aggregation of hazard/exposure (EI_i) and vulnerability (VI_i) was conducted to determine relative risk for each township in a risk index (RI_i) (Eq. 6). Natural break method was used to visualize spatial distribution of exposure, vulnerability, and risk indices.

$$RI_i = \frac{EI_i + VI_i}{2} \tag{6}$$

2.3.4 Spatial analysis

Spatial differences of dynamic pressures were calculated in the raster calculator in ArcGIS to understand changes between the start and end of the available period. Resulting differences in terms of decrease, increase, and no changes were again calculated for each township using zonal statistics as Table tool in ArcGIS. For flood plain calculations, raster data of dynamic pressures were clipped with flood plain boundaries. Then, the similar processes of calculation of spatial differences were carried out for those dynamic pressures in the flood plains.

To understand the spatial relationships of dynamic pressures and flood risk, a modified-t test was carried out in R studio using SpatialPack library. Similarly, for the dynamic pressures on the flood plains, the same approach was conducted to understand their spatial relationships with flood risk index.

3. Results

3.1 Root causes and dynamic pressures

In response to the first research question, root causes and dynamic pressures of flood risk in Myanmar are explained as a result of the literature review. Social flood vulnerability in Myanmar is expressed in the form of multidimensional inequalities, and widespread poverty, while ecological flood vulnerability is expressed in the form of enormous environmental degradation. Multidimensional inequalities can be in terms of gender, ethnicities, class, and spatial divides such as urban, rural areas, bordered areas [53]. Those socio-economic and ecological vulnerabilities are connected with each other and also shape flood exposure and hazard.

Inadequate natural resource governance is a major root cause that deepens flood risk in Myanmar. Corruption at different levels allows most resource-based activities to happen in the black economy, resulting in widespread deforestation [54]. Particularly in sectors such as forestry and agribusiness, corruption can result in increased deforestation from land conversions [54]. Economic dependency on resource extraction can increase corruption [55]. Myanmar was ranked 136 out of 176 countries in terms of corruption index [56]. Widespread deforestation causes amplified erosion, limited water retention capacity and high sedimentation, worsening flood hazards. Studies proved that deforestation increases flood frequency [57, 58] and flood risk in nonlinear way [59]. As forests serve as a safety nets for forest-dependent communities, deforestation has limited their coping and adaptive capacities after floods. Biodiversity loss can have a cascading effect on the ecosystems and its regulation services for disasters.

An example can be seen in the Ayeyarwaddy delta, the rice bowl of the country, where wetlands and mangrove forests were massively transformed into rice fields. Mangrove forests were again degraded during insurgency periods [60]. Later, additional conversions into commercial fish and shrimp farming result in massive deforestation of mangrove forests in the delta [61]. As a result, risk reduction services and provision services of mangrove ecosystems were declined seriously affecting livelihood base and adaptive capacities of society and ecosystems to adapt to flooding and other coastal hazards such as cyclones, salt water intrusion, and sea level rise.

Poor resource governance contributes to environmental degradation and increased flood risk. Development projects such as natural gas extraction and dam construction are often carried out without proper environmental regulations or local participation [55]. Corruption results in such projects with violations of safeguards. These projects threaten the environmental security on which local communities depend [55]. One of the examples of inadequate natural resource governance which increases flood risk can be seen in the mining sector. The townships where many of mines existed, Hpa Khant, experience deterioration of its environment where mountains become valleys, and *vice versa* [62] increasing landslide risk and collapse of mining sites. Illegal and large-scale mining practices, such as exceeding permitted tailings height and dumping tailings near river banks, also contribute to high risk of collapse, resulting in loss of life and releasing large amounts of soil into rivers [62]. This in turn raises river bed levels, increasing flood hazards for nearby and downstream populations and villages, potentially washing them away. Mining also causes water pollution, producing toxic water that can be highly dangerous after flooding. Additionally, mining projects often result in land grabbing at the expense of local communities, weakening their ability to respond to flooding.

Inadequate governance of natural resources can lead to inequalities that exacerbate flood vulnerability [63]. This is because natural resources are often exploited in ways that benefit only a small group, while the pollution, deforestation, and other negative consequences affect the entire population [55]. It was worsened as foreign investment places immensely toward resource extractive sector [63]. For example, in the energy sector, even though energy resources are exported, the areas where the gas pipelines are located, Tanintharyi region and Rakhine State, have the lowest per capita electricity consumption in Myanmar [64]. Limited access to electricity relates to reliance of fuel wood on forests including mangroves and underdevelopment in those coastal regions. Deforestation, underdevelopment, and dependency on fuel wood contribute to flood vulnerability. Other injustices can also be witnessed in local protests against dam construction, which increase flood vulnerability. Moreover, dams in Myanmar are not designed for flood control [65].

Inadequate resource governance can also be witnessed in widespread land confiscations, which increases flood vulnerability through poverty and inequality. In 1999 and 2000, transition to market-based economy led to land appropriations, for the sake of agribusiness such as oil palm plantations and industrial production [55]. Therefore, millions of land acres were transferred to private companies. This made millions of people to depend only on labor wage alone [66]. By 2030, there was a national mandate to convert 10 million acres [67]. While violent land grabbing especially for agribusinesses makes many people landless, there is no systematic policy for those conflict resolutions of their concerns [15]. Those landless and displaced labors were less likely to be reabsorbed in the manufacturing industries [15] creating enormous inequalities. Finally, poor, landless, and jobless communities have limited capacities to cope and severely vulnerable to large-scale flooding after damaging their livelihood base.

Those flood vulnerabilities were exacerbated by continuous conflicts currently and dynamically. Myanmar has experienced a long history of civil war [16]. Conflicts in ethnic areas make developments of those areas seriously lacking behind. Conflicts with ethnic armed groups, and shared resource exploitation with private actors in those conflicted and ethnic people areas, caused marginalization of ethnic populations [68]. Current conflicts and instabilities in Myanmar have wider spill-over effects. Current instabilities disrupted humanitarian assistance, climate action, food production, employment, public services, and development activities. As many international projects and companies left Myanmar, many people faces difficulties and uncertainties to secure food, job, income, and safety. The problems are exacerbated by economic downturn of the political instability and the COVID-19 pandemic [16]. The consequences are the souring food and commodity prices, which make it difficult to be affordable by poor population. Despite some progress in the environmental governance in the recent years in terms of local recognition and transparency, now current political development has put all of those efforts to a halt.

A consequence of conflicts and land appropriations, displacement, is worth to mention for its contribution to flood risk. Besides increasing flood vulnerability, displacement due to the land grabbing, and conflicts, can lead to high exposures to flood-prone areas and flood plains and other sensitive areas such as urban slums. Displacement was used as a counter-insurgency measures, leaving many people vulnerable. Dramatically high figures of rural-to-rural migration [69] highlight that the displaced people are not absorbed by urban jobs [69]. Moreover, urban jobs are so inadequate that the income differences between urban and rural jobs were not apparent. In addition, out migration to Thailand has continued quickly [70], despite unsafe conditions there [71]. Increased migration and displacements disrupted traditional social networks of care and support, eroding community bonds [72]. Therefore, displacement can decrease social adaptive capacities to floods. Finally, displacement cycle begins when those people have to displace again from floods.

Despite those widespread inequalities and poverty in Myanmar, there is limited expenditure and funding on public services and infrastructure such as education and health facilities. Total government expenditure on education is still very low compared to the ASEAN average [73]. Moreover, there are huge disparities in infrastructure and provision of public services between rural and urban areas, especially lower in bordered areas [53]. Those regional differences become more apparent due to the conflicts faced in the bordered ethnic people's regions [53]. Inadequate health facilities can lead to major fatalities in case of large-scale flooding and other disasters. Levels of education are directly related to risk awareness of flooding and preparatory actions. Deficient public services cannot facilitate well to vulnerable people to escape out of the underemployment cycle. Particularly in health and education sector, the poor people have to subsist with under-qualified and under-funded public options. Moreover, lack of road networks and communication infrastructure leave people in remote areas of Myanmar isolated until humanitarian aid could reach them.

In the disaster risk management, and climate change adaptation actions, there are certain factors that drive to insufficiency and ineffectiveness in flood risk reduction in Myanmar. Despite widespread deforestation, ecosystem-based climate change adaptation is still yet to be developed in Myanmar. Similarly, social protection is at its infancy, in terms of coverage and level of institutionalization, with remarkable exclusion errors [17]. The coverage on problems related to health, child, and elderly support is extremely inadequate. The share of government expenditure allocated to social protection is limited only at 0.8% of National GDP, making Myanmar the lowest in its spending for social protection in the region, compared to regional average of 2% [17]. Moreover, the current social protection system has focused on the wrong demographics, with only 0.1 percent of existing social protection spending emphasized on the most vulnerable and the poorest [17], making major population to depend on informal forms of safety nets. Community-based actions are crucial for the success of adaptation and risk reduction efforts; however, the current political climate makes it challenging for international actors to support these efforts [16]. Consequently, progress in climate action has been hindered. Moreover, development aid providers are struggling with issues of increasing domestic and regional inequalities [53]. Given these challenges, it is likely that international assistance for response and recovery efforts would be difficult in the event of a large-scale flooding.

Importantly, root causes were interconnected reinforcing each other. For example, Rustad et al. (2008) found a link between Myanmar's forest resources and armed conflicts [74]. Woods (2019) also supported the relationship between Myanmar's natural resources and history of armed conflicts in ethnic areas [75]. Additionally, countries affected by conflicts are at risk of corruption in their natural resource sector due to weakened and resource-deprived bureaucratic systems [76]. Poverty in Myanmar is related to low levels of education [73] as a result of inadequate funding for education infrastructure. Households head with primary or lower education are likely to be poor [73].

Based on the described root causes and their consequences on socio-ecological systems and associated flood risk, the following dynamic pressures were identified. They are deforestation, agricultural expansion, urban extension, mining, conflicts, and wetland loss. Other disasters such as heatwaves are also considered here as dynamic pressures in this study.

3.2 Temporal distribution of dynamic pressures

The figure shows changes in dynamic pressures of flood risk in comparison with flood impacts in Myanmar. **Figure 4(a)** and **(b)** shows consistently symmetric trends between deforestation and expansion of agricultural areas. While forests were constantly declined between the period between 1988s and 2000s (from 68 percent to 57 percent), agriculture was consistently flourished during that period. Ochards and planations (**Figure 4(c)**) mirrored the same trend of agriculture. In that period, forest areas of around 767,000 km^2 were declined while agriculture had increased by 38,000 km^2. When comparing to deforestation and agriculture increase with flood impacts during that period (**Figure 4(g)** and **(h)**), it shows that frequent relatively small-scale floods that affected up to 359,000 populations and brought up to 68 deaths during the same period. Later after years of 2000, rates of deforestation were slow down to further loss from around 404,000 km^2 to 387,00 km^2. The same rate applied for increase in agricultural and orchard plantation areas. At the same time after 2000s, flood events were more intense and rarer with less frequent smaller-scale floods. Therefore, the gaps in the intensities and frequencies of flood impacts become wider after 2000s.

When detecting at the figures in urban, wetland and mining areas (**Figure 4(e)**), there were substantial declines in wetland areas across Myanmar. Urbanization has increased steadily from 587 km^2 to 958 km^2 similar to sluggish increase in mining areas. Urban areas have also larger rate of increase before 2000s, similar to agricultural trends. Compared to agriculture, urbanization was less a dynamic pressure due to its minor spatial extent. Small fluctuation in mining areas might indicate that old mines were demolished and new mines were produced. Overall, looking at reduction trends in forests and wetland areas, the flood risk reduction capacities were diminished over time, while flood-sensitive and flood-induced and flood-sensitive agriculture, orchard plantations, and mining areas have proliferated over the period.

Average annual land surface temperature (**Figure 4(d)**) shows a fluctuated trend over the period from 2000 to 2022. Land surface temperature had its peaks in the years of 2005, 2010, 2013, 2015, 2019, and 2020. In 2010, there were extreme meteorological droughts occurred in Myanmar, which saw record-breaking temperatures of

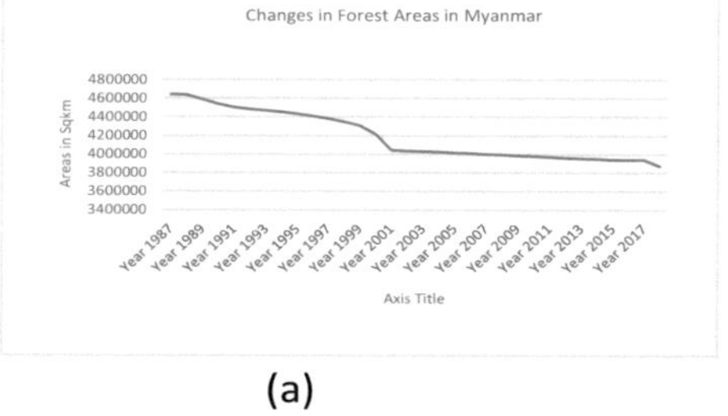

(a)

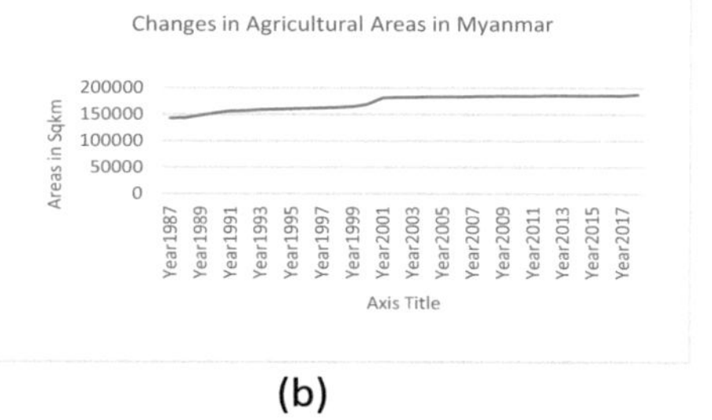

(b)

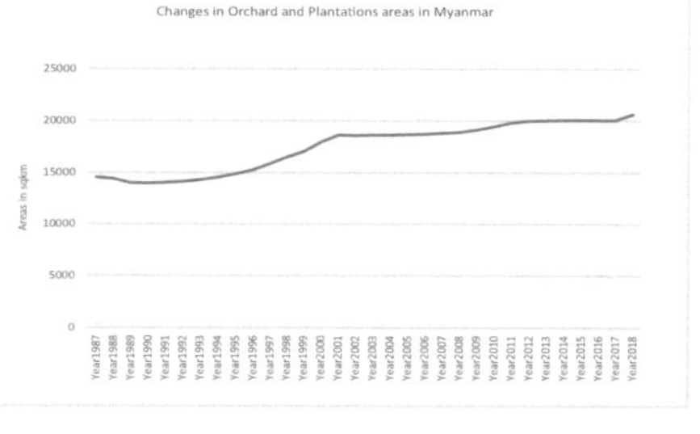

(c)

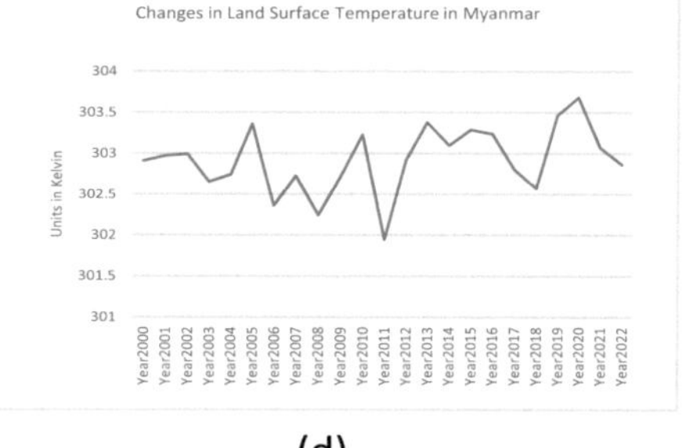

(d)

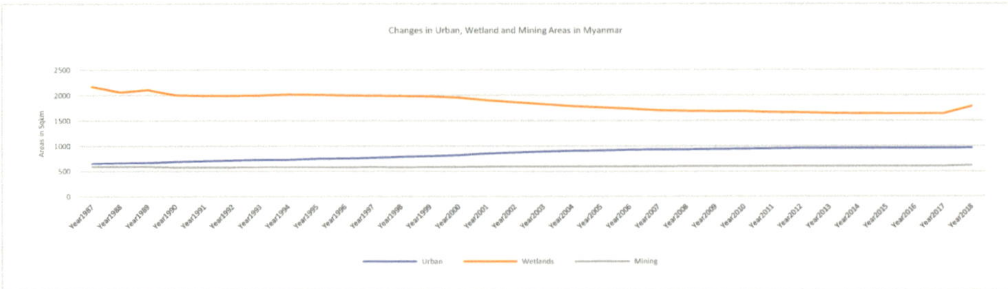

(e)

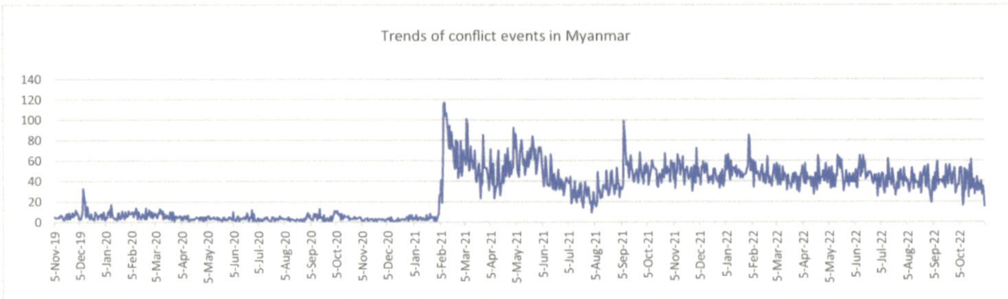

(f)

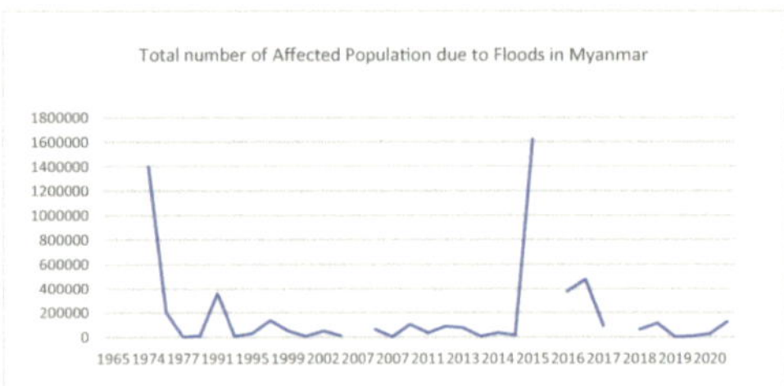

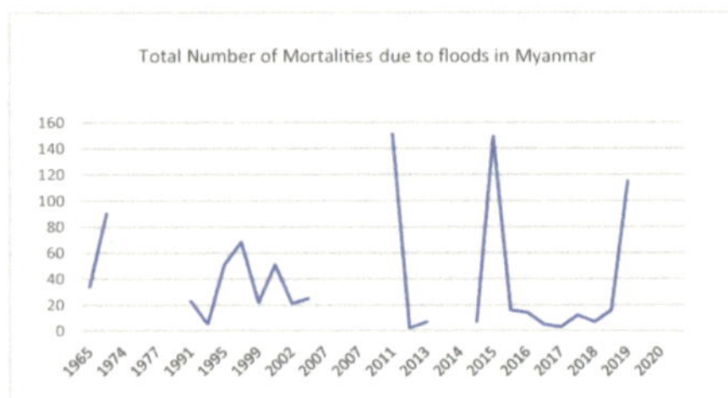

(g) (h)

Figure 4.
Temporal changes in flood risk and its dynamic pressures: (a) changes in forests areas (b), changes in agricultural areas, (c) changes in orchards and plantation areas, (d) changes in land surface temperature, (e) changes in urban, wetlands, and mining areas, (f) trends of conflict events, (g) trends of affected population due to floods, and (h) trends of mortalities due to floods in Myanmar.

three times higher than normal summer months. When compared to floods, nationwide large-scale flooding occurred with 1.6 million affected populations and 149 deaths in 2015, which is also a peak year in average annual land surface temperature. However, in 2011 when there was a decline in land surface temperature, it coincided with a large-scale flood which caused 151 deaths of people.

When it comes to added pressures by conflicts (**Figure 4(f)**), there was a significant increase in the number of conflict events due to recent political changes in Myanmar. Despite existences of dynamic conflicts in Myanmar, data coverage was only limited to the last three years. Fortunately, only small-scale flood events were reported so far while experiencing the high nationwide conflicts. However, there was a deadly flood occurred in Mon State in 2019, while 33 conflicts were recorded. To understand the spatial overlap of those pressures, spatial dimension is required to considered. Therefore, the spatial distributions of dynamic pressures were discussed in Section 4.5.

Overall, the results show that significant changes in dynamic pressures over the period from 1987 to 2018. Forest cover reduced from 68 percent in 1987 to 57 percent in 2018. In 1987, coverage of agricultural area was 21 percent, which increases to 27 percent in 2018. For orchards and plantations, total plantation areas increase from 6 percent to 8 percent in 2018. Urban areas increase from 0.09 percent of the total country areas in 1987 to 0.14 percent in 2018. Mining areas increase from 0.08 to 0.09 percent, while wetland areas reduced from 0.32 percent to 0.26 percent in 2018 (**Figure 5**).

3.3 Temporal evolution of dynamic pressures in highest risk townships

Figure 6 shows the temporal evolution of dynamic pressures in the 10 highest risk townships (See those township locations in **Figure 5**). It is important to note that the values are not relative values to township areas but rather absolute values in km^2. In trends of forest loss (**Figure 6(a)**), Paletwa, Myitkyina, and Minbya townships have substantially high forest areas compared to other high-risk townships, followed by Kyauktaw. Townships such as Myitkyina and Paletwa have highest forest loss of around 500 km^2 among them during the period from 1987 to 2018, while Myinbya and Kyauktaw have less loss of around 100 km^2. Most of those losses happened between 1999s and 2000s indicating a significant decrease during those periods.

In trends of agricultural area increase (**Figure 6(b)**), extent of agricultural areas ranges from 51 km^2 to 1600 km^2. Townships with large agricultural extent are Bogale, Mawlamyinekyun, Wakema, Mrauk-U, Pauktaw, Minbya, Rathedaung, Kyauktaw, Myitkyina, and Palawe in ascending order. While most of the trends in agricultural extent are stable, Myitkyina has increase sharply from around 295 km2 to 579 km2 over the period, while it is assessed as the highest risk township in Myanmar. It sharply increased in years between 2000s and 2001s, and slowly increased in the later years until it saw again a sharp increase between 2017 and 2018. Bogale also experienced its increase in agricultural areas steadily after 2001.

In trends of orchard and plantation area increase (**Figure 6(c)**), there is a widespread range from around 1.1 km^2 to 900 km^2 in the year 2018. Townships with large plantation extent are Paletwa, Kyauktaw, Myinbya, Mrauk-U, Pauktaw, and Myitkyina in ascending order. Most townships with high plantation areas such as Myitkyina, Kyauktaw, Paletwa, and Minbya experienced a steep increase in 1994s till 2001s where the trend became stable one. Among them, Paletwa and Myitkyin experienced a steep rise during the whole period. Pauktaw and Mrauk-U experienced a level-off trend over the whole period from 1987 to 2018 with little raise between 1995s and 2000s. Wakema and Mawlamyinekyun have almost no plantation areas.

In trends of urban settlement areas (**Figure 6(d)**), all townships have similar urban extents and have little differences in area extents. Interestingly like other drivers, they saw similar changes during some points of times over the periods. Those sudden and slight increase happened in three steps of the times: in 1990, 2003, and 2017.

In trends of wetland areas (**Figure 6(e)**), only Myitkyina has relatively high wetland areas while other townships have only 0 to 3 km^2. Surprisingly, Myitkyina shows slight increase in wetland areas from approximately 25 km^2 to 34 km^2 over the period from 1987 to 2018. Similarly, only Myitkyina has mining areas (**Figure 6(f)**) among other high-risk townships. Mining areas of 0.4 km^2 have started to increase in 2000 till it has demolished in 2003, and it increased again in 2017.

Looking at the temperature trends in high-risk townships (**Figure 6(g)**), land surface temperature values range from 297 to 303 kelvin. Over the period from 2000

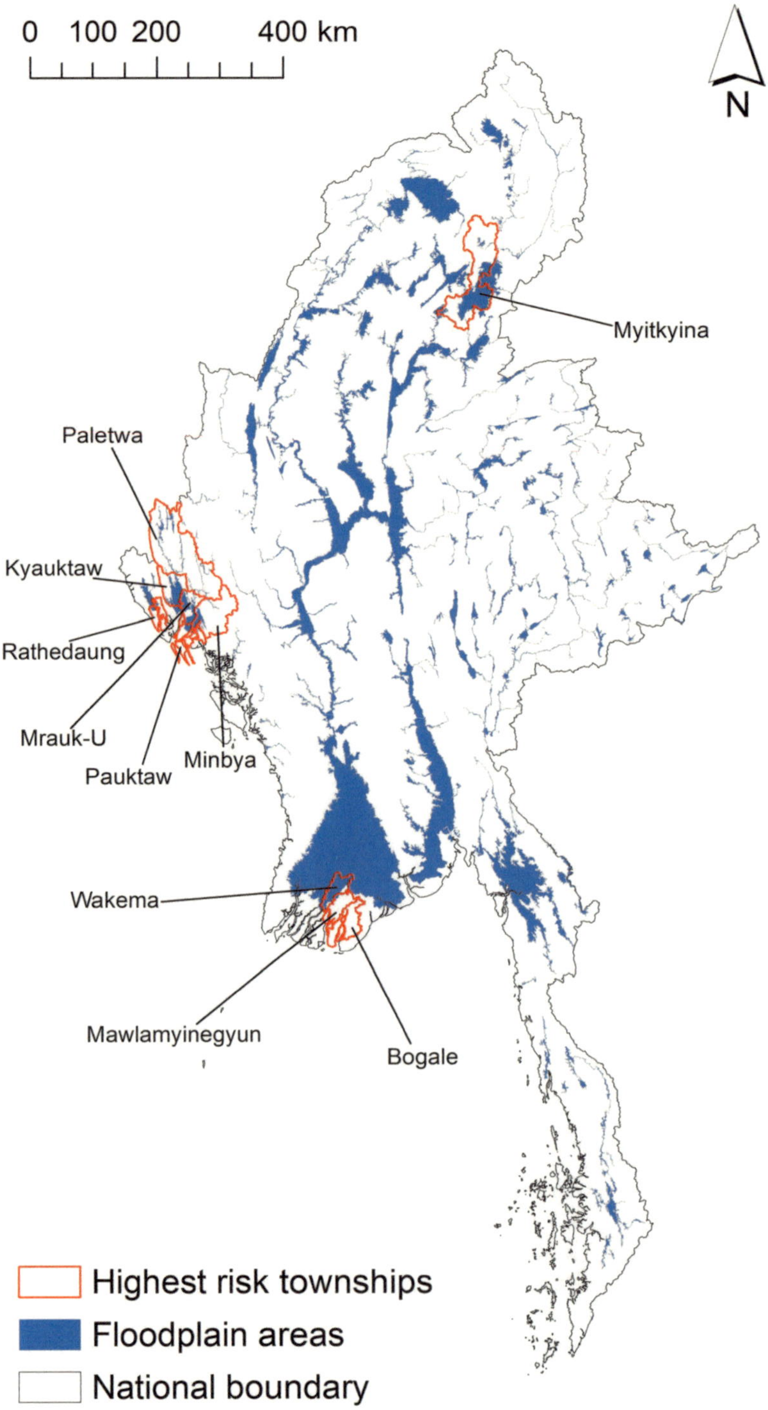

Figure 5.
Maps showing flood plain areas and ten highest (flood) risk townships in Myanmar.

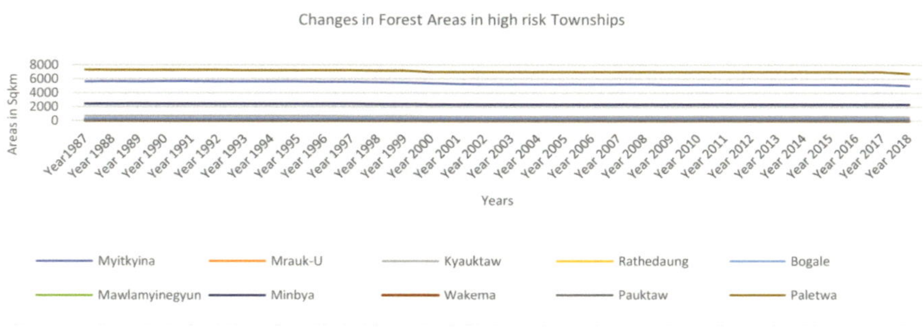
Changes in Forest Areas in high risk Townships
Areas in Sqkm
Years
Myitkyina
Mrauk-U
Kyauktaw
Rathedaung
Bogale
Mawlamyinegyun
Minbya
Wakema
Pauktaw
Paletwa

(a)

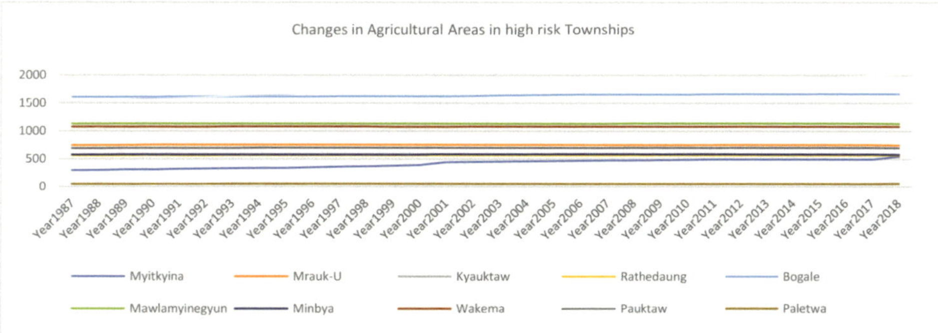
Changes in Agricultural Areas in high risk Townships
Myitkyina
Mrauk-U
Kyauktaw
Rathedaung
Bogale
Mawlamyinegyun
Minbya
Wakema
Pauktaw
Paletwa

(b)

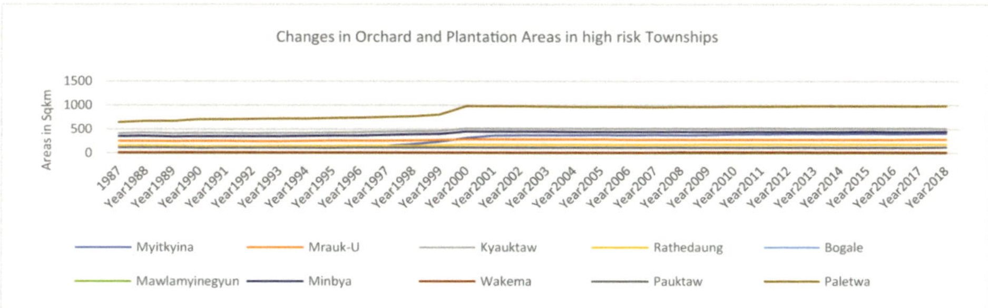
Changes in Orchard and Plantation Areas in high risk Townships
Areas in Sqkm
Myitkyina
Mrauk-U
Kyauktaw
Rathedaung
Bogale
Mawlamyinegyun
Minbya
Wakema
Pauktaw
Paletwa

(c)

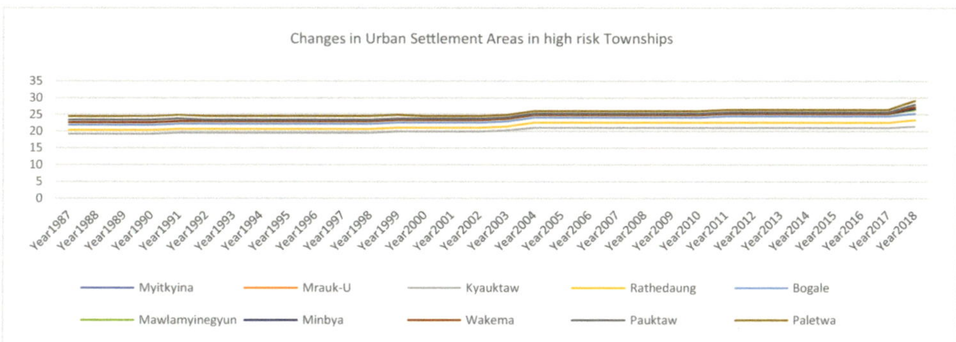
Changes in Urban Settlement Areas in high risk Townships
Myitkyina
Mrauk-U
Kyauktaw
Rathedaung
Bogale
Mawlamyinegyun
Minbya
Wakema
Pauktaw
Paletwa

(d)

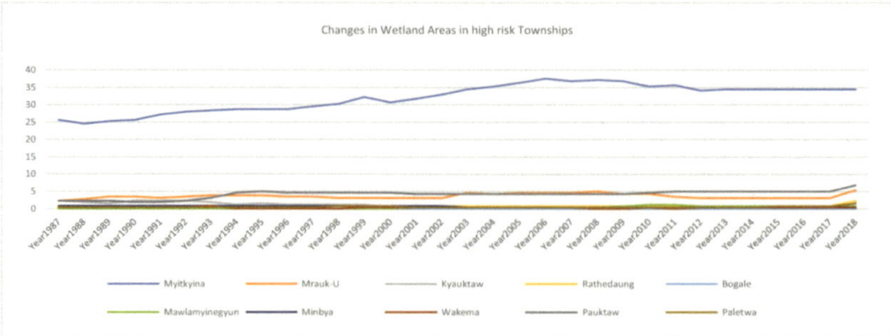

(e)

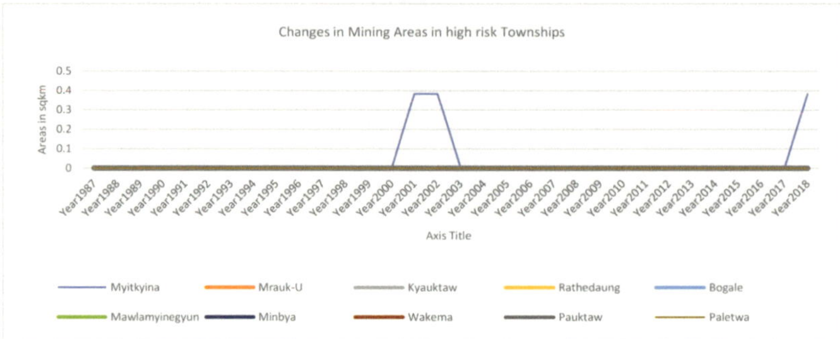

(f)

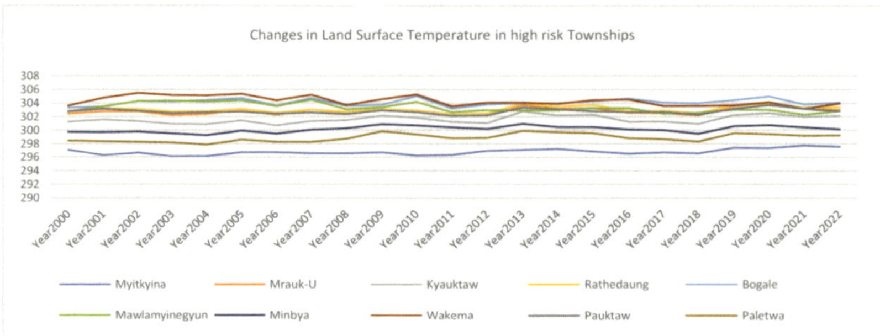

(g)

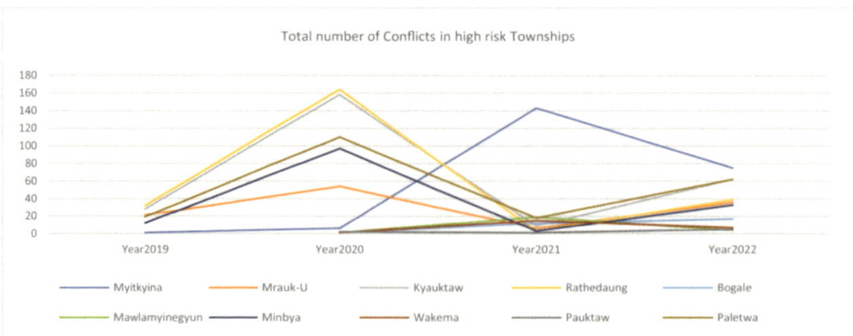

(h)

Figure 6.
Temporal changes in dynamic pressures in high risk townships: (a) changes in forests areas, (b) changes in agricultural areas, (c) changes in orchards and plantation areas, (d) changes in urban settlement areas, (e) changes in wetlands areas, (f) changes in mining areas, (g) changes in land surface temperature, and (h) trends of conflicts in high-risk townships.

to 2022, temperature values saw fluctuated trends. It was found that temperatures have slightly downward trends during those critical years of large-scale and high-impacted floods in 2008, 2011, and 2015 except for Bogale township. That means temperature was slightly higher before those years except for the case in 2015, which has stable and lower temperature before.

When it comes to changes in conflicts over the year from 2019 to 2022 in high-risk townships (**Figure 6(h)**), it was uncovered that most of the townships (except Myitkyina) have high frequency of conflict events (54 events to 164 events) in 2020 and the numbers dropped in 2021 when the usurpation of national power took place. This connotes that those townships likely experienced pre-existing conflicts over their history. However, it required data from previous years to draw explicit conclusions. A reverse trend was occurred in Myitkyina, where there is stable trend before 2021, where it reaches its peak at 143 events.

3.4 Temporal evolution of dynamic pressures in flood plains

Figure 7 shows changes in dynamic pressures in the flood plains (See area distribution of flood plains in Figure [5]). It was apparent that many of the dynamic pressures experienced a remarkable drastic change (increase or decrease) in the period before 2001s followed by steady changes after 2001s except for mining. Among those drastic changes, it saw a steep slope especially in period between 1994s and 2001s indicating that some substantial root causes systematically existed to drive those dynamic pressures. Moreover, those temporal patterns also followed major trends of dynamic pressures in the whole Myanmar.

The results underscore significant dynamic changes within the flood plain areas. Forest area coverage (**Figure 7(a)**) was 3 percent in the flood plains in 1987, which reduced to 2 percent in 2018. For agriculture (**Figure 7(b)**), the total percent of agricultural flood plains compared to the country areas increased from 7 percent to 8 percent in 2018. Plantations in the flood plains (**Figure 7(c)**) also rised double from 0.8 percent of the total country area to 1.16 percent. Urban settlements (**Figure 7(d)**) in the flood plain areas increased from 0.05 percent of the total country areas to 0.07 percent. However, 56 percent of total urban settlements within flood plains reduced to 53 percent of total urban settlements in 2018. Wetland areas (**Figure 7(e)**) also saw similar raise with similar figures. Mining in the flood plains (**Figure 7(f)**) increased from 0.03 percent of country areas to 0.04 percent. Surprisingly, 44 percent of total mining areas existed in the flood plains in both 1987 and 2018.

3.5 Spatial distribution of dynamic pressures in relation to flood risk

Figure 8 below shows spatial distribution of changes of dynamic pressures and flood risk. **Figure 8(a)** represents area amount of deforestation for the period from 1987 to 2018. Spatially, deforestation was mainly occurred in Shan States, with forest losses of more than 1000 square kilometers (**Figure 8(j)**). It can be seen that the majority of forest conversions were realized in the periphery areas of dry zone region and in the middle of Shan State. It was clear in the maps that forests were converted to agricultural areas except for the lower part of Bago region (**Figure 8(a)** and **(c)**). Highest forest loss occurred in Hsipaw, Kanbulu, Kyethi, Nansang, and Tanitharyi townships, where majority of those losses were converted to agricultural areas, except for the loss in Tanintharyi township which transformed to orchards and oil palm

Figure 7.
Temporal changes of dynamic pressures in flood plains: (a) changes in forests areas, (b) changes in agricultural areas, (c) changes in orchards and plantation areas, (d) changes in urban settlement areas (e) changes in wetlands areas (f) changes in mining areas in flood plains.

plantations. Forests were consistently declined in the bordered areas to the neighboring countries. Ayeyarwady delta experienced remarkable loss in mangrove forests compared to Tanintharyi Region which have higher proportion of orchards and plantations (**Figure 8(e)**). It is interesting to note that Myikyina Township is the township, which ranked as the second highest townships in terms of its forest loss and agricultural increase in flood plains (**Figure 8(k)** and **(m)**) while being the township with the highest flood risk of socio-ecological systems (**Figure 8(aa)**). Forest area loss within flood plains mainly occurred in the Kachin and upper part of Sagaing Region, which are mainly upland forests of Ayeyarwady River in Myanmar (**Figure 8(k)**). Spatial distribution of forest loss and flood risk may not match, except for mangrove forest loss in the Ayeyarwady delta (**Figure 8(j)** and **(aa)**).

Agricultural areas were mainly concentrated in the middle Myanmar: dry zone area, Ayeyarwaddy Delta, and Rakhine State which are highly prone to numerous climate risks such as floods, cyclones, and droughts (**Figure 8(c)** and **(d)**). However, increase in agriculture occurred likely in the form of a circle in the middle Myanmar during the period between 1987 and 2018 (**Figure 8(c)** and **(i)**). Over the period, the increase was the highest in Shan State and Sagaing Region with the largest in Hsipaw and Kanbulu of approximately 2000 km^2 each (**Figure 8(i)**). Many agricultural areas in the flood plains were mainly concentrated in the Ayeyarwady delta (**Figure 8(d)**). However, in the **figure 8(m)**, increase in agriculture within flood plains mostly occurred in the Kachin, upper part of Sagaing region, and Shan State with amount up to 294 km^2 per township. Changes are insignificant in Ayeyarwady delta after 1987 compared to those regions. Areas with high agricultural extension were spatially coincident with the second highest risk classes (**Figure 8(i)** and **(aa)**).

Orchards and plantations areas were mainly concentrated in the Mon, Tanintharyi, Kayin, and Rakhine State (**Figure 8(n)**). Apart from those areas, orchards and plantation areas were limited in other states and regions. Mon State and Tanintharyi Region saw the highest increase in orchards and plantation during the period between 1987 and 2018, while the increase in the flood plain areas occurred mostly in Kachin State and Tanintharyi Region (at least 110 km^2 per townships) (**Figure 8(o)**). Some of spatial patterns of general increase of orchards (i.e., in Kachin and Rakhine state) were overlapped with flood exposure patterns (**Figure 8(y)**). When compared to the flood risk, increase in orchards and plantations in Mon State and Tanintharyi Region spatially matched with the second highest risk townships in those regions (**Figure 8(n)** and **(aa)**).

Urbanization rates were minor in Myanmar compared to other dynamic pressures such as deforestation and agricultural increase (**Figure 8(i)**). As shown in **Figure 8(i)**, it is invisible at all at the National-scale map. In the zonal classification map that shows changes per townships (**Figure 8(p)**), it was apparent that urbanization mainly occurs near major cities such as Yangon and Mandalay, and in Kachin State to some extent. However, even though the extent of increase was small (at most 25 km^2 per township), the increase mainly happened within the floodplains (**Figure 8(p)** and **(q)**). Compared to flood risk, the distribution of urbanization was widespread throughout the country (**Figure 8(p)** and **(aa)**).

Unlike urbanization, mining and wetland areas were only limited to a few states and regions (**Figure 8(r)** and **(t)**). Decrease in wetland areas mainly occurred near Yangon and Ayeyarwady regions, while most of those losses located on flood plains (**Figure 8(r)** and **(s)**). Moreover, areas of wetland loss partly fall in some highest risk townships in Mon State. Similar to the wetland areas, major increase in mining areas realized in upper part of Myanmar (**Figure 8(t)**). Most of the increase (6.51 km^2 out

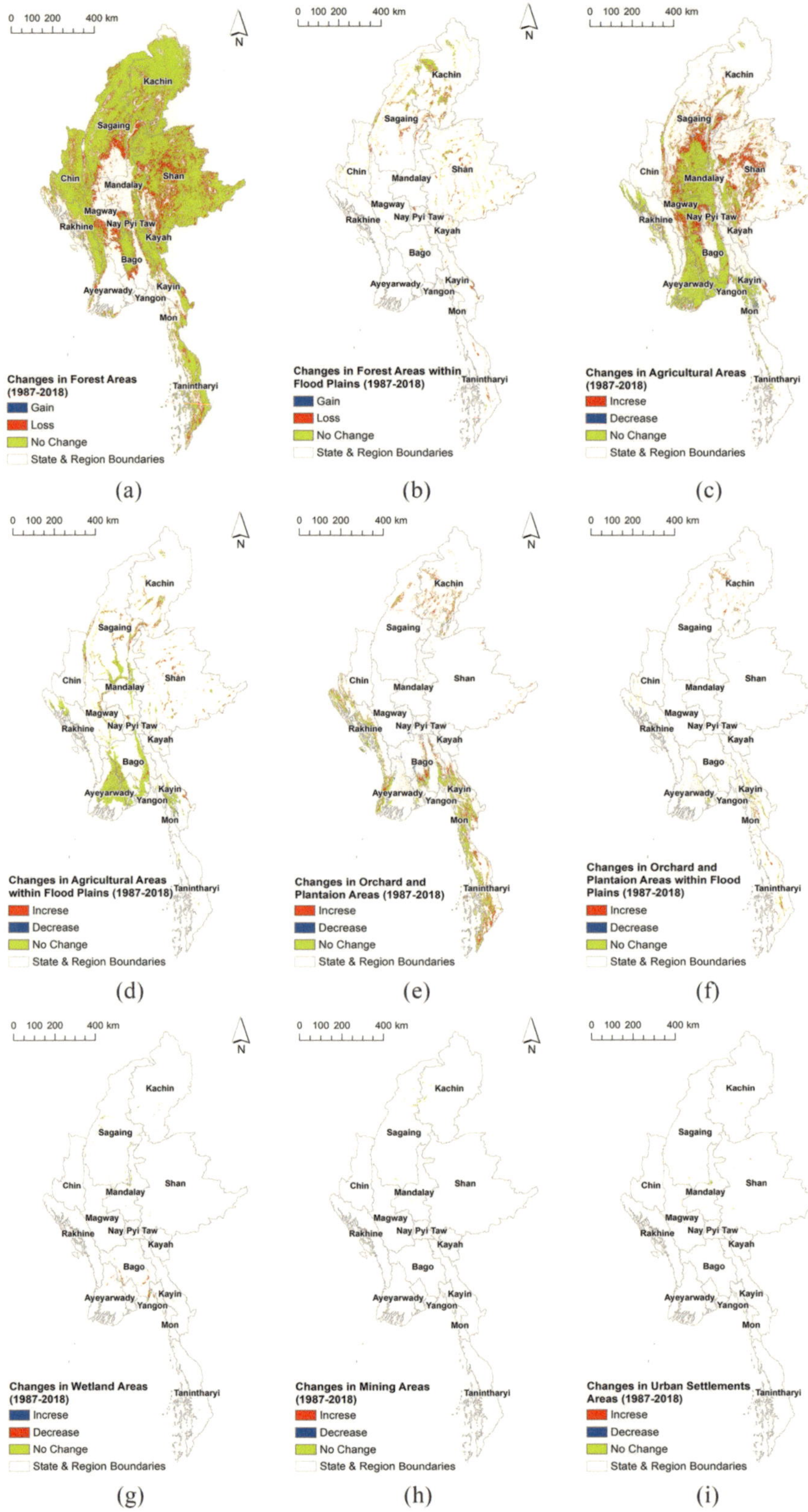
0 100 200 400 km
N
Kachin
Sagaing
Chin
Mandalay
Shan
Magway
Rakhine
Nay Pyi Taw
Kayah
Bago
Ayeyarwady
Kayin
Yangon
Mon
Tanintharyi
Changes in Forest Areas (1987-2018)
Gain
Loss
No Change
State & Region Boundaries
(a)
Changes in Forest Areas within Flood Plains (1987-2018)
Gain
Loss
No Change
State & Region Boundaries
(b)
Changes in Agricultural Areas (1987-2018)
Increse
Decrease
No Change
State & Region Boundaries
(c)
Changes in Agricultural Areas within Flood Plains (1987-2018)
Increse
Decrease
No Change
State & Region Boundaries
(d)
Changes in Orchard and Plantaion Areas (1987-2018)
Increse
Decrease
No Change
State & Region Boundaries
(e)
Changes in Orchard and Plantaion Areas within Flood Plains (1987-2018)
Increse
Decrease
No Change
State & Region Boundaries
(f)
Changes in Wetland Areas (1987-2018)
Increse
Decrease
No Change
State & Region Boundaries
(g)
Changes in Mining Areas (1987-2018)
Increse
Decrease
No Change
State & Region Boundaries
(h)
Changes in Urban Settlements Areas (1987-2018)
Increse
Decrease
No Change
State & Region Boundaries
(i)

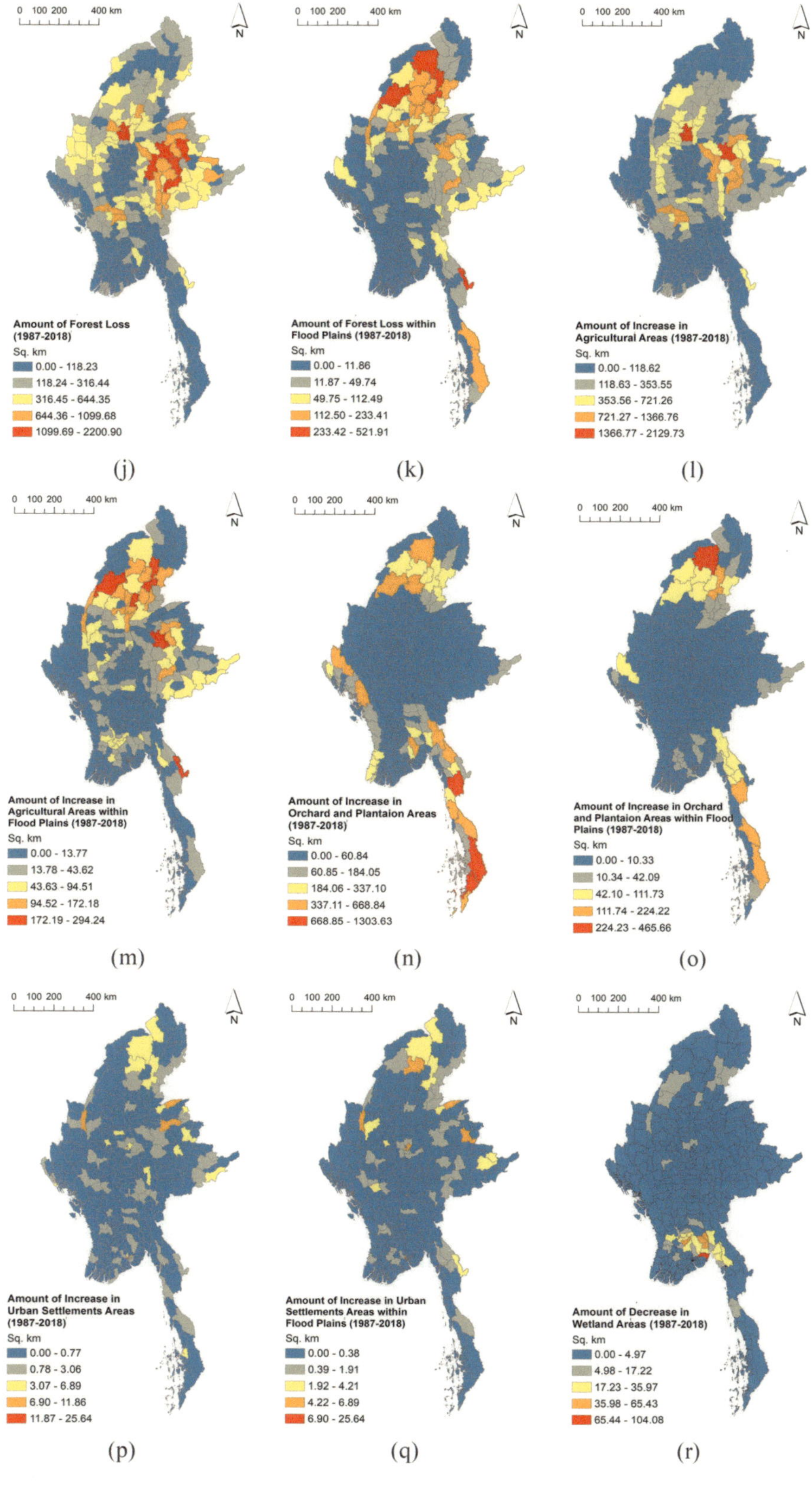
0 100 200 400 km
N
Amount of Forest Loss (1987-2018)
Sq. km
0.00 - 118.23
118.24 - 316.44
316.45 - 644.35
644.36 - 1099.68
1099.69 - 2200.90
(j)
Amount of Forest Loss within Flood Plains (1987-2018)
Sq. km
0.00 - 11.86
11.87 - 49.74
49.75 - 112.49
112.50 - 233.41
233.42 - 521.91
(k)
Amount of Increase in Agricultural Areas (1987-2018)
Sq. km
0.00 - 118.62
118.63 - 353.55
353.56 - 721.26
721.27 - 1366.76
1366.77 - 2129.73
(l)
Amount of Increase in Agricultural Areas within Flood Plains (1987-2018)
Sq. km
0.00 - 13.77
13.78 - 43.62
43.63 - 94.51
94.52 - 172.18
172.19 - 294.24
(m)
Amount of Increase in Orchard and Plantaion Areas (1987-2018)
Sq. km
0.00 - 60.84
60.85 - 184.05
184.06 - 337.10
337.11 - 668.84
668.85 - 1303.63
(n)
Amount of Increase in Orchard and Plantaion Areas within Flood Plains (1987-2018)
Sq. km
0.00 - 10.33
10.34 - 42.09
42.10 - 111.73
111.74 - 224.22
224.23 - 465.66
(o)
Amount of Increase in Urban Settlements Areas (1987-2018)
Sq. km
0.00 - 0.77
0.78 - 3.06
3.07 - 6.89
6.90 - 11.86
11.87 - 25.64
(p)
Amount of Increase in Urban Settlements Areas within Flood Plains (1987-2018)
Sq. km
0.00 - 0.38
0.39 - 1.91
1.92 - 4.21
4.22 - 6.89
6.90 - 25.64
(q)
Amount of Decrease in Wetland Areas (1987-2018)
Sq. km
0.00 - 4.97
4.98 - 17.22
17.23 - 35.97
35.98 - 65.43
65.44 - 104.08
(r)

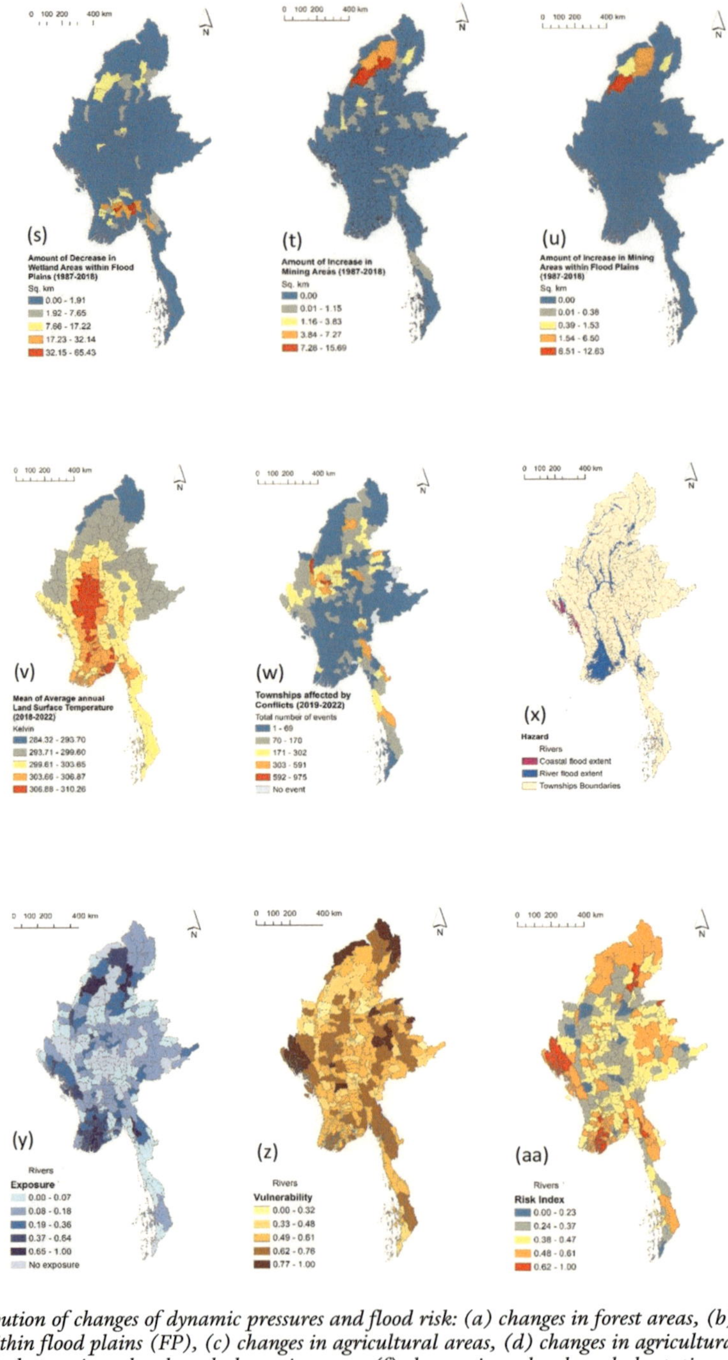

Figure 8.

Spatial distribution of changes of dynamic pressures and flood risk: (a) changes in forest areas, (b) changes in forest areas within flood plains (FP), (c) changes in agricultural areas, (d) changes in agricultural areas within FP, (e) changes in orchards and plantation areas (f) changes in orchards and plantation areas within FP (g) changes in wetland areas, (h) changes in mining areas, (i) changes in urban areas, (j) amount of forest losses, (k) amount of forest losses within FP, (l) amount of agricultural areas increase, (m) amount of agricultural areas increase within FP, (n) amount of increase in plantation areas, (o) amount of plantation areas increase within FP, (p) amount of increase in urban areas, (q) amount of urban areas increase within FP, (r) amount of wetland losses, (s) amount of wetland losses within FP, (t) amount of mining area increase, (u) amount of mining areas increase within FP, (v) mean of average annual land surface temperature, (w) townships affected by conflict events, (x) hazard, (y) exposure, (z) vulnerability, and (aa) risk index.

of 7.26 km^2 in the highest class) occurred in the flood plains, indicating the potential unsafe conditions (**Figure 8(u)**). Some of the townships with highest mining increase fall in the second highest risk categories (e.g., Homalin, Tanai, Hkamti, Injangyang), in the highest exposure (e.g., Homalin) and in the township with the highest forest loss (e.g., Hsipaw).

Interestingly, average annual land surface temperature patterns match consistently with agricultural area patterns where the highest values exist in the central Myanmar (**Figure 8(v)** and **(c)**). This implies that almost all agricultural areas in Myanmar (1987 to 2018) face extreme land surface temperature (over 301 kelvin). When comparaed to flood risk with temperature values, some townships with the highest risk and those with second highest temperature matches each other (e.g., in Ayeyarwady delta and Rakhine State) (**Figure 8(v)** and **(aa)**). Townships with the second highest temperature values (over 301 kelvin) overlapped with some townships of highest flood vulnerability (e.g., in Ayeyarwady delta: Bogale, Maubin, Mawlamyinekyun) (**Figure 8(z)**) and highest flood exposure (e.g., in Rakhine State: Mrauk-U, Rathedaung, and Kyauktaw) (**Figure 8(y)**).

In terms of distribution of conflict events (**Figure 8(w)**), conflicts took place highly in Sagaing and Magway regions. It was followed by Kayah, Tanintharyi, Kachin, and Kayin states. Some townships with the highest conflicts such as Kale, Monywa, Yesagyo, Khin-U fall in the second highest risk classes (**Figure 8(aa)**), eventuating in compound risk. In addition, other townships that have the highest conflicts such as Demoso, Yinmarpin, Salingyi, and Myaing are also near to second highest flood risk with risk index of over 0.45. Similarly, those areas with high conflicts rates also have the second highest flood vulnerability indices, except for Kale.

The analysis detected that the 100-year return period river and coastal flood hazard predictably followed the contours of the rivers and coast lines in Myanmar (**Figure 8(x)**). The highly exposed populations and ecosystems appeared in the Ayeyarwady region, Rakhine, and Kachin states. Populations and ecosystems in 73 townships were not exposed (**Figure 8(y)**). Interestingly, two townships with the highest flood exposure (i.e., Myitkyina and Homalin) have high forest loss of 624 km^2 and 940 km^2, respectively. It was followed by Toungup, Pyapon, and Laputta with forest loss of at least 150 km^2 each where most of those forests are expected to be mangrove forests (Jones 2022).

Vulnerability was widespread through Myanmar (**Figure 8(z)**). Our analysis shows that some vulnerability indicators were more critical than others. For the six most critical indicators, most townships had an index score between 1.0 and 0.9. The total number of townships is 331 townships in Myanmar. Access to healthcare was poor with a maximum of one doctor and 51 hospital beds per 10,000 people in 222 and 312 townships, respectively. Extent of wetland areas for flood water storage and reduction capacities was limited below 2 km^2 for 307 townships. Accessibility was also low for 290 townships with the density of roads between 0 and 2.82 (road kernel density: km road/km2, search radius = 5 km). The percentage of households owning a boat, which are important for saving lives and transportation during flood events, was between zero and 4.86 percent in 242 townships. Poverty was also widespread. The average income for 312 townships was between 324,225 to 3,720,021 Kyat (US$ 251 to US$2884), and 14 townships had an average income below the 2015 poverty line of 475,595 Kyat (US$ 369) [23].

Among the remaining indicators, some contributed to vulnerability more than others. For access to alternative electricity sources such as solar panels and generators that are useful during the power shortages during floods, 199 townships had maximum of 20 percent of households with these facilities. Extent of forest areas has less than 9 km^2 for 127 townships. However, rate of loss in wetland areas, expansion rate in urban areas, and vector-borne diseases contributed less to vulnerability for most townships.

Spatially, the most vulnerable townships were concentrated in bordered areas of Myanmar (**Figure 8(z)**). The majority of townships in the highest vulnerable classes were located in the Rakhine state, followed by Shan, Chin, and Kachin states. Highly vulnerable areas overlapped with high forest loss areas and high agricultural increase areas to some extent. Logically, the least vulnerable townships were concentrated in Yangon region, where the former capital of Myanmar, Yangon, is located. This was followed by the Mandalay region which has the second largest city, Mandalay, and the Nay Pyi Taw which is the current capital city. Therefore, urban areas were less vulnerable to flood risk compared to rural bordered areas in Myanmar.

Figure 8(aa) shows the spatial distribution of risk. Risk was highly concentrated in townships in the Rakhine and Ayeyarwady regions, followed by Kachin State. The elements contributing to risk for townships in the highest class were multifaceted. Common critical elements included high levels of poverty, limited capacities of wetlands, and a low number of hospital beds and doctors per 10,000 people. Rural areas were mainly present in the highest risk class with the travel time to the nearest city ranging from 52 to 521 minutes.

3.6 Spatial relations of flood risk and its driving factors (dynamic pressures)

Table 3 below shows spatial relationships between dynamic pressures and flood risk. It shows some of the significant and non-significant spatial relations between dynamic pressures and flood risk. Agricultural areas' increase in flood plains, wetland loss in flood plains, wetland loss, forest loss in flood plains, forest loss, increase in plantation areas, and increase in plantation areas in flood plains have statistically significant positive spatial relationships with flood risk. Therefore, many of the dynamic pressures, especially those in the flood plains, show positive spatial relationships, highlighting that their increase contributes to the flood risk. Among them, wetland loss, wetland loss in flood plains, and agricultural areas' increase in flood plains have relatively stronger positive relations than other drivers. It was followed by drivers such as plantation areas' increase in flood plains, forest loss, and forest loss within flood plains. Interestingly, forest loss over the whole of Myanmar has stronger positive relationships with flood risk than forest loss within flood plains, indicating the need to consider the impacts of scales and downstream effects on flood risk.

However, urban areas' increase and land surface temperature have statistically significant negative spatial relationships with flood risk. Among them, land surface temperature has stronger negative relationships with flood risk. This means that low land surface temperature areas have strong flood risk to some degree. Urban areas' increase within flood plains has almost neutral relationships with flood risk, while urban areas' increase also has weak negative relationships. This highlights that urbanization is rather not a pressing problem for flood risk in Myanmar according to the provided data.

Non-significant but positive relationships with flood risk are detected in drivers such as agricultural areas' increase, mining areas' increase, mining areas' increase within flood plains, and conflicts. Therefore, although not statistically significant,

	ftl	ftlfp	agrg	agfp	wtl	wtlfp	ming	minfp	urb	urbfp	plt	pltfp	con	Lst5
Correlation Coefficient	0.1561*	0.1294*	0.1041	0.1907***	0.2061**	0.218***	0.0335	0.0627	−0.1137*	−0.0072	0.1437*	0.173**	0.0446	−0.1944
P values	0.0212	0.0405	0.1006	0.0012	0.0018	0.0008	0.5727	0.2798	0.041	0.8949	0.0296	0.0049	0.5048	0.001

**represent significant levels.*

Table 3.
Spatial relationships between dynamic pressures and flood risk.

they are contributing to flood risk spatially. Here, it is again the questions of scales over development of flood risk.

4. Discussion

Our study aimed to analyze spatial-temporal relations of flood risk and its potential drivers. The main finding in our analysis is that dynamic pressures of flood risk such as agricultural increase in flood plains, forest loss, forest loss in flood plains, wetland loss, wetland loss in flood plains, increase in plantation areas, and increase in plantation areas in flood plains have statistically significant positive spatial relationships with flood risk. Therefore, their escalation contributes to the flood risk.

Reinforcing those statistical results, their spatial distribution shows that some townships with highest agricultural extension, with highest conflicts, highest mining, and highest plantations, were spatially coincident with the second highest risk classes, indicating compound risks of socio-ecological disasters. This kind of spatial overlap of dynamic pressures can be extremely dangerous for vulnerable populations of Myanmar.

These spatial overlaps and positive relations with flood risk are on account of poor governance in landuse and environment management. We found large proportion (44 percent) of total mining areas exist in the upper flood plains, which are linked to inadequate environmental governance. This can result in poor water qualities, and rising flood water levels deepen risk in downstream areas due to the rain-induced disposal of mine tailings in the upstream rivers. Moreover, deforestation in the upland flood plain areas of rivers can lead to high flood risk in the downstream areas. High deforestation in the upper peripheral areas of dry zone may increase flood and drought risk, especially to agrarian populations in the middle Myanmar. Despite the extent is small, wetland area loss is materialized within the flood plain areas. These dynamic increases can have serious implications in case of large-scale disasters.

More importantly, increase in urban settlements mostly occurs in flood plain areas although urbanization has shown weak negative spatial relationship. The fact that 53 percent of urban areas are within the flood plains in 2018, indicating high risk to those people. There might be many underlying reasons behind the existence on the high-risk flood plain areas. In Myanmar, most of the townships and villages are located near the rivers and streams. The reasons can also be some push and pull factors of migration or original existence in the alluvial areas. It may be voluntary relocations due to high fertility of soils in the flood plains. It may be forced displacements due to land grabbing, conflicts, or development activities. Under the conditions of structural constraints such as environmental degradation, land tenure insecurity, and limited access to the infrastructure, the agency of people has little choices to response and adapt to those pressures.

However, it is essential to notice the importance of scales in expressing the impacts. Impacts of mining and deforestation on flood risk are not just realized in the areas where mining and deforestation accommodate. Downstream effects, and spatial and temporal knock-on effects that cascade beyond the local areas should not be ignored. Although major increase in agriculture is not in the delta, they are mostly realized in the upland areas of Myanmar's rivers, which are critical for flood risk of the downstream communities in the rest of the country. It is assumed that the delta already has limited space for the agricultural expansion. In addition, increase of plantations areas in Kachin State can exacerbate the flood risk for those downstream communities.

As a result, a history of poor environmental governance is related to flood risk in Myanmar. Ongoing deforestation, farmland expansion, and unstructured or poorly planned land use change can be linked back to natural resource mismanagement and state inefficiencies [37]. This has contributed to decreased flood plain areas and high levels of sedimentation and erosion in waterways, worsening flood hazard and increasing exposure of people and assets to floods [22, 77]. Moreover, the associated environmental degradation and poverty increase flood vulnerability of many Myanmar people, whose livelihood base and survival depend on the ecosystems, resulting reduced capacities of both society and ecosystems. The more the people are poor, and they depend more on ecosystems. Degraded ecosystems and environments make them more susceptible to flood risk.

Our results also highlighted the importance of recognition of compound and cascaded disasters in disaster risk management. The fact that over one-third (31 percent) of the current agricultural areas (in 2018) in Myanmar are within floodplains, showing potentially high financial agricultural losses in case of large-scale flood disasters. Ayeyarwady delta belongs to most agricultural areas within the flood plains. Being that Ayeyarwady delta produce more than 30 percent of country's rice production and a biodiversity hotspot area, damaging this critical region, will have wider and long-term impacts millions of people in the rest of the country. In addition, almost all of agricultural area distributions spatially mirror with the highest land surface temperature areas. This indicates that agriculture in Myanmar is highly susceptible to heatwaves, droughts, and flooding.

Floods, heatwaves, and droughts are reinforcing each other in terms of increasing hazard intensities, even without considering vulnerabilities. Floods that follow droughts and heatwaves can be extreme due to damaged soil moisture and ability of soil to absorb water. This was epitomized in the case in Magway region where flooding occurs after serious drought and extreme temperature in the dry zone areas in 2011.

Overlap of disasters and their drivers in terms of time and space can lead to compound disasters. Another visible overlap can be seen in Myitkyina. Among other highest risk townships, Myitkyina experiences its peculiarities, representing an example of compound drivers. Among those townships, it has highest forest loss of 500 km^2, and experiences sharp increase in agriculture (2000–2001s), and a steep rise in orchards and plantation areas (1994–2001s). It also possessed mining areas and saw a reverse trend in conflicts which increases in 2021. Surprisingly, 100% of urban settlements of Myikityina in 2018 are in the flood plain area, which is critical for the residents to be aware of flood risk. Most importantly, Myitkyina has the highest socio-ecological flood risk in Myanmar.

Another important finding is that many dynamic pressures accelerated during the period between 1995s and 2000s. We also find that dynamic pressures of flood risk are temporally related to each other. Changes in pressures that contribute to flood risk (deforestation, agriculture and plantations increase, urbanization) were rapid and remarkable in the first-half period between 1987 and 2000s, especially between 1995s and 2000s. This highlighted that some socio-economic and political reforms during that critical period substantially and dramatically cause to increase flood risk. In the first-half period (1987–2000s), floods were more frequent but less intense, after 2000s, flood becomes more intense and less frequent.

It is alarming that dynamic pressures are increasing over time. On the one hand, global anthropogenic climate change is accelerating worryingly. On the other hand, forests and wetlands that will absorb and provide capacities to adapt climate impacts were decreasing while the agriculture, mining, and urban areas that increase climate vulnerabilities were increasing over time.

It is important to include indirect drivers that drive direct drivers of risk in the risk assessments. Indeed, those indirect drivers are root causes of risk. For example, while deforestation drives flood risk, drivers that lead to deforestation should be considered in flood risk assessments. If risk assessments only consider indicators that directly related to flooding, risk assessments will only recognize symptoms of flood risks, leading to prioritization of treatment of symptoms.

Similar to the identified root causes in the literature review, flood vulnerability is driven by poverty, limited health care such as insufficient doctors and hospitals, and wetland area loss. The consequence of mismanagement of natural resources, corruption in resource allocation, and social and economic insecurity is continued poverty and vulnerability in Myanmar [78]. Moreover, persistent high levels of poverty have resulted in a growing population living in unsafe areas, with poor infrastructure and housing conditions, contributing to compounding impacts. Widespread inadequate access to services that assist people in everyday life such as health care and electricity sources has also reduced the potential for economic participation that enables people to escape from poverty, while further limiting capacities to cope and adapt to flooding [79].

Spatially, the highest vulnerability occurs in ethnically dominated bordered areas such as Rakhine, Shan, Chin, and Kachin states where poverty and lack of access to services are especially severe due to the continuing conflicts and violence [15, 79]. This has resulted in a situation where it is difficult for people to pursue livelihoods and make it uncertain for their future due to the lack of security [15, 79].

One of our limitations is the resolution of the data, because most of the dynamic pressures were based on the datasets that have resolution of 618.74 m^2. Data paucity was also an issue for other datasets such as longer frame of conflicts, shrimp and fish farming, landslides, droughts, waste management, salt water intrusion, water and air pollution, and sea level rise, which might help to explain the bigger complex figure of flood risk in Myanmar. Other high-risk townships, for example, in Rakhine State, likely have long history of conflicts. However, due to inaccessibility of long-term data, it could not be proved.

Due to the differences in classification methods and aggregation methods, determination of risk for each township can be different. We purposely use natural-break method over quantile methods as it better reflects reality in terms of spatial differences. Moreira et al. (2021) also recommended the natural break method as the best performed classification approach in risk assessments [80] .

The results of our study are essential for disaster risk management and climate change adaptation in Myanmar because the study considers comprehensive socio-ecological flood risk that is compared to and incorporated its dynamic pressures and root causes. Only through addressing those underlying factors of flood risk, it is possible to reduce vulnerabilities and increase resilience of Myanmar people, while also facilitating flood exposure and hazard patterns. Our findings also stress the need of effective land use and environmental governance that consider compound and cascaded flood risk, and prioritized investment in public services and infrastructure such as health, and education, due to their links with factors increasing vulnerability and flood risk.

5. Conclusion

This chapter presents spatial temporal relation of flood risk and its dynamic pressures in Myanmar. It illustrates the first, conceptually sound risk assessment for major river and coastal flooding in Myanmar. The finding highlights many socio-ecological dynamic pressures driven by economic- and governance-related root causes were positively and spatially correlated with flood risks. We call attentions to critical concerns about existence of those dynamic pressures in the flood plains and their potential compound and complex disasters. Effective land use and environmental management is essential to prevent deforestation, agricultural extension, urban extension, mining, and wetland loses in the critical areas such as upstream areas and flood plain areas of Myanmar's major rivers and to protect key ecosystems such as mangrove forests in the coastal areas.

Acknowledgements

The authors appreciate much to Ms. Alexandra Dudley for her contribution in the first version of river flood risk assessment, which was later further developed to socio-ecological risk assessment for river and coastal floods. We are very thankful to Dr. Michael Hagenlocher, Dr. Yvonne Walz, and Dr. Mostapha Harb for their guidance in the first version of river flood risk assessment.

Conflict of interest

The authors declare no conflict of interest.

A. Indicators

Table A1 provides details of all the relevant indicators for flooding in Myanmar. Indicators that were not included due to lack of data are highlighted red. Direction refers to the impact of the indicator on vulnerability and risk. A positive (+) sign indicates that a higher score increases vulnerability and risk. A negative (−) sign indicates that a higher score decreases vulnerability and risk.

Datasets:

WorldPop: http://www.worldpop.org.ukCensus 2014: http://themimu.info/census-data (spreadsheet: BaselineData_Census Dataset - Sr., District & Township_MIMU 16Jun2016 ENG.xlsx).

General Administration Department Survey 2016–2017:

Income: https://data.opendevelopmentmekong.net/dataset/general-administration-department-gad-2016-17–module-6-individual-income-section-prod2?type=dataset

Population: https://data.opendevelopmentmekong.net/dataset/general-administration-department-gad-2016-2017—module-4-population-population-section-prod2?type=dataset

Category	Indicator	Code	Direction	References	Data Source	Justification	Comments
Social Exposure							
Exposure	The population exposed to floods (%)	e_pop	+	[40]	Hazard: JRC and the WorldBank Catalog WorldPop	Without population exposure to a hazard there is no risk to the population.	Hazard return period: 100 years
	Exposed forests (%)	e_fl	+	[52]	Servir Mekong	Without forest exposure to a hazard there is no risk to ecosystems	
	Exposed Agriculture (%)	e_agri	+	[52]	Servir Mekong	Without agriculture exposure to a hazard there is no risk to the ecosystems	
Social susceptibility							
Poverty and inequality	Dependency ratio (%)	s_dep	+	[81–83]	Census 2014	Young and old people are economically dependent on the state or their family. They may need assistance evacuating, and they are more prone to succumb to illnesses such as those from water borne diseases.	Calculated by summing the population under 15 years and over 65 years and dividing by the population 16–64 years.
	Average annual income	s_pov	+	[81, 84]	General Administration Department Survey 2016–2017	Low average income means that the township is likely to be deprived of goods, services, and opportunities. They have little buffer for shocks and will struggle to recover after a flood.	Proxy for: Population below the poverty line (%)
	Households with existing debt (%)	s_deb	+	[81, 82]	—	Existing debt restricts ability to access more credit after a flood. Can reflect financial stress.	
	Minority population (%)	s_min	+	[85]	—	It is harder for minorities to access education, employment, credit, due to structural inequality.	

Social Exposure							
Category	**Indicator**	**Code**	**Direction**	**References**	**Data Source**	**Justification**	**Comments**
Disability and health status	Proportion of population with a disability (%)	s_dis	+	[81]	Census 2014	High numbers of disabled people increase pressure on communities due to dependency, and more people will need assistance during flood.	Disability defined by MIMU (2018) as seeing, hearing, walking, remembering.
	Number of cases of malaria per 10,000 people	s_vec	+	[82, 83, 85, 86]	General Administration Department Survey 2016–2017	Vector borne diseases increase the health burden. Flooding increases breeding grounds for the vectors, increasing the likelihood further of contracting the disease.	Proxy for: Population with a vector borne disease (%)
	Percent share of the incidence of tuberculosis, dysentery, and hepatitis (%)	s_chr	+	[82]	General Administration Department Survey 2016–2017	Chronic diseases reduce the ability of people to pursue capital building activities and increase the need for assistance and specialized care during a flood.	Proxy for: Population with chronic illness (%)
	Population undernourished (%)	s_nou	+	[81]	—	Undernourishment impacts education, employment, and health outcomes.	
Gender	Female-headed households (%)	s_fhh	+	[81, 87]	Census 2014	Females have less access to education, employment, land, and credit which affects the whole household here.	
Stability/ conflict	Number of conflict events per 10,000 people in the last year	s_con	+	[81]	Conflict: The Armed Conflict Location & Event Data Project Population: WorldPop	Conflict is extremely relevant for Myanmar. It affects people's ability to achieve livelihood goals, and ongoing instability undermines development activities.	Proxy for: Population affected by conflict in townships (%) Time: June 1, 2019–June 1, 2020. Includes: battles, explosions, riots, and violence against civilians.
	Proportion of internally displaced persons (%)	s_idp	+	[81]	—	This population has no safety net, limited shelter, and emotional trauma.	

Social Exposure							
Category	**Indicator**	**Code**	**Direction**	**References**	**Data Source**	**Justification**	**Comments**
WASH	Households with safe sanitation (%)	s_san	—	[81, 88]	Vulnerability assessment 2018 based on Census 2014	Poor sanitation leads to poor health outcomes. Unsealed sanitation will contaminate waterways during a flood.	Defined by MIMU (2018) as flush and improved pit latrine.
	Households access to improved drinking water supply (%)	s_dri	—	[81, 82, 85, 86, 89]	Vulnerability assessment 2018 based on Census 2014	Poor drinking water source leads to poor health outcomes. During and after a flood, water from non-improved sources is more likely to be contaminated.	Defined by MIMU (2018) as tap, tube well, protected well, and bottled water.
Settlement and Housing	Households with walls or floors made from leaf, bamboo or earth (%)	s_wfl	+	[81, 84, 87, 90–92]	Census 2014	These houses are likely to be damaged or washed away in a flood reducing quality shelter for populations.	Proxy for: population living in poorly constructed housing. Calculated by average the % of houses with walls made from leaf, bamboo, or earth and the % of houses with floors made from leaf, bamboo, or earth.
	Population living in informal settlements (%)	s_ins	+	[88]	—	Are less likely to be considered by aid agencies/government, have limited access to services.	
	Population living in unofficial wards (%)	s_unw	+	[88]	—	Are less likely to be considered by aid agencies/government, have limited access to services.	
Remoteness	Average travel time to the nearest city (minutes)	s_ttc	+	[83, 93]	The Malaria Atlas Project 2015	This defines how easily services normally only found in cities can be accessed, for example most health care. Also indicates access to markets, creditors etc.	
Employment	Daily wage workers (% of working population)	s_daw	+	[81, 84, 94]	General Administration Department Survey 2016–2017	When disruption occurs, daily wage workers will be the first to lose employment and therefore income.	

Social Exposure							
Category	**Indicator**	**Code**	**Direction**	**References**	**Data Source**	**Justification**	**Comments**
Environment and land use	Extent of forest loss (%)	s_ftl	+	[52]	Servir Mekong 1987–2018	Loss of forest areas reduces capacities of ecosystems to reduce flood risk	
	Extent of Wetland area loss (%)	s_wetl	+	[52]	Servir Mekong 1987–2018	Loss of wetland areas reduces capacities of ecosystems to reduce flood risk	
	Extent of urban extensions in the flood plains	s_ure	+	[52]	Servir Mekong 1987–2018	Urban extension in the flood plains increases risk for the people living in those settlements	
	Extent of mining area extension in the flood plains (%)	s_age	+	[52]	Servir Mekong 1987–2018	Mining area extension can increase water pollution and sedimentation	
	Extent of Agricultural extension in the flood plains (%)	s_mine	+	[52]	Servir Mekong 1987–2018	Agricultural extension in the flood plains reduces rooms for rivers and increases financial losses	
Adaptative capacity							
Information/ early warning	Households with access to radio, television, internet, or mobile (max %)	c_ati	—	[81, 82, 89]	Census 2014	People can be warned, find more information on the flood and find services to help cope.	Proxy for: Households with access to information (%)
	Number of rain gauges per 10,000 km	c_rai	—	[95]	—	Provides accurate input to early warning systems.	
Ecosystems	Extent of forest areas (%)	c_ft	+	[52]	Servir Mekong 1987–2018	Forest provides ecosystem services to reduce flood risk.	
	Extent of wetland areas (%)	c_wet		[52]	Servir Mekong 1987–2018	Wetlands provide ecosystem services to reduce flood risk.	

Social Exposure							
Category	**Indicator**	**Code**	**Direction**	**References**	**Data Source**	**Justification**	**Comments**
	Presence early warning system (yes/no)	c_ear	—	[86, 89, 91, 96]	—	Early warning allows timely evacuation.	Potential proxy: Households that received early warning in last event (%)
Social capital	Population that has lived in the township for more than 1 year (%)	c_ltr	—	[83, 87, 97]	—	Indicates knowledge of the area, evacuation routes, social capital that would help during and after a flood.	
Education	Literacy rate (%)	c_lit	—	[83, 87, 97]	Census 2014	Literate people can read information and warning pamphlets. They can fill in forms to get aid.	
Institutional capacity	Money spent on disaster risk reduction per person.	c_res		[86, 89, 91, 96, 98]	—	Indicates the institutional capacity to assist population and the systems in place to help populations cope.	Potential proxy: Number of disaster risk reduction projects per 10,000 people
Transportation	Households owning a boat (%)	c_boa	—	[87, 91, 94]	Census 2014	Boats are useful in floods to evacuate people and important goods.	
Health	Number of doctors per 10,000 people	c_doc	—	[81–83, 86, 88, 89]	General Administration Department Survey 2016–2017	A population needs doctors to treat illnesses and injuries after a flood.	This was not aggregated with c_bed, as even if there are enough hospital beds, without doctors no treatment is available.
	Number of hospital beds per 10,000 people	c_bed	—	[81–83, 86, 88, 89]	General Administration Department Survey 2016–2017	Populations need access to hospital beds to access health care	This was not aggregated with c_doc as without hospital beds, it is challenging for populations to access doctors and treatment.
Access to energy	Households with access to alternative electricity sources (%)	c_aes	—	[82]	Census 2014	After floods it is likely that there will be damage to electricity transmission wires. Alternative sources can help communities stay safe with light, charge or use devices to maintain contact with others and authorities.	Solar energy and generators were considered alternative energy sources. The number of sources per township were summed and divided by the number of households.

Social Exposure							
Category	**Indicator**	**Code**	**Direction**	**References**	**Data Source**	**Justification**	**Comments**
Accessibility	Density of roads (road kernel density: km road/ km2, search radius = 5 km)	c_dens	—	[81, 88, 94]	Open Street Map 2020	Presence of roads increases accessibility for emergency vehicles and provides paths of evacuation for populations.	Trunk, primary, secondary, and tertiary roads were considered. Data downloaded 6/6/2020.
Settlement and housing	Houses with more than one floor (%)	c_flo	—	[92]	—	Can move furniture out of flood, can evacuate or live on top floor.	
Financial capital	Households with diverse incomes (%)	c_div	—	[81, 82, 87]	—	If one income is disrupted due to flooding, they can rely on the other.	
	Households with savings (%)	c_sav	—	[81]	—	If income is disrupted, they can rely on savings.	
	Households with insurance (%)	c_ins	—	[81, 87]	—	Provides financial security. Can rebuild after a flood.	
Public infrastructure	Average distance to nearest shelter place.	c_dts	+	[81, 85, 89, 90, 96]	—	Determines how easily populations can evacuate and find safety.	Potential proxy: Number of shelter places per 10,000 people. Includes schools, monasteries, and evacuation shelters.
Food security	Households that save food (%)	c_sfo	—	[82, 86]	—	Flooding disrupts agriculture which impacts food availability. Saving food means families will be able to eat after a flood.	

Table A1.
List of desired indicators with references, data sources (where applicable), and justification.

Malaria/chronic disease: https://data.opendevelopmentmekong.net/dataset/general-administration-department-gad-2016-2017-module-7-health-most-disease-in-region-section?type=datasetDoctors: https://data.opendevelopmentmekong.net/dataset/general-administration-department-gad-2016-2017-module-7-health-personal-health-care-section?type=dataset

Hospital beds: https://data.opendevelopmentmekong.net/dataset/general-administration-department-gad-2016-2017—module-7-health-hospital-section-prod2?type=dataset

The Armed Conflict Location & Event Data project (ACLED): https://acleddata.com/data-export-tool/

Vulnerability assessment 2018: http://themimu.info/vulnerability-in-myanmar (Spreadsheet: Datasets_Vulnerability Analysis in Myanmar_09Jul2018.xlsx).

The Malaria Atlas Project: https://malariaatlas.org/research-project/accessibility_to_cities/

Open Street Map: https://export.hotosm.org/en/v3/

B. Data analysis

Appendix B provides additional information on outlier analysis and treatment, and multicollinearity assessment.

Outliers and winzorization: Box plots based on the interquartile range and skewness and kurtosis were used to identify outliers (where skewness >1 and kurtosis >3.5 indicate potential outliers [52]. There was a high number of data points outside the interquartile range, and indicators with high skewness and kurtosis. However, only nine points in total were determined to be outliers and errors based on the local expert knowledge of one of the authors. These were treated with winzorization which is a common method to manage outliers [50]. As the data was not treated with the intention to create normally distributed data (as this would alter the characterization of vulnerability and risk beyond what is the true situation in Myanmar), the median as a measure of central tendency for each indicator. **Table B1** outlines which data points were treated and which value they were given.

Indicator	Treated value	From	To	Value
s_pov	34,303,567	Seikkan	Yebyu	12,357,626
s_pov	26,322,559	Kyeemyindaing	Yebyu	12,357,626
s_pov	23,616,513	Kyauktada	Yebyu	12,357,626
s_pov	18,114,882	Magway	Yebyu	12,357,626
s_vec	509.32	Paletwa	Tanai	231.98
s_chr	2.281641	Palaw	Aunglan	1.558724
s_fhh	55.05	Thayetchaung	Amarapura	40.97
s_fhh	48.41	Patheingyi	Amarapura	40.97
c_doc	248.09	Oke Ta Ra Thi Ri	Cocokyun	8.97

Table B1.
Treated indicators and their values.

c_wetl	s_ure	s_age	c_ftl	c_ft	c_wet	s_dep	s_pov	s_dis	s_vec	s_chr	s_fhh	s_con	s_san	s_dri	s_wal	s_ttc	s_daw	c_ati	c_lit	c_den	c_boa	c_doc	c_bed	c_aes	e_pop	e_agri	e_ft
0.021893	0.089094	0.096741	0.152707	-0.21637	-0.02347	0.041097	0.037434	-0.08146	0.190188	0.006665	0.023159	0.153874	-0.03394	-0.00991	-0.07466	0.163338	-0.03165	0.005122	0.04065	0.1875	-0.00338	-0.1686	-0.14633	-0.170929588	-0.01596	-0.005610202	0.143404
1	0.100974	0.248728	-0.1387	0.144707	-0.60037	0.101711	-0.02654	0.090724	-0.00879	0.252606	0.082416	0.05221	0.061466	-0.0219	0.07683	-0.02342	0.060086	0.150304	-0.11968	-0.04939	-0.4375	0.08401	0.175607	0.012119106	0.354158	0.421978275	-0.09777
0.100974	1	0.249729	-0.03898	0.000576	-0.06919	-0.13208	-0.07185	-0.16047	-0.0538	0.035471	0.061602	0.098126	-0.11921	-0.11621	0.003324	-0.04292	-0.06304	-0.15857	-0.04771	-0.13257	-0.01082	-0.05468	-0.14105	0.088553983	0.239707	0.117320123	0.076188
0.248728	0.249729	1	0.205012	-0.09162	-0.20986	0.015493	0.070749	-0.07556	0.152311	0.128088	0.018859	0.075853	0.050692	-0.07361	0.166337	0.064259	-0.06195	0.031684	0.016837	0.041686	-0.15697	-0.04219	0.076946	-0.183348394	0.277059	0.296449814	0.085363
-0.1387	-0.03898	0.205012	1	-0.38661	0.133229	0.209146	0.192126	0.000957	0.289268	-0.03562	-0.09448	0.015127	0.212166	0.114416	0.052318	0.280866	-0.11972	0.11423	0.310443	0.267689	0.152839	-0.05406	-0.03371	-0.359071845	-0.18257	-0.125234455	0.231533
0.144707	0.000576	-0.09162	-0.38661	1	-0.18605	-0.34862	-0.37877	-0.09201	-0.47258	0.129707	0.171773	0.089416	-0.22446	-0.21059	-0.01014	-0.65647	0.226518	-0.18761	-0.38998	-0.53705	-0.12564	0.169999	0.211049	0.328088037	0.2426	0.200050214	-0.48773
-0.60037	-0.06919	-0.20986	0.133229	-0.18605	1	-0.0448	0.094441	-0.02743	0.031938	-0.23687	-0.09607	-0.13064	-0.09162	0.039057	-0.11359	0.114117	-0.14729	-0.0841	0.12572	0.069125	0.423994	-0.06169	-0.14891	0.03894435	-0.36213	-0.390744203	0.071341
0.101711	-0.13208	0.015493	0.209146	-0.34862	-0.0448	1	0.368031	0.271043	0.308453	0.034387	-0.0964	-0.05373	0.412436	0.343979	0.105578	0.449796	-0.04168	0.438924	0.481896	0.423436	-0.09448	0.034873	0.035124	-0.253642386	-0.03084	0.035349211	0.261411
-0.02654	-0.07185	0.070749	0.192126	-0.37877	0.094441	0.368031	1	0.119902	0.267831	-0.02407	-0.17536	-0.14303	0.238953	0.250373	0.1098	0.373775	-0.11047	0.241915	0.390053	0.338651	0.01921	0.02323	0.002377	-0.148278548	-0.07116	-0.052976804	0.276419
0.090724	-0.16047	-0.07556	0.000957	-0.09201	-0.02743	0.271043	0.119902	1	0.1226	0.094494	-0.08507	-0.06672	0.167862	0.233418	0.082536	0.230126	0.012133	0.269126	0.192746	0.130002	-0.18308	0.014089	0.043558	-0.020066952	0.075817	0.134357637	0.054744
-0.00879	-0.0538	0.152311	0.289268	-0.47258	0.031938	0.308453	0.267831	0.1226	1	0.091254	-0.10024	0.024521	0.259431	0.176518	0.054509	0.449974	-0.13955	0.207613	0.282262	0.44477	-0.04502	-0.10491	-0.11723	-0.330234973	-0.09843	-0.059633644	0.376913
0.252606	0.035471	0.128088	-0.03562	0.129707	-0.23687	0.034387	-0.02407	0.094494	0.091254	1	0.068651	0.195336	-0.01143	-0.02394	0.052966	-0.06564	0.084295	0.052779	-0.0518	-0.07812	-0.19566	0.064989	0.141193	-0.041244907	0.212507	0.230448333	-0.12868
0.082416	0.061602	0.018859	-0.09448	0.171773	-0.09607	-0.0964	-0.17536	-0.08507	-0.10024	0.068651	1	0.333877	-0.23293	-0.24564	-0.15851	-0.24475	0.006083	-0.25908	-0.19065	-0.18538	0.080262	-0.00311	-0.01482	0.037685408	-0.02631	-0.046534534	-0.13934
0.05221	0.098126	0.075853	0.015127	0.089416	-0.13064	-0.05373	-0.14303	-0.06672	0.024521	0.195336	0.333877	1	-0.14324	-0.17562	-0.05773	-0.10562	0.037125	-0.14083	-0.11472	-0.09723	0.024829	-0.00672	-0.02089	-0.047101879	0.07439	0.06000348	-0.06223
0.061466	-0.11921	0.050692	0.212166	-0.22446	-0.09162	0.412436	0.238953	0.167862	0.259431	-0.01143	-0.23293	-0.14324	1	0.412058	0.30921	0.342903	-0.02569	0.405693	0.428427	0.314557	-0.11752	0.13037	0.118355	-0.188949834	0.032219	0.081991352	0.21441
-0.0219	-0.11621	-0.07361	0.114416	-0.21059	0.039057	0.343979	0.250373	0.233418	0.176518	-0.02394	-0.24564	-0.17562	0.412058	1	0.151676	0.311843	-0.00352	0.30083	0.387339	0.365373	-0.15763	0.040235	0.052989	-0.146564923	-0.0228	0.033299014	0.234428
0.07683	0.003324	0.166337	0.052318	-0.01014	-0.11359	0.105578	0.1098	0.082536	0.054509	0.052966	-0.15851	-0.05773	0.30921	0.151676	1	0.089063	0.027715	0.28947	0.165088	0.056041	-0.06695	0.049382	0.193449	-0.120653599	0.128532	0.140706558	-0.03071
-0.02342	-0.04292	0.064259	0.280866	-0.65647	0.114117	0.449796	0.373775	0.230126	0.449974	-0.06564	-0.24475	-0.10562	0.342903	0.311843	0.089063	1	-0.14384	0.339491	0.409826	0.580709	-0.03966	-0.08044	-0.14234	-0.280063725	-0.1412	-0.073381032	0.415323
0.060086	-0.06304	-0.06195	-0.11972	0.226518	-0.14729	-0.04168	-0.11047	0.012133	-0.13955	0.084295	0.006083	0.037125	-0.02569	-0.00352	0.027715	-0.14384	1	-0.00737	-0.07944	-0.11271	-0.09943	0.032073	0.118947	0.053887139	0.136228	0.148630494	-0.16409
0.150304	-0.15857	0.031684	0.11423	-0.18761	-0.0841	0.438924	0.241915	0.269126	0.207613	0.052779	-0.25908	-0.14083	0.405693	0.30083	0.28947	0.339491	-0.00737	1	0.289786	0.308781	-0.19749	0.026801	0.118823	-0.103838192	0.112757	0.20358276	0.117414
-0.11968	-0.04771	0.016837	0.310443	-0.38998	0.12572	0.481896	0.390053	0.192746	0.282262	-0.0518	-0.19065	-0.11472	0.428427	0.387339	0.165088	0.409826	-0.07944	0.289786	1	0.366166	0.107831	0.028203	-0.01331	-0.223031338	-0.1434	-0.127002937	0.306698
-0.04939	-0.13257	0.041686	0.267689	-0.53705	0.069125	0.423436	0.338651	0.130002	0.44477	-0.07812	-0.18538	-0.09723	0.314557	0.365373	0.056041	0.580709	-0.11271	0.308781	0.366166	1	-0.07719	-0.12785	-0.0992	-0.316364748	-0.15744	-0.073665876	0.414036
-0.4375	-0.01082	-0.15697	0.152839	-0.12564	0.423994	-0.09448	0.01921	-0.18308	-0.04502	-0.19566	0.080262	0.024829	-0.11752	-0.15763	-0.06695	-0.03966	-0.09943	-0.19749	0.107831	-0.07719	1	-0.03689	-0.13363	0.020234857	-0.43675	-0.528997734	-0.05737
0.08401	-0.05468	-0.04219	-0.05406	0.169999	-0.06169	0.034873	0.02323	0.014089	-0.10491	0.064989	-0.00311	-0.00672	0.13037	0.040235	0.049382	-0.08044	0.032073	0.026801	0.028203	-0.12785	-0.03689	1	0.383644	0.092333872	0.098167	0.086815351	-0.11997
0.175607	-0.14105	0.076946	-0.03371	0.211049	-0.14891	0.035124	0.002377	0.043558	-0.11723	0.141193	-0.01482	-0.02089	0.118355	0.052989	0.193449	-0.14234	0.118947	0.118823	-0.01331	-0.0992	-0.13363	0.383644	1	-0.023433755	0.168639	0.210192938	-0.23181
0.012119	0.088554	-0.18335	-0.35907	0.328088	0.038944	-0.25364	-0.14828	-0.02007	-0.33023	-0.04124	0.037685	-0.0471	-0.18895	-0.14656	-0.12065	-0.28006	0.053887	-0.10384	-0.22303	-0.31636	0.020235	0.092334	-0.02343	1	0.159425	0.109962099	-0.20394
0.354158	0.239707	0.277059	-0.18257	0.2426	-0.36213	-0.03084	-0.07116	0.075817	-0.09843	0.212507	-0.02631	0.07439	0.032219	-0.0228	0.128532	-0.1412	0.136228	0.112757	-0.1434	-0.15744	-0.43675	0.098167	0.168639	0.159424524	1	0.791828767	0.030967
0.421978	0.11732	0.29645	-0.12523	0.20005	-0.39074	0.035349	-0.05298	0.134358	-0.05963	0.230448	-0.04653	0.060003	0.081991	0.033299	0.140707	-0.07338	0.14863	0.203583	-0.127	-0.07367	-0.529	0.086815	0.210193	0.109962099	0.791829	1	0.0155
-0.09777	0.076188	0.085363	0.231533	-0.48773	0.071341	0.261411	0.276419	0.054744	0.376913	-0.12868	-0.13934	-0.06223	0.21441	0.234428	-0.03071	0.415323	-0.16409	0.117414	0.306698	0.414036	-0.05737	-0.11997	-0.23181	-0.203947769	0.030967	0.01550033	1

Table B2.
Results of multicollinearity assessment using Kendall's tau using SPSS (IBM SPSS statistics).

Multicollinearity assessment: A multicollinearity assessment was conducted to determine if there were any redundant indicators. Using Kendall's Tau suitable for non-normal data (Puth et al. 2015), and two tailed significance, no issue of collinearity was detected based on a threshold of ±0.9. **Table B2** outlines the results from the analysis.

Author details

Hnin Wuit Yee Kyaw[1,2]* and Phone Pyai Tun[2]

1 Konstanz University of Applied Sciences, Konstanz, Germany

2 University of Forestry and Environmental Sciences, Nay Pyi Taw, Myanmar

*Address all correspondence to: hninwuityeekyaw95@gmail.com

References

[1] Zschau J. Where are we with multihazards, multirisks assessment capacities?. In: Science for disaster risk management 2017: Knowing better and losing less. European Union. 2017. pp. 98-115

[2] Pescaroli G, Alexander D. Understanding compound, interconnected, interacting, and cascading risks: A holistic framework. Risk analysis. 2018;**38**(11): 2245-2257

[3] Goldin I, Vogel T. Global governance and systemic risk in the 21st century: Lessons from the financial crisis. Global Policy. 2010;**1**(1):4-15

[4] UNDRR & UNU-EHS. Understanding and managing cascading and systemic risks: Lessons from COVID-19. Geneva, UNDRR; Bonn: UNU-EHS; 2022

[5] Helbing D. Globally networked risks and how to respond. Nature. 2013; **497**(7447):51-59

[6] Schweizer P-J. Systemic risks–concepts and challenges for risk governance. Journal of Risk Research. 2021;**24**(1):78-93

[7] Wisner B, Blaikie P, Cannon T, Davis I. At Risk: Natural Hazards, people's Vulnerability, and Disasters. Second ed. Abingdon, Oxon; New York, New York: Routledge; 2004. p. 492

[8] Blaikie P, Cannon T, Davis I, Wisner B. At Risk. London: Routledge; 1994

[9] Pelling M. Toward a political ecology of urban environmental risk. Political ecology: An integrative approach to geography and environment-development studies. 2003:73-93

[10] Global assessment report on disaster risk reduction: Making Development Sustainable: The Future of Disaster Risk Management. 4th ed. ed. Geneva, Switzerland: United Nations International Strategy for Disaster Reduction; 2015. xx, 266, cccxi pages p. 2015

[11] Eckstein D, Künzel V, Schäfer L. Global Climate Risk Index: Who Suffers Most Extreme Weather Events? Weather-Related Loss Events in 2019 and 2000–2019. Bonn: Germanwatch Nord-Süd Initiative e.V; 2021 1 online resource p

[12] Marin-Ferrer M, Vernaccini L, Poljanšek K. Index for Risk Management - INFORM: Concept and Methodology, Version 2017. Luxembourg: Publications Office; 2017. p. 85

[13] Simpson A, Farrelly N, Holliday I, editors. Routledge Handbook of Contemporary Myanmar. London; New York: Routledge/Taylor & Francis Group; 2018

[14] Simpson A. Environment and natural resources. In: Simpson A, Farrelly N, Holliday I, editors. Routledge Handbook of Contemporary Myanmar. Routledge Handbooks. London; New York: Routledge/Taylor & Francis Group; 2018

[15] Prasse-Freeman E, Latt PW. Class and inequality. In: Simpson A, Farrelly N, Holliday I, editors. Routledge Handbook of Contemporary Myanmar. Routledge Handbooks. London; New York: Routledge/Taylor & Francis Group; 2018

[16] Kim K. How Military Rule Makes Myanmar more Vulnerable to Climate Change 2022. Available from: https://www.thethirdpole.net/en/climate/how-military-rule-makes-myanmar-more-vulnerable-to-climate-change/ [Accessed: June 22, 2022]

[17] Niño-Zarazúa M, Tarp F. Poverty, Inequality and Social Protection in Myanmar. Development Economics Research Group Working Paper Series. 2021 Jul 27:11-2021

[18] Hamlet AF, Lettenmaier DP. Effects of 20th century warming and climate variability on flood risk in the western US. Water Resources Research. 2007;**43**: W06427

[19] Winsemius HC, Aerts JCJH, van Beek LPH, Bierkens MFP, Bouwman A, Jongman B, et al. Global drivers of future river flood risk. Nature Climate Change. 2016;**6**(4):381-385

[20] EM-DAT. EM-DAT | The international Disasters Database 2022. Available from: https://www.emdat.be/ [Accessed: May 12, 2022]

[21] Hirabayashi Y, Mahendran R, Koirala S, Konoshima L, Yamazaki D, Watanabe S, et al. Global flood risk under climate change. Nature Climate Change. 2013;**3**(9):816-821

[22] Taft L, Evers M. A review of current and possible future human–water dynamics in Myanmar's river basins. Hydrology and Earth System Sciences. 2016;**20**(12):4913-4928

[23] World Bank Group. An Analysis of Poverty in Myanmar (English). Washington, D.C. Available from: https://documents.worldbank.org/en/publication/documents-reports/documentdetail/829581512375610375/an-analysis-of-poverty-in-myanmar; 2017

[24] Nguyen MT, Sebesvari Z, Souvignet M, Bachofer F, Braun A, Garschagen M, et al. Understanding and assessing flood risk in Vietnam: Current status, persisting gaps, and future directions. Journal of Flood Risk Management. 2021;**14**(2):e12689

[25] Zischg AP, Hofer P, Mosimann M, Röthlisberger V, Ramirez JA, Keiler M, et al. Flood risk (d)evolution: Disentangling key drivers of flood risk change with a retro-model experiment. Science of The Total Environment. 2018; **639**:195-207

[26] Hamlet AF, Lettenmaier DP. Effects of 20th century warming and climate variability on flood risk in the western US. Water Resources Research. 2007;**43** (6)

[27] Garschagen M, Surtiari GAK, Harb M. Is Jakarta's new flood risk reduction strategy transformational? Sustainability. 2018;**10**(8):2934

[28] Abbas HB, Routray JK. Assessing factors affecting flood-induced public health risks in Kassala state of Sudan. Operations Research for Health Care. 2014;**3**(4):215-225

[29] Vojinovic Z, Hammond M, Golub D, Hirunsalee S, Weesakul S, Meesuk V, et al. Holistic approach to flood risk assessment in areas with cultural heritage: A practical application in Ayutthaya, Thailand. Natural Hazards. 2016;**81**(1):589-616

[30] Odemerho FO. Building climate change resilience through bottom-up adaptation to flood risk in Warri, Nigeria. Environment and Urbanization. 2015;**27**(1):139-160

[31] Stewart RM, Rashid H. Community strategies to improve flood risk communication in the Red River Basin,

Manitoba, Canada. Disasters. 2011;**35**(3): 554-576

[32] Treby EJ, Clark MJ, Priest SJ. Confronting flood risk: Implications for insurance and risk transfer. Journal of environmental management. 2006; **81**(4):351-359

[33] Abbas HB, Routray JK. Vulnerability to flood-induced public health risks in Sudan. Disaster Prevention and Management. 2014;**23**(4):395-419

[34] Li C-j, Chai Y-q, Yang L-s, Li H-r. Spatio-temporal distribution of flood disasters and analysis of influencing factors in Africa. Natural Hazards. 2016; **82**(1):721-731

[35] Yang W, Hu D, Jiang X, Dun X, Hou B, Zheng C, et al. Framework for Spatio-temporal distribution of disasters and influencing factors: Exploratory study of Tianjin, China. Sustainability. 2022;**14**(17):10488

[36] International Panel on Climate Change (IPCC). Summary for policymakers. In: Edenhofer O, Pichs-Madruga R, Sokona Y, Farahani E, Kadner S, Seyboth K, et al. editors. Climate Change 2014: Mitigation of Climate Change. Contribution of Working Group III to the Fifth Assessment Report of the Intergovernmental Panel on Climate Change. Cambridge, United Kingdom and New York, NY, USA: Cambridge University Press; 2014

[37] Oo TN, Hlaing EES, Aye YY, Chan N, Maung NL, Phyoe SS, et al. The Context of REDD+ in Myanmar: Drivers, Agents and Institutions. Bogor, Indonesia: CIFOR; 2020

[38] Nixon H, Joelene C, Saw KPC, Lynn TA, Arnold M. State and Region Governments in. Myanmar: MDRI-CESD; 2013

[39] Kummu M, Taka M, Guillaume JH. Gridded global datasets for gross domestic product and Human Development Index over 1990–2015. Scientific Data. 2018 Feb 6;**5**(1):1-5

[40] IPCC. Climate change 2014: Impacts, adaptation, and vulnerability. Part a: Global and sectoral aspects. In: Field CB, Barros VR, Dokken DJ, Mach KJ, Mastrandrea MD, Bilir TE, Chatterjee M, Ebi KL, Estrada YO, Genova RC, Girma B, Kissel ES, Levy AN, MacCracken S, Mastrandrea PR, White LL, editors. Contribution of Working Group II to the Fifth Assessment Report of the Intergovernmental Panel on Climate Change. Cambridge, United Kingdom and New York, NY, USA: Cambridge University Press; 2014. p. 1132

[41] Servir Mekong. Regional Land Cover Monitoring System 2018. Available from: https://servir.adpc.net/tools/regional-land-cover-monitoring-system [Accessed: February 9, 2022]

[42] The Armed Conflict Location & Event Data (ACLED). ACLED | Bringing Clarity to Crisis: The Armed Conflict Location & Event Data (ACLED). 2022. Available from: https://acleddata.com/ [Accessed: January 1, 2022]

[43] Nardi F, Annis A, Di Baldassarre G, Vivoni ER, Grimaldi S. GFPLAIN250m, a global high-resolution dataset of Earth's floodplains. Scientific data. 2019;**6**(1): 1-6

[44] Dottori F, Alfieri L, Bianchi A, Lorini V, Feyen L, Salamon P, Cartographers. River flood Hazard maps for Europe: Version 1. European Commission, Joint Research Centre

(JRC) [dataset]. PID: http://data.europa.eu/89h/8e49997c-ba99-4ed1-9aec-059bb440001b2016

[45] World Bank Group. A Country on the Move : Domestic Migration in Two Regions of Myanmar. Yangon: © World Bank; 2016. https://openknowledge.worldbank.org/handle/10986/24981 License: CC BY 3.0 IGO

[46] Myanmar Information Management Unit (MIMU). The 2014 Myanmar Population and Housing Census | MIMU 2014. Available from: http://themimu.info/census-data [Accessed: May 12, 2022]

[47] Open Development Mekong. The Asia Foundation - Organizations OD Mekong Datahub 2016. Available from: https://data.opendevelopmentmekong.net/organization/the-asia-foundation [Accessed: May 12, 2022]

[48] Weiss DJ, Nelson A, Gibson HS, Temperley W, Peedell S, Lieber A, et al. A global map of travel time to cities to assess inequalities in accessibility in 2015. Nature. 2018;**553**(7688):333-336

[49] OpenStreetMap. HOT Export Tool 2022. Available from: https://export.hotosm.org/en/v3/ [Accessed: October 31, 2022]

[50] Damioli G. The Identification and Treatment of Outliers. Italy: European Commission Ispra; 2017

[51] Puth M-T, Neuhäuser M, Ruxton GD. Effective use of Spearman's and Kendall's correlation coefficients for association between two measured traits. Animal Behaviour. 2015;**102**:77-84

[52] Hagenlocher M, Renaud FG, Haas S, Sebesvari Z. Vulnerability and risk of deltaic social-ecological systems exposed to multiple hazards. The Science of the total environment. 2018;**631-632**:71-80

[53] Fumagalli M, Kemmerling A. Development aid and domestic regional inequality: The case of Myanmar. Eurasian Geography and Economics. 2022:1-30

[54] Barbier EB, Damania R, Léonard D. Corruption, trade and resource conversion. Journal of Environmental Economics and Management. 2005;**50**(2):276-299

[55] Simpson A. Corruption, investment and natural resources. In: International Natural Resources Law, Investment and Sustainability. London: Routledge; 2017. pp. 416-434

[56] Transparency International. 2017 Corruption Perceptions Index - Transparency.Org. 2017. Available from: https://www.transparency.org/en/cpi/2017

[57] Peptenatu D, Grecu A, Simion AG, Gruia KA, Andronache I, Draghici CC, et al. Deforestation and frequency of floods in Romania. In: Water Resources Management in Romania. Cham: Springer; 2020. pp. 279-306

[58] Bradshaw C, Sodhi N, Peh K, Brook B. Global evidence that deforestation amplifies flood risk and severity in the developing world. Global Change Biology. 2007;**13**:2379-2395

[59] Brookhuis BJ, Hein LG. The value of the flood control service of tropical forests: A case study for Trinidad. Forest Policy and Economics. 2016;**62**:118-124

[60] Oo N. Present state and problems of mangrove management in Myanmar. Trees. 2002;**16**(2):218-223

[61] Scott JC. The erosion of patron-client bonds and social change in rural Southeast Asia. The Journal of Asian Studies. 1972;**32**(1):5-37

[62] Fishbein E, Lamung AM. How A Beloved Gemstone Became A Symbol Of Environmental Tragedy In Myanmar. 2020. Available from: https://pulitzercenter.org/stories/how-beloved-gemstone-became-symbol-environmental-tragedy-myanmar

[63] Gylfason T, Zoega G. Inequality and economic growth: Do natural resources matter? Inequality and Growth: Theory and Policy Implications. 2003;**1**:255

[64] Simpson A. The environment in Southeast Asia: Injustice, conflict and activism. Contemporary Southeast Asia: The politics of change, contestation, and adaptation. 2017 Oct 25:164-180

[65] Brakenridge GR, Syvitski JPM, Niebuhr E, Overeem I, Higgins SA, Kettner AJ, et al. Design with nature: Causation and avoidance of catastrophic flooding, Myanmar. Earth-Science Reviews. 2017;**165**:81-109

[66] Brown I. BURMA's ECONOMY in the Twentieth Century. Cambridge: Cambridge University Press; 2013

[67] Andersen KE. Study of Upland Customary Communal Tenure in Chin and Shan States. Yangon, Myanmar: Livelihoods and Food Security Trust Fund (LIFT). Retrieved ...; 2015

[68] Brenner D. Ashes of co-optation: From armed group fragmentation to the rebuilding of popular insurgency in Myanmar. Conflict, Security & Development. 2015;**15**(4):337-358

[69] Internal labour migration in Myanmar: Building an evidence-base on patterns in migration, human trafficking and forced labour; International Labour Organization, ILO Liaison Officer for Myanmar. Yangon: ILO; 2015. p. 106

[70] Rhoden TF. No Burmese returning: Economics across Myanmar-Thailand border. Thai Journal of East Asian Studies. 2015;**19**(2):51-70

[71] Arnold D, Pickles J. Global work, surplus labor, and the precarious economies of the border. Antipode. 2011;**43**(5):1598-1624

[72] Boutry M. From British to humanitarian colonization: The 'EarlyRecovery'Response in Myanmar after Nargis. South East Asia Research. 2013;**21**(3):381-401

[73] Thein ATZ, Akita T. Education and expenditure inequality in Myanmar: An analysis with the 2006 and 2012 household income and expenditure survey in an urban and rural setting. Regional Science Policy & Practice. 2019;**11**(1):55-70

[74] Rustad SCA, Rød JK, Larsen W, Gleditsch NP. Foliage and fighting: Forest resources and the onset, duration, and location of civil war. Political Geography. 2008;**27**(7):761-782

[75] Woods K. CP maize contract farming in Shan State, Myanmar: A regional case of a place-based corporate agro-feed system. BRICS initiative for critical agrarian studies (BICAS). 2015 May:1-25

[76] Le Billon P. Natural resources and corruption in post-war transitions: Matters of trust. Third World Quarterly. 2014;**35**(5):770-786

[77] Khaing TW, Tantanee S, Pratoomchai W, Mahavik N. Coupling flood hazard with vulnerability map for flood risk assessment: A case study of Nyaung-U township in Myanmar.

Greater Mekong Subregion Academic and Research Network International Journal. 2021;**15**:127-138

[78] Thang HNC, Uyen NTT, wa Mungai N. What constitutes poverty in Myanmar? The social work appraisal. Development. 2014;**2**(2):10-15

[79] Hudson-Rodd N, Nyunt M, Tun ST, Htay S. State Induced Violence and Poverty in Burma. June, Bangkok: Federation of Trade Unions-Burma (FTUB); 2004

[80] Moreira LL, Brito MM, Kobiyama M. Effects of different normalization, aggregation, and classification methods on the construction of flood vulnerability indexes. Water. 2021;**13**(1):98

[81] The Government of the Union of Myanmar - Post-disaster needs assessment of floods and landslides. Washington, D.C.: World Bank Group; 2015. Available from: http://documents.worldbank.org/curated/en/646661467990966084/Myanmar-Post-disaster-needs-assessment-of-floods-and-landslides-July-September-2015

[82] Oo AT, van Huylenbroeck G, Speelman S. Assessment of climate change vulnerability of farm households in Pyapon District, a delta region in Myanmar. International Journal of Disaster Risk Reduction. 2018;**28**:10-21

[83] Phongsapan K, Chishtie F, Poortinga A, Bhandari B, Meechaiya C, Kunlamai T, et al. Operational flood risk index mapping for disaster risk reduction using Earth Observations and cloud computing technologies: A case study on Myanmar. Frontiers in Environmental Science. 2019 Dec 11;**7**:191

[84] Win S, Zin WW, Kawasaki A, San ZMLT. Establishment of flood damage function models: A case study in the Bago River basin, Myanmar. International Journal of Disaster Risk Reduction. 2018;**28**:688-700

[85] Burki T. Floods in Myanmar damage hundreds of health facilities. The Lancet. 2015;**386**(9996):843

[86] Oo AT, van Huylenbroeck G, Speelman S. Characterising households' vulnerability to climate change in Pyapon District in the delta region of Myanmar. International Journal of Global Warming. 2018;**16**(4):365

[87] Kawasaki A, Kawamura G, Zin WW. A local level relationship between floods and poverty: A case in Myanmar. International Journal of Disaster Risk Reduction. 2020;**42**:101348

[88] Boutry M. Migrants Seeking out and Living with Floods: A Case Study of Mingalar Kwet Thet Settlement. Yangon, Myanmar. Living with Floods in a Mobile Southeast Asia: Routledge; 2017. pp. 42-62

[89] Htein MK, Lim S, Zaw TN. The evolution of collaborative networks towards more polycentric disaster responses between the 2015 and 2016 Myanmar floods. International Journal of Disaster Risk Reduction. 2018;**31**:964-982

[90] Kawasaki A, Ichihara N, Ochii Y, Acierto RA, Kodaka A, Zin WW. Disaster response and river infrastructure management during the 2015 Myanmar floods: A case in the Bago River basin. International Journal of Disaster Risk Reduction. 2017;**24**:151-159

[91] Otsuyama K, Aung SP, Maki N. Adaptive strategies and transformation for community recovery – A case study of villages in Hinthada, Ayeyarwady Region, Myanmar. International Journal of Disaster Risk Reduction. 2019;**34**:75-93

[92] Zin W, Kawasaki A, Hörmann G, Acierto RA, San Z, Thu A. Multivariate flood loss estimation of the 2018 Bago flood in Myanmar. Journal of Disaster Research. 2020;**15**:300-311

[93] Sritarapipat T, Takeuchi W. Land cover change simulations in Yangon under several scenarios of flood and earthquake vulnerabilities with master plan. Journal of Disaster Research. 2018; **13**:50-61

[94] Food and Agriculture Organization (FAO). Agriculture and Livelihood Flood Impact Assessment in Myanmar. Rome: FAO; 2015. p. 60

[95] Yuan F, Zhang L, Soe K, Ren L, Zhao C, Zhu Y, et al. Applications of TRMM- and GPM-era multiple-satellite precipitation products for flood simulations at sub-daily scales in a sparsely gauged watershed in Myanmar. Remote Sensing. 2019;**11**(2):140

[96] Reeder G. Urban governance of flooding in Myanmar: A case study of Bago. Urban Book Series. 2019:103-126. DOI: 10.1007/978-3-319-98968-6_6

[97] Jones L, Ballon P. Tracking changes in resilience and recovery after natural hazards: Insights from a high-frequency mobile-phone panel survey. Global Environmental Change. 2020;**62**:102053

[98] Zaw TN, Lim S. The military's role in disaster management and response during the 2015 Myanmar floods: A social network approach. International Journal of Disaster Risk Reduction. 2017; **25**:1-21

Section 4

From Climate Changes to Hurricanes: Exploring the Range of Climatological Hazards

Chapter 15

Vulnerability Assessment of Process Vessels in the Event of Hurricanes

Nima Khakzad

Abstract

Hurricanes are multi-hazard natural hazards that can cause severe damage to chemical and process plants *via* individual or combined impact of strong winds, torrential rainfall, floods, and hitting waves especially in coastal areas. To assess and manage the vulnerability of process plants, failure modes and respective failure probabilities both before and after implementing safety measures should be assessed. However, due to the uncertainties arising from interdependent failure modes and lack of accurate and sufficient historical data, most conventional quantitative risk assessment techniques deliver inaccurate results, which in turn lead to inaccurate risk assessment and thus ineffective or non-cost-effective risk management strategies. Bayesian network (BN) is a probabilistic technique for reasoning under uncertainty with a variety of applications is system safety, reliability engineering, and risk assessment. In this chapter, applications of BN to vulnerability assessment and management of process vessels in the event of hurricanes are demonstrated and discussed.

Keywords: Na-techs, Hurricanes, probabilistic risk assessment, Bayesian network, atmospheric storage tanks

1. Introduction

The impact of natural disasters such as earthquakes, hurricanes, and floods on chemical and process plants can be quite catastrophic due to the possibility of damage to process units and subsequent release of large quantity of hazardous chemicals which may lead to disastrous environmental pollution [1] or fires and explosions [2–8]. Such technological accidents, which involve release of chemical substances into the environment, due to the impact of natural hazards are also known as Na-techs [9]. Katrina and Harvey are the hurricanes, which hit the U.S. in 2005 and 2017, respectively, are the first and second costliest events in the U.S. history and considered as noteworthy examples for Na-techs and their substantial damage to the environment and the industry. Katrina (2005) caused extensive damage to chemical and process plants and resulted in release of $\sim$ 8 million gallons of oil, making it the second largest environmental pollution in the U.S after the Deep Horizon disaster in the Gulf of Mexico in 2010. Similarly, Harvey (2017) caused substantial damage to refineries and petrochemical plants in the Huston Greater Area, leading to release of $\sim$ 2000 tons of chemicals into the environment as of September 2017 [10–12]. **Table 1** includes a list of industrial plants that suffered direct damages during Hurricane Harvey [10].

Among the process units, storage tanks – and atmospheric storage tanks in particular – have reportedly been the most vulnerable type of industrial vessels during hurricanes both due to their thin shell, which makes them very susceptible to lateral forces exerted by wind and flood, and due their high volume-weight ratio, which makes them susceptible to buoyancy force exerted by flood and heavy rainfall as components of hurricanes [11–14]. During Harvey in 2017, a total of 47 incidents of tank damage was reported with a total spillage 2 million liters of gasoline, crude oil, and other chemicals [15].

Damage to storage tanks can cause considerable environmental pollutions. During Katrina, ruptured storage tanks released several millions of gallons of oil while during Harvey, more than two dozen storage tanks ruptured, spilling $\sim$ 550 cubic meters of crude oil and gasoline into surrounding areas [10]. **Figure 1** shows the release of crude oil from to storage tanks, which have been removed from their foundations due to flood forces during Hurricane Katrina (2005).

Environmental pollution, however important, is not the only concern with regard to Na-techs. Release of petrochemicals – which in most cases are highly flammable and explosive – from damaged storage tanks can readily result in fire or explosion particularly in the presence of a variety of ignition sources and hot surfaces in the chemical and process plants. Such fires and explosions can easily go out of control due to large quantities of released chemicals and vicinity of storage tanks, capable of resulting in a sequence of fires and explosions – known as domino effects [16, 17].

Company name	Type of industry	Type of toxic chemicals
Invista	Plastic Manufacturer	Polytetramethylene ether glycol, Tetrahydrofuran, and 1,4 Butane Diol
LyondellBasell	Plastic Manufacturer	Polymers
LyondellBasell	Refinery	Gasoline, Diesel, Benzene, Paraxylene, Orthoxylene
Celanese Chemical	Chemicals Manufacturer	Methanol
Dow	Plastic Manufacturer	Polyurethanes
Eastman	Plastic Manufacturer	Non-phthalate plasticizers
Indorama Ventures	Chemicals Manufacturer	Ethylene oxide
Indorama Ventures	Plastics Manufacturer	Glycols
Arkema	Chemicals Manufacturer	Sulfuric substances
ExxonMobil	Refinery	Acrylic acid
Enterprise	Refinery	Hydrocarbon fuels
Chevron Phillips	Petrochemical	Hydrocarbon fuels
Shell	Refinery	NGL fractionators

Table 1.
Type of industrial plants damaged during the Hurricane Harvey [10].

Figure 1.
Release of crude oil due to displacement of storage tanks from their bases due to storm surges during Hurricane Katrina, 2005 [10].

Compared to conventional technological accidents that are caused by random failures or human error, risk assessment and management of Na-techs are more challenging [9, 18] due to the presence of more uncertain parameters and interdependent and complex failure mechanisms. The uncertainty consists of aleatory uncertainty arising from the randomness of natural disasters, resistance and vulnerability of process vessels to combined forces – wind, flood, rainfall, lightning – during hurricanes, and epistemic uncertainty arising from lack of sufficiently accurate and large historical databases.

Probability theory has effectively been used to account for uncertainties arising from environmental and operational random variable involved in risk assessment of Na-techs [19–25]. Even given the probability of random variables (depth of flood inundation, speed of wind, etc.), conventional quantitative risk assessment (QRA) techniques such as fault tree analysis fall short in modeling interdependent failure modes and common root causes arising from concurrent impact of hurricane forces on process vessels.

Bayesian network (BN) is a probabilistic tool for handling uncertainty and modeling dependencies [26, 27]. Previous studies demonstrated that BN can effectively accommodate and handle a variety of probabilities in the form of point probabilities [28], probability distributions [29, 30], fuzzy probabilities [31], and imprecise probabilities [32, 33].

In the present chapter, we demonstrate how BN can be employed to assessing the vulnerability of process vessels to the forces of hurricanes. Section 2 reviews the constituent natural hazards in the event of a hurricane and the resulting failure modes with emphasis on atmospheric storage tanks as the most vulnerable type of process vessels. Section 3 reviews the fundamentals of BN. Section 4 demonstrates the application of BN to modeling, combining, and calculating the failure probabilities of an illustrative storage tank. The main outcomes of the study are summarized in Section 5.

2. Failure mechanisms during hurricanes

According to the Hurricane Hazard Mitigation Handbook for Public Facilities [34], winds, rains, and flood (or storm surge in coastal areas) are the three natural hazards or forces of hurricanes responsible for damages, including Na-techs. In line with [34], in the present study, hurricanes are considered as multi-hazard natural disasters consisting of strong winds, floods, and heavy rainfalls.

2.1 Flood

Compared to seismic or wind related Na-techs, the one caused by floods have received relatively less attention; this has mainly been due to the scarcity of experimental or high-resolution field observations. Only for a limited number of floods were the flood inundation depths registered and even for fewer floods were the flood speeds recorded.

According to [14], 272 flood related Na-techs were reported in Europe and the U.S. from 1960 to 2007, with the aboveground storage tanks as the most frequently damaged equipment (74% of cases), including atmospheric storage tanks, floating roof tanks, and pressurized tanks. Displacement of storage tanks (due to rigid sliding or floatation) and subsequent disconnection or rupture of attached pipelines, shell rupture due to lateral forces or debris impact, and the collapse of equipment were reportedly the main failure modes during floods. Godoy [13] reported similar failure modes while investigating the process plants affected by the Hurricanes Katrina in Louisiana and Texas, U.S. These failure modes are depicted in **Figure 2**.

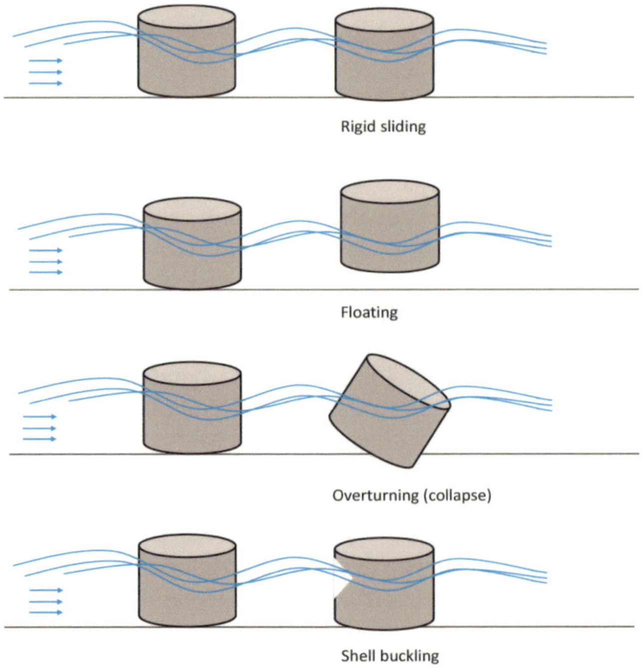

Figure 2.
Flood related failure modes for storage tanks.

Due to a lack of sufficiently accurate and reliable historical data, a majority of previous studies has employed analytical or numerical techniques to model the abovementioned failure modes and estimate the respective failure probabilities [12, 20, 22–25].

In these studies, limit state equations (LSEs) that were developed based on "physics of failure" models were used to calculate the failure probabilities. Such physics-of-failure models consider forces (loads) exerted by floods on target vessels (e.g., buoyancy force, which tends to float the tank) and the resistance of the vessel (e.g., total weight of the tank, which tend to resist the floatation). The developed LSEs can then be coupled with Monte Carlo simulation to generate artificial databases that in turn can be used to develop fragility curves [23] or be combined with other QRA techniques, such as BN [12, 24] for probabilistic reasoning.

2.2 Wind

Shell buckling is the main failure mode caused by wind [35] especially if the storage tank is empty or less than half-filled. According to Godoy [13], several storage tanks were reported to suffer shell buckling during Hurricane Katrina, mostly due to strong winds rather than flood. High winds can enforce critical pressures on tank shells. The magnitude of this pressure depends on several parameters, the most important of which being the wind speed and the shape and size of the storage tank.

Considering cylindrical storage tanks, however, the height of the tank does not play a key role [36, 37] as does the tank's circumference. As such, the wind pressure is usually considered as constant along the tank's height [38]. Similar to the flood related failure modes, it has been a common practice to develop LSEs for wind related failure mode – i.e., shell buckling and shell damage due to windborne debris – based on the load-resistance relationships. Considering the shell buckling, for instance, external pressure exerted by wind (i.e., load) and the internal pressure of the tank (i.e., resistance) were taken into account to develop the related LSE [12].

2.3 Rainfall

Heavy rainfalls before, during and after passage of hurricanes were rarely accounted for as a separate natural hazard directly capable of causing damage to process vessels. In that sense, contribution of heavy rainfalls to Na-techs has been indirectly taken into account via the runover of water bodies and rivers due to heavy rainfall and consequent flooding of process plants.

Hurricane Harvey, however, demonstrated the potential of heavy rainfall as a natural hazard, which can cause significant damage to process vessels standalone or in combination with other forces of the hurricane. Large amount of rainfall that fell during Hurricane Harvey caused damage to 400 "floating roof storage tanks", including sinking of 14 roofs. A floating-roof storage tank is a tank the roof of which is not fixed and rather floats on the surface of chemicals stored inside the tank.

3. Bayesian network

Bayesian network (BN) [26] is a directed acyclic graph consisting of random variable nodes and directed edges connecting the nodes, with the edges directed from

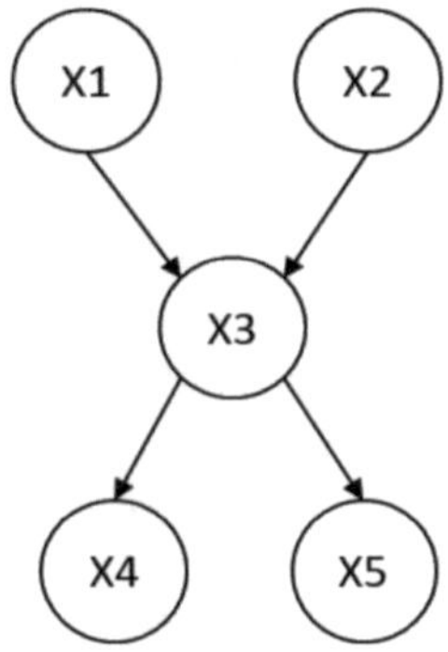

Figure 3.
BN consisting of 5 random variables (nodes). X1 and X2 are the root nodes and parents of X3. X3 is an intermediate node, the child of X1 and X2 and the parent of X4 and X5. X4 and X5 are the leaf nodes and the children of X3.

parents to children. A BN can mathematically be defined as BN = (G, θ), where G is the network structure – nodes and edges – and θ is the network parameters – conditional probabilities of child nodes given their parents. **Figure 3** shows an illustrative BN consisting of 5 nodes. X1 and X2 are the root nodes (equivalent to the basic events of a fault tree), X3 is an intermediate event (equivalent to an intermediate event of a fault tree), and X4 and X5 are the leaf nodes (equivalent to the top event of a fault tree). The root nodes are assigned marginal probability distributions while the other nodes are assigned conditional probability distributions. Such conditional probabilities reflect the type (usually causal) and strength of the impact parent nodes have on their child nodes.

Considering the local dependencies and chain rule, the joint probability distribution of the random variables in a BN can be presented as the product of marginal and conditional probabilities considering only the immediate parents of a child node:

$$P(X_1, X_2, \dots, X_n) = \prod_{i=1}^{n} P(X_i | pa(X_i)) \tag{1}$$

For example, for the BN in **Figure 3**, P (X1, X2, X3, X4, X5) = P (X1) P(X2) P(X3| X1, X2) P(X4|X3) P(X5|X3). BN can be used for either forward or backward reasoning. In forward reasoning, for example, the knowledge about X1 can be used to deduce about X5, whereas in backward reasoning, the knowledge about X5 can be used to infer about X1. The main advantages of BN over conventional techniques such as fault tree analysis are its capability of considering conditional dependencies, handling multistate variables, and applying Bayes' rule for belief updating (backward reasoning) [28]. Using the Bayes' theorem, BN can update the probabilities assigned a priori to the random variables in the presence of new information about the states of its nodes:

$$P(X|E) = \frac{P(X, E)}{P(E)} \tag{2}$$

The parameters of the BN—the marginal and conditional probabilities needed to quantify the model—can be assigned by subject matter experts or be estimated from a dataset by applying machine learning techniques such as the maximum likelihood estimation. **Table 2** summarizes some of the advantages of BN against fault tree.

Fault tree	Bayesian network
Has only one top event	Can have more than one leaf node
Handles only binary variables	Handles a variety of multi-state variable
Assumes variables are independent	Can consider dependencies
N/A	Conducts probability updating
Can be used only for forward reasoning	Can be used for both forward and backward analyses

Table 2.
Modeling advantages of BN as opposed to fault tree.

4. Vulnerability assessment of storage tanks

As previously discussed, one way to analyze the failure mechanism and to generate failure data required for probabilistic vulnerability assessment of process vessels is application of "physics of failure" to develop LSEs. The benefit of LSEs is twofold: From one side, they help better understand random variables that contribute to the failure mechanisms and should thus be considered in the BN, and from the other side, they can be coupled with, for instance, Monte Carlo simulation to generate datasets required for estimating the parameters—especially, the conditional probabilities—of the BN [12, 23, 24].

To demonstrate the abovementioned methodology, we consider floatation as one of the frequently reported failure modes (failure mechanism) for above-ground atmospheric storage tanks. Since such storage tanks are usually unanchored (not bolted or cemented to their foundations), the buoyancy force of flood in Eq. (3) can make the storage tank float should it overcome the total weight of the tank—i.e., the weight of the tank in Eq. (4) plus the weight of chemical in the tank in Eq. (5)—as the only resisting force. Considering the bouncy as the only moving force and the total weight of the tank as the only resistance, the LSE of the tank due to the floatation can be modeled as in Eq. (6):

$$F_B = \rho_w g.\pi \frac{D^2}{4} S \tag{3}$$

$$W_T = \rho_T g.\left(\pi DH + 2\pi \frac{D^2}{4}\right).t \tag{4}$$

$$W_L = \rho_L g.\pi \frac{D^2}{4} h \tag{5}$$

$$LSE = F_B - W_L - W_T \tag{6}$$

where F_B is the buoyancy force; W_T and W_L are the weight of the tank and the contained chemical, respectively; ρ_w, ρ_T, and ρ_L are, respectively, the density of flood water, of the tank structure's material, and of the chemical; g is the gravitational acceleration; D and H are, respectively, the diameter and height of the tank; S is the depth of flood; h is the depth of chemical in the tank; t is the shell thickness of the tank. Given the LSE in Eq. (6), if LSE > 0, or simply $F_B > W_L + W_T$, the tank floats. The parameters in Eqs. (3)–(6) are depicted in **Figure 4**. Considering the depth of flood (S) and the depth of chemical (h) as the only random variables (the other parameters can reasonably be considered as constants), $F_B > W_L + W_T$ can be further simplified as:

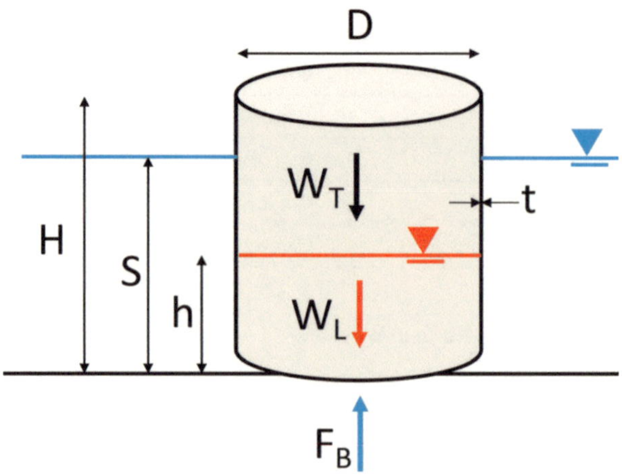

Figure 4.
Parameters used to develop the LSE required for floatation failure mechanism of an unanchored atmospheric storage tank [23].

$$S > ah + b \tag{7}$$

where a and b are constants:

$$a = \frac{\rho_L}{\rho_w} \tag{8}$$

$$b = \frac{W_T}{\rho_w . g . \pi . \frac{D^2}{4}} \tag{9}$$

According to Eq. (7), the probability of floatation P_{float} can be calculated as:

$$P_{float} = P(S > ah + b) = 1 - P(S < ah + b) \tag{10}$$

Given the probability distribution functions of S and h as $f_S(S)$ and $f_h(h)$, and considering that the level of chemical in the tank cannot exceed H (in practice, 15-20% of the tank's height is usually left for safety purposes), Eq. (10) can be expanded as:

$$P_{float} = 1 - P(S < ah + b) = 1 - \int_0^H \left(\int_0^{ah+b} f_S(S).dS \right) f_h(h)\, dh \tag{11}$$

4.1 Analytical approach

For some types of probability distributions, P(S < ah +b) can easily be calculated with no need for calculating the double integral in Eq. (11). One type of such probability distributions is Normal distribution. Given two normally distributed random variables as X $\sim$ Normal (μ_x, σ_x) and Y $\sim$ Normal (μ_y, σ_y), any linear function of X and Y is a normally distributed random variable as Z $\sim$ Normal (μ_z, σ_z). For instance, if Z = mX + nY, the mean value μ_z and standard deviation σ_z of Z can be calculated as:

$$\mu_z = m\,\mu_x + n\,\mu_y \tag{12}$$

$$\sigma_z = \sqrt{m^2.\sigma_x^2 + n^2.\sigma_y^2} \tag{13}$$

Therefore, if S $\sim$ Normal (μ_s, σ_s) and h $\sim$ Normal (μ_h, σ_h), then Q = S – ah is a linear function of S and h and thus normally distributed as:

$$\mu_Q = \mu_s - a\,\mu_h \tag{14}$$

$$\sigma_Q = \sqrt{\sigma_s^2 + a^2.\sigma_h^2} \tag{15}$$

As a result:

$$\mathrm{P(S < ah + b) = P(S - ah < b) = P(Q < b)} = \Phi_Q(b) \tag{16}$$

where $\Phi_Q(.)$ is the cumulative distribution function for normal distribution. Consequently:

$$P_{float} = P(S > ah + b) = 1 - P(S < ah + b) = 1 - \Phi_Q(b) \tag{17}$$

To demonstrate the application of Eq. (17), consider an unanchored crude oil storage tank with diameter of D = 91m, height of H = 6m, and shell thickness of t = 15mm. The characteristics of the tank, crude, and flood are presented in **Table 3**.

Given the values in **Table 3**, the values of a and b are calculated via Eqs. (8) and (9) as a = 0.88 and b = 0.26. Subsequently, using Eqs. (14) and (15):

$$\mu_Q = \mu_s - a\,\mu_h = 1 - 0.88 \times 1 = 0.62$$

$$\sigma_Q = \sqrt{0.1^2 + 0.88^2 \times 1^2} = 0.89$$

Knowing that Q is a normal random variable Q $\sim$ Normal (0.62, 0.89), using Eq. (17), $P_{float} = 1 - \Phi_Q(b) = 1 - \Phi_Q(0.26) = 0.66$.

4.2 Numerical approach

Given the probability distribution of S and h, the probability of flotation in Eq. (10) can be approximated numerically using, among others, Monte Carlo

Parameter	Symbol (unit)	Value or probability distribution
Tank diameter	D (m)	91
Tan height	H (m)	6
Shell thickness	t (mm)	15
Density of shell material	ρ_T (kg/m^3)	7900
Density of flood water	ρ_w (kg/m^3)	1024
Density of crude oil	ρ_L (kg/m^3)	900
Depth of crude in the tank	h (m)	$f_s(S)$ = Normal (1.0, 1.0)
Depth of flood water	S (m)	$f_h(h)$ = Normal (1.5, 0.1)

Table 3.
Parameters used for the illustrative storage tank and hitting flood.

Sample No	h ~ N(1.0, 1.0)	S ~ N(1.0, 0.1)	Q = S – 0.88h	1 if Q > 0.26; 0 otherwise
120	0.089	1.608	1.529	1
203	0.724	1.592	0.955	1
311	1.410	1.599	0.359	1
457	1.657	1.508	0.049	0
590	1.768	1.745	0.189	0
602	1.481	1.454	0.150	0
740	1.129	1.417	0.423	1

Table 4.
Sample of dataset generated by Monte Carlo simulation given the probability distributions of S and h.

simulation. **Table 4** show part of the dataset generated for S and h given their probability distributions. Obviously, for each pair of S-h dataset, if Q > 0.26 the tank floats. The last column of **Table 4** assign 1 to each pair if Q > 0.26 (the tank floats) and 0 if Q < 0.26 (the tank does not float). Having a sufficiently large dataset, the mean value of the last column denotes the probability of floatation. Given 1000 samples, P_{float} = 0.676, which is close enough to the probability calculated using the analytical approach.

4.3 Bayesian network approach

Instead of solving Eq. (10) or Eq. (11) analytically or numerically, the BN in **Figure 5** can be used to model the tank flotation and estimate its probability. Knowing that *S* and *h* are the only random variables that contribute to the flotation, the BN should consist of two root nodes "S" and "h", and a leaf node, "Floatation".

S and *h* in Eq. (11) are continuous variables, but their corresponding nodes in the BN are discrete nodes. As such, *S* and *h* need to be discretized into a finite set of states. The probabilities of the states can then be calculated using the probability

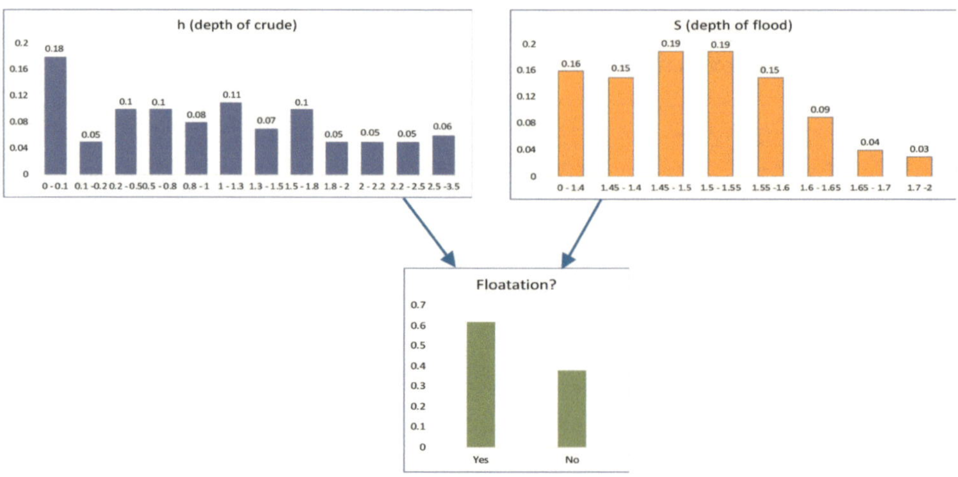

Figure 5.
BN for modeling and assessing the probability of tank floatation.

distributions of S and h, that is, $f_S(S)$ and $f_h(h)$. Subsequently, the conditional probability table for node "Floatation" can readily be populated with 1s and 0s given Eq. (7), that is, 1 if S > ah + b, and 0 otherwise. Given the BN in **Figure 5**, the probability of floatation is estimated as 0.62, sufficiently close to the results obtained by solving the integrals analytically (Section 4.1) or numerically (Section 4.2).

While by increasing the resolution of nodes h and S (increasing their states) the calculated probability for the node Floatation can further be improved, such refinement comes at a substantially increased modeling time and effort. It is because the size of the conditional probability table for the node Floatation increases exponentially with the size of the probability tables (number of states) for nodes h and S.

Following the same procedure for the other failure modes, the final BN for assessing the combined failure probability can be developed as in **Figure 6**. In **Figure 6**, the nodes "V_wind" and "V_flood" denote the speed of wind and flood, respectively; "H_ac" is the height of accumulated rainfall on the roof of storage tank; "Wind_buckle" and "Flood_buckle" represent shell buckling failure mode due to lateral forces of wind and flood, respectively; "Slide" denotes rigid sliding of the tank

Figure 6.
Final BN for assessing the total failure probability considering all the failure modes [12].

Methodology	Feasibility	Accuracy
Analytical (Section 4.1)	Low	Exact solution
Numerical (Section 4.2)	High	Depends on sample size
BN (Section 4.3)	High	Depends on discretization resolution

Table 5.
A comparison between the discussed methodologies.

due to lateral force of flood, and "Roof-sinking" denotes the failure mode due to accumulation of rainfall on the top.

5. Conclusions

In this chapter, we discussed failure modes and vulnerability of process vessels – with emphasis on atmospheric storage tanks – in the event of hurricanes. It was demonstrated how physics-of-failure models and Bayesian network can be used to assess the failure probabilities of storage tanks due to the impact of flood, wind, and rainfall.

Although the physics-of-failure models can directly be used to develop mathematical relationships for estimating the failure probabilities (see Eq. (11) as an example), such mathematical relationships cannot seem to efficiently account for dependencies when modeling and combining interdependent failure modes. These interdependencies may arise when one root node contributes to several failure modes or when one failure mode excludes or contributes to another failure mode. An example for such interdependencies includes the accumulation of rainfall on the rooftop of a floating- roof tank ("H_ac" in **Figure 6**), which causes the roof to sink ("Roof_sinking") but reduces the likelihood of tank's floatation ("Floatation") and sliding ("Slide") due to its added weight.

The application of BN, on the other hand, enables the analyst to consider the dependencies among the failure modes, which would otherwise result in inaccurate failure probabilities. **Table 5** summarizes the modeling features of the methodologies discussed in this chapter.

Acknowledgements

The financial supports by the Natural Sciences and Engineering Research Council of Canada (NSERC) via Discovery Grant, and by the Faculty of Community Services, Ryerson University, Canada, via Publication Grant is much appreciated.

Author details

Nima Khakzad
Ryerson University, Toronto, ON, Canada

*Address all correspondence to: nima.khakzad@ryerson.ca

References

[1] Fernandes R, Necci A, Krausmann E. Model(s) for the Dispersion of Hazardous Substances in Floodwaters for RAPID-N, EUR 30968 EN. Luxembourg: Publications Office of the European Union; 2022

[2] Showalter PS, Myers MF. Natural disasters in the United-States as release agents of oil, chemicals, or radiological materials between 1980 and 1989. Risk Analysis. 1994;**14**:169-181

[3] Rasmussen K. Natural events and accidents with hazardous materials. Journal of Hazardous Materials. 1995;**40**: 43-54

[4] Young S, Balluz L, Malilay J. Natural and technologic hazardous material releases during and after natural disasters: A review. The Science of the Total Environment. 2004;**322**:3-20

[5] Cruz AM, Okada N. Consideration of natural hazards in the design and risk management of industrial facilities. Natural Hazards. 2008;**44**:213-227

[6] Antonioni G, Bonvicini S, Spadoni G, Cozzani V. Development of a frame work for the risk assessment of Na-Tech accidental events. Reliability Engineering and System Safety. 2009; **94**:1442-1450

[7] Krausmann E, Renni E, Campedel M, Cozzani V. Industrial accidents triggered by earthquakes, floods and lightning: Lessons learned from a database analysis. Natural Hazards. 2011;**59**:285-300

[8] Marzo E, Busini V, Rota R. Definition of a short-cut methodology for assessing the vulnerability of a territory in natural–technological risk estimation. Reliability Engineering and System Safety. 2015;**134**:92-97

[9] Salzano E, Basco A, Busini V, Cozzani V, Marzo E, Rota R, et al. Public awareness promoting new or emerging risks: Industrial accidents triggered by natural hazards (NaTech). Journal of Risk Research. 2013;**16**(3-4):469-485

[10] Sebastian T, Lendering K, Kothuis B, Brand N, Jonkman B, van Gelder P, et al. Hurricane Harvey Report: A Fact-finding Effort in the Direct Aftermath of Hurricane Harvey in the Greater Houston Region. Delft, The Netherlands: Delft University Publishers; 2017

[11] Qin R, Khakzad N, Zhu j. An overview of the impact of Hurricane Harvey on chemical and process facilities in Texas. International Journal of Disaster Risk Reduction. 2020;**45**:101453

[12] Qin R, Zhu J, Khakzad N. Multi-hazard failure assessment of atmospheric storage tanks during hurricanes. Journal of Loss Prevention in the Process Industries. 2020;**68**:104325

[13] Godoy LA. Performance of storage tanks in oil facilities damaged by Hurricanes Katrina and Rita. Journal of Performance of Constructed Facilities. 2007;**21**(6):441-449

[14] Cozzani V, Campedel M, Renni E, Krausmann E. Industrial accidents triggered by flood events: Analysis of past accidents. Journal of Hazardous Materials. 2010;**175**:501-509

[15] Bernier C, Padgett JE. Fragility and Risk Assessment of Aboveground Storage Tanks Subjected to Concurrent Surge, Wave, and Wind Loads. Reliability Engineering and System Safety. 2019;**191**:106571

[16] Khakzad N, Khan F, Amyotte P, Cozzani V. Domino effect analysis using

Bayesian networks. Risk Analysis: An International Journal. 2013;**33**(2): 292-306

[17] Khakzad N. Application of dynamic Bayesian network to risk analysis of domino effects in chemical infrastructures. Reliability Engineering and System Safety. 2015;**138**:263-272

[18] Salzano E, Basco A, Cozzani V, Renni E, Rota R. The risks related to the interaction between natural hazards and technologies. In: Proceedings of 8th World Congress of Chemical Engineering (WCCE8). Ontario: Canadian Society for Chemical Engineering; 2009. pp. 119-201

[19] Hauptmanns U. A decision-making framework for protecting process plants from flooding based on fault tree analysis. Reliability Engineering and System Safety. 2010;**95**:970-980

[20] Landucci G, Antonioni G, Tugnoli A, Cozzani V. Release of hazardous substances in flood events: Damage model for atmospheric storage tanks. Reliability Engineering and System Safety. 2012;**106**:200-216

[21] Necci A, Antonioni G, Cozzani V, Krausmann E, Borghetti A, Nucci CA. A model for process equipment damage probability assessment due to lightning. Reliability Engineering and System Safety. 2013;**115**:91-99

[22] Mebarki A, Jerez S, Prodhomme G, Reimeringer M. Natural hazards, vulnerability and structural resilience: Tsunamis and industrial tanks. Geomatics, Natural Hazards and Risk. 2016;**7**(S1):5-17

[23] Khakzad N, Van Gelder P. Fragility assessment of chemical storage tanks subject to floods. Process Safety and Environment Protection. 2017;**111**:75-84

[24] Khakzad N, van Gelder P. Vulnerability of industrial plants to flood-induced natechs: A Bayesian network approach. Reliability Engineering and System Safety. 2018; **169**:403-411

[25] Kameshwar S, Padgett JE. Storm surge fragility assessment of above ground storage tanks. Structural Safety. 2018;**70**:48-58

[26] Pearl J. Probabilistic Reasoning in Intelligent Systems. San Francisco, CA: Morgan Kaufmann; 1988

[27] Jensen F. An Introduction to Bayesian Networks. London, UK: UCL Press; 1996

[28] Khakzad N, Khan F, Amyotte P. Safety analysis in process facilities: Comparison of fault tree and Bayesian network approaches. Reliability Engineering and System Safety. 2011; **96**(8):925-932

[29] Khakzad N, Khan F, Paltrinieri N. On the application of near accident data to risk analysis of major accidents. Reliability Engineering and System Safety. 2014;**126**:116-125

[30] Abrishami S, Khakzad N, Hosseini SM, van Gelder P. BN-SLIM: A Bayesian Network methodology for human reliability assessment based on Success Likelihood Index Method (SLIM). Reliability Engineering and System Safety. 2020;**193**:106647

[31] Zarei E, Khakzad N, Cozzani V, Reniers G. Safety analysis of process systems using Fuzzy Bayesian Network (FBN). Journal of Loss Prevention in the Process Industries. 2019;**57**:7-16

[32] Simon C, Weber P. Evidential networks for reliability analysis and

performance evaluation of systems with imprecise knowledge. IEEE Transactions on Reliability. 2009;**58**(1):69-87

[33] Khakzad N. System safety assessment under epistemic uncertainty: Using imprecise probabilities in Bayesian network. Safety Science. 2019;**116**: 149-160

[34] FEMA. Hurricane mitigation: A handbook for public facilities. 2005. Available online from: https://www.fema.gov/media-library-data/20130726-1715-25045-932 4/hurricane_mitigation_handbook_for_public_facilities.pdf. (Accessed 2 November 2022).

[35] Zhao Y, Lin Y. Buckling of cylindrical open-topped steel tanks under wind load. Thin-Walled Structures. 2014;**79**:83-94

[36] Eurocode C. 3–Design of Steel Structures–Part 1–6: Strength and Stability of Shell Structures. Brussels: European Committee for Standardization (CEN); 2007

[37] Bu F, Qian C. A rational design approach of intermediate wind girders on large storage tanks. Thin-Walled Structures. 2015;**92**:76-81

[38] Mayorga S et al. Development of parametric fragility curves for storage tanks: A Natech approach. Reliability Engineering and System Safety. 2019; **189**:1-10

Chapter 16

Climate Change-Related Natural Hazards and Risk Communication: Incorporating Traditional Indigenous Knowledge

Muhammad Arshad K. Khalafzai

Abstract

This chapter explores the incorporation of traditional Indigenous knowledge into climate change-related natural hazard risk communication. Ample research has been conducted on climate change-related risk communication, and a significant body of literature exists on the role of traditional Indigenous knowledge in reducing climate change impacts. However, even in the face of mounting climate change-related risks, little effort has been made to incorporate traditional Indigenous knowledge into climate change-related natural hazard risk communication. Scientific knowledge and traditional Indigenous knowledge pertain to different knowledge systems; however, in terms of methods and content, many aspects exist where both systems converge or follow similar patterns. Rather than focusing on points of divergence, researchers, policymakers, and decision-makers, and risk-communication experts should focus on common features of both systems. Points of convergence may provide common ground for knowledge integration and co-production, enabling Indigenous and scientific understandings of climate change to be reconciled. This may help improve risk communication processes between disaster risk management practitioners, agencies, and Indigenous Peoples. However, it is also important to recognize that traditional Indigenous knowledge may not fit with every scientific model; therefore, a more in-depth research is needed to learn which forms of traditional Indigenous knowledge can help scientific researchers improve climate change-related natural hazard risk communication processes.

Keywords: climate change adaptation, natural hazards, disaster risk reduction, risk communication, scientific knowledge, traditional Indigenous knowledge, integration, co-production

1. Introduction

Climate change is a global phenomenon, which impacts different people unevenly [1]. Scientists, politicians, special interest groups, and members of the general public are engaged in extensive debate on climate change issues, including their potential impacts [2]. Research suggests that public beliefs on climate change vary significantly [3, 4], and different people view the risks associated with climate change differently. Addressing public perceptions of climate-change risk can be challenging, due to the sociocultural construction of risk and its multi-dimensional complexity [5]. Differing perspectives on climate change and associated risks must be understood within specific contexts of climate change—and within interconnected socioeconomic and cultural settings [6]. Because climate change is a complex phenomenon, its risks can be difficult for laypeople to comprehend. Moreover, climate-change awareness and understanding are also influenced by environmental values and political and economic agendas. While climate-change skeptics have accused climate scientists of "alarmism"—of over-interpreting or overreacting to evidence of climate change—recent evidence suggests that scientists have been conservative in their projections of risk [7]. For example, Brysse et al. [7] found that the Intergovernmental Panel on Climate Change (IPCC) underestimated the expected impacts of climate change; similarly, Rahmstorf et al. [8] found that experts have underestimated expected climatic impacts. Along with opposing perspectives on climate change, problems of "poor (mis)communication," "misinformation," and implicit presumptions have complicated climate-change discourses and risk-communication efforts [5].

Many populations have already started to experience the impacts of climate change [9–11]. As a result of climate change, experts project that populations will experience increasingly extreme weather events, rising sea levels, and flooding and drought events in the future [10]. The intensity of such events is expected to increase over time—and some populations are at greater risk than others. For example, indigenous Peoples across the globe are expected to be among the communities most heavily affected by climate change [1, 12]. In part, this reflects the dependence and close proximity of many indigenous Peoples to their natural environments and resources [1, 12]. In northern Canada, indigenous communities are already experiencing the effects of significant climate changes [13]. For example, members of the Kashechewan First Nation community near the Albany River in the southwestern James Bay region have been evacuated 14 times to 22 different host communities across the Ontario province due to actual flooding or flooding risk because of the potential failure of the dike [14].

The uneven impacts of climate change raise questions of fairness [15]. Given the disproportionate effects of climate change on indigenous communities, it is vital for climate-change risk-communication experts to consider the needs and concerns of indigenous Peoples. It is also important for experts to consider and incorporate tradition into their efforts to address climate change and mitigate its risks. While ample research has been conducted on climate-change risk communication, as well as the role of traditional indigenous knowledge in reducing climate change impacts, little effort has been made to integrate traditional indigenous knowledge into climate-change risk communication—even in the face of mounting climate-change risks. This chapter addresses a gap in the literature by exploring the potential contributions of traditional indigenous and local spatial knowledge systems to climate-change risk communication processes and efforts to understand and mitigate the impacts of climate change.

2. Materials and method

In this chapter, I synthesize information from a range of secondary sources to consider the potential contributions that traditional indigenous knowledge systems can make to risk communication processes. Specifically, I explore how three of the "five convergence areas" developed by Riedlinger and Berkes [16] and "five areas of complementarity" presented by Moller et al. [17] may facilitate the incorporation and co-production of knowledge across scientific and traditional indigenous knowledge systems. Riedlinger and Berkes' five convergence areas (1) local scale expertise; (2) climate history; (3) research hypothesis; (4) community adaptation; and (5) community-based environmental monitoring; I restrict my analysis to climate history; community adaptation, and community-based environmental monitoring. Moller and colleagues' give areas of complementarity include: (1) synchronic-diachronic complementarity; (2) complementary foci on averages versus extremes; (3) interplay between quantitative and qualitative information; (4) science for a better test of mechanisms, traditional indigenous knowledge for better hypothesis; and (5) complementing objectivity with subjectivity.

2.1 Traditional Indigenous knowledge recognition

European conquests, colonization, and subsequently forced assimilation of Indigenous Peoples into the dominant Eurocentric Canadian culture have had serious and far-reaching implications. Indigenous communities have been relegated to systematic poverty, isolation, and powerlessness as a result of Eurocentric "cultural imperialism and racism" [18]. During the colonization and modernization eras, traditional indigenous knowledge and values were devalued as an "impediment to development" [19]. Unequal power dynamics between western science and traditional knowledge have undermined the potential role that traditional indigenous knowledge-holders can play as active participants in risk-communication decision-making processes.

Advocates, however, have been arguing that Indigenous Peoples and their knowledge should not be treated as "clients" or "stakeholders" in efforts to mitigate the risks of climate change. Instead, their meaningful participation should be engaged at all levels of decision-making, and traditional indigenous knowledge should be treated as coequal and complementary to western scientific knowledge [20]. Indeed, the value of traditional indigenous knowledge systems is increasingly recognized by many scientists, managers, policy-makers, and lawmakers, and traditional indigenous knowledge has become a subject of national and international law [21]. Presently, international institutional structures—such as TRIPS (Trade Related Intellectual Property Rights) and other legal instruments—do not provide patent protection to traditional indigenous knowledge [22]. However, initiatives such as the UN Permanent Forum on Indigenous Peoples (UNPFIP), Intergovernmental Forum on Forests (IFF), and Arctic Environmental Protection Strategy (AEPS) are working toward protecting Indigenous Peoples' rights and traditional knowledge. The IPCC's 32nd Session [10] also recognized traditional indigenous knowledge as an important guiding principle for the Cancun Adaptation Framework (CAF), which was adopted by its members at the 2010 FCCC Conference in Cancun [23]. This framework highlights the contributions of traditional indigenous knowledge to broader understanding of global climate change—including climate-change observations, impacts, and opportunities for adaptation.

While traditional indigenous knowledge has been historically devalued in many fields of scientific and social science research, traditional indigenous knowledge contributions have also been well documented in many fields—including fields of agro-forestry, traditional medicine, biodiversity conservation, customary resource management, applied anthropology, impact assessment, and risk reduction [24]. For example, traditional indigenous knowledge contributions have been made to biological information and ecological insights, resource management, biodiversity conservation, environmental assessment, and ethics [25]. The IPCC [9] has also recognized the value of traditional indigenous knowledge as a basis for developing adaptation and natural-resource-management strategies while responding to environmental change and other forms of change. However, scientists are often skeptical of its importance. Moreover, they tend to value traditional indigenous knowledge after recasting it in scientific terms to make it more rational, empirical, and objective [26]. It is important that one not lose the essence of traditional indigenous knowledge while recasting indigenous indigenous content into scientific.

2.2 How can traditional Indigenous knowledge be integrated into climatic risk communication?

This section will discuss various methods that can help in the integration of indigenous traditional knowledge systems with scientific knowledge. These methods include "scientization," co-management, and co-production by scientization indigenous knowledge systems and science. Traditional indigenous and scientific knowledge systems can be integrated to help address issues of climate-change risk communication. These systems have been in "intimate interaction" with each other for at least 50 years and cannot remain "untouched" by each other [27]. Traditional indigenous knowledge systems comprise the process of observing, discussing, and making sense of new information—in other words, Indigenous ways of knowing [28]. There needs to be a paradigm shift in the system of power between Indigenous Peoples and mainstream power in the contemporary risk communication decision-making process.

The process of "scientization" will separate useful traditional indigenous knowledge from other knowledge, practices, context, and cultural beliefs [29]. While filtering traditional indigenous knowledge for climate-change risk communication, only functional content should be selected. Agarwal [29] describes scientization as the process of particularization, validation, and generalization. Particularization involves selecting useful traditional indigenous knowledge that can be tested and validated. After it is validated, that knowledge is generalized through the steps of cataloging, archiving, and circulating it for use in risk-communication decision-making. Scientization can facilitate the integration of scientific and traditional indigenous knowledge systems for the co-production of climate-change risk communication.

Knowledge co-production can be instrumental for decision-making by sharing information. Armitage et al. [30] assert that knowledge co-production is not merely knowledge integration. It can offer more opportunities to improve the risk communication process through joint gathering, sharing, integration, interpretation, production, and application of knowledge. Knowledge co-production may be facilitated through collaborative research projects between Indigenous Peoples and scientists. Such collaboration may help members of the mainstream scientific community recognize that indigenous observations and assessments offer valuable insights into, and provide local verification of, global scientific models; moreover, taking indigenous observations and assessments seriously may help climate-change experts ensure that adaptation measures align with local needs and priorities [31]. Collaboration between scientific and Indigenous communities can promote innovative and effective adaptation action, and relevant traditional knowledge can inform cost-effective and sustainable climate change adaptation measures.

2.2.1 Partnership and governance for knowledge co-production

Collaborative research projects are bringing Indigenous Peoples and scientists closer to knowledge co-production. To facilitate the co-production process, the collaboration between the two parties should be strengthened by focusing on common features and characteristics of scientific knowledge and traditional indigenous knowledge systems. Anderson David [32] has identified several common features (**Figure 1**): (1) the two systems consider the universe to be unified; (2) their organizing principle is based on a body of knowledge systems that is stable, but subject to modification; (3) both require honesty, inquisitiveness, and perseverance; (4) they record empirical observations in natural settings, with verification through repetition; (5) their inference and prediction are based on skills and procedures; (6) their knowledge base originates from plant and animal behavior, cycles, habitats needs, and interdependence; and (7) both focus on cycles and changes in earth and sky. The following figure explores the similarities and differences between the two systems.

Collaboration between the two parties can be facilitated through resource management governance approaches, including: (1) co-management; (2) adaptive management; and (3) adaptive co-management [28]. These governance approaches can contribute to the climatic and environmental risk management process by involving Indigenous communities, their local institutions, networks, and knowledge systems.

Berkes [28] argues that co-management—as knowledge production and as social learning—is becoming increasingly important, as demonstrated in the literature over the past two decades. Co-management evolves adaptively as a result of deliberate problem-solving. To compare co-management and adaptive co-management, first I provide Berkes's definition in this respect. Adaptive co-management, as defined by Berkes & Turner [33], is "a process by which institutional arrangements and ecological knowledge are tested and revised in a dynamic, ongoing, self-organized process of

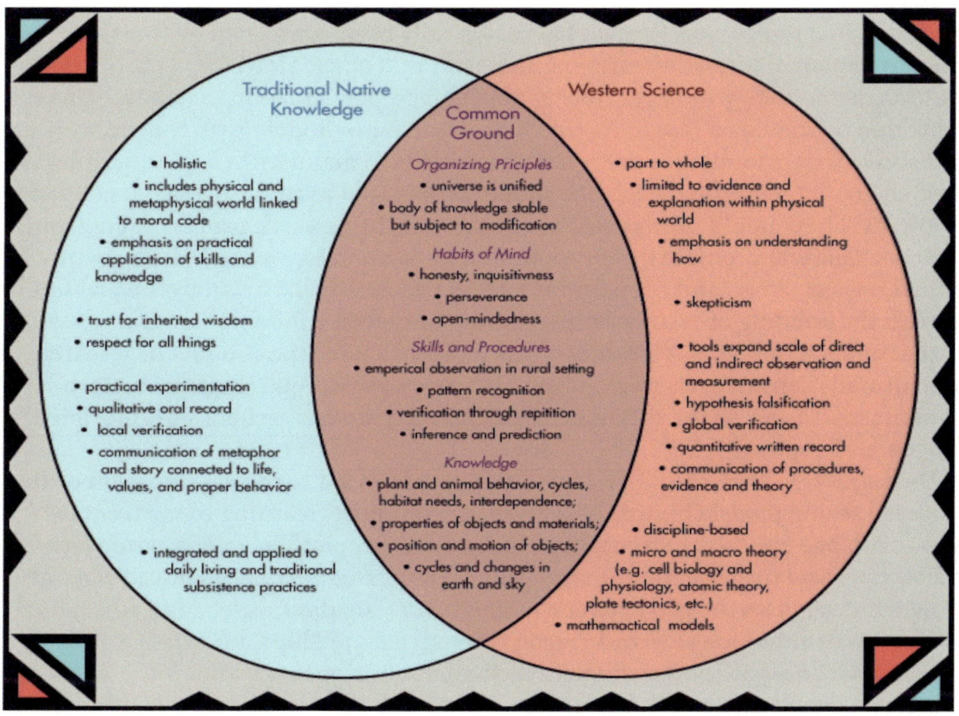

Figure 1.
Traditional Indigenous and scientific knowledge: common ground. (Source: http://www.beliefnet.com/columnists/healingandtransformation/).

learning-by-doing" (2). According to Anderson David [32], generally, co-management's primary focus remains on making vertical institutional linkage to create connections between both systems. While its temporal scope is short to medium-term, the gap between indigenous (local) and government levels is bridged at the organizational level while building the capacity of Indigenous communities as resource users. Co-management essentially includes power sharing, institutional building, and trust building between parties and social capital, as a process, as problem-solving, and as people-centered governance with shared management and responsibility. In comparison, adaptive management requires collaborative processes to build consensus among the parties. However, co-management and adaptive management can complement one another.

Adaptive management develops links between science and resource management [25], and adaptive management's multiple cycles of learning and adaptation have medium to long-term temporal scope [32]. Adaptive management essentially focuses on management's needs and relationships at multi-level—with self-organized networks—and capacity building of resource managers and decision-makers. Berkes [34] notes that learning-based approaches were primarily developed to deal with environmental uncertainty. Therefore, adaptive management relies upon "learning by doing" through interactive practice, evaluation, and activity modification.

Contrary to the first two approaches, Anderson David [32] argues that adaptive co-management targets both horizontal and vertical linkages for joint "learning-by-doing". In adaptive co-management, cycles of learning and adaptation are multiple and have medium- to long-term temporal scope. The organizational level is generally multi-level with self-organized networks. Adaptive co-management focuses on the needs of and relationships among all partners for capacity building.

According to the Canadian Institute of Health Research (CIHR), research involving Aboriginal Indigenous Peoples has historically been conducted by "outsiders"—which is a colonial way of researching Indigenous Peoples. However, a participatory research methodology requires Aboriginal Peoples' participation in research projects. right from conceiving a research project to research methodology to collection of data to analysis of data to sharing and validating results to manuscript/report writing As affected and interested parties, they should be engaged as active partners, not passive recipients. They should be involved at all stages of the research process, right from the start. This will promote the development of partnership within a framework of mutual respect, trust, and cooperation; it will promote collaboration on equal footing through the building of relationships, sharing of power, equitable resourcing, and mutual understanding. Furthermore, this will also enable the conduct of research in a culturally sensitive, relevant, respectful, responsive, equitable, and reciprocal manner—as well as the sharing of benefits between researchers and Aboriginal Indigenous Peoples [35, 36].

During 2009–2010, the CIHR consultative process led to the development of the two-eyed seeing model (**Figure 2**). This model promotes learning to see from two eyes—with one eye on the strengths of Indigenous Peoples' knowledge and ways of knowing, and the other on the strengths of mainstream knowledge and ways of knowing. It suggests the gap between mainstream Canadian society and Aboriginal Indigenous communities can be bridged through partnerships, knowledge exchange, and research on significant challenges facing First Nations and other Indigenous Aboriginal Peoples. This collaborative approach may provide more opportunities to improve climate-change risk communication.

According to Berkes [28], traditional indigenous knowledge is evolving all the time. It involves adaptive management or "learning-by-doing," experimenting, and

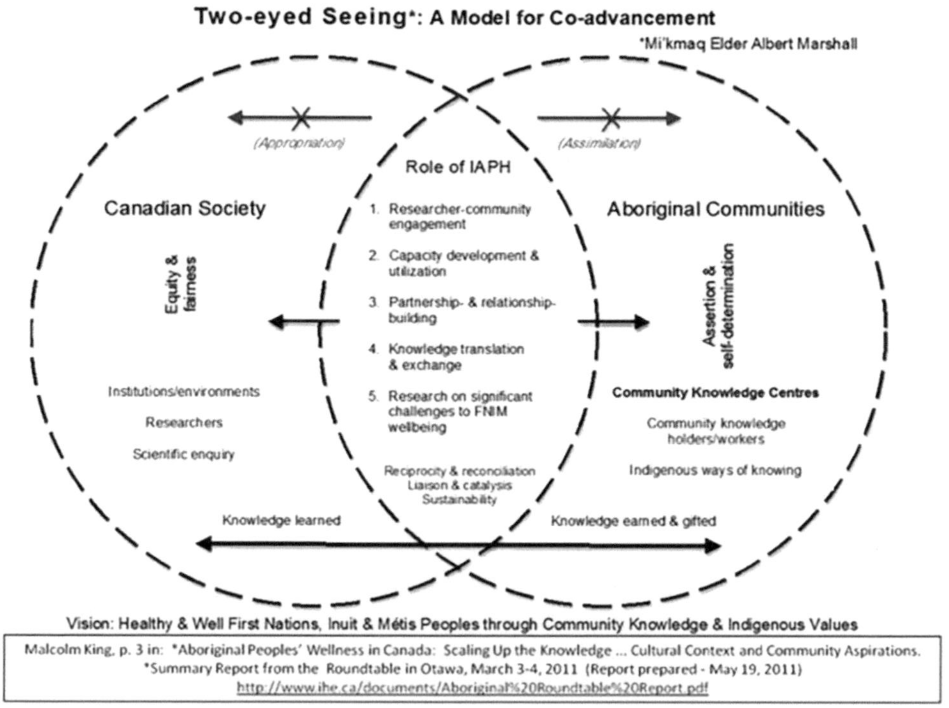

Figure 2.
Two-eyed seeing–a model for knowledge co-production. (Source: http://www.cihr-irsc.gc.ca/e/43686.html).

knowledge-building; the evolving process depends upon community members' ability to constantly observe the climate and environmental changes occurring around them. Such observations have been crucial for climate history, community adaptation, and community-based environmental monitoring. These are three of the "five convergence areas" outlined by Reidlinger and Berkes. These areas can provide a starting point for climate-change-related risk mitigation knowledge integration and co-production.

2.2.2 Climate history

Riedlinger and Berkes [16] identify a range of scientific and non-scientific sources of information that can help climate researchers establish a historical record of change. For example, these sources include meteorological data, proxy data (ice-core samples), and historical documentation (fur trade companies and expedition records). In northern Canada, the sequential scope of scientific data is limited, since the process of scientific meteorological data collection only began after 1946. In contrast, Indigenous Peoples have been closely observing changes in weather, sea ice, permafrost, fish and wildlife, and seasons for a long time. In this respect, traditional indigenous knowledge systems can offer insights into past climate fluctuation and variability. For example, research has shown the accuracy, consistency, and precision of Inuit historical recall regarding the Caribou population in the Canadian Arctic [16]. Current and projected climate change conditions may differ from those experienced in the past; however, traditional indigenous knowledge remains the basis for Indigenous Peoples' local-level responses to climatic and environmental change.

2.2.3 Community adaptation

Community adaptation is fundamental to mitigating climate change impacts. According to IPCC [9], adaptation refers to "the adjustment in natural or human systems in response to actual or expected climatic stimuli or their effects, which moderates harm or exploits beneficial opportunities" (6). Adapting to climate essentially includes adaptation to variability [37]. Understanding how Indigenous communities experience and respond to changes will guide the scientific community in articulating possible coping and adaptive options for managing climatic risks. Adaptation and resilience to climate change impacts are intertwined; communities that are able to quickly adapt to changes are more resilient to emerging climatic challenges and threats. Resilience refers to the capacity of a community or system to anticipate and cope with, respond to, and recover from climate change impacts as well as avail the opportunities created due to the change; in other words, a community's climate resilience is the capacity of a socio-ecological system to absorb stresses, maintain function, and adapt and evolve into a condition that improves the sustainability of the system, thereby preparing it for future climatic risks [38–40]. I have added the term anticipation to the definition above. In believe the ability of a community to anticipate future hazards' occurrence caused due to existing and/or expected uncertainties/climate change impacts can also be considered a community's resilience, as the ability to anticipate allows the community (with the capacity) to anticipate in advance an expected hazardous event in short-, medium-, and/or long-term future.

Berkes [34] argues that "building resilience into human–environment systems is an effective way to cope with change characterized by surprises and unknowable risks" (1). His definition of resilience is based on learning to live with change and uncertainty and promoting various kinds of ecological, social, and political diversity to increase adaptation choices for climatic risk mitigation. Berkes's analysis brings the community's shared communal resources for its wellbeing under the spotlight. Indigenous Peoples' socioeconomic and cultural organization is based on communal natural resources, and their collective wisdom in the shape of traditional indigenous knowledge helps them address emerging climate challenges.

2.2.4 Community-based environmental monitoring

Community-based environmental monitoring is another convergence area, where both scientific and traditional indigenous knowledge systems can complement each other [17]—particularly in managing environmental risks and adapting to climate change. Indigenous communities are natural resource users; therefore, their knowledge originates from close observation of the environment and resources over a long period of time. For example, Indigenous Peoples in northern Canada have been closely observing changes in the natural environment [16]. Their climate-related environmental-change-monitoring knowledge can be useful for natural resource and environmental management as well as mitigating natural hazard risks by focusing on signs and indicators of the changes and information specific to small local areas. This can help bridge the gap between scientific knowledge and traditional indigenous knowledge systems while incorporating synchronic and diachronic information [17]. In addition to the cost-effectiveness of indigenous monitoring, traditional indigenous knowledge can also help identify and differentiate natural variation from "non-natural" or unexpected changes. According to Moller et al. [17], climate change monitoring is politically significant because the causes and implications of environmental risk management issues are defined within a particular set of power relations. Climate change monitoring is also ecologically significant because it requires such issues to be understood and dealt with in the context of environmental complexity and uncertainty. In this respect, Indigenous Peoples are better placed in dealing with ecological complexity and uncertainty. However, such issues are important as power dynamics interplay between politics and environment and resource management decision-making.

2.3 Contributions traditional Indigenous knowledge offers for climate-change risk communication

In this section, I will discuss traditional indigenous knowledge contributions that it can offer to the scientific research community while focusing on five "areas of complementarity" given by Moller et al. [17] to demonstrate traditional indigenous systems' practicability. Traditional indigenous knowledge systems can contribute to climate-change risk communication in many ways: one is adding value to the risk communicators' knowledge. Indigenous Peoples' understanding of changes and variability occurring in their environments guide their responses to environmental changes. This knowledge can help climate-change adaptation policymakers and practitioners in articulating possible coping and adaptive options, priorities, and opportunities to manage climatic risks. In this respect, the following five "areas of complementarity" demonstrate the application of traditional indigenous knowledge that it can offer to science by integrating scientific knowledge and traditional indigenous knowledge systems.

2.3.1 Synchronic-diachronic information

In climate change research, data collection typically relies on a small sample size over a large area, with short temporal scope, which may likely miss vital information due to the small sample size, sampling technique used, and focus on a large area. Moller et al. [17] call this process "synchronic". In contrast, indigenous observations are based on smaller areas over a longer time period, which provides more opportunities to collect "diachronic" information, thereby minimizing the chances of any change being overlooked. Using synchronic and diachronic approaches together will provide complete information on both temporal and spatial scales [17]. However, developing a mechanism that provides a practical tool to combine both techniques can be challenging. Nevertheless, diachronic information offered by traditional indigenous knowledge will be a valuable resource, particularly when scientists are facing decreasing resources for research projects.

2.3.2 Average-extreme events

Science primarily depends on numerical and quantitative data, and statistical analyses [17]; therefore, scientific knowledge is based on averages, which makes the system vulnerable to overlooking critical information. Conversely, traditional indigenous knowledge, however, captures extreme events in a better way by closely observing both changes and variations. The two techniques can be integrated, which can also help narrow the trust deficit between Indigenous Peoples and non-Indigenous researchers.

2.3.3 Quantitative-qualitative data

According to Moller et al. [17], science examines quantitative aspects of parts of a system, while relying on numbers or numerical data. Conversely, traditional indigenous knowledge deals with a qualitative understanding of the whole system. As the two perspectives are complementary, their combination will enable scientists to use qualitative measures that are less costly and more rapid; however, this may come at the "cost of precision". Combining both quantitative and qualitative approaches will require a framework that allows the two data sources to be integrated without compromising the value or integrity of each. Qualitative data for climate and environmental risk management research offers valuable insights, particularly in mixed research methods as qualitative data can be collated with quantitative one for a better understanding of climate-change risk communication issues.

2.3.4 Testing-hypothesis formulation

In testing a hypothesis, scientific statistical methods have an edge over traditional indigenous knowledge: they involve powerful tools for testing "why?". However, scientific statistical methods may also waste resources, such as time and man-hours, on testing insignificant hypotheses. Traditional indigenous knowledge can contribute to the process of formulating a scientific hypothesis, as another way of identifying, understanding, and assessing climate change risks. Right from setting off a research project, a collaboration between the two systems will expand the scope of inquiry and establish a role for Indigenous communities and their knowledge in research planning. Using both approaches together takes advantage of their relative strengths [17].

2.3.5 Objective-subjective perspectives

Moller and colleagues [17] assert that science does not account for people's emotions, feelings, and sentiments, as it remotely monitors populations and is essentially objective in nature. Conversely, traditional indigenous knowledge is subjective in essence; therefore, it explicitly includes people, their feelings, and the relationships among various groups. It considers human beings and their relationships with the land, the natural environment, and resources as "sacred". Instead of making either-or choices, objective and subjective views should be combined by researchers. However, it may be challenging to determine which emotions, feelings, and relationships should be included after recasting them into scientific language.

2.3.6 Collective wisdom

Traditional indigenous knowledge has cultural and local meaning, such as the Inuit way of doing things based on past, present, and future knowledge, experiences and value of the Inuit society, or the collective wisdom of Cree First Nations Communities. The knowledge is contextualized as a system connected with the long-term use of a certain place. It consists of all the experiences and knowledge of a social group (a social and mental construction), which guide, organize, and regulate a community's way of life and make sense out of it [41].

Traditional indigenous knowledge is diverse and complex, capturing the histories, cultures, and lived realities of Indigenous Peoples. Indigenous cultural beliefs position individuals as part of the natural system. Their worldview respects community elders' collective wisdom, value those living, the dead, and future generation by sharing responsibility, wealth, community resources, and by embracing spiritual values, traditions and practices, and social values connected to their land and natural resources [41]. Such belief in the sacredness of land guides respect for ecology, the natural environment, and resources [42], which can help inform natural resource and environmental protection agencies in formulating risk messages and designing communication in a manner that addresses the Indigenous Peoples' concerns. In short, the collective wisdom of scientific and indigenous systems together can facilitate the risk communication process between non-Indigenous experts and Indigenous communities, particularly in adapting to the changing climate.

2.3.7 Other areas of mutual interests and benefits

Experts' understanding of how Indigenous communities experience and react to climate change can guide them in developing coping and adaptive strategies to manage environmental risks. Therefore, scientific and traditional indigenous knowledge-holders can meaningfully collaborate in many areas, including issues related to food, health, fisheries, water, and ecosystems management for risk mitigation.

In the face of climate change, the linkage between wildlife and food security calls for collaborative efforts in monitoring and managing wildlife in northern Canada. Such efforts can contribute to effectively dealing with variability and change by ensuring the supply of traditional food. In some cases, relevant traditional indigenous knowledge has drawn the attention of scientists in long-term efforts to monitor wildlife populations and their status. For example, Chronic Wasting Disease (CWD) in northern Canada is threatening traditional food sources, including caribou and moose populations. According to Parlee et al. [43], Indigenous Elders' knowledge, harvester observations, and harvest and consumption data indicate a decline in the availability of the caribou population. While resources for scientific monitoring are becoming scarce, co-management approaches incorporating traditional indigenous knowledge and Indigenous Peoples' experiences can help experts monitor changes in Arctic ecosystems and caribou populations. Although, recent tests found no CWD cases; however, wildlife disease experts recommend surveillance on a regular basis. Similarly, Indigenous harvesters' experiences and observations of variability in berry production may prove useful to scientists [44]. An example drawn from African experience also shows that ample opportunity and a growing need for co-production of rain forecasts exists, which can increase crop production capacity to deal with food insecurity. In this connection, "Indigenous meteorological" insights on temperature, wind, and rainfall may help explain meteorology-based forecasts [45].

The role of traditional indigenous knowledge in watershed and fisheries management presents another avenue for collaboration. The "Mackenzie River Basin Board Traditional Knowledge and Strengthening Partnerships Committee" project is a case in point, wherein indigenous traditional knowledge holders demonstrate an understanding of social and ecological change in the river basin and contribute to regional, territorial/provincial, and federal decisions about its continued sustainability. However, in case of Kashechewan First Nation in northern Ontario, their traditional indigenous knowledge has been undervalued in dealing with the spring flooding risk of the Albany River [46]. However, empirical evidence indicates that despite traditional indigenous environmental knowledge coupled with climate data establishes temporal relationships between extreme climate events in 2005 and fish die-offs [47].

Fishers' interannual, seasonal, lunar, diel, and tide- and habitat-related ecological knowledge can also be valuable sources of information for managing marine fisheries. Johannes et al. [24] argue that older fishers and their traditional indigenous knowledge are often the only source of information when other long-term datasets are not available on historical changes in local marine stocks and marine environmental conditions. Ignoring such information can lead to wrong decision-making. The Canadian government's ban on hunting codfish on the north east coast during the 1980s is a classic example as experts did not consider and value local spatial knowledge,[1] which resulted in losing thousands of jobs by local population in the area. In this respect, community-based co-management between experts and traditional indigenous knowledge-holders can help government agencies to learn more about patterns of change in fish habitat and water resources and their influence over human livelihoods and wellbeing. Understanding health and environmental risks that exist among northern Indigenous communities in Canada demand recognition of communities' unique place-based circumstances and characteristics for risk mitigation [6].

In northern Canada, as discussed earlier, Indigenous Peoples have been closely observing changes occurring and affecting them, such as extreme weather conditions, melting sea ice and permafrost, rivers' breakup ice jamming, and threatened fisheries and wildlife resulting in food insecurity. As a result of global warming, access to sparsely populated Indigenous communities in Arctic Canada is becoming restricted as winter ice roads are becoming more dangerous. Climate change is also affecting forestry in Canada. The example of the mountain pine beetle (MPB) problem in in British Columbia and Alberta is a case in point. Global warming is considered to be one of the main causes of the problem. In the past, the insect population was controlled by (1) two weeks of continuous temperatures below −40°C; and (2) traditional burning of trees in the affected areas by Indigenous Peoples. However, in recent years, the use of modern technology to extinguish wildfires has contributed to the increase and expansion of the MPB population. Traditional indigenous climate and environmental change monitoring knowledge can help researchers address such emerging issues. To do so, a paradigm shift requires institutional sociocultural and political change by taking traditional indigenous knowledge into account for understanding and improving the knowledge base for decision-making [46, 48].

3. Discussion

The literature review indicates that climate-change risk communication has not adequately incorporated traditional knowledge. In spite of increasing climate-change risks, no serious effort has been made to incorporate traditional knowledge into relevant risk communication. The two knowledge systems pertain to different domains; while there are many areas of divergence, there are also many areas that exist where the systems converge or follow analogous patterns [45]. Exploring points of convergence may help experts better understand and deal with climate change risks. By reconciling scientific and traditional indigenous knowledge systems, experts can help improve the risk communication process.

Three of the "five convergence areas" given by Riedlinger and Berkes [16] and the "five areas of complementarity" presented by Moller et al. [17] are discussed in detail above. This previous discussion has demonstrated how the "areas of complementarity" can facilitate climate-change risk communication through integration of

[1] Importance of local spatial knowledge in South Asia is highlighted by sharing a personal example. During childhood, my parents used "Neem Tree" (or Azadirachta indica, also called Margosa) leaf as a medicine to treat my boils. Today, traditional use of neem for a variety of reasons includes dental and health hygiene by millions of people in the region.

scientific and traditional indigenous knowledge systems. However, while traditional indigenous knowledge offers value addition potential, there are limitations to integrating it with scientific knowledge. One of the limitations is that there is a lack of research on incorporating traditional indigenous knowledge into climate-change risk communication, which suggests that it is a neglected area.

Until recently, disaster risk reduction researchers and practitioners did not recognize the important role that traditional indigenous knowledge and practices can play in reducing risks and improving preparedness. In spite of more recent recognition, traditional indigenous knowledge and practices are yet to be employed by experts, practitioners, and policymakers. In many cases, strategies to improve risk communication have targeted non-Indigenous people [49]. However, a set of distinct challenges are also experienced by experts while communicating risks to Indigenous communities, due to pre-existing histories of distrust between the two parties.

In Canada, Aboriginal Peoples have been subject to Eurocentric and colonial discriminatory laws and oppressive regimes since the arrival of European settlers. As a result, they have faced a set of unique vulnerabilities and risks, which are not just a function of climate change but rather arise from multiple social conditions—including environmental, socioeconomic, and political factors. Consequentially, Indigenous Peoples are the most vulnerable (in general and due to climatic and environmental change impacts) communities in Canada. However, despite the many serious issues that Indigenous Peoples have faced in Canada, they have demonstrated resiliency, which needs to be capitalized while building mutual trust and respecting their diverse cultures, traditions, and ways of life. This can be achieved by creating opportunities and spaces where the Indigenous languages, cultural beliefs, and social fabric of Indigenous societies not only revive, but also thrive. The unlearning of colonial memories—to heal the social, emotional, and psychological traumas of colonization—and the relearning of Indigenousness will greatly help in curing the wounded soul [18].

The resurgence of dormant traditional adaptive capacities will not only reduce the vulnerabilities and risk exposure of Indigenous communities; it will also contribute to the development of lost trust between Indigenous communities and mainstream Canadian society. Vertical and horizontal communication and coordination are critical to rebuild trust between Indigenous communities and risk management agencies. To achieve this, effective risk communication demands empowering Indigenous communities as active partners and not just passive stakeholders in risk decision-making processes [50–52]. In this respect, localized and customized (context-based and situation-focused) dialogic strategies should be developed to guide when informed decision-making will be an appropriate goal for risk communication among "at-risk" Indigenous communities. The strategy must be just and fair for all [53].

Non-participatory risk management process often results in poor risk communication, which in turn can cause more harm than good—particularly in the case of Indigenous Peoples [54, 55]. Instead of taking a top-down, unidirectional, or one-way approach to communication, participatory communication involves a two-way dialogue [56] between risk communication experts and Indigenous communities. Moreover, frequent interaction between both parties will legitimatize the risk decision-making process as democratic by addressing the legitimate concerns of Indigenous Peoples.

However, power dynamics between the risk management agencies and Indigenous communities may very likely undermine processes of participation and engagement, as risk communicators may attempt to influence and manipulate the participatory and democratic process [57]. To promote trust between Indigenous communities and non-Indigenous agencies and risk communicators, a balanced, democratic, and transparent consultation process is necessary. In the process, the differing perspectives (subjective, experiential, social, and cultural values) of Indigenous Peoples and the technical standpoints (rational, efficiency, and quantitative data) of scientists [53] will congregate on convergence areas—thereby resolving differences in a manner that makes the process workable, just, and fair. Incorporating traditional knowledge for informed decision-making in climate-change risk communication can be made viable only if Indigenous communities are meaningfully engaged and involved in the process right from the start. The integration and co-production process will promote participation, thereby making the process democratic as Indigenous Peoples are actively included as affected and interested parties in the risk communication process.

Trust is central to the risk communication process: as the interplay between risk and trust can either significantly amplify [58] or attenuate risks [59, 60] among Indigenous communities. Trust-destroying events are more evident than trust-building events in the history of Aboriginal Peoples in Canada [61]. In this respect, the risk management guiding principle of "do more good than harm" [62] is crucial for regaining lost trust. A unique approach is needed to communicate risks to Indigenous Peoples, as trust and risk relationship becomes more complex and sensitive when dealing with sociocultural, political, and psychological factors [61]. These factors significantly influence Indigenous Peoples' trust in government agencies and/or risk communication experts. Similarly, risk encompasses both objective and subjective factors; therefore, the role of indigenous social values in risk perception and risk acceptance is fundamental [63] in accomplishing and maintaining mutual trust [64].

Winning buy-in among Indigenous Peoples is crucial and very difficult to achieve. However, this can be achieved by intensely involving Indigenous community members in the risk decision-making process and conveying both overt and covert messages, while ensuring that the process is participatory and democratic [54]. Perception of Indigenous Peoples that whether or not they have been meaningfully involved in the process will also largely depend upon their satisfaction and trust in the process than in the outcome [54]. However, since satisfaction, trust (antecedents vary across Indigenous communities), and confidence [65] are subjective and heuristic, the dilemma is that winning buy-in among Indigenous community members will depend upon their perceptions—whether or not the process was just and fair, their voices heard, and their concerns addressed. Furthermore, some members of Indigenous communities may challenge the process while others accept it since individual risk perception can differ from the wider community risk perception.

The trust deficit between Indigenous communities and risk management agencies can complicate the process of risk communication, including risk management efforts to change attitudes and behaviors among Indigenous Peoples when dealing with climate risks. Trust between Indigenous communities and non-Indigenous experts appears to be already fragile; lost trust can take a long time to rebuild—and among some of the indigenous communities, empirical evidence suggests it may never be regained. Trust building is very challenging, particularly in the context of a long history of distrust between Indigenous Peoples and mainstream Canadian society. As a trust-building strategy, risk managers working with indigenous communities need to engage traditional indigenous knowledge holders and Indigenous Elders to capitalize on their skills and knowledge. This would help them develop trust with community members while demonstrating respect for Indigenous Peoples' unique cultures, traditions, and way of life. However, the trust-building task is challenging due to: (a) past colonial oppressions and (b) the misconception among scientists that traditional indigenous knowledge does not meet scientific criteria.

The integration and co-production process will bring many opportunities and benefits, not just for the scientists but also for Indigenous communities. Such collaboration between both will create opportunities for funding, information and knowledge sharing, and capacity building among traditional knowledge holders and gatekeepers. Scientific knowledge can be transferred to traditional indigenous knowledge-holders at the local level, particularly among those settled in remote areas. Human resource development of traditional indigenous knowledge holders and gatekeepers will support the documentation of traditional knowledge, which can contribute to efforts to preserve that knowledge while preventing further erosion. Such collaboration efforts will attract more investment for joint research projects on different aspects of traditional knowledge. It will also help to promote the development of skills and capacity of Indigenous communities to promote and disseminate their knowledge for the benefit of the general public.

4. Conclusion

A relevant literature review indicates that climate-change risk communication has not adequately integrated traditional indigenous knowledge, despite the value that it can offer. There appears to be a lack of serious effort in this respect, as demonstrated by a major gap in the literature. For integration and co-production of scientific knowledge and traditional indigenous knowledge systems, methods, and content as well as several aspects exist where both systems converge [57]. Researchers, policy and decision-makers, risk communication experts, and practitioners need to focus on areas and aspects of convergences rather than divergences. By focusing on common ground, both knowledge systems can be reconciled to improve the risk communication process to benefit Indigenous as well as non-Indigenous communities. As a result of deliberate reconciliation between the two, the process will help promote power sharing and trust building between parties. Furthermore, collective wisdom, the sharing of climate-change risk information, and the democratization of the knowledge-production process will help empower Indigenous communities through the valuing and mobilization of traditional indigenous knowledge for risk management. This may also help promote the development of trust between indigenous Peoples and mainstream Canadian society. Traditional indigenous knowledge may not necessarily fit with every scientific model; however, a more in-depth research is warranted to identify points of convergence where traditional indigenous knowledge may help improve the climate-change risk communication process.

Author details

Muhammad Arshad K. Khalafzai
Royal Roads University, Victoria, Canada

*Address all correspondence to: khalafzai29@gmail.com; khalafza@ualberta.ca

References

[1] Vinyeta K, Lynn K. Exploring the role of traditional ecological knowledge in climate change initiatives. In: Gen. Tech. Rep. PNW-GTR-879. Portland, OR: US Department of Agriculture, Forest Service, Pacific Northwest Research Station; 2013

[2] Brody SD, Zahran S, Vedlitz A, Grover H. Examining the relationship between physical vulnerability and public perceptions of global climate change in the United States. Environment and Behavior. 2008;**40**(1):72-95

[3] Smith JB, Hans-Joachim Schellnhuber M, Mirza MQ, Fankhauser S, Leemans R, Lin Erda L, et al. Vulnerability to climate change and reasons for concern: A synthesis. In: Climate Change. Cambridge University Press; 2001. pp. 913-967

[4] Taylor AL, Dessai S, Bruine W, de Bruin. Public perception of climate risk and adaptation in the UK: A review of the literature. Climate Risk Management. 2014;**4**:1-16

[5] Etkin D, Ho E. Climate change: Perceptions and discourses of risk. Journal of Risk Research. 2007;**10**(5):623-641

[6] Jardine CG, Boyd AD, Furgal CM. Gender and place influences on health risk perspectives in northern Canadian aboriginal communities. Gender, Place and Culture. 2009;**16**(2):201-223

[7] Brysse K, Oreskes N, O'Reilly J, Oppenheimer M. Climate change prediction: Erring on the side of least drama? Global Environmental Change. 2013;**23**(1):327-337

[8] Rahmstorf S, Cazenave A, Church JA, Hansen JE, Keeling RF, Parker DE, et al. Recent climate observations compared to projections. Science. 2007;**316**(5825):709-709

[9] Intergovernmental Panel on Climate Change (IPCC). Climate Change Working Group II: Impacts, Adaptation and Vulnerability. IPCC Fourth Assessment Report; 2007

[10] Intergovernmental Panel on Climate Change (IPCC). Impacts on indigenous peoples: Contribution of working group II chapter 16. In: Polar Regions (Arctic and Antarctic) Impacts, Adaptation, and Vulnerability in the Third Assessment Report of the Intergovernmental Panel on Climate Change. IPCC; 2001

[11] Intergovernmental Panel on Climate Change (IPCC). Vulnerability to climate change and reasons for concern. A Synthesis. 2001;**19**(3):4.2

[12] Green D, Raygorodetsky G. Indigenous knowledge of a changing climate. Climatic Change. 2010;**100**(2):239-242

[13] Furgal C, Seguin J. Climate change, health, and vulnerability in Canadian northern aboriginal communities. Environmental Health Perspectives. 2006;**114**:1964-1970

[14] Khalafzai M-AK, McGee TK, Parlee B. Flooding in the James Bay region of Northern Ontario, Canada: Learning from traditional knowledge of Kashechewan First Nation International Journal of Disaster Risk Reduction. 2019:36. DOI: 10.1016/j.ijdrr.2019.101100

[15] Adger WN, Paavola J, Huq S, Mace MJ. Toward justice in adaptation to climate change. In: Fairness in Adaptation to Climate Change. MIT Press; 2006. pp. 1-19

[16] Riedlinger D, Berkes F. Contributions of traditional knowledge to understanding climate change in the Canadian Arctic. Polar Record. 2001;**37**(203):315-328

[17] Moller H, Berkes F, Lyver OP, Kislalioglu M. Combining science and traditional ecological knowledge: Monitoring populations for co-management. Ecology and Society. 2004;**9**(3):2

[18] Battiste M. Nourishing the learning Spirit: Living our way to new thinking. Education Canada. 2010;**50**(1):14-18

[19] Payyappallimana U, Koike O. Traditional Knowledge for Sustainable Development: A Case from the Health Sector in Kerala, India. Biodiversity and Community Health; 2010

[20] Mauro F, Hardison PD. Traditional knowledge of indigenous and local communities: International debate and policy initiatives. Ecological Applications. 2000;**10**(5):1263-1269

[21] Anaya J. Indigenous Peoples in International Law. New York, New York, USA: Oxford University Press; 1996

[22] Arewa Olufunmilayo B. TRIPS and Traditional Knowledge: Local Communities, Local Knowledge, and Global Intellectual Property Frameworks. Social Science Research Network Electronic Paper Collection; 2006

[23] United Nations Framework Convention on Climate Change (UNFCCC). Reducing emissions from deforestation in developing countries: Approaches to stimulate action. In: Conference of the Parties, Bali, 3-14 December 2007. 2007

[24] Johannes RE, Freeman MMR, Hamilton RJ. Ignore fishers' knowledge and miss the boat. Fish and Fisheries. 2000a;**1**(3):257-271

[25] Berkes F. Sacred Ecology: Traditional Ecological Knowledge and Management Systems. Philadelphia: Taylor & Francis; 1999

[26] Scott J. Seeing like a State: How Certain Schemes to Improve the Human Condition Have Failed. New Haven, Connecticut, USA: Yale University Press; 1998

[27] Dove MR. Indigenous people and environmental politics. Annual Review of Anthropology. 2006;**35**:191-208

[28] Berkes F. Indigenous ways of knowing and the study of environmental change. Journal of the Royal Society of New Zealand. 2009;**39**(4):151-156

[29] Agrawal A. Indigenous knowledge and the politics of classification. International Social Science Journal. 2002;**54**(173):287-297

[30] Armitage D, Berkes F, Dale A, Kocho-Schellenberg E, Patton E. Co-management and the co-production of knowledge: Learning to adapt in Canada's Arctic. Global Environmental Change. 2011;**21**(3):995-1004

[31] United Nations University International (UNU). Seeing traditional Technologies in a new Light: Using traditional approaches in water Management in Drylands. In: United Nations University International Network on Water, Environment and Health (UNU-INWEH). A Joint Undertaking of UNESCO and UNV (UNU-INWEH):12-14. 2009

[32] Anderson David G. The Mistake of Western Science. Healing and Transformation; 2015

[33] Berkes F, Turner NJ. "Knowledge, Learning and the Resilience of Social-Ecological Systems." Panel "Knowledge

for the Development of Adaptive Co-Management". Vol. 4. Mexico: IACSP; 2004

[34] Berkes F. Understanding uncertainty and reducing vulnerability: Lessons from resilience thinking. Natural Hazards. 2007;**41**(2):283-295

[35] Canadian Institute of Health Research (CIHR). CIHR Guidelines for Health Research Involving Aboriginal People. Canada: Government of Canada; 2007

[36] Canadian Institute of Health Research (CIHR). Internal Assessment for 2011International Review - CIHR Institute of Aboriginal Peoples' Health. Canada: Government of Canada; 2009

[37] Smit B, Wandel J. Adaptation, adaptive capacity and vulnerability. Global Environmental Change. 2006;**16**:282-292

[38] Folke C. Resilience: The emergence of a perspective for social–ecological systems analyses. Global Environmental Change. 2006;**16**(3):253-267

[39] Nelson DR, Neil Adger W, Brown K. Adaptation to environmental change: Contributions of a resilience framework. Annual Review of Environment and Resources. 2007;**32**(1):395

[40] Palenchar MJ, Heath RL, Orberton EM. Terrorism and industrial chemical production: A new era of risk communication. Communication Research Reports. 2005;**22**(1):59-67

[41] Dei GJS, Hall BL, Rosenberg DG. Indigenous Knowledges in Global Contexts: Multiple Readings of our World. University of Toronto Press; 2000. Corpus ID: 140510638

[42] Audrey WT. A Shining Trail to the Sun's Lodge: Renewal through Blackfoot Ways of Knowing. Lethbridge, Alta: University of Lethbridge, Faculty of Education; 1997

[43] Parlee BL, Goddard E, Łutsël KÉ, Nation DF, Smith M. Tracking change: Traditional knowledge and monitoring of wildlife health in northern Canada. Human Dimensions of Wildlife. 2014;**19**(1):47-61

[44] Parlee Brenda, Fikret Berkes. "Teetl'it Gwich'in Renewable Resources Council. Indigenous knowledge of ecological variability and commons management: A case study on berry harvesting from northern Canada". Human Ecology. 2006;34(4):515-528

[45] Roncoli C, Ingram K, Kirshen P. Reading the rains: Local knowledge and rainfall forecasting in Burkina Faso. Society & Natural Resources. 2002;**15**(5):409-427

[46] McCarthy D, Crandall D, Whitelaw G, General Z, Tsuji L. A critical systems approach to social learning: Building adaptive capacity in social. Ecology and Society. 2011;**16**(3):18

[47] Hori Y, Tam B, Gough WA, Ho-Foong E, Karagatzides JD, Liberda EN, et al. Use of traditional environmental knowledge to assess the impact of climate change on subsistence fishing in the James Bay region of northern Ontario, Canada. Rural and Remote Health. 2012;**12**:1878

[48] McCay B, J, Finlayson AC. The political ecology of crisis and institutional change: The case of the northern cod. In: Annual Meeting of the American Anthropological Association, Washington, DC. 1995. pp. 15-19

[49] Alcántara-Ayala I, López-Mendoza M, Melgarejo-Palafox G, Borja-Baeza RC, Acevo-Zarate R. Natural hazards and risk communication strategies among indigenous communities: Shedding light

on accessibility in Mexico's mountains. Mountain Research and Development. 2004;**24**(4):298-302

[50] Heath RL, Seshadri S, Lee J. Risk communication: A two-community analysis of proximity, dread, trust, involvement, uncertainty, openness/accessibility, and knowledge on support/opposition toward chemical companies. Journal of Public Relations Research. 1998;**10**(1):35-56

[51] Rich RC, Edelstein M, Hallman WK, Wandersman AH. Citizen participation and emprowerment: The case of local environmental hazards. American Journal of Community Psychology. 1995;**23**(5):657-676

[52] Rich RC, Edelstein M, Hallman WK, Wandersman AH. Citizen participation and empowerment: The case of local environmental hazards. American Journal of Community Psychology. 1995;**23**(5):657-676

[53] Gurabardhi Z, Gutteling JM, Kuttschreuter M. An empirical analysis of communication flow, strategy and stakeholders' participation in the risk communication literature 1988-2000. Journal of Risk Research. 2005;**8**(6):499-511

[54] Bostrom A, Löfstedt RE. Communicating risk: Wireless and hardwired. Risk Analysis. 2003;**23**(2):241-248

[55] Fischhoff MB, Granger W, Bostrom A, Lave L, Atman C. ES&T Features. Communicating risk to the public. First, learn what people know and believe. Environmental Science & Technology. 1992;**26**(11):2048-2056

[56] Jardine CG. Role of risk communication in a comprehensive risk management approach. In: Encyclopedia of Quantitative Risk Analysis and Assessment. Wiley; 2008

[57] Santos SL, Chess C. Evaluating citizen advisory boards: The importance of theory and participant-based criteria and practical implications. Risk Analysis. 2003;**23**(2):269-279

[58] Kasperson RE, Renn O, Slovic P, Brown HS, Emel J, Goble R, et al. The social amplification of risk: A conceptual framework. Risk Analysis. 1988;**8**(2):177-187

[59] Knox JC. Large increases in flood magnitude in response to modest changes in climate. Nature. 1993;**361**(6411):430-432

[60] Lewis RE, Tyshenko MG. The impact of social amplification and attenuation of risk and the public reaction to mad cow disease in Canada. Risk Analysis. 2009;**29**(5):714-728

[61] Slovic P. Trust, emotion, sex, politics, and science: Surveying the risk-assessment battlefield. Risk Analysis. 1999;**19**(4):689-701

[62] Jardine CG, Hrudey S, Shortreed J, Craig L, Krewski D, Furgal C, et al. Risk management frameworks for human health and environmental risks. Journal of Toxicology and Environmental Health, Part B: Critical Reviews. 2010;**6**(6):569-718

[63] Slovic P. Perceived risk, trust, and democracy. Risk Analysis. 1993;**13**(6):675-682

[64] Jardine CG. Considerations in planning for successful risk communication. In: Encyclopedia of Quantitative Risk Analysis and Assessment. Wiley; 2008c

[65] Earle TC. Trust in Risk Management: A model-based review of empirical research. Risk Analysis. 2010;**30**(4):541-574

Chapter 17

New Remote Sensing Technologies Applied to the Prediction of Spontaneous Forest Fires

Emilio Ramírez-Juidías and Emilio José Cabello-Franco

Abstract

One of the causes of a forest fire is a combination of environmental variables such as temperature and relative humidity, as well as wind speed. When environmental conditions are favorable, chemical reactions occur at the forest mass level, resulting in pre-ignition. A big-data analysis of three spontaneous forest fires that occurred in Spain between 2015 and 2019 was performed to determine the chemical compound from which the forest fire starts. After analyzing satellite data, it was discovered that the critical temperature at which a fire starts spontaneously is 51.27°C, a value that coincides with the maximum limit of decrease in environmental sulfur dioxide concentration ([SO_2]), presumably due to sulfur capture by part of the wood, an element that is released into the environment after the fire occurs. The Spontaneous Forest Fire Process (SFFP) and fAPAR have a close relationship because pre-ignition occurs when the critical temperature determined by environmental conditions is reached.

Keywords: sulfur dioxide, thermal plugging, Broglie Law, Iberian Peninsula, remote sensing technologies

1. Introduction

Hitherto, when a forest fire occurs, the habitat is destroyed, without it being possible to avoid the devastation of the environment when the fire is due to natural causes.

According to Government of Spain [1], Spanish forest ecosystems occupy just over 26 million hectares (26,280.281 ha), of which almost 15 million (14,717.898 ha) are wooded and some 12 (11,562.382 ha) treeless, which respectively represent 29 and 23% of the national territory. The coniferous forests have a similar extension than the hardwood ones (5.7 and 5.2 million hectares, respectively) while the mixed ones have a somewhat smaller population (3.9 million hectares).

In another vein, and based on the latest study published by [2], it is of great interest to take into account the deforestation linked to international trade, especially since Spain causes the deforestation of 32,900 hectares of forest each year (the production and consumption of soybeans, palm oil and beef are the main items that explain this situation). However, at the national level, a notable increase in the area of wooded mountains has been detected at the cost of a decrease in the cultivated area, as well as that corresponding to land without forest mass [1]. In general terms, the Government of Spain has action policies to control both deforestation and forest conversion to other uses through the sustainable use of its resources. In this regard, and although deforestation occurs as a result of the mutual interaction of various factors, the problem of a forest fire, as an independent variable, should not be ignored, since as consequence of the destruction of areas of both biological and ecological interest, it is difficult to adapt the damaged ecosystem to the situation prior to the fire.

In relation to the above, the trend that climate-change represents in the medium term is of enormous importance, since the increase in temperatures as well as the increase in the incidence of prolonged and intense drought periods will result in a decrease in relative humidity. in the atmosphere close to the forest mass, and therefore the appearance of suitable conditions for spontaneous forest fires to take place much more often than at present [3].

Likewise, it is important to take into account the relationship between the dynamics of the lower atmosphere and the movement of air resulting in large fires that favor the development of enormous plumes, generating a series of processes that give rise to a vertical transfer of matter and energy in turbulent movements and rapid [4]. On the other hand, it is interesting to know that, during a forest fire, high-risk situations are frequently generated when specific fire sources are created, associated or not, with sudden changes in direction or speed of the wind. It is evident that, when temperature, relative humidity and local wind speed take place together, the forest fire expansion danger increases, making extinction work difficult [5].

The importance of the type of fuel existing in the forest mass should not be forgotten. For this reason, it should be noted that in Spain the predominant hardwood species are *Quercus ilex*, followed by *Quercus pyrenaica, Quercus pubescens, Eucalyptus spp., Quercus suber, Fagus sylvatica, Quercus faginea, Quercus canariensis, Quercus robur, Quercus petraea and Castanea sativa*. Regarding conifers, *Pinus halepensis, Pinus pinaster, Pinus sylvestris, Pinus nigra* and *Pinus radiata* stand out. The great richness in scrub, bush and grass species existing in Spanish forests should not be forgotten, mainly due to the amount of dry matter that they can contribute to the fire.

According to [6], in this country, around 5% of fires are due to natural causes (adequate relationship between relative humidity and temperature, lightning, etc.), being the cause of the remaining 95% very varied, highlighting among them accidents or negligences, intentional fire and unknown reasons.

In another vein, and as a result of the technological advances that have occurred in recent decades, it is evident that remote sensing can be a tool of greater interest, than it is currently, in the evaluation and monitoring of forest fires. To date, satellite sensors collect data about the location of the fire, as well as those characteristics of the forest mass that may influence its duration and/or its extension [7]. As it is known, there are several satellites that inform us in almost real time of forest fires. One of the best known is NASA's MODIS (Moderate-Resolution Imaging Spectroradiometer), which is a scientific instrument on board both the Terra satellite and the Aqua satellite. Another of them is the VIIRS (Visible Infrared Imaging Radiometer Suite) also from NASA, a sensor on board of Suomi National Polar-orbiting (Suomi NPP) and NOAA-20, and which together with the MODIS provide data on forest fires within the NASA program, Fire Information for Resource Management System, dedicated exclusively to giving us timely and very precise information in near real time (NRT) of forest fires, within 3 hours after the passage of the satellite. The Sentinel satellites of the Copernicus program cannot be forgotten, a great help in this field of study carried out jointly by the European Space Agency (ESA) and by the European Union through the European Environment Agency. ESA is currently developing seven missions, and has five underway, of which, each Sentinel mission is made up of two satellites to ensure total coverage in each mission.

As is well known, in many parts of the world most forest monitoring is still carried out using traditional methods, no existing information on the advantages of new and accurate remote sensing techniques. For this reason, it is very important to point out that there is currently no procedure, based on environmental information, capable of predicting those areas with a high probability of occurrence of a forest fire with time enough for the firefighting services can intervene before a major disaster occurs. Based on the comments, this study focuses on obtaining an algorithm, based on environmental information obtained through remote sensing, capable of predicting the moment in which the pre-ignition of a spontaneous forest fire occurs. The resulting algorithm can be useful for decision making, as well as preventing spontaneous forest fires.

2. Study area

The municipality of Lújar (36°47′16″N; 3°24′10″W), Grenada (Spain), is located at south of the Iberian Peninsula (**Figure 1**). In general, the study area has a temperate and warm climate, with rainfall concentrated mainly in winter. According to the Köppen-Geiger classification, Lújar is located in the Csa climatic zone, with an

Figure 1.
Location of the study area.

average annual temperature of 16.8°C, while its annual rainfall amounts to an average of 416 mm.

Of a total of 36.83 km2 (3683 ha) of surface that Lújar has, on July 30, 2015, as a result of a spontaneous fire, 1034.96 ha (28.1%) were burned (**Table 1**).

Lújar is located in the Sierra of Lújar, one of the most outstanding natural spaces in the region of the Grenada tropical coast. The affected area is made up of deep and steep ravines distributed in all geographical orientations and has a vegetation that, although it has been greatly damaged in large part by fires in the last 30 years, preserves remnants of the original natural formations. It must be emphasized, as a result of its fauna and flora characteristics, that the Sierra of Lújar is entirely a hunting ground, most of it managed by local hunting societies and some areas by the owners of the farms.

As a result of the above, over time there has been a process of renewal of the traditional landscape, characterized by the mining and agricultural sectors as the main drivers of the municipal economy, in which the maintenance of natural vegetation stands out, resulting in an improvement in the quality of the landscape, and therefore, in an increase of the ecological value of the study area. This aspect has been of great importance for the existing sustainable rural tourism in the municipality of Lújar, although after the fire that occurred it was radically damaged, causing a terrible impact at all levels.

On the other hand, according to [8], the study area is characterized by soils with a pH close to neutrality (mean value of 7.15), with a Base Saturation Percentage (BSP) of 77.80%, a Cation Exchange Capacity (CEC) of 8.58 $meq \cdot 100g^{-1}$, with an average value in Active Limestone (AL) of 0.179% and, an Organic Matter (OM) value of 2.766%.

Land uses	Value (€)	Burned area (ha)
Cork oak	2,733,155.04	138.01
Irrigated almond tree	75,767.88	12.79
Non-irrigated almond tree	1,002,130.83	171.77
Subtropical fruit trees	30,894.42	0.70
Irrigated fruit tree	21,364.36	0.67
Intensive farmland	171,583.18	3.88
Non-productive farmland	14,418.64	8.99
Horticultural greenhouse	292,915.52	6.62
Dry lands	45,496.84	5.61
Shrubs	480,809.30	299.76
Non-irrigated olive grove	1,039,887.76	28.32
Irrigated olive grove	43,288.80	1.07
Pastureland	497,723.32	310.30
Timber pine forest	878,343.08	44.50
Non-irrigated vineyard	42,442.75	1.98
Total sum →	**7,370,221.73**	**1034.96**

Table 1.
Affected area by the spontaneous forest fire, and their economic value.

With regard to the species, or type of canopy, affected by the fire, it should be noted that, before the disaster occurred, there were different typology of agricultural crops, noble Mediterranean scrub, vegetation classified as other Mediterranean scrub, a mixture of *Pinus* and *Quercus* in a similar proportion, cork oak and areas with scarce vegetation.

One of the main economic bastions of the municipality of Lújar is natural or green tourism. In general, this type of tourism refers to those activities that tourists can develop in a completely natural environment, such as hunting tourism (one of those mainly affected by the fire), ecotourism, ornithological tourism, agro-tourism and rural tourism. To a greater or lesser extent, and to date, the recovery of these activities has been supervised by a series of active policies, resulting in both an economic and social resurgence of the affected area, and therefore in a possibility that the Sierra of Lújar is declared a protected natural area.

3. Materials and methods

To carry out this study, a total of 503 images were downloaded from the Earth Explorer archive, belonging to the U.S. Geological Service (USGS), which covered the period between January 2015 and December 2020, both inclusive. In order to adequately characterize the affected area by the fire, from an environmental point of view, each one of the images was examined and classified in one of three determined periods of time (January 1, 2015-July 29, 2015; July 30, 2015- August 31, 2015; September 1, 2015-December 31, 2020) depending on how the study area was (before, immediately after, and after the spontaneous forest fire). All images (Path 200 Row 34 and 35 "WRS") were acquired during January 2021. Subsequently, and with the use of the Geographic Information System (GIS) Ilwis, all the images (Sentinel-2 L2A, Sentinel-2 L1C both of them with 10 m spatial resolution and Landsat 8-C "LC8" Operational Land Imager "OLI", with a spatial resolution of 30 m, and Thermal Infrared Sensor "TIRS" of 100 m spatial resolution rescaled by cubic convolution to 30 m in order to match the multispectral bands of the OLI sensor) were improved. In order to keep the original brightness of the pixel values unchanged, the nearest neighbor algorithm was used.

As a result of the existence of a process of assimilation, and dis-assimilation, of sulfur by the wood of forest mass [9, 10], as well as the need for the existence of sulfur, in trace amounts, in wood for spontaneous combustion, an extensive search for sulfur isotopologues was carried out for the purpose of finding out the proper wavelength at which sulfur can be found in the atmosphere, and for this reason, to be able to select the appropriate satellite to analyze the possible variation of said element in the atmosphere near the forest mass.

For that reason, and as a result of the fact that when a forest fire occurs, the sulfur dioxide concentration ($[SO_2]$) in the atmosphere immediately in contact with the forest mass increases, data on this pollutant were downloaded from [11] between January 2015 and December 2019. Subsequently, 50 images were downloaded from the Sentinel-5P satellite (7×3.5 km spatial resolution) from May 6, 2018 to December 31, 2020 to correlate [11] data with satellite data using multiple triangulation. The objective of this correlation was to find out which of the three specified pollutants is related to the pre-ignition of spontaneous forest fires, based on the conditions of temperature (greater than 30°C) and relative humidity (less than 30%) that must occur during the days prior to the fire.

From the analysis of data obtained from satellite images, as well as those downloaded from [11], a linear model was obtained that allows obtaining, in any forest area and through remote sensing, the moment in which the pre-ignition of forest mass occurs.

In order to adequately contrast the obtained algorithm, a bibliographical review was carried out to determine the possible existence of a temporary record of climatic data (temperature and precipitation) of the affected area by the spontaneous fire, which is broad and significant enough to be able to obtain results, discussions and conclusions consistent with this research. In this regard, the information provided by the Global Climate Monitor (University of Seville, Spain) was of great importance, as it provided a temporal record of data, with significance ≤ 0.05, between 1901 and 2020 (120 years).

Last but not least, it was decided to test the algorithm obtained, as well as the results of this study, in two other spontaneous fires that occurred in Spain (1.- Doñana Natural Park "Matalascañas, Huelva" on June 24, 2017; 2.- Municipality of La Torre de l'Espanyol "Catalonia" on June 26, 2019), obtaining results consistent with those achieved in the study area.

4. Results and discussion

As has been confirmed in this study, remote sensing plays a fundamental role in the prevention of spontaneous forest fires. For this reason, the historical climatic variables of the study area (mean, maximum and minimum temperatures, as well as precipitation) should not be ignored, since when they exist in significant quantity they can help explain certain reflectivity characteristics produced at the pixel level [12].

In this type of study, the atmospheric contaminant amount in the study area is also of great importance, which will affect the optical depth of the atmosphere, and therefore the luminance of the analyzed pixel.

Therefore, in the analysis carried out using satellite images, it has been taken into account that the amount of radiation reflected by each object, and receiving by the satellite sensor, will result in a different reading per pixel than the one that would be in the case of atmosphere clean. This is an aspect of great interest, in addition to the relationship between the radiation reflected by the object in question, and the increase in atmosphere temperature, since when there are extremely reflective surfaces, a Thermal Plugging (TP) is caused in the atmosphere that leads to a decrease in Photosynthetically Active Radiation (fAPAR) in the study area [13], and therefore, a pronounced decrease in both Forest Mass Production (FMP) and Dry Matter Productivity (DMP).

As a result of the fact that in a forest fire the concentration of carbon monoxide (CO), carbon dioxide (CO_2) and sulfur dioxide (SO_2) increases in the affected area, it is logical to think that the temperature also increases, and therefore the fAPAR radiation is affected [14]. This fact is due to the fact that at higher atmospheric temperatures, a TP is generated (as long as there is wind speed ≥ 5 m/s that causes a situation of turbulent regime, and therefore, a mixture of the different air layers), thanks to which the fAPAR radiation, a consequence of the mixture of air layers, will be reflected or refracted, resulting in a lower amount of photosynthetically active radiation that will reach the surface of the plant canopy. The relationship between fAPAR radiation and the Average Environmental Temperature (AET), in Celsius degrees, is presented in the following equation (Eq. (1)):

$$fAPAR = 0.564 - 0.011 \cdot AET \left(r = 0.83; R^2 = 0.7; p \leq 0.001\right) \quad (1)$$

If the Eq. (1) is analyzed based on the TP, the critical temperature for a spontaneous pre-ignition to take place coincides with the moment in which the TP is maximum. At this time, and as a result of the increase in $[SO_2]$ in the atmosphere immediately close to the forest mass, the air particules movement amount (Broglie's Law) increases, giving rise to successive collisions between these particules up to a point where, due to the elastic collision at high speed, an air particules coupling is produced that causes a wave front that prevents the fAPAR radiation reaches the plant canopy, since it is reflected towards the outside of the atmosphere close to the forest mass. This fact causes the successive heating of the air layers towards the surface, implying a successive decrease in the environmental Relative Humidity (RH in %), and as a final consequence another increase in $[SO_2]$. According to what was mentioned above, the fAPAR radiation decreases to zero, and therefore the accumulated temperature necessary for a spontaneous pre-ignition process to start is 51.27°C according to Eq. (1).

It is evident that the sulfur dioxide concentration depends on the Forest Mass Production (FMP), mainly due to the fact that the quantity and type of forest mass existing in the fire area is what gives rise, after its ignition, to certain levels of environmental $[SO_2]$, in mol·m^{-2}, or others. In this sense, a linear relationship has been found between $[SO_2]$ and the FMP (variable dimensionless spectral index between 0 "zero Forest Mass Production" and 1 "maximum Forest Mass Production"), which is presented below (Eq. (2)):

$$[SO_2] = 2.84 \cdot 10^{-5} + FMP \cdot 4.35 \cdot 10^{-4} \left(r = 0.92; R^2 = 0.873; p \leq 0.001\right) \quad (2)$$

In order to facilitate, as far as possible, obtaining the FMP, it is important to comment that a direct relationship has been found with the Normalized Difference Humidity Index (NDMI), which is logical if one takes into account that greater forest mass humidity, greater will be the probability that this mass will be more productive (see Eqs. (3) and (4)):

$$NDMI = -0.353 + 0.791 \cdot FMP \left(r = 0.913; R^2 = 0.833; p \leq 0.001\right) \quad (3)$$

$$NDMI = \frac{NIR - SWIR}{NIR + SWIR} \quad (4)$$

where:

NIR = Near Infrared spectral band.

SWIR = Short wave infrared spectral band.

Likewise, a relationship has been found to be able to obtain, in kg per hectare, the Dry Matter Productivity (DMP) (see Eqs. (5) and (6)):

$$DMP = -636.35 + 276.32 \cdot NDMI + 3798.18 \cdot GNDVI \left(r = 0.89; R^2 = 0.79; p \leq 0.001\right) \quad (5)$$

$$GNDVI = \frac{NIR - Green}{NIR + Green} \quad (6)$$

where:

GNDVI = Green Normalized Difference Vegetation Index.

NIR = Near Infrared spectral band.

Green = Green spectral band.

However, apart from the above, after analyzing the data obtained, it has been possible to verify the decrease, from the 9th day before the spontaneous fire occurred (**Figure 2**), in $[SO_2]$. This fact is the key to being able to predict whether or not the conditions that give rise to a spontaneous forest fire will occur. As can be inferred from **Figure 2**, during the summer months mainly, seems to be a variation in sulfur dioxide concentration levels in the atmosphere immediately close to the forest mass, which begins to be more pronounced 9 days prior to spontaneous fire. This decrease in $[SO_2]$ will be more (**Figure 2a** and **c**) or less (**Figure 2b**) visible depending on the predominant type of vegetation in the affected area. However, it must be taken into account that 2 or 3 days before the spontaneous fire occurs, the decrease in $[SO_2]$ is more pronounced, which may be due to the existence of a process of rapid assimilation of environmental sulfur by forest mass similar to that described by [9, 10]. Logically, and as a result of the decrease in fAPAR radiation, the increase in environmental temperature, as well as the decrease in relative humidity, favors the appearance of a process, which will be called Spontaneous Forest Fire Process (SFFP) (Eq. (7)). Thanks to this process, and according to [15], the sulfur dioxide concentration plays an extremely important role, since when the mean value of the concentrations corresponding to the 9 days prior to the fire ($[SO_2]_{\text{mean value 9 days before}}$) is greater, positive difference, than the estimated value of the day of the fire ($[SO_2]_{\text{estimated value}}$), a spontaneous fire will take place. Otherwise, negative difference, the spontaneous fire will not occur (**Table 2**).

$$SFFP = [SO_2]_{mean\ value\ nine\ days\ before} - [SO_2]_{estimated\ value} \tag{7}$$

It is evident that the assumption of the SFFP is of great importance when it comes to monitoring the possible potential risk of spontaneous forest fires occurrence, and

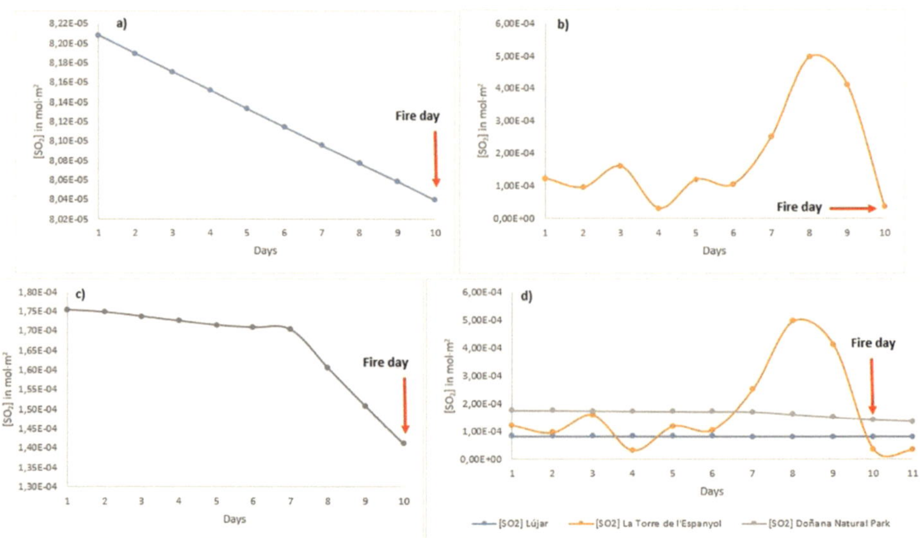

Figure 2.
Variation of sulfur dioxide concentration in Lújar "(a)", La Torre de l'Espanyol "(b)" and Doñana Natural Park "(c)" during the 9 days prior to the spontaneous fire. Picture "(d)" shows the comparison of the previous three.

	$[SO_2]$ Lújar	$[SO_2]$ La Torre de l'Espanyol	$[SO_2]$ Doñana Natural Park
Mean value days 1 to 9 ($mol \cdot m^{-2}$)	8.13×10^{-5}	1.98×10^{-4}	1.69×10^{-4}
Fire day value ($mol \cdot m^{-2}$)	8.04×10^{-5}	3.44×10^{-5}	1.41×10^{-4}
Percentage difference	1.15	82.67	16.51
Difference type	Positive	Positive	Positive
SFFP	Occurs	Occurs	Occurs

Table 2.
Occurrence of the spontaneous forest fire process (SFFP).

therefore, of great interest for the sustainable environmental maintenance of all forests. From the point of view of land use planning, control, evaluation and environmental development of forests around the world, it is of great interest to alleviate the effects of climate change, whose relationship with the effect of Thermal Capping (TP), as well as with the decrease in fAPAR, must be taken into account.

There is no doubt that the incorporation of the proposed methodology to determine the spontaneous forest fire occurrence is simple, as well as easy to implement in any type of airborne or satellite platform.

5. Conclusions

Methodology proposed in this study provides a useful tool to carry out summer monitoring of environmental sulfur dioxide concentration, through satellite images, in any forest area. Likewise, the finding called Spontaneous Forest Fire Process (SFFP) should be highlighted, since it allows a better understanding of what happens in the atmosphere immediately close to the forest mass before the spontaneous fire takes place. It should not be forgotten that, associated with SFFP, there is a decrease in fAPAR radiation which has as a consequence, when it becomes equal to zero, the appearance of a critical temperature (51.27°C) at which the pre-ignition of the spontaneous forest fire occurs. It is necessary to emphasize the importance of Relative Humidity (RH), which decreases in relation to the increase in environmental temperature. Without this decrease in RH, it is difficult for SFFP to take place.

In another vein, it may be of great interest to take into account the cumulative effect of temperature on the soil-forest mass-atmosphere system, as well as its relationship with the assimilation, and dis-assimilation, of sulfur by the forest mass. However, the time dimension in which climatic variables can be forecast is still too short. For this reason, continuous monitoring of them is of vital importance, especially in forest areas where access is difficult in order to avoid natural disasters.

Regarding the affected area, Lújar, it is evident that both the environmental and landscape deterioration to which it was doomed after the fire, whose consequences may be the result of climate change [16].

However, as can be seen in **Figure 3**, the sulfur dioxide concentration before the fire (June 30, 2015) is related, on the surface, to the type of existing vegetation. On the other hand, this concentration varies completely once the fire has been controlled (August 1, 2015), giving rise to a situation very similar to what occurs in a desert area, that is, due to thermal accumulation, or thermal inertia, existing sulfur dioxide levels

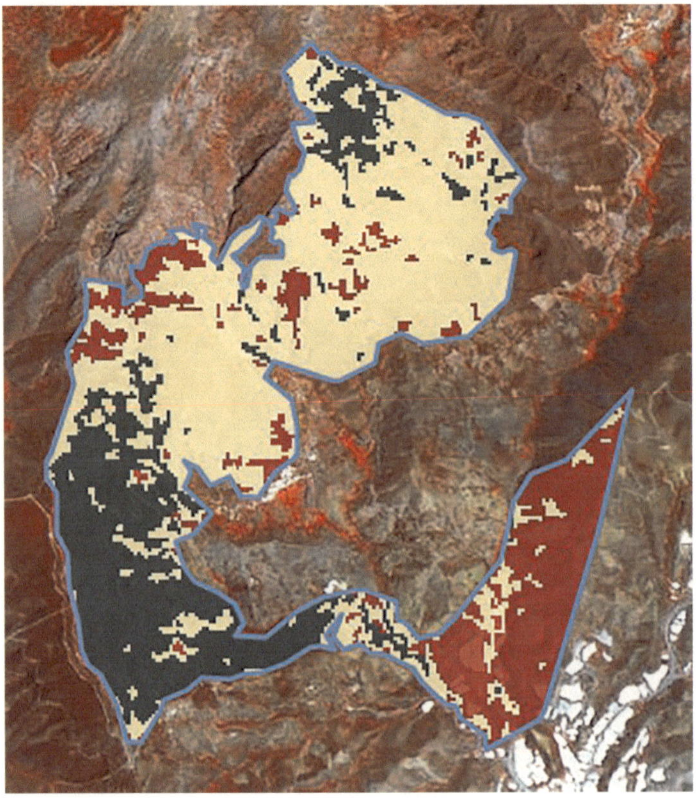

Figure 3.
Environmental SO_2 concentration, in Lújar, 1 month before (June 30, 2015) of the spontaneous fire. The red color is equivalent to 0 $mol \cdot m^{-2}$, the area in yellow has values $<1.25 \times \cdot 10^{-3}$ $mol \cdot m^{-2}$, while the area in green has values $>1.25 \times \cdot 10^{-3}$ $mol \cdot m^{-2}$.

drop in those areas without vegetation due to thermal plugging (**Figure 4**). Over time, and as a result of the restoration policies carried out by the municipal government, the affected area has been recovering its initial status, giving rise to a situation (**Figure 5**), on August 14, 2020, similar to the one shown in **Figure 3**.

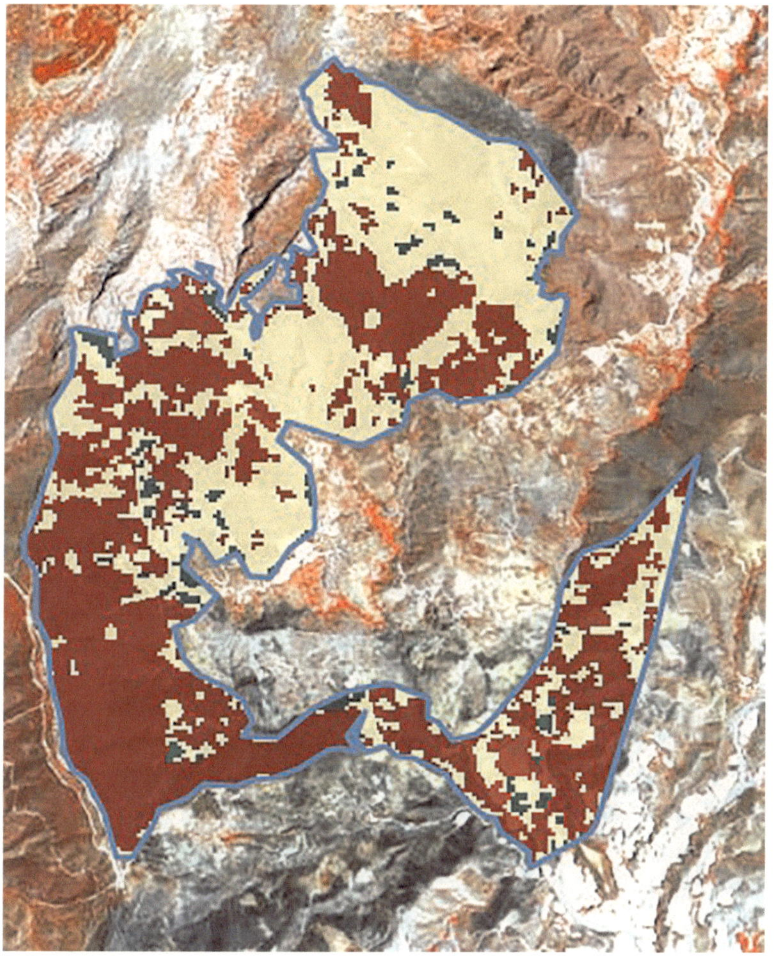

Figure 4.
Environmental SO_2 concentration, in Lújar, when the fire was controlled (August 1, 2015). The red color is equivalent to 0 mol · m^{-2}, the area in yellow has values <1.25 × ·10^{-3} mol · m^{-2}, while the area in green has values >1.25 × ·10^{-3} mol · m^{-2}.

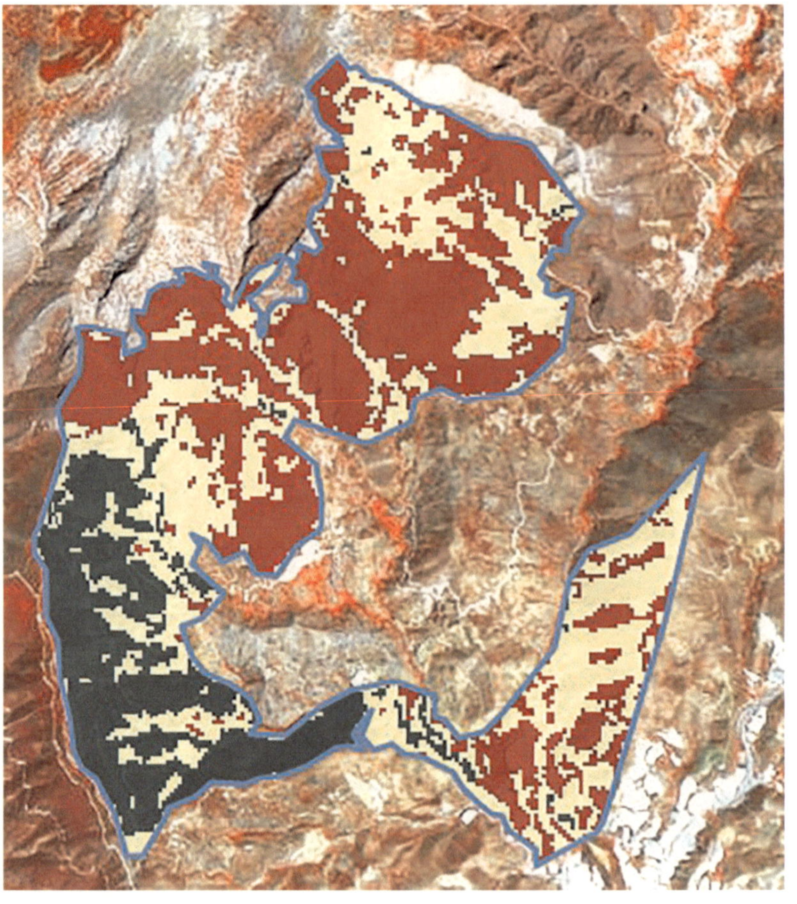

Figure 5.
Environmental SO_2 concentration, in Lújar, on August 14, 2020. The red color is equivalent to 0 $mol \cdot m^{-2}$, the area in yellow has values $<1.25 \times \cdot 10^{-3}$ $mol \cdot m^{-2}$, while the area in green has values $>1.25 \times \cdot 10^{-3}$ $mol \cdot m^{-2}$.

Acknowledgements

Authors appreciate the collaboration of Technology-Based Company RS3 Remote Sensing S.L. in this work.

Conflict of interest

The authors declare no conflict of interest.

Author details

Emilio Ramírez-Juidías[1]* and Emilio José Cabello-Franco[2]

1 University of Seville, Seville, Spain

2 School of Agricultural Engineering, University of Seville, Seville, Spain

*Address all correspondence to: erjuidias@us.es

References

[1] Ministry for Ecological Transition. Bosques españoles y su evolución [Internet]. 2022. Available from: https://www.miteco.gob.es/es/biodiversidad/temas/default.aspx [Accessed: 2022-12-01]

[2] World Wild Fund for Nature. Stepping up?: the continuing impact of EU consumption on nature worldwide [Internet]. 2021. Available from: https://wwfes.awsassets.panda.org/downloads/stepping_up___the_continuing_impact_of_eu_consumption_on_nature_worldwide_fullreport.pdf?56780/denuncia-que-el-consumo-de-la-UE-es-responsable-del-16-de-la-deforestacion-tropical-importada-y-Espana-es-el-tercer-pais-europeo-con-mayor-impacto [Accessed: 2022-12-02]

[3] Molina-Terrén DM, Xanthopoulos G, Diakakis M, Ribeiro L, Caballero D, Delogu GM, et al. Analysis of forest fire fatalities in Southern Europe: Spain, Portugal, Greece and Sardinia (Italy). International Journal of Wildland Fire. 2019;**28**:85-98. DOI: 10.1071/WF18004

[4] Jaffe DA, O'Neill SM, Larkin NK, Holder AL, Peterson DL, Halofsky JE, et al. Wildfire and prescribed burning impacts on air quality in the United States. Journal of the Air & Waste Management Association. 2020;**70**(6): 583-615. DOI: 10.1080/10962247.2020.1749731

[5] Liu N, Lei J, Gao W, Chen H, Xie X. Combustion dynamics of large-scale wildfires. Proceedings of the Combustion Institute. 2021;**38**:157-198. DOI: 10.1016/j.proci.2020.11.006

[6] National Statistics Institute of Spain. Estadística de incendios forestales [Internet]. 2022. Available from: https://www.ine.es/dyngs/IOE/es/operacion.htm?numinv=04002 [Accessed: 2022-12-03]

[7] Wu J, Baartman JEM, Nunes JP. Comparing the impacts of wildfire and meteorological variability on hydrological and erosion responses in a Mediterranean catchment. Land Degradation & Development. 2021;**32**: 640-653. DOI: 10.1002/ldr.3732

[8] Andalusian Environmental Information Network (REDIAM). Geoportal: Visualizadores [Internet]. 2022. Available from: https://www.juntadeandalucia.es/medioambiente/portal/web/guest/acceso-rediam/geoportal [Accessed: 2022-12-05]

[9] Barrelet T, Ulrich A, Rennenberg H, Krähenbühl U. Seasonal profiles of sulphur, phosphorus, and potassium in Norway spruce wood. Plant Biology. 2006;**8**(4):462-469. DOI: 10.1055/s-2006-924044

[10] Zhang Y, Wu FZ, Tan B, Xu ZF, Li H, Liu JC, et al. Interception of sulfate deposition from a closed canopy to a forest gap edge canopy in a subalpine dragon spruce plantation. Journal of Mountain Science. 2019;**16**(12):2856-2866. DOI: 10.1007/s11629-019-5537-1

[11] Andalusian Environmental Statistics Viewer (VEM). Red de vigilancia y control de emisiones a la atmósfera [Internet]. 2022. Available from: https://www.juntadeandalucia.es/medioambiente/vem/?c=Menu/tema/584 [Accessed: 2022-12-05]

[12] Shaharyar A, Cetin M. Compressed sensing ISAR reconstruction considering highly maneuvering motion. Electronics. 2017;**6**(1):2-29. DOI: 10.3390/electronics6010021

[13] Ramírez-Juidías E. Cómo predecir las cosechas a partir de imágenes por

satélite. The Conversation. 2021;**7**:1-4. Available from: https://theconversation.com/como-predecir-las-cosechas-a-partir-de-imagenes-por-satelite-162684

[14] Ramírez-Juidías E. A mini-review of remote sensing applied to salt-marshes. Academia Letters. 2022;**1**:1-5. Article 4662. DOI: 10.20935/AL4662

[15] Ramírez-Juidías E, Pozo-Morales L, Galán-Ortiz L. Procedimiento para obtener una imagen teledetectada a partir de fotografía. Patent n° ES2537783B2 (29/09/2015), international extension PCT-WO2014198974. University of Seville, Spain. 2015. Available from: https://consultas2.oepm.es/pdf/ES/0000/000/02/53/77/ES-2537783_B2.pdf

[16] Keeley JE, Syphard AD. Large California wildfires: 2020 fires in historical context. Fire Ecology. 2021; **17**(1):1-11. DOI: 10.1186/s42408-021-00110-7

Chapter 18

Evolution of Agroclimatic Indicators in Senegal Using CMIP6 Simulations

Cheikh Modou Noreyni Fall, Adama Faye, Mbaye Diop, Babacar Faye and Amadou Thierno Gaye

Abstract

Climate has a strong influence on agriculture, which is considered the most dependent human activity on climate variations. The future performance of the Senegalese agricultural sector will depend on its ability to adapt to the negative impacts of climate change. This study demonstrated that the impact of three climate change scenarios (ssp126, ssp245 and ssp585) on the evolution of 14 agro-climatic indicators is already evident in Senegal in the near and distant future. Indeed, the results obtained show a generalized decrease over the whole country in seasonal rainfall totals of about −10% in the near future (2020–2049; PSE horizon) up to −40% in the distant future (2070–2099) for the ssp585 scenario. This decrease in precipitation will be associated with two phenomena, namely a shortening of the rainy season due to increasingly late starts and an increase in dry spells, particularly the DSl and DSxl. The other trend observed is an increase in the frequency and intensity of extreme rainfall events (R99 and R20), which illustrates an increasingly chaotic distribution of rain in the future. Finally, this characterization of agroclimatic indicators made it possible to evaluate and classify the sensitivity of four global models corrected by the CFD-t method in order to run agronomic simulations and to explore adaptation strategies for farmer management in the future.

Keywords: agroclimatic indicators, climate change, shared socioeconomic pathways (SSP), near future, far future

1. Introduction

Climate change is now considered one of the major challenges facing the world's populations [1]. Most sectors of human activity on which development efforts are focused are climate-sensitive. The agricultural sector is one of the most vulnerable sectors to climate change [2–6]. Indeed, the Sahelian zone is already subject to high spatiotemporal variability in rainfall. The Sahel experienced continuous drought from the late 1960s to the mid-1990s and early 2000s [7–10]. During these decades of drought, the isohyets lost between 20 and 50% of their annual cumulative rainfall compared to the 1950s and 1960s. Precipitation has also decreased between 20 and 40%, as well as average temperatures have increased by 1.3°C [11]. This has been described as the most negative trend in precipitation in the contemporary world [12]. The impacts on the populations of sub-Saharan Africa have been very significant with per capita food production falling sharply from 98% in the 1960s to about 86% in the mid-1980s [13, 14]. On average, each inhabitant had lost 12% of the food grown in the 1980s, when drought was at its peak [15]. Climate change will exacerbate these impacts on agriculture. There is growing evidence that changes in climate parameters, namely precipitation, temperature and the intensity and frequency of extreme events, will affect the agricultural sector in several ways [16]. Increased temperatures and the possibility of more extreme thermal events will impact crop productivity. Heading is one of the most sensitive phenological stages to temperature extremes for all cereal species, and during this stage of development, temperature extremes will significantly affect crop production [16, 17]. While increased pockets of drought during the rainy season will limit water availability, thereby promoting soil and groundwater salinization ultimately reducing crop area [18, 19]. Studies have also shown a proliferation of parasites harmful to plants and an increasing incidence of pest attacks which could be detrimental to agricultural production [20]. Given this context of high exposure to the impacts of climate change, the implementation of adaptation measures is an imperative issue for the agricultural sector now and in the future. Thus, several agro-climatic indicators have been created so that agricultural actors can better understand the effects of climate change on their production systems and prepare the necessary adaptations. However, most of the studies that evaluate these agro-climatic indicators focus on global or regional scales. The objective of this paper is to quantify the impacts of global warming on 14 selected agroclimatic indicators in Senegal. Among these indicators, we have the start of the rainy season, different categories of dry sequences and the frequency and intensity of extreme rainfall events. The study was carried out through a spatiotemporal assessment of changes in these agro-climatic indicators in the near (2020–2049) and distant (2070–2099) future in relation to the historical period (1985–2014) for the Shared Socioeconomic Pathways scenarios (ssp126, ssp245 and ssp585).

2. Data and methods

2.1 Data

In this study, four CMIP6 general circulation models (**Table 1**; [25]) and three greenhouse gas emission scenarios known as Shared Socioeconomic Pathways (SSPs; [26, 27]) (SSP1-2.6, SSP2-4.5, and SSP5-8.5,) were used. The CMIP6 GCM runs were developed in support of the Intergovernmental Panel on Climate Change Sixth Assessment Report (IPCC AR6). Each of the climate projections includes daily precipitation and temperature variables for the periods 1950–2014 ("hindcast") and 2015–2100 ("forwardcast"). A bias reduction and correction procedure, described in the following paragraph, was applied to the historical and future climate projections to provide a set of climate change projections at high resolution (0.25°) relative to their native resolution that can be used to assess the impacts of climate change on

No	Models	Institutions	Horizontal resolution Lon × Lat	Reference
1	CNRM-CM6-1	Centre National de Recherches Météorologiques (CNRM); Centre Européen de Recherches et de Formation Avancée en Calcul Scientifique, France	1.41° × 1.41°	Voldoire et al. [21]
2	CNRM-ESM2-1	Centre National de Recherches Météorologiques, Toulouse, France	1.41° × 1.41°	Séférian et al. [22]
3	IPSL-CM6A-LR	Institut Pierre Simon Laplace, Paris, France	2.5° × 1.26°	Boucher et al. [23]
4	MRI-ESM2-0	Meteorological Research Institute (MRI), Japan	1.13° × 1.13°	Yukimoto et al. [24]

Table 1.
List of CMIP6 models used in this study, modeling centers, horizontal resolution and references.

processes that are sensitive to smaller-scale climate gradients and the effects of local topography on climate conditions.

2.2 Bias correction

A bias correction procedure with the CDF-t method was applied on each grid point with the GCM data via the historical Global Meteorological Forcing Dataset (GMFD; [28]). For each daily parameter, the algorithm generates the cumulative distribution function (CDF) for the GMFD data and for the retrospective GCM simulations. The corresponding source values (day of year +/− 15 days) are clustered and sorted over the period from 1960 to 2014 at various probability thresholds to produce a quantile map between models and observations. Based on this map, the model values in the different quantiles (e.g., p = 90%) can be translated into corresponding GMFD values for the same quantile. Assuming that the CDF of the simulations is stable during the retrospective and prospective periods, the algorithm simply searches for the probability quantile associated with the predicted climate variations from the estimated CDF of the historical data and then accepts these as the adjusted climate projections. The climate projections adjusted in this way have the same CDF as the GMFD data; therefore, possible biases in the statistical structure (variance, in particular) of the original GCM outputs are eliminated by this procedure. At the end of the bias correction step, the previously extracted parameter climate trends (precipitation and temperature) are added to the fitted model climate fields.

2.3 Agroclimatic indicators

Table 2 provides details on the 14 agroclimatic indicators calculated in this study. These include seasonal rainfall characteristics (cumulative seasonal rainfall, number of rainy days, rainfall intensity, extreme rainfall, dry sequences, start, end and length of the rainy season). A dry sequence is obtained by counting the number of consecutive days without rainfall that lie between two rainy days (a day is rainy when

Indicator	Details	Definitions	Units
PRCPTOT	Total annual daily precipitation	Total annual rainy days >1 mm	mm
SDII	Daily intensity index	Ratio of annual total to number of rainy days	mm/day
R95 (HIP)	Rather rainy days	Total daily rainfall >95th percentile	mm
R99 (HIP)	Extreme daily rainfall	Total daily rainfall >99th percentile	mm
R10 (FIP)	Number of rainy days	Number of days with rain >10 mm	days
R20 (FIP)	Number of very rainy days	Number of days with rain >20 mm	days
WD	Rainy days	Number of days with rain >1 mm	days
DSs	Dry spells of short duration	Dry spell between 1 and 3 days	days
DSm	Dry spells of medium duration	Dry spell between 4 and 7 days	days
DSl	Dry spells of long duration	Dry spell between 8 and 15 days	days
DSxl	Dry spells of extreme duration	Dry spell >15	days
Onset	Start date of the agronomic season	Sivakumar Definition	days
Cessation	End date of the agronomic season	Sivakumar Definition	days
Length	Length of the agronomic season	Sivakumar Definition	days

Table 2.
Definitions of the agroclimatic indicators used.

its cumulative rainfall is greater than or equal to 1 mm). These are intra-seasonal droughts, also called rainfall breaks. We have defined four types: DSs, DSm, DSl and DSxl (see **Table 2**). The dry sequences that determine the potential risks of reseeding are the longest rainfall breaks observed over the 30 days after sowing [29]. Here, DSl and DSxl present themselves as effective indicators for detecting false starts to the season [18]. Next, in order to assess the ability of global models to adequately represent the agronomic start of the rainy season, we used Sivakumar's [30] definition of local rainy season start. Based on an agro-nomic criterion, he considers the start of the rainy season in the Sahelian and Sudanian regions as the date, starting on May 1, when there is a rainfall of at least 20 mm on three consecutive days, with no dry spells of more than seven days in the following 30 days. The end of the agronomic season is defined by considering the date after September 1 when a long dry period of 20 days occurs. The duration of the agronomic season is defined for a given season as the difference between the start and end date of the rainy season according to the agronomic criterion. Finally, indicators illustrating the intensity (R95 and R99) and frequency (R10 and R20) of rainfall events are used in addition to the cumulative season (PRCTOT).

3. Results

3.1 Changes in agroclimatic indicators in the near (2020–2049) and far future (2070–2099)

To study the impacts of climate change on the evolution of these indicators on each grid point of Senegal, the changes between the historical period (1985–2014) and the climate simulations SSP126, SSP245 and SSP585 for the near future (2020–2049) and the far future (2070–2099) are represented in **Figure 1**. The middle of the whisker diagram represents the spatial average over Senegal, and it is important to note that here the multi-model average is considered. An amplification of the changes in the evolution of the agroclimatic indicators is observed when moving from the near future to the distant future and from the sustainability scenario SSP126 to the fossil fuel development scenario, SSP585. In the near future, seasonal precipitation totals will be broadly the same or even slightly lower than what is observed in the historical model simulations. This slight decrease is more a consequence of a high occurrence of dry sequences and a shortening of the rainy season illustrated by a late start to the season. This evolution will become more pronounced in the distant future with a decrease in average over the country of −10% for SSP126, −20% for SSP245 and over −35% for SSP585.

By analyzing the distribution of precipitation over the season, it is clear that the trend observed over the period 2020–2049 is confirmed and amplified in the distant future. Indeed, a strong increase in dry sequences (DSs and DSxl) will be observed, which will increase from +15% for SSP 126 to +50% for SSP585. This situation will favor the risks of reseeding and post-flowering water stress. Although at the beginning of the rainy season (June to July), the occurrence of rainfall breaks of 8 to 14 days is a fairly frequent event, and the observation of a DSl or DSxl after the semi flowering is not something that farmers want. Indeed, even if young millet plants show a certain capacity to adapt to such pockets of drought, the same cannot be said for maize, sorghum or certain legumes (groundnuts, cowpeas, etc.). The water requirement of crops during the heading-maturity phase (or reproductive phase) is one of the main factors conditioning their final yield. In general, this period corresponds to the critical reproductive phase for non-photoperiodic crops with a cycle length of about 90 days. A late start of the agronomic season, from 10 days for SSP126 to about 20 days for SSSP585, will also be observed, whereas the end of the season seems to be less sensitive to the effects of climate change. Paradoxically, extreme rainfall events, notably R99 (mm) and R20 (days), will increase by about 10–15% and 20–25%, respectively, by 2050 before decreasing slightly in the distant future. These results are perfectly consistent with studies [31–33] demonstrating the installation of a new rainfall regime over the Sahel characterized by false starts, a shortening of the rainy season, an increased frequency of intense rainfall and an occurrence of dry sequences, which is directly associated with global warming.

3.2 Spatial distribution of the change in agroclimatic indicators

In order to better analyze the spatial distribution of these indicators, the average change of these indicators on each grid point is represented in **Figure 2**. This spatial distribution of changes on these agroclimatic indicators allows to measure the spatial heterogeneity of the impact of global warming.

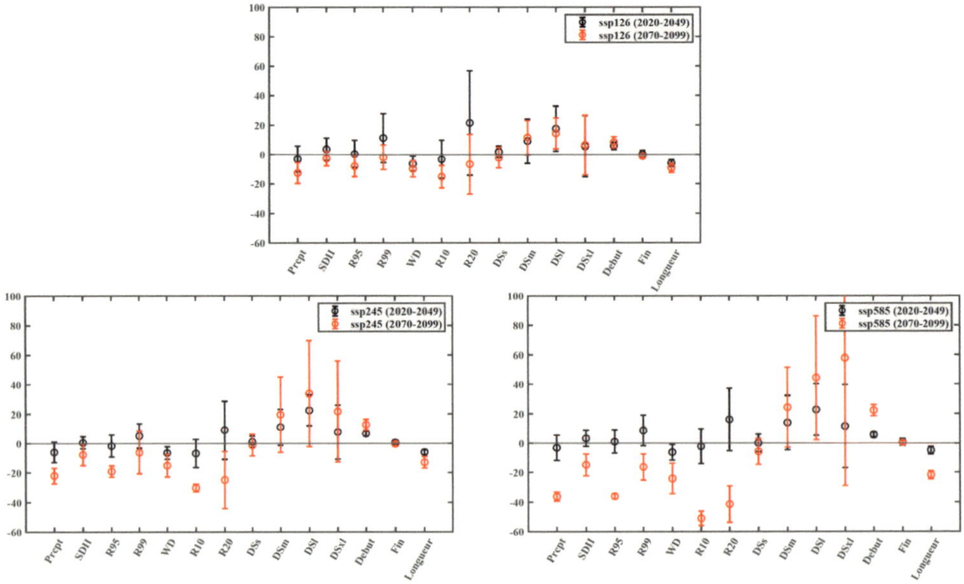

Figure 1.
Projected changes in the evolution of the 14 agroclimatic indicators. Changes are defined as the difference between the historical period (1985–2014) and the climate simulations SSP126 (top panel), SSP245 (bottom panel left) and SSP585 (bottom panel right) for the near future (2020–2049; boxplot in black) and the distant future (2070–2099; boxplot in red).

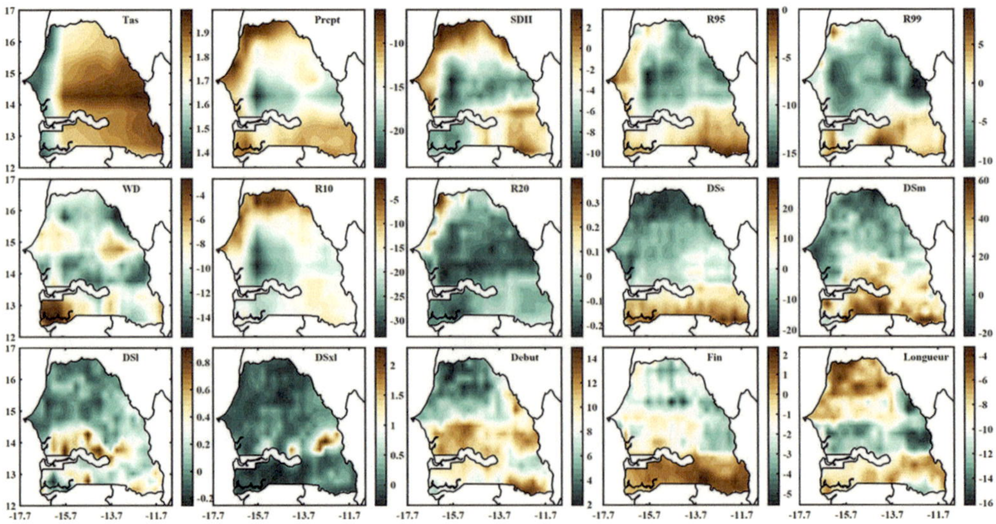

Figure 2.
Spatial distribution of changes in the 14 agroclimatic indicators in addition to the mean surface temperature (Tas). The changes are defined as the difference between the historical period (1985–2014) and the SSP126 climate simulations for the near future (2020–2049).

Indeed, despite a general decline in cumulative seasonal rainfall on all pixels of the country, some areas of the country will be more impacted than others. The center of the country and even the Central East will suffer the largest decreases in terms of cumulative seasonal exceeding −20% followed by the south with rates between −15

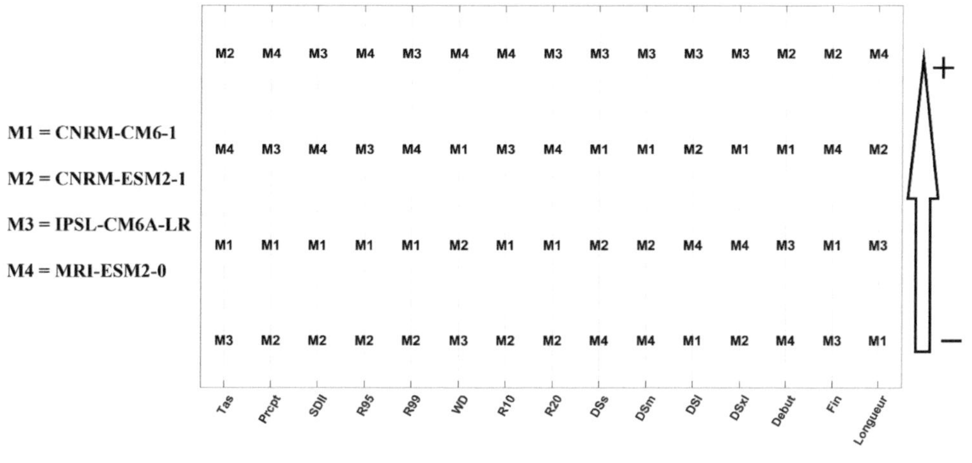

Figure 3.
Ranking of the four models used according to the 14 agroclimatic indicators. The sign (+) of the arrow indicates an increase in the near future of the indicator and the sign (−) a decrease. For example, M2, which is the model where warming is stronger, tends to capture later season starts.

and − 10%. The north and the coastal area of the country will record the smallest decreases (less than −10%). Precipitation intensity indicators will increase in the coastal areas, the southeast and the north except for R99 which will increase only in the south of the country. The 8-to-15-day dry spells (DSl) will increase more in the south of the groundnut basin, while DSxl will increase almost throughout the country. Finally, the delays in the start of the season observed in **Figure 1** affect the groundnut basin (Fatick, Kaolack and Kaffrine), the center-east and the southeast of the country more, so the seasons are shorter in these areas of the country. This characterization of agroclimatic indicators allowed us to set up this ranking matrix of four global climate models (GCMs), namely CNRM-CM6-1, CNRM-ESM2-1, IPSL-CM6A-LR and MRI-ESM2-0 (see **Figure 3**).

This diagram combining temperature changes, and precipitation characteristics relative to the 1985–2014 history allow us to determine, in terms of trend, whether the models are warm/dry/extreme/early-onset, warm/moist/extreme/early-onset, cool/moist/moderate/early-onset, cool/dry/moderate/early-onset for the near-future SSP126 (2020–2049). For example, temperatures are expected to increase in this period by +0.98°C with IPSL-CM6A-LR (M3), +1.07°C with CNRM-CM6-1 (M1), +1.09°C with MRI-ESM2-0 (M4) and + 1.1°C with CNRM-ESM2-1 (M2). With regard to precipitation, the changes are variable: With 3/4 models predicting a decrease in precipitation on average over Senegal ranging from −1 to −13%, the largest decrease is recorded by the warmest model, namely CNRM-ESM2-1 (M2). While the second warmest model MRI-ESM2-0 (M4) predicts an increase in precipitation explained mainly by a high occurrence of events such as R95, WD, R10 and longer seasons compared to the historical period. Finally, the IPSL-CM6A-LR, which is the model with less warming compared to the other three models, predicts a climate with an increase in extreme rainfall events such as R99, SDII, R20 and all categories of rainfall breaks (DSs, DSm, DSl and DSxl). This approach could allow for sensitivity testing with agronomic impact models. In addition, it will allow us to better understand the response of crops to different degrees of warming and extreme climate in the different agroclimatic indicators defined (**Table 2**).

4. Discussions and conclusions

This assessment of the evolution of 14 agroclimatic indicators shows that the effects of climate change that will affect all of Africa are already visible at the local level. Knowledge about future climate trends and their impacts has increased in recent years in Africa. According to the IPCC, monsoon rainfall is expected to increase over the central Sahel and decrease over the extreme western Sahel (Senegal area). The monsoon season is expected to have a delayed start. However, at a more localized scale, the question of the reliability of this information arises. The ability of global models to represent the Sahelian climate faces difficulties related to the limited availability of quality climate data and the many uncertainties intrinsic to the physics of the models. In order to better understand the impact of climate change on key indicators for the agricultural sector, four corrected global models are used to characterize their evolution in the near (2020–2049) and distant (2070–2099) future.

The results corroborate with the conclusions of the latest IPCC report that predict a decrease in rainfall in the western Sahel, agricultural droughts and extreme rainfall events. Indeed, we show that climate change will lead to an almost constant decrease for the different socioeconomic scenarios, of the seasonal cumulative rainfall in the near future below −10% for the ssp126 scenario and about −40% in the distant future for the ssp585 scenario. This decrease in rainfall is rather explained by an increase in rainfall breaks combined with a shortening of the rainy season as illustrated by the results. The frequency of the different categories of dry sequences will tend to increase to reach +20% for the DSl. This trend will persist in the distant future and could exceed +50% with the ssp585 scenario. The shortening of the rainy season will be mainly explained by a delay in the start of the rainy season which could be 1 month compared to the average observed over the 1985–2014 history. Furthermore, despite a decrease in precipitation, extreme rainfall events will increase in intensity (R99) and frequency (R20) in the near future for all scenarios before decreasing in the distant future. This increase in the frequency of heavy rainfall corroborates the results of Taylor et al. [34], Panthou et al. [35] and Chagnaud et al., [36] who attribute this intensification of rainfall to global warming, which particularly affects temperatures in the Sahara. Thus, a warmer Sahara intensifies convection in Sahelian storms through increased wind shear and changes in the Saharan air layer [37]. We also analyzed the strong spatial disparities behind these future trends. The results indicate that the central part of the country and the Central East will experience the largest decreases in terms of seasonal accumulation. Precipitation intensity indicators will increase most strongly in the coastal areas, the southeast and the north. The increase in dry sequences of the DSl and DSxl type and the delays in the start of the season in the southern part of the groundnut basin constitute a source of vulnerability for the agricultural sector, with significant consequences for food security, diseases, farm capital, loss of livestock, etc. [38]. In conclusion, our work could be the starting point for simulations of yields, biomass and several relevant indicators for the agricultural sector in order to quantify the impact of climate change on agriculture in Senegal. These studies could be carried out on various spatial scales (field, commune, department or region). This will allow for the exploration of adaptation strategies such as scaling up investments in irrigation, fertilization or pesticides and accelerating the adoption of climate-proof agricultural technologies such as drought-resistant crop varieties and collaboration between farmers and breeders for the mass use of organic fertilization.

Acknowledgements

We would like to thank the Projet d'Appui au Plan National d'Adaptation du Sénégal (PNA-FEM) for funding this research program to assess the impact of climate change on the Senegalese agricultural sector. The PNA-FEM is funded by UNDP and the Global Environment Facility to strengthen the capacity of sectoral ministries and local governments to better assess the impacts of climate change and to adapt decision-making to the risks posed by climate change.

Author details

Cheikh Modou Noreyni Fall*, Adama Faye, Mbaye Diop, Babacar Faye and Amadou Thierno Gaye
Laboratoire de Physique de l'Atmosphère et de l'Océan Siméon Fongang (LPAOSF), École Supérieure Polytechnique (ESP), University of Cheikh Anta Diop, Dakar, Senegal

*Address all correspondence to: cheikhmodou.fall@ucad.edu.sn

References

[1] IPCC. Summary for policymakers. In: Climate Change 2021: The Physical Science Basis: Contribution of Working Group I to the Sixth Assessment Report of the Intergovernmental Panel on Climate Change. Masson-Delmotte V, Zhai P, Pirani A, Connors SL, Péan C, Berger S, editors. New York, NY, USA: In press and Cambridge, United Kingdom: Cambridge University Press; 2021. DOI: 10.1017/9781009157896

[2] Faye B, Webber H, Naab J, Maccarthy D, et al. Impacts of 1.5 versus 2.0°C on cereal yields in the West African Sudan Savanna. Environmental Research Letters. 2018;**13**:034014. DOI: 10.1088/1748-9326/aaab40

[3] Schmidhuber J, Tubiello F. Global food security under climate change. Proceedings of the National Academy of Sciences of the United States of America. 2008;**104**:19703-19708

[4] Sultan B, Defrance D, Iizumi T. Evidence of crop production losses in West Africa due to historical global warming in two crop models. Scientific Reports. 2019;**9**:12834. DOI: 10.1038/s41598-019-49167-0

[5] Van den Berg J, Greyvenstein B, du Plessis H. Insect resistance management facing African smallholder farmers under climate change. Current Opinion in Insect Science. 2022;**2022**:100894

[6] Hansen JE. A brighter future. Climatic Change. **2002**;52:435-440. DOI: 10.1023/A:1014226429221

[7] Nicholson S. The West African Sahel: A review of recent studies on the rainfall regime and its interannual variability. Meteorology. 2013;**2013**:1-32. DOI: 10.1155/2013/453521

[8] Nicholson S, Tucker C, Ba M. Desertification, drought, and surface vegetation: An example from the West African Sahel. Bulletin of The American Meteorological Society. 1998;**79**:815-829

[9] Hatfield JL, Takle G, Grothjahn R, Holden P, Izaurral RC, Mader T, et al. Agriculture. In: Melillo JM, Richmond TC, Yohe GW, editors. Climate Change Impacts in the United States: The Third National Climate Assessment. US: Global Change Research Program. 2014. pp. 150-174. DOI: 10.7930/J02Z13FR

[10] Sultan Benjamin GM. Agriculture in West Africa in the twenty-first century: Climate change and impacts scenarios, and potential for adaptation. Frontiers in Plant Science. 2016;**7**:1262

[11] Maranz S. Tree mortality in the African Sahel indicates an anthropogenic ecosystem displaced by climate change. Journal of Biogeography. 2009;**36**:1181-1193. DOI: 10.1111/j.1365-2699.2008.02081.x

[12] Nicholson S. On the question of the "recovery" of the Ravis in the West African Sahel. Journal of Arid Environments. 2005;**63**:615-641. DOI: 10.1016/j.jaridenv.2005.03.004

[13] Clover J. Food security in sub-Saharan Africa. African Food Review. 2010;**12**(1):5-15. DOI: 10.1080/10246029.2003.9627566

[14] Hugon P. L'agriculture en Afrique subsaharienne restituée dans son environnement institutionnel. In: Tradition et modernisation des économies rurales: Asie-Afrique-Amérique latine: Mélanges en l'honneur de Gilbert Étienne [en ligne]. Genève: Washington, DC., USA: Graduate Institute Publications; 1998

[15] Terence E, Chehbouni A, Dhiba D. Recent climate change adaptation strategies in the Sahel: A critical review. The nature, causes, Effects and Mitigation of Climate Change on the Environment. IntechOpen; 2022;**2022**. DOI: 10.5772/intechopen.100110

[16] Hatfield JL, Prueger JH. Agroecology: Implications for plant response to climate change. In: Yadav SS, Redden RJ, Hatfield JL, Lotze-Campen H, Hall AE, editors. WestSussex: Wiley-Blackwell; 2011. pp. 27-43

[17] Hatfield JL, Boote KJ, Fay P, Hahn L, Izaurralde RC, Kimball BA, et al. Agriculture. In: The Effects of Climate Change on Agriculture, Land Resources, Water Resources, and Biodiversity in the United States. A report by the U.S. Climate Change Science Program and the Subcommittee on Global Change Research. Washington DC, USA: U.S. Environmental Protection Agency. 2008. pp. 362

[18] Noreyni FCM, Lavaysse C, Kerdiles H, Dramé MS, Roudier P, Gaye AT. Performance of dry and wet spells combined with remote sensing indicators for crop yield prediction in Senegal. Climate Risk Management. 2021;**33**:100331. DOI: 10.1016/j.crm.2021.100331

[19] Libutti A, Gatta G, Gagliardi A, Vergine P, Pollice A, Beneduce L, et al. Agro-industrial wastewater reuse for irrigation of a vegetable crop succession under Mediterranean conditions. Agricultural Water Management. 2018;**196**:1-14. DOI: 10.1016/j.agwat.2017.10.015

[20] Skendzic S, Zovko M, Zivkovic IP, Lesic V, Lemic D. The impact of climate change on agricultural insect pests. Insects. 2021;**12**:440. DOI: 10.3390/insects12050440

[21] Voldoire A, Saint-Martin D, Sénési S, Decharme B, Alias A, Chevallier M, et al. Evaluation of CMIP6 DECK experiments with CNRM-CM6-1. Journal of Advances in Modeling Earth Systems. 2019;**11**:2177-2213. DOI: 10.1029/2019MS001683

[22] Séférian R, Nabat P, Michou M, Saint-Martin D, Voldoire A, Colin J, et al. Evaluation of CNRM earth-system model, CNRMESM2-1: Role of Earth system processes in present-day and future climate. Journal of Advances in Modeling Earth Systems. 2019;**11**:4182-4227. DOI: 10.1029/2019MS001791

[23] Boucher O, Denvil S, Levavasseur G, Cozic A, Caubel A, Foujols MA, et al. IPSL IPSL-CM6A-LR model output prepared for CMIP6 CMIP historical. Earth System Grid Federation. Journal of Advances in Modeling Earth Systems. 2018;**12**:1-52. DOI: 10.1029/2019MS002010

[24] Yukimoto S, Koshiro T, Kawai H, Oshima N, Yoshida K, Urakawa S, et al. MRI MRI-ESM2.0 model output prepared for CMIP6 CMIP. Earth System Grid Federation. 2019;**2019**. DOI: 10.22033/ESGF/CMIP6.621

[25] Eyring V, Bony S, Meehl GA, Senior CA, Stevens B, Stouffer RJ, et al. Overview of the coupled model intercomparison project phase 6 (CMIP6) experimental design and organization, Geoscientific Model Development. 2016;**9**:1937-1958. DOI: 10.5194/gmd-9-1937-2016

[26] O'Neill BC, Tebaldi C, van Vuuren DP, Eyring V, Friedlingstein P, Hurtt G, et al. The scenario model intercomparison project (ScenarioMIP) for CMIP6, Geoscientific Model Development. 2016;**9**:3461-3482. DOI: 10.5194/gmd-9-3461-2016

[27] Meinshausen M, Nicholls ZRJ, Lewis J, Gidden MJ, Vogel E, Freund M, et al. The shared socio-economic pathway (SSP) greenhouse gas concentrations and their extensions to 2500, Geoscientific Model Development. 2020;**13**:3571-3605. DOI: 10.5194/gmd-13-3571-2020

[28] Thrasher B, Maurer EP, McKellar C, Duffy PB. Technical note: Bias correcting climate model simulated daily temperature extremes with quantile mapping. Hydrology and Earth System Sciences. 2012;**16**(9):3309-3314. DOI: 10.5194/hess-16-3309-2012

[29] Sivakumar MVK. Empirical analysis of dry spells for agricultural applications in West Africa. Journal of Climate. 1992;5:532-539. DOI: 10.1175/1520-0442

[30] Sivakumar M. Predicting rainy season potential from the onset of rains in southern Sahelian and Sudanian climatic zones of West Africa. Agricultural and Forest Meteorology. 1988;**42**:295-305. DOI: 10.1016/0168-1923(88)90039-1

[31] Salack S, Klein C, Giannini A, Sarr B, Worou N, Belko N, et al. Global warming induced hybrid rainy seasons in the Sahel. Environmental Research Letters. 2016;**11**:104008. DOI: 10.1088/1748-9326/11/10/104008

[32] Trenberth KE. Attribution of climate variations and trends to human influences and natural variability. WIREs Climate Change. 2011;**2**:925-930. DOI: 10.1002/wcc.142

[33] Trenberth KE, Dai A, Rasmussen R, Parsons D. The changing character of precipitation. British American Meteorological Society. 2003;**84**:1205-1217. DOI: 10.1175/BAMS-84-9-1205

[34] Taylor CM et al. Frequency of extreme Sahelian storms tripled since 1982 in satellite observations. Nature. 2017;**544**:475-478

[35] Panthou G, Vischel T, Lebel T. Recent trends in the regime of extreme rainfall in the Central Sahel. International Journal of Climatology. 2014;**34**:3998-4006. DOI: 10.1002/joc.3984

[36] Chagnaud G, Panthou G, Vischel T, Lebel T. A synthetic view of rainfall intensification in the west African Sahel. Environmental Research Letters. 2022;**17**:044005. DOI: 10.1088/1748-9326/ac4a9c

[37] Berg A, de Noblet-Ducoudre N, Sultan B, Lengaigne M, Guimberteau M. Projections of climate change impacts on potential C4 crop productivity over tropical regions. Agricultural and Forest Meteorology. 2013;**170**:89-102. DOI: 10.1016/j.agrformet.2011.12.003

[38] Agossou DSM et al. Perception des perturbations climatiques, savoirs locaux et stratégies d'adaptation des producteurs agricoles béninois. African Crop Journal. 2012;**20**(Supplement):565-588

Section 5

Slopes in Motion: Exploring Landslides as Natural Hazards

Chapter 19

The Role of Polarization Analysis in Reducing Natural Hazard

Mohsen Kazemnia Kakhki and Webe João Mansur

Abstract

The complexity of the regions where the landslide occurred requires a detailed survey of the site response properties of these regions. Ambient noise analysis is a common and nondestructive approach that provides more detailed information on site resonance properties characterized by directional variations. Obtaining geological information from seismic data motivates researchers to innovate and improve efficient tools for seismic wave processing. Polarization-based methods have received much attention regarding their capability to discriminate between different phases of the seismic wave based on their polarities. Using polarization filtering to extract Rayleigh wave ellipticity provides more detailed information on site resonance properties characterized by directional variations. The outputs showed better performance of the method in terms of stability and reliability of the results. Indeed, it enabled the detection of site resonance characteristics that were previously undetectable by classic Nakamura method.

Keywords: site response, landslide, Rayleigh wave ellipticity, ambient noise, polarization analysis, time-frequency decomposition

1. Introduction

The annual destruction caused by landslides inspires near-surface engineers to discover strategies to prevent them. Among the mechanisms responsible for slope instability, earthquakes require special attention due to their ability to trigger a large number of simultaneous mass movements, complicating the management of seismic crises. Slopes susceptible to landslides exposed to earthquakes can be more easily mobilized in the presence of ground motion amplification due to subsoil physical properties [1] or topography [2], characterized by directional variations of resonance phenomenon. The impacts of directional resonance on soil slopes have been investigated by analyzing seismic ground motion at landslide-prone slopes [3]. For instance, Burjánek et al., in [4] found that the largest ground motion amplification was directed along the steepest slope in a survey conducted at the Randa rock slope in Switzerland.

The standard spectral ratio (SSR) approach, which records seismic events at research sites and a nearby reference site to compare site response characteristics, is the most straightforward method for studying site response. The lack of long-term accelerometers on the rough topography of landslide-prone areas to record ground motion has increased the need for alternative approaches to studying site effects. The high expense and limitations of studying subsurface structure by boreholes and other geophysical methods motivated the researchers to choose ambient noise analysis as a feasible and cost-effective alternative tool for monitoring landslide-prone slopes [5–7]. The ambient noise analysis provides valuable information regarding the effect of subsurface structure on site response at observation sites [8]. Nakamura method [9], also known by the acronym HVSR (horizontal-to-vertical spectral ratio), is the most popular method for ambient noise analysis. It analyzes the spectral ratios, H/V, between the horizontal and vertical components of ambient noise recordings, disclosing site-specific resonance frequencies via H/V curve peaks. Directional resonance can also be defined by the analysis of azimuthal changes in spectral ratios [10].

It has been shown that Nakamura method cannot precisely define resonance frequencies when amplification occurs at several frequencies, particularly in sites with complex geology [7, 8, 11]. To effectively analyze local site amplification, a complete understanding of local soil characteristics such as shear-wave velocities and sediment thickness is required. Surface wave dispersion curves generated from active or passive seismic array data can provide valuable information on soil properties. However, it often renders reliable information for frequencies higher than the resonance frequency, and the inverted velocity profiles do not give information on deeper structures. Several studies have shown that the S-wave velocity profile may be calculated by inverting the H/V spectra [12, 13] or by inverting the H/V spectra and dispersion curves [13, 14]. Since both Rayleigh and Love waves contribute to the H/V spectrum, the extraction of these waves can be effective in the inversion process [15, 16].

Polarization methods are the efficient approaches that have the ability to discriminate between different phases of the waves and extract Rayleigh waves depending on their polarities [17]. Polarization is a seismic wavefield property that describes the particle motion, such as linear polarization in P-waves or the elliptical particle motion in Rayleigh waves. The ellipticity curve calculated by the polarization methods is more similar to the theoretical curve than the classic Nakamura H/V curve [18]. In the following, some of these polarization analysis methods are explained with the aim of improving the reliability of the results in seismic hazard assessments.

2. Polarization analysis methods

Polarization techniques are mostly used in the time or frequency domains, depending on the wave content and event type. Time-domain approaches are more useful in the event of the near arrival of waves in time, such as wave conversions at crustal discontinuities. The direction of particle motion and rectilinearity of the three-component (3C) earthquake data were defined using the correlation approach in time-domain polarization filtering proposed by Montalbetti and Kanasewich in [19]. Although their nonlinear polarization filter enhanced the shear and compressional phases, it ignored the fact that the waveforms have distinct polarizations for various frequencies. Addressing this problem, Samson in [20] studied particle polarization as a function of frequency, and after that, Samson and Olson in [21] developed a frequency domain polarization filter. In real-world seismograms, events with different frequencies and polarization characteristics may overlap, while events with identical frequencies but distinct polarization content may occur at different times. Therefore, the researchers became interested in time-frequency (TF) methods that consider polarization variations of 3C signals [22–26].

2.1 Time-domain polarization analysis

According to certain research (e.g., [27, 28]), ambient noise contains a mix of several types of waves (body, Rayleigh, and Love waves). Their proportional percentage is determined by site circumstances and source characteristics, which may differ from one example to the next. As a result, assumptions about this fraction make the inference of the site resonance features from H/V curves more challenging. Generally, extracting Rayleigh waves from a recording can yield the following results: (I) an estimation of the S-wave resonance frequencies; (ii) identification of site response directivity and orientation from preferential polarization directions; and (iii) some constraints for velocity modeling using the ellipticity curves. Since horizontal components of ambient noise can contain a large or even dominating fraction of Love waves (cf. [29]), the H/V ratios can be different from the anticipated one for Rayleigh waves. Therefore, some studies tried to separate Rayleigh waves from Love waves in noise recordings by selecting signal parts with considerable energy in the vertical component [30, 31]. On the same lines of inquiry, Del Gaudio in [11] provided a technique for identifying wave packets with Rayleigh-type particle motion inside ambient noise recordings by identifying instantaneous polarization features.

2.1.1 Methodology

Considering the multicomponent signal, its analytic transformation has the form

$$\mathbf{u}_c(t) = \mathbf{u}(t) + j\hat{\mathbf{u}}(t) = \mathbf{A}(t)e^{j\boldsymbol{\varphi}(t)} \tag{1}$$

where $\mathbf{u}(\mathrm{t})$ represents the real component of a vector on a complex plane, j is the imaginary unit, and $\hat{\mathbf{u}}(t)$ is the Hilbert transform of $\mathbf{u}(\mathrm{t})$. The vector modulus $\mathbf{A}(\mathrm{t})$ is the instantaneous amplitude, and phase $\boldsymbol{\varphi}(\mathrm{t})$ is the real phase. Consider $\mathbf{a}(\mathrm{t})$ and $\mathbf{b}(\mathrm{t})$ as the major and minor semi-axes of the instantaneous elliptical trajectories, respectively: these two vectors can be derived by calculating the phase shifts, Φ_0, as defined in [23].

$$\begin{aligned} \mathbf{a}_i(t) &= \operatorname{Re}\left[\mathbf{A}_i(t).e^{j(\boldsymbol{\varphi}_i(t)-\phi_0(t))}\right] \quad for i = E, N, Z \\ \mathbf{b}_i(t) &= \operatorname{Re}\left[\mathbf{A}_i(t).e^{j\left(\boldsymbol{\varphi}_i(t)-\phi_0(t)-\frac{\pi}{2}\right)}\right] \quad for i = E, N, Z \end{aligned} \tag{2}$$

In 3D, the orientation of the instantaneous elliptical trajectory can be characterized by the planarity vector, $\mathbf{p}(\mathrm{t})$, which is defined as the vector product of $\mathbf{a}(\mathrm{t})$ and $\mathbf{b}(\mathrm{t})$ [32]. When the $\mathbf{a}(\mathrm{t})$ amplitude exceeds the $\mathbf{b}(\mathrm{t})$ amplitude, the elliptical trajectory degenerates into a rectilinear ground motion. According to Schimmel and Gallart in [33], the rectilinearity (Eq. 3) can be used to assess the proximity to such a state.

$$\mathbf{rl}(t) = 1 - \frac{|\vec{\mathbf{b}}(t)|}{|\vec{\mathbf{a}}(t)|} \tag{3}$$

The rectilinearity takes values ranging from 0 to 1, which relate to precisely circular and linear polarizations, respectively.

To define Rayleigh wave packets and their features (ellipticity and polarization direction), an instantaneous polarization analysis can be employed on time series of ambient noise recordings. The analytic transformation can be applied to time series by filtering them through several narrowband filters with varying central frequencies (f_c) to see the variation of Rayleigh wave characteristics with frequency.

The analytic transformation of the two horizontal components permits calculating the maximum amplitude Hmax of ground motion on the horizontal plane and its orientation for each frequency. The analytic transformation of the vertical component yields a measure of its amplitude V, and dividing Hmax by V yields an estimate of the ratio between the amplitudes of the horizontal and vertical components of ground motion, as well as the relative azimuth, at each instant and for the analyzed frequency.

Generally, a considerable dispersion of H/V values may be predicted as a result of the overlapping of various types of waves arriving from a geographically spread distribution of distinct noise sources. However, if Rayleigh waves can be extracted from the noise recording, they should exhibit more coherent characteristics, which can be correlated to site response properties. To detect such signals, the analytic transform is applied to all 3C, yielding the vectors $\mathbf{a}(t)$ and $\mathbf{b}(t)$, as well as the planarity vector $\mathbf{p}(t)$ and the rectilinearity vector $\mathbf{rl}$. Parts of a noise recording containing Rayleigh waves are then recognized based on the presence of a Rayleigh-type polarization of ground motion, that is, if (i) the instantaneous trajectories lie on a vertical plane (which indicates a horizontal $\mathbf{p}(t)$) and (ii) one of $\mathbf{a}(t)$ and $\mathbf{b}(t)$ is horizontal and the other is vertical. If a(t) is horizontal, H/V will be more than 1; otherwise, H/V will be less than 1.

Since Rayleigh waves can be contaminated by noise effects from other wave types, it is important to consider deviations of $\mathbf{p}(t)$, $\mathbf{a}(t)$, and $\mathbf{b}(t)$, which should not surpass angular thresholds. Theoretically, instantaneous polarization analysis can recognize single Rayleigh wave samples, and the existence of isolated samples matching the identification requirements among a substantial amount of data can be completely coincidental. Therefore, if a coherent form of polarization is observed in more consecutive samples, more reliable detection of Rayleigh waves may be achieved.

Finally, HVIP values derived from Rayleigh-type wave packets (denoted as $HVIP_R$) are averaged within each filtered time series, and a curve of such HVIP values as a function of filtering central frequency fc represents an estimate of Rayleigh wave ellipticity as a function of frequency. The scatter of instantaneous $HVIP_R$ values around their average represents the uncertainty impacting ellipticity estimations. Analyzing the distribution of $HVIP_R$ azimuths reveals the presence of preferred signal polarization, which might indicate the presence of site response directivity.

2.1.2 Synthetic example

A 900-s synthetic ambient noise recording with a sampling interval of 0.01 s on a model introduced in [34] was generated by the synthetic seismogram codes [35]. The recording was analyzed using classic Nakamura and HVIP methods to compare the H/V curves derived from these methods. The code gpell [36] was applied to calculate the theoretical Rayleigh wave ellipticity through 1D modeling and hence to verify the reliability of the results. Based on the model used in this example, it is expected to observe two main peaks on the H/V curve for the fundamental and first higher mode of Rayleigh waves at 2 and 6 Hz.

Comparing the results represents that both methods detect the fundamental mode of the Rayleigh waves, while only HVIP detects the first higher mode peak. Multiple peaks can be seen on the curves derived from HVIP, which can be related to different vibration modes. It can be inferred that H/V curves obtained by the use of the

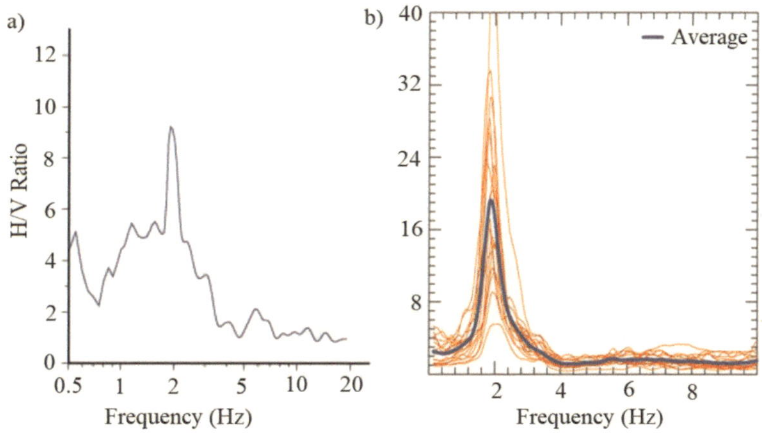

Figure 1.
a) Synthetic H/V curves derived from a) HVIP, and b) Nakamura methods.

	H (km)	V_p (km/s)	V_s (km/s)	RHO (GM/CC)	Q_p	Q_s
Layer 1	0.025	1.35	0.2	1.9	50	25
Layer 2	0	2.0	1.0	2.5	100	50

Table 1.
Synthetic crustal model.

geometric mean, similar to HVSR method, can underestimate the H/V ratio, and ignore the secondary peaks (**Figure 1** and **Table 1**).

It is known that Nakamura method can work properly under the condition that both site response and noise source are distributed isotropically. In some cases where noise sources insufficiently energize the frequency band of maximum site amplification, HVIP outperforms HVSR since the former at least can show some reliable peaks in some particular directions, while Nakamura method shows small peaks not only at frequencies that are not significantly amplified but also with the underestimate of amplification. The main discrepancy between HVIP and Nakamura method is that the HVSR averages the H/V values over the length of the recording, in which different types of waves can come along with Rayleigh waves.

The main objective of applying HVIP is to study the site response more accurately due to the ambiguous nature of ambient noise. There are some similarities between the HVIP method and other techniques of polarization analysis. For instance, Jurkevics in [22] applied polarization analysis to site response studies based on the analysis of earthquake recordings. The instantaneous properties of the particle motion are calculated in [37], which show the most similarities to the HVIP approach, to analyze earthquake recordings, while Burjánek et al., performed the method on ambient noise data [4]. Applying polarization analysis simultaneously to the 3C of a recording to identify Rayleigh wave packets makes the HVIP approach different from [4].

2.1.3 Site response directivity

Site response directivity is a directional resonance characteristic that has been shown to accurately reflect resonance frequencies associated with topographic amplification, anisotropy in slope materials, and mechanical features [38]. In studying the site response directivity in landslide areas, Del Gaudio et al., in [8] were the pioneers who presented a comprehensive directional analysis of ambient noise data to study the differences in H/V ratios estimated along orthogonal directions. Despite the numerous case studies in landslide areas, the origin of directional resonance has remained uncertain due to the difficulty in recognizing the involved factors. The anisotropy of shear wave velocity generated by landslide mobilization might represent directed amplification; however, the directional amplification is not always in the direction of the landslide, and other factors, such as topography, geology, lithology, and tectonic features, can be effective in generating this phenomenon [7, 39–41]. It is important to note that some sites are more likely to experience stronger shaking in a specific direction and frequency range, which is of interest to earthquake engineers. For instance, the similar direction of slope and maximum amplification can amplify the influence of directivity on mass movement [7, 41]. For these reasons, studying site response directivity has become increasingly important in seismic hazard assessments.

2.1.4 Real example

On June 1, 2014, an earthquake with a magnitude of 3.6 followed by five aftershocks occurred on two sides of an active fault in the vicinity of the Nargeschal village, located in the north of Iran, and then a landslide occurred 3 days after this seismic sequence [7]. Based on the landslide hazard zonation map of the province given by the Geological Survey and Mineral Exploration of Iran, Nargeschal village lies in a high-risk zone, which makes landslides the most serious geological hazard in the study region. Slope steepness (nearly 30°) and the presence of soft and sticky soils covering relatively hard and brittle rocks create conditions for gravity-driven motions. Furthermore, the region's tectonic activity, which retains the dynamic stresses discharged by seismic activity, can operate as both a triggering and predisposing factor for slope failures.

To evaluate the landslide and prevent future damages, ambient noise data were recorded and analyzed by the HVIP method. The results confirmed the role of landslides and structural features in the directional resonance detected in the study region. As an example, the polar diagrams that show the results of HVIP analysis at stations located on the head and foot of the landslide are depicted in **Figure 2**. The results represent a preferential directivity that is perpendicular to the tension cracks. The reason for the observed directivity is a mass movement, which induces the opening of fissure or fracture systems transverse to the sliding direction and makes the material more deformable in such a direction.

This example confirms the necessity of an advanced ambient noise analysis technique on sites with complex lithology, where the classic Nakamura method could have difficulty revealing site resonance properties. The detailed HVIP analysis allows us to have a more reliable hazard zonation map to identify the areas that are more susceptible to future seismically induced mobilization. More details about this example can be found in [41].

2.2 TF polarization analysis

TF decompositions have a significant role in designing TF polarization filters. Although several TF decompositions are available in the literature [42–47], the resolution of these decompositions is remarkably effective in designing polarization filters.

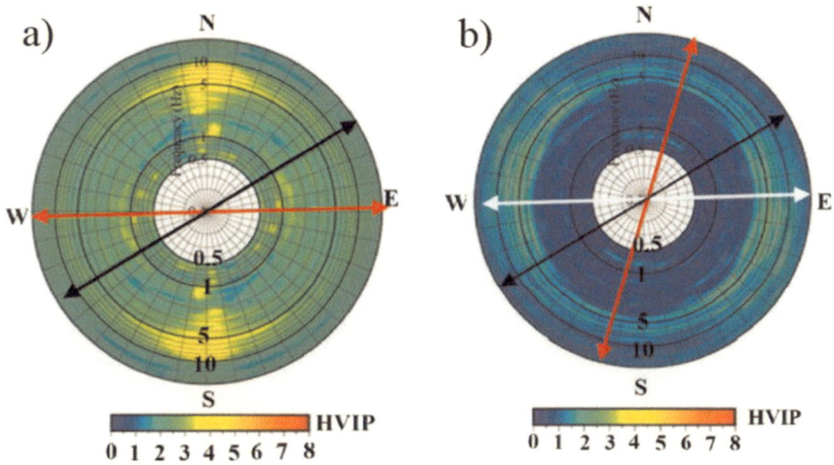

Figure 2.
Polar diagrams depicting the results of the ambient noise analysis processed via the HVIP method for the stations located on a) the head and b) the foot of the landslide region (the black arrow denotes the slope direction, the red and white arrows are the direction of tension cracks and old landslides, respectively).

Wavefields are typically recorded in geophysical research using 3C sensors that transfer the same information with various energies. Since real geophysical data are typically contaminated by noise, signals with lower energy may be weakened while being decomposed by high-resolution methods. This is critical in analyzing all 3C of a signal at the same time since the support for any specific part of a wavefield should be consistent in all 3C TF maps and just vary by a scaling factor. Kakhki et al., in [48] proved that employing a single-component sparse adaptive S transform for 3C data results in suboptimal decompositions in the TF domain. Therefore, they proposed the 3C sparse adaptive ST (3C-SAST), which employs the group sparse constraint to lock the TF supports of all 3C of a signal together and retrieves the spectral paths of weak parts of waveforms without deformation. The invertibility and resolution flexibility make the proposed method a viable tool for integrating the TF relationship in polarization analysis. Then, the polarization attributes can be calculated directly from the TF maps derived from the 3C-SAST without using Eigen analysis or other complicated methods [25, 49]. These attributes, which have a higher resolution compared with other available decompositions, can be used to construct an efficient TF filter to remove the undesired waves.

2.2.1 Methodology

Polarization has an intuitive and straightforward description of a monochromatic wave propagating in 3D space. Assume the 3C seismic data, $[N(t), E(t), Z(t)]$, and the TF transform of these components are $[N(t,f), E(t,f), Z(t,f)]$, whose real and imaginary parts are $N_R(t,f)$, $N_I(t,f)$, $E_R(t,f)$, $E_I(t,f)$, $Z_{R,}(t,f)$, and $Z_I(t,f)$, respectively. The ellipse parameters, schematically depicted in **Figure 3**, can be calculated directly from the real and imaginary parts of the 3C TF spectra. For the sake of brevity, only the results are presented, and more explanation can be found in [49].

$$a = \frac{1}{\sqrt{2}}\sqrt{A + \sqrt{B^2 + C^2}}$$

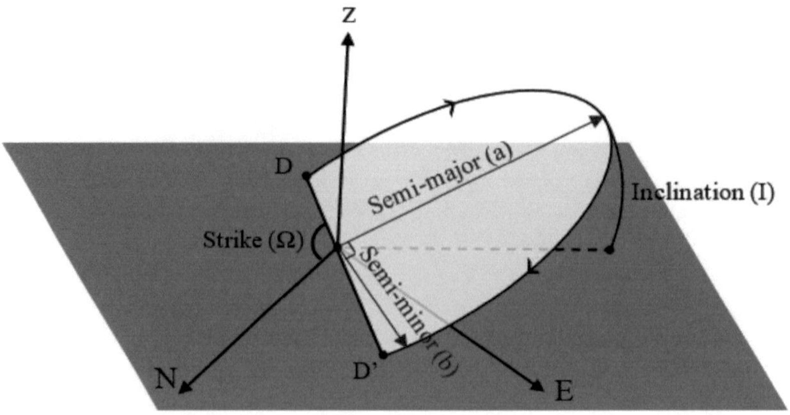

Figure 3.
The ellipse of prograde motion is plotted together with the polarization attributes. D *and* D′ *are the intersection of the ellipse with the horizontal plane,* a *and* b *are the ellipse major and minor semi-axes, respectively, the angle* I *is the inclination of the ellipse plane concerning the horizontal plane. The angle* Ω *is that between* DD' *and the* x *axis, measured on the horizontal plane.*

$$b = \frac{1}{\sqrt{2}}\sqrt{A - \sqrt{B^2 + C^2}}$$

$$I = \arctan\left\{\frac{\left[(Z_R Y_I - Z_I Y_R)^2 + (Z_R X_I - Z_I X_R)^2\right]^{1/2}}{(Y_R X_I - Y_I X_R)}\right\}$$

$$\Omega = \arctan\left\{\frac{(Z_R Y_I - Z_I Y_R)}{(Z_R X_I - Z_I X_R)}\right\} \tag{4}$$

where

$$\begin{aligned} A &= X_R^2 + X_I^2 + Y_R^2 + Y_I^2 + Z_R^2 + Z_I^2 \\ B &= X_R^2 - X_I^2 + Y_R^2 - Y_I^2 + Z_R^2 - Z_I^2 \\ C &= -2(X_R X_I + Y_R Y_I + Z_R Z_I) \end{aligned} \tag{5}$$

2.2.2 Earthquake example

To evaluate the role of a high-resolution TF decomposition in calculating the polarization attributes and to represent the efficiency of the TF polarization method in detecting different wave types, we chose an earthquake example that occurred on June 3, 2003, and was recorded at the PEMO station, in Canada (45.68° N, 77.25° W), with a magnitude of 6.9 and an epicentral distance of 7124 km (**Figure 4**).

Figure 5 depicts the polarization attributes estimated by employing the 3C-SAST and SAST, which represent the higher resolution of the 3C-SAST in providing polarization attributes in the TF domain. The resolution of the attributes is an important

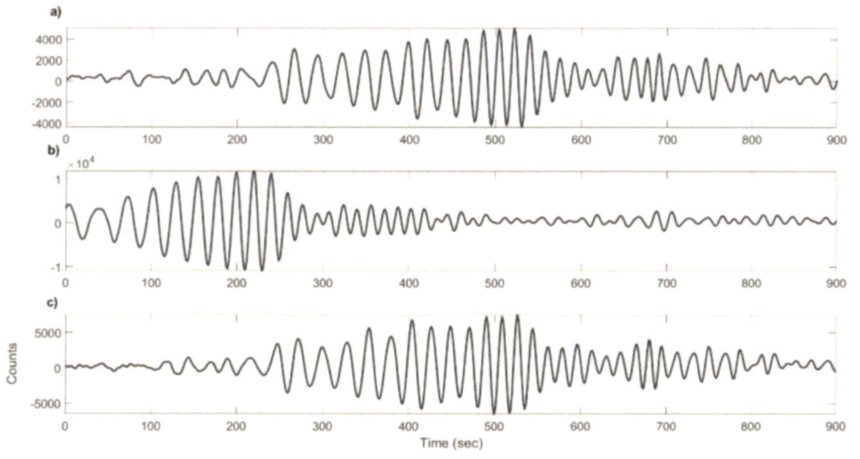

Figure 4.
The earthquake recorded at PEMO station. a) Radial, b) transverse, and c) vertical components.

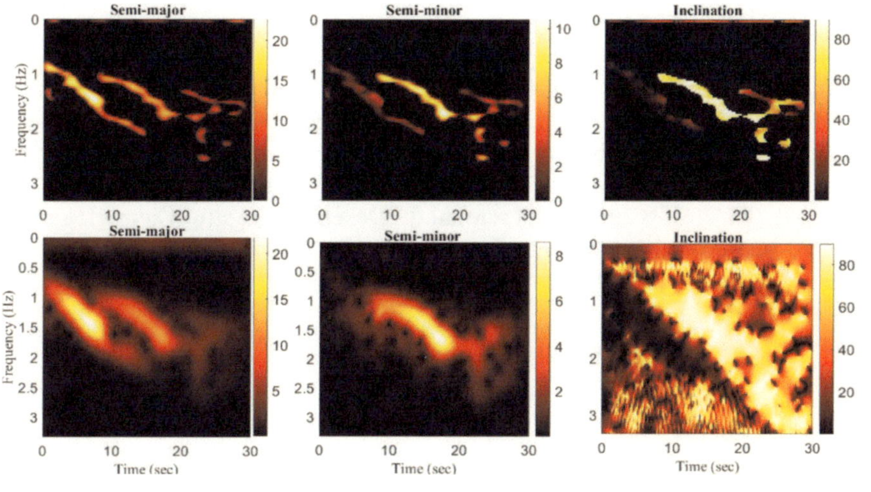

Figure 5.
The spectra of semi-major, semi minor, and the inclination of the seismogram in ***Figure 4*** *by 3C-SAST transform (upper plots) and SAST (lower plots).*

factor in designing the polarization filters since their lower resolution leads to inaccuracies in the distinction between various wave types, particularly in the presence of noise. In this example, the polarization attributes were used to detect Rayleigh and Love waves. For instance, Rayleigh waves can be observed more obviously in the semi-minor axis since this attribute is representative of the circular motion. On the other hand, the dip attribute shows a stable value of $\pi/2$ for the Rayleigh waves, while it has an unstable value between 0 and π for the Love waves, which is 0 in this example. Therefore, the initial high-energy arrival has Love wave features, but the second one has Rayleigh wave characteristics. Finally, after detecting the aimed wave type, which is Rayleigh wave in this example, a filter can be designed based on the polarization attributes to extract it.

3. Conclusions

Evaluating seismically induced vulnerability to slope failure has been a critical problem for geotechnical engineers, especially in metropolitan areas near landslide-prone zones. Slopes with a deep-seated landslide can be an amplification factor for seismic ground motion, with a maximum along the sliding direction. In such cases, slopes with high impedance contrast of layers throughout the 3D geometry of a landslide body may trap seismic waves and prolong shaking, making slopes more susceptible to seismically induced landslide reactivation. This phenomenon is more possible if shaking energy is polarized along the hill slope facing direction. The presence of more factors generating directional spectral peaks is a common characteristic of site response directivity phenomena. Ambient noise analysis is a practical and cost-effective method for studying site response on landslide-prone slopes. Despite the popularity of Nakamura method for ambient noise analysis, it has limitations at sites with complex geology. This limitation can be addressed by extracting Rayleigh waves from ambient noise recordings using polarization methods in the time, frequency, or TF domain. The synthetic and real-world examples proved the superiority of polarization methods over classical methods in studying site response in landslide areas.

Acknowledgements

The authors appreciate the funding for this study from the Fundação de Amparo à Pesquisa do Estado do Rio de Janeiro (FAPERJ) with grant numbers E-204.210/2021, and E-26/201.124/2021, Conselho Nacional de Desenvolvimento Científico e Tecnológico (CNPq) with grant numbers 305949/2019-5, and 422589/2021-7, and Centro de Pesquisas Leopoldo Américo Miguez de Mello (CENPES) under grant number SAP 4600571711.

Conflict of interest

The authors declare no conflict of interest.

Author details

Mohsen Kazemnia Kakhki[1,2*] and Webe João Mansur[1,2]

1 Modelling Methods in Engineering and Geophysics Laboratory (LAMEMO), COPPE, Federal University of Rio de Janeiro, RJ, Brazil

2 Department of Civil Engineering, COPPE, Federal University of Rio de Janeiro, RJ, Brazil

*Address all correspondence to: kazemnia@coc.ufrj.br

References

[1] Bourdeau C, Havenith HB. Site effects modelling applied to the slope affected by the Suusamyr earthquake (Kyrgyzstan, 1992). Engineering Geology. 2008;**97**(3–4):126-145. DOI: 10.1016/j.enggeo.2007.12.009

[2] Sepúlveda SA, Murphy W, Jibson RW, Petley DN. Seismically induced rock slope failures resulting from topographic amplification of strong ground motions: The case of Pacoima canyon, California. Engineering Geology. 2005;**80**(3–4):336-348. DOI: 10.1016/j.enggeo.2005.07.004

[3] Del Gaudio V, Wasowski J. Advances and problems in understanding the seismic response of potentially unstable slopes. Engineering Geology. 2011;**122** (1–2):73-83. DOI: 10.1016/j.enggeo.2010.09.007

[4] Burjánek J, Gassner-Stamm G, Poggi V, Moore JR, Fäh D. Ambient vibration analysis of an unstable mountain slope. Geophysical Journal International. 2010;**180**(2): 820-828. DOI: 10.1111/j.1365-246X.2009.04451.x

[5] Vella A, Galea P, D'Amico S. Site frequency response characterisation of the Maltese islands based on ambient noise H/V ratios. Engineering Geology. 2013;**163**:89-100. DOI: 10.1016/j.enggeo.2013.06.006

[6] Kakhki MK, Mansur WJ, Bamani B. Comparison of microtremor and electrical resistivity in detecting sliding surface. 78th EAGE Conference and Exhibition 2016. 2016;**2016**(1):1-5

[7] Kakhki MK, Peters FC, Mansur WJ, SadidKhoii A, Rezaei S. Deciphering site response directivity in landslide-prone slopes from ambient noise spectral analysis. Engineering Geology. 2020; **269**:105542. DOI: 10.1016/j.enggeo.2020.105542

[8] Del Gaudio V, Coccia S, Wasowski J, Gallipoli MR, Mucciarelli M. Detection of directivity in seismic site response from microtremor spectral analysis. Natural Hazards and Earth System Sciences. 2008;**8**(4):751-762. DOI: 10.5194/nhess- 8-751-2008

[9] Nakamura Y. A method for dynamic characteristics estimation of subsurface using microtremor on the ground surface. Quarterly Report of the Railway Technical Reasearch Institute. 1989;**30**:25-33

[10] Kakhki MK, Peters F, Mansour JW, Ghazvini SH. Detection of landslide direction based on HVNR method. 80th EAGE Conference and Exhibition 2018. 2018;**2018**(1):1-3

[11] Del Gaudio V. Instantaneous polarization analysis of ambient noise recordings in site response investigations. Geophysical Journal International. 2017;**210**(1):443-464. DOI: 10.1093/gji/ggx175

[12] Fäh D, Kind F, Giardini D. Inversion of local S-wave velocity structures from average H/V ratios, and their use for the estimation of site-effects. Journal of Seismology. 2003;7(4):449-467. DOI: 10.1023/B:JOSE.0000005712.86058.42

[13] Arai H, Tokimatsu K. S-wave velocity profiling by joint inversion of microtremor dispersion curve and horizontal-to-vertical (H/V) spectrum. Bulletin of the Seismological Society of America. 2005;**95**(5):1766-1778. DOI: 10.1785/0120040243

[14] Picozzi M, Parolai S, Richwalski SM. Joint inversion of H/V ratios and

dispersion curves from seismic noise: Estimating the S-wave velocity of bedrock. Geophysical Research Letters. 2005;**32**(11). DOI: 10.1029/ 2005GL022878

[15] Boore D, Toksöz M. Rayleigh wave particle motion and crustal structure. Bulletin of the Seismological Society of America. 1969;**59**(1):331-346

[16] Malischewsky FS, Peter G. Love's formula and H/V-ratio (ellipticity) of Rayleigh waves. Wave Motion. 2004; **40**(1):57-67

[17] Du Z, Foulger GR, Mao W. Noise reduction for broad-band, three-component seismograms using data-adaptive polarization filters. Geophysical Journal International. 2000;**141**(3):820-828. DOI: 10.1046/j.1365-246X.2000. 00156.x

[18] Kakhki MK, Peters F, Webe J, Ghazvini SH. Separation of Rayleigh wave from ambient noise data by instantaneous polarization. In: 24th European Meeting of Environmental and Engineering Geophysics. Vol. 2018, No. 1. European Association of Geoscientists and Engineers; 2018. pp. 1-5

[19] Montalbetti JF, Kanasewich ER. Enhancement of teleseismic body phases with a polarization filter. Geophysical Journal International. 1970;**21**(2):119-129. DOI: 10.1111/j.1365-246X.1970. tb01771.x

[20] Samson JC. Matrix and Stokes vector representations of detectors for polarized waveforms: theory, with some applications to teleseismic waves. Geophysical Journal International. 1977;**51**(3):583-603. DOI: 10.1111/ j.1365-246X.1977.tb04208.x

[21] Samson JC, Olson JV. Some comments on the descriptions of the polarization states of waves. Geophysical Journal International. 1980;**61**(1):115-129. DOI: 10.1111/j.1365-246X.1980. tb04308.x

[22] Jurkevics A. Polarization analysis of three-component array data. Bulletin of the seismological society of America. 1988;**78**(5):1725-1743

[23] Morozov IB, Smithson SB. Instantaneous polarization attributes and directional filtering. Geophysics. 1996; **61**(3):872-881. DOI: 10. 1190/1.1444012

[24] Diallo MS, Kulesh M, Holschneider M, Scherbaum F. Instantaneous polarization attributes in the time-frequency domain and wavefield separation. Geophysical Prospecting. 2005;**53**(5):723-731. DOI: 10.1111/ j.1365-2478.2005.00500.x

[25] Kakhki MK, Mansur WJ, Peters FC. Rayleigh wave separation using high-resolution time-frequency polarization filter. Geophysical Prospecting. 2020; **68**(7):2104-2118. DOI: 10.1111/1365-2478.12994

[26] Schimmel M, Stutzmann E, Gallart J. Using instantaneous phase coherence for signal extraction from ambient noise data at a local to a global scale. Geophysical Journal International. 2011; **184**(1):494-506. DOI: 10.1111/j.1365-246X.2010.04861.x

[27] Bonnefoy-Claudet S et al. H/V ratio: A tool for site effects evaluation. Results from 1-D noise simulations. Geophysical Journal International. 2006;**167**(2): 827-837

[28] Lunedei E, Albarello D. On the seismic noise wavefield in a weakly dissipative layered earth. Geophysical Journal International. 2009;**177**(3): 1001-1014

[29] Bonnefoy-Claudet S, Köhler A, Cornou C, Wathelet M, Bard P-Y. Effects of love waves on microtremor H/V ratio. Bulletin of the Seismological Society of America. 2008;**98**(1):288-300

[30] Fäh D, Kind F, Giardini D. A theoretical investigation of average H/V ratios. Geophysical Journal International. 2001;**145**(2):535-549. DOI: 10.1046/j.0956-540X.2001.01406.x

[31] Poggi V, Fäh D, Burjanek J, Giardini D. The use of Rayleigh-wave ellipticity for site-specific hazard assessment and microzonation: application to the city of Lucerne, Switzerland. Geophysical Journal International. 2012;**188**(3):1154-1172. DOI: 10.1111/j.1365-246X.2011.05305.x

[32] Schimmel M, Gallart J. The use of instantaneous polarization attributes for seismic signal detection and image enhancement. Geophysical Journal International. 2003;**155**(2):653-668. DOI: 10.1046/j.1365-246X.2003.02077.x

[33] Schimmel M, Gallart J. Degree of polarization filter for frequency-dependent signal enhancement through noise suppression. Bulletin of the Seismological Society of America. 2004;**94**(3):1016-1035. DOI: 10.1785/0120030178

[34] Asten MW. Comment on 'microtremor observations of deep sediment resonance in metropolitan Memphis, Tennessee' by Paul Bodin, Kevin smith, Steve Horton and Howard Hwang. Engineering Geology. 2004;**72**(3–4):343-349. DOI: 10.1016/j.enggeo.2003.09.001

[35] Herrmann RB. Computer programs in seismology: An evolving tool for instruction and research. Seismological Research Letters. 2013;**84**(6):1081-1088. DOI: 10.1785/0220110096

[36] Wathelet M. Array Recordings of Ambient Vibrations: Surface-Wave Inversion, PhD thesis. Belgium: Li'ege University; 2005

[37] Vidale JE. Complex polarization analysis of particle motion. Bulletin of the Seismological society of America. 1986;**76**(5):1393-1405

[38] Chávez-García FJ, Sánchez LR, Hatzfeld D. Topographic site effects and HVSR. A comparison between observations and theory. Bulletin of the Seismological Society of America. 1996;**86**(5):1559-1573

[39] Del Gaudio V, Luo Y, Wang Y, Wasowski J. Using ambient noise to characterise seismic slope response: the case of Qiaozhuang peri-urban hillslopes (Sichuan, China). Engineering Geology. 2018;**246**:374-390. DOI: 10.1016/j.enggeo.2018.10.008

[40] Kleinbrod U, Burjánek J, Fäh D. Ambient vibration classification of unstable rock slopes: A systematic approach. Engineering Geology. 2019;**249**:198-217. DOI: 10.1016/j.enggeo.2018.12.012

[41] Kakhki MK, Del Gaudio V, Rezaei S, Mansur WJ. Directional variations of site response in a landslide area using ambient noise analysis via Nakamura's and polarization-based method. Soil Dynamics and Earthquake Engineering. 2021;**141**:106492. DOI: 10.1016/j.soildyn.2020.106492

[42] Lu W, Li F. Seismic spectral decomposition using deconvolutive short-time Fourier transform spectrogram. Geophysics. 2013;**78**(2):V43-V51

[43] Cohen L. Time-Frequency Analysis. Vol. 778. New Jersey: Prentice Hall; 1995

[44] Daubechies I. The wavelet transform, time-frequency localization and signal analysis. IEEE Transactions on Information Theory. 1990;**36**(5): 961-1005

[45] Stockwell RG, Mansinha L, Lowe RP. Localization of the complex spectrum: the S transform. IEEE transactions on signal processing. 1996;**44**(4):998-1001. DOI: 10.1109/78.492555

[46] Khakhki MK, Moghaddam PP, Yazdanpanah H, Mansur WJ. High-resolution time-frequency hilbert transform using sparsity-aware weighting function. Earth Science Informatics. 2021;**14**(3):1197-1212. DOI: 10.1007/s12145-021-00628-z

[47] Kazemnia Kakhki M, Aghazade K, Mansur WJ, Peters FC. Seismic Attributes via robust and high-resolution seismic complex trace analysis. Acta Geophysica. 2020;**68**(6):1689-1701. DOI: 10.1007/s11600-020-00499-w

[48] Kakhki MK, Mokhtari A, Mansur WJ. Three-Component Sparse S Transform. In: IEEE Transactions on Geoscience and Remote Sensing. Vol. 60. 2022. pp. 1-7. Art no. 5922907. DOI: 10.1109/TGRS.2022.3219420

[49] Pinnegar CR. Polarization analysis and polarization filtering of three-component signals with the time—frequency S transform. Geophysical Journal International. 2006;**165**(2): 596-606. DOI: 10.1111/j.1365-246X.2006.02937.x

Chapter 20

Landslide Assessment and Hazard Zonation in the Birbir Mariam District, Gamo Highlands, Rift Valley Escarpment, Ethiopia

Yonas Oyda and Hailu Regasa

Abstract

The current research focused on landslide assessment and hazard zonation in the Birbir Mariam district of the Gamo highlands. The study examined landslide causative factors and used the slope susceptibility evaluation parameter to create a landslide hazard zonation covering an area of 110 km^2. The landslide hazard zonation was classified using facet-wise observation. As a result, the intrinsic and external causal parameters of score schemes have been held responsible for slope instability. Inherent causative elements consist of slope geometry, slope material (rock/soil), structural discontinuities, land use/land cover, and groundwater conditions. Rainfall and human interest have seemed as external elements. The intrinsic and external triggering elements for every facet (a total of 106) were rated for their contribution to slope instability. Finally, an evaluated landslide hazard value was calculated and classified into three landslide hazard classes. According to the findings, the area has a high hazard zone of 18.87% (20.76 km^2), a moderate hazard zone of 54.72% (60.19 km^2), and a low hazard zone of 26.41% (29.05 km^2).

Keywords: Ethiopia, landslide, hazard zonation, landslide evaluation, slope stability

1. Introduction

Landslides are a series of events in which a mass of rocks, soil, or debris slides down a slope due to gravitational pull, the mechanisms include sliding, falling, or flowing material down a slope [1–3]. Landslides are one of the most common geological hazards in the world, with a high incidence, a wide range of distribution, and catastrophic severity, resulting in numerous fatalities each year [3–9].

Landslides are caused by inherent causative parameters such as slope geometry, slope material, structural discontinuities, land use, land cover, and groundwater conditions, which define the unfavorable stability conditions within the slope [4, 10, 11]. External causative factors, such as rainfall, volcanism, seismic motion, and human activities, are also relatively variable or dynamic, temporary, and forced by upcoming events [11–14], the slope will be prone to instability when the slope morphometry is steep, and a chance of landslide increases with an increase in slope steepness through landslide occurs in all slope. Landslides such as rock-fall, toppling, and rockslides/ avalanches are common in the area because the slope material has been covered by highly fractured bedrock such as basalt and ignimbrite [15–17].

Discontinuities in the slope, such as bedding, joints, and faults, are potentially weak planes that affect slope stability [9, 15, 18]. As a result, fractured rocks have lower strength than intact rocks. They are the most vulnerable component of slope geology, and knowledge of their orientation, spacing, continuity, roughness, separation, and type of filling material, as well as slope angle, slope direction, and strength along such potential weak planes, is essential. Slope stability is influenced by land use/land cover, and the material's shear strength is reduced by subsoil conditions in the slope [17, 19, 20].

Landslides are one of Ethiopia's most common natural disasters, and the current study area of the Birbir Miriam district is one of the most vulnerable. A variety of factors affect slope stability, including slope angle, lithology, soil type, and hydrologic conditions [21]. Deforestation, changes caused by the construction of engineering structures on the slope, road construction undercutting the toe of the slope, and other human activities all contribute to potential factors. Variations in human activity on the cliff-side can make the slope less stable [4, 22–25]. Landslides harm infrastructure (houses, roads, buildings, irrigation, canals, and so on), as well as cause geomorphological damage, serious injuries, and the death of humans and animals. For minimizing damage to infrastructure, homes, cultivated lands, and human lives, well-organized landslide hazard zonation is critical. These slope failures zonation maps become significant when they are used by decision-makers in land use planning, landslide prevention, and mitigation methods [8, 15, 26–30].

The goal of a landslide study is to determine the essence of predisposition as well as the outcomes for human life, land, roads, buildings, and other resources [4, 31]. As a result, locating landslide-prone areas is critical for ensuring human safety and avoiding negative consequences for regional and national economies [1, 9, 32–36].

Landslides are common in the Birbir Mariam district along riverbanks, slope toes, and slope faces. Landslides are responsible for any losses and also affect much of the farmland and farmers' income. Moreover, landslides in the study area cause significant damage to properties and massive destruction, particularly in Zala Gutisha and Waro localities, among the heights-prone areas.

2. Materials and methods

2.1 Study area

The current study area is in the Birbir Mariam district of Ethiopia's Gamo highlands, in the rift valley escarpments. It is approximately 450 km from Addis Ababa's capital city and 47 km from the Zonal capital of Arba Minch town. It has a total area of 110 km^2 and is geographically bounded (UTM Zone 37 N) by latitudes ranging from 692,000 to 702,000 m N and longitudes ranging from 342,000 to 360,000 m E. The area can be reached from Arba Minch town via the Chencha-Ezo main road and the Birbir Mariam gravel road (**Figure 1**).

The study area's geological setting has a significant impact on the occurrence of landslides. Since the Oligocene, volcanoes have been active in the southern part of the

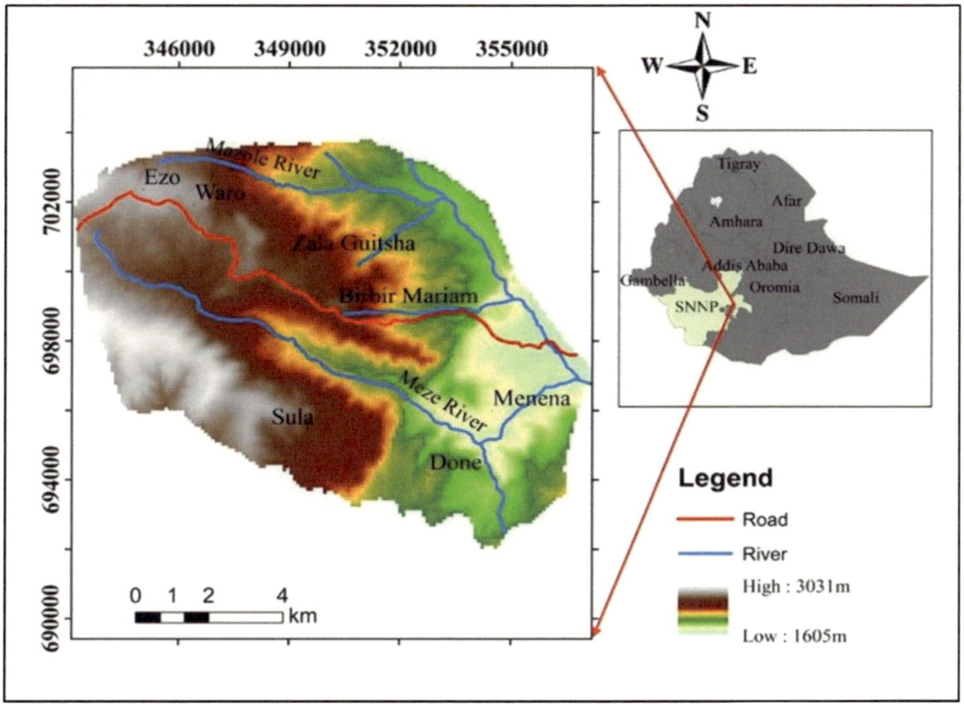

Figure 1.
Location map of Birbir Mariam district.

Main Ethiopian Rift (MER), which includes the Ganjuli graben (Lake Abaya), and the western side of Lake Abaya, which includes the plateau and the Chencha escarpment [37–39]. The geology of the study area is dominated by pre-rift and post-rift deposits. Thereby, the stratigraphy [38, 40] has the following significant units in the Galena basin and north Abaya basin. These are ignimbrite, unwelded tuff units, early flood basalts, alkaline basalt intermediate flows, pyroclastic rocks, Pleistocene basalt, trachyte, and rhyolites.

The geology of the research area can be mapped using field investigations. Basalt, tuff, and Ignimbrite are among the geologic units found in the study area (rock units). The measurement's rock units can be present along rivers, road cuts, and natural hillsides. Overall, the dominance of destructive materials, basalt, and ignimbrite is a crucial feature of the lithologies of this region (**Figure 2**).

2.2 Materials

The Ethiopia Mapping Agency's toposheet number (0637 D1) at a scale of 1:50,000 was used in the current study to demarcate the study area, and a land facet map was created using the topographic map. The slope morphometry and relative relief map were extracted using a DEM (digital elevation model) with a spatial resolution of 12.5 m from the Alaska Satellite Facility site. On January 10, 2019, cloud-free optical satellite data acquired by Landsat-8 Operational Land Imager (OLI) with path-row numbers 170-051 was also entered into the following portal (http://earthexplorer.usgs.gov/) and used to develop land use and land cover map of the current study area (**Figure 3**).

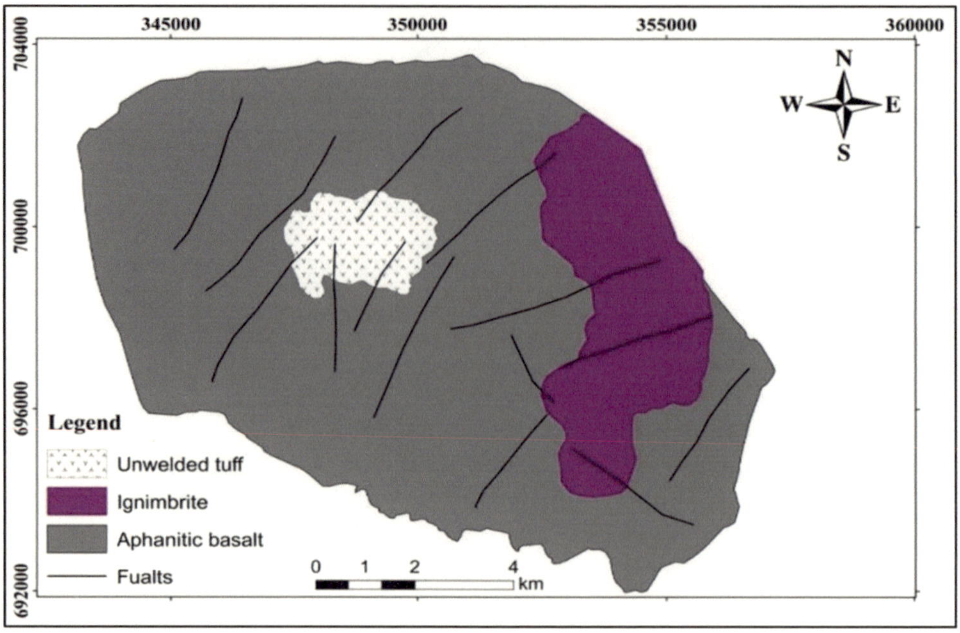

Figure 2.
Geological map of the study area.

2.3 Methods

The entire research area was initially divided into 106 slope land facets (**Figure 4**). In terms of slope inclination and slope direction, a land facet is defined as a land unit with more or less uniform slope geometry [4]. The slope facets were delineated using topographical maps. Facets were defined by major and minor hill ridges, primary and secondary streams, and other topographical undulations [4, 11]. The prepared facet map was then used as a base map for collecting data on the various causative factors. Lithology, structural discontinuity with slope, slope morphometry, relative relief, land use/land cover, groundwater-surface drops, rain-induced index, and human activities are among the factors that contribute to slope instability [4, 11, 12, 41]. As a result, the susceptibility evaluation parameter rating was assigned for each causative factor to get evaluated landslide hazard.

The total maximum susceptibility evaluation parameter rating for the various causative factors is 15. The maximum value was accounted for slope morphometry, groundwater situation, and seismicity 2.0. Land use land cover, rainfall-induced surface index, and human activity contributed to a maximum susceptibility evaluation parameter rate of 1.5. Furthermore, relative relief and slope geo-material each contribute at a rate of 1.0. Structured discontinuity, on the other hand, contributes a maximum susceptibility evaluation parameter rate of 2.50 (**Table 1**). The evaluated landslide hazard is the total sum of susceptibility evaluation parameter ratings for all causative factors; therefore, the greater the value of the susceptibility evaluation parameter, the greater the degree of hazard.

The slope susceptibility evaluation parameter was developed by [11] by taking intrinsic (inherent) and external landslide causative factors. Furthermore, the total prospect of instability was established by assessing landslide hazard, and it was determined facet-by-facet for which observations and investigations were made during

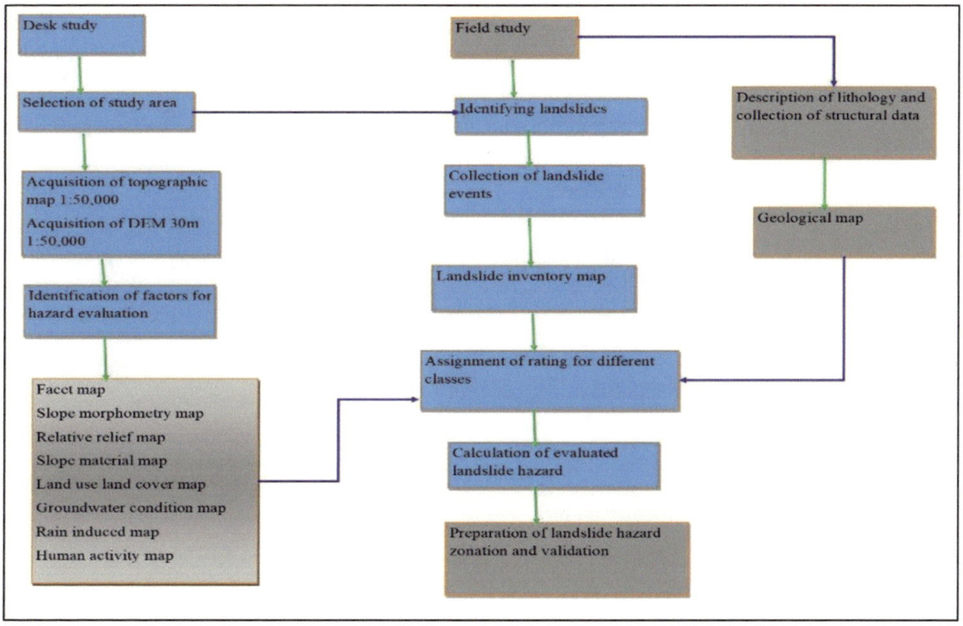

Figure 3.
General methodology flow chart for the study.

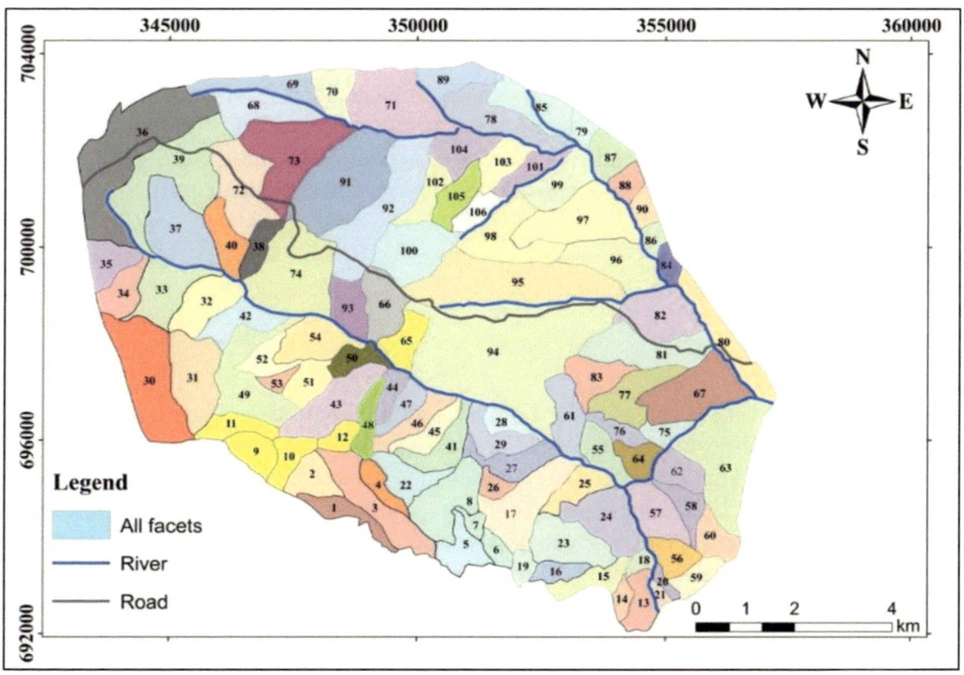

Figure 4.
Land facet map of the study area.

the fieldwork. As a result, ratings from **Table 1** have been assigned and evaluated landslide hazard is the total sum of susceptibility evaluation parameter ratings for the various causative factors for each facet. Each causative factor was assigned a rate

Code	Susceptibility evaluation parameter (SEP) parameters		Maximum rate assigned (R)	Landslide hazard zone	Landslide hazard class	Evaluated landslide hazard
	Intrinsic parameters			High hazard zone	IV	12-8
R1	Slope geometry	Relative relief	1.0	Moderate hazard zone	III	7.9-5
R2		Slope morphometry	2.0	Low hazard zone	II	4.9-2
R3	Slope geo-material		1.0	Very low hazard zone	I	<2
R4	Structural discontinuity		2.5			
R5	Land use land cover		1.5			
R6	Groundwater condition		2.0			
	External parameters					
R7	Seismicity		2.0			
R8	Rainfall		1.5			
R9	Man-made activity		1.5			
	Total parameters		15.00			

Table 1.
Rating methods for both intrinsic, external causative factors and evaluated landslide hazard.

based on subjective decisions attained from past research on intrinsic and external causing factors and their relative contribution to slope instability. Field and literature review data on intrinsic and extrinsic causative factors were incorporated. Finally, each causative parameter was rated on a facet-by-facet basis (**Figure 4**).

3. Results and discussion

3.1 Landslide inventory

Landslide inventory is the base for landslide hazard zonation mapping [5, 28, 33, 35, 42]. A clear understanding of landslide conditions and a more detailed assessment of the landslide hazard of the area concerned are essential to make a systematic landslide inventory. The landslide inventory map depicts the location and characteristics of prior landslides. The geologic, topographic, and climatic conditions that are overcome at the site give an important indication regarding the causes and triggering mechanisms of the past slope failures. Hence, landslide inventory mapping provides helpful insight regarding the potential for future landslide occurrences. Based on relevant literature, historical sources, and classical field survey and mapping, landslide inventory on sliding processes was compiled. The inventory includes 46 different types of mass movements from the past and present that are dispersed throughout the area. The landslide inventories (**Figure 5a**) were collected through the direct field survey method. The density of landslides is very high in the central part of the area which is covered by highly weathered basalt.

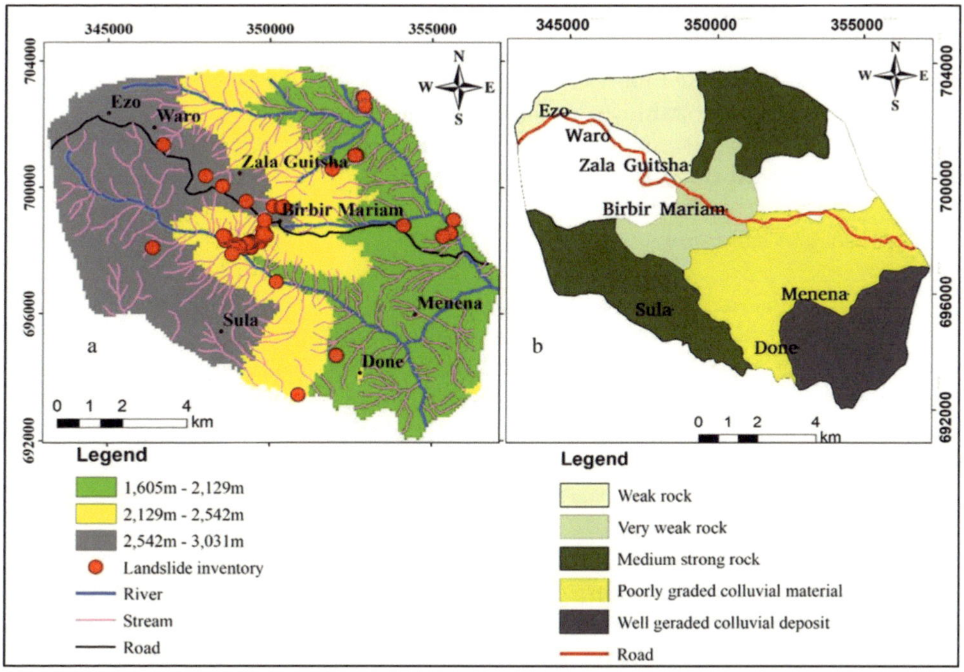

Figure 5.
a. Landslide inventory map of Birbir Mariam district, b. slope geo-material map of Birbir Mariam district.

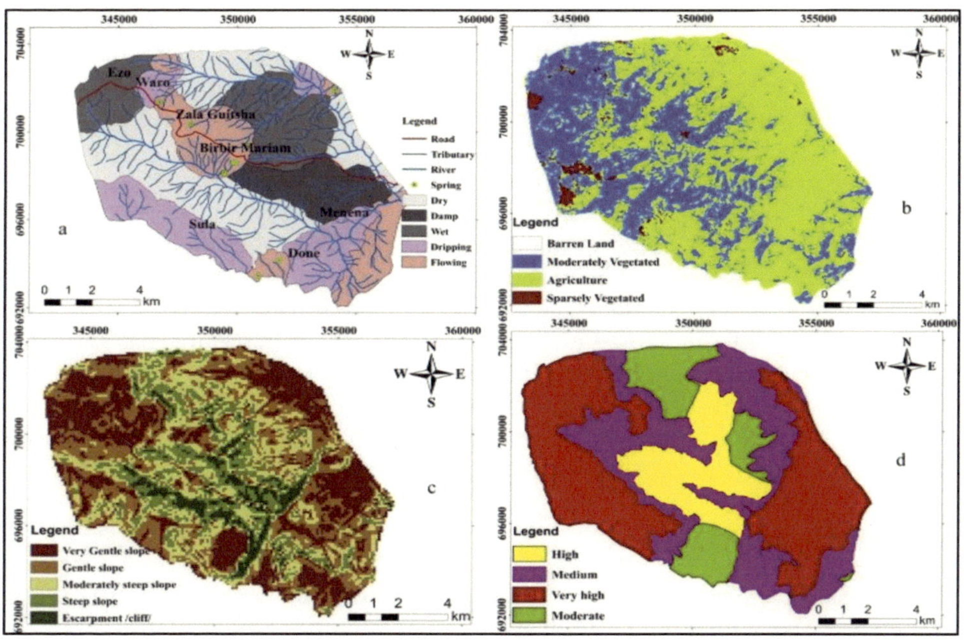

Figure 6.
Landslide causative factor maps, a. groundwater condition manifestation, b. land use land cover, c. slope morphometry, and d. relative relief.

3.2 Landslide causative factors

Intrinsic factors are the inherent or static causative parameters that define the favorable or unfavorable stability conditions within the slope [4, 11]. The main landslide causative factors selected for the Birbir Mariam area are summarized in **Table 1** and their brief descriptions are presented hereunder.

3.2.1 Groundwater condition

Groundwater has a major influence on slope stability. In hilly terrains, groundwater does not follow a consistent pattern and is generally channeled along rock structural discontinuities. It is difficult and time-consuming to assess the behavior of groundwater in hilly terrains over large areas. Groundwater surface indicators (**Figure 6a**) such as damp, wet, dripping, and flowing provide valuable information on the stability of hill slopes and are helpful for the rating [4].

3.2.2 Land use and land cover

Landslides are caused by a variety of factors, including land use and land cover [25]. Thereby, vegetation covers play an important role in slope stabilization through a different mechanism. The land use land cover of the study area (**Figure 6b**) is characterized by agricultural land, bare land, sparsely vegetated land, and moderately vegetated land categories.

3.2.3 Slope morphometry

Slope morphometry is the steepness of the slope [4]. Slope morphology brings to bear major controls on landslide types and the severity of associated damages to life and property. It is classified into five classes (**Figure 6c**) such as escarpment/cliff (>45°), steep slope (36–45°), moderately steep slope (26–35°), gentle slope (16–25°), and very gentle slope (<15° [11]).

3.2.4 Relative relief

The difference between a facet's maximum and minimum elevation is known as relative relief. Inside the apiece facet, the relative relief map depicts the local relief of maximum height between the ridge top and valley floor (**Figure 6d**). This relative relief lies on very high, high, medium, and moderate classes in the area based on the slope geometry classification system, and very high-class value range from elevation greater than 300 m, high 201–300 m, medium 101–200 m, and moderate has an elevation of 51–100 m. Therefore, an area with high relative relief is more susceptible to slope failures compared with low relative relief [15, 34, 35].

3.2.5 Lithology

Types of lithologies are the main influencing factors that contributed to the occurrences of landslides in the study area. The dominant lithological units are highly weathered basalts, ignimbrite, tuff, colluvium, alluvial soil deposits, and residual soils overlying the bedrocks (**Figure 5b**). The main criterion in the allocation of the rating for lithological subclasses is the response of the rocks to the weathering processes and erosion. The degree of weathering may vary based on the rock types and mineralogical compositions. Basalt and ignimbrites form steep slopes because they are hard, massive, and resistant to erosion. Soft rocks, on the other hand, such as tuff, are more susceptible to weathering and erosion, as well as slope instability (**Figure 5b**).

3.2.6 Structural discontinuities

The association between structural discontinuity and rock slopes is a critical factor. The orientation of structural discontinuities was determined by collecting data facet-by-facet from the exposed rock mass and determining its affiliation to slope proclivities. Based on field observations and preliminary analysis joints and faults were considered the main geologic structures for further hazard/susceptibility/evaluations. Facet wise the structural data were collected and ratings were assigned based on the proposed slope susceptibility evaluation factors.

3.2.7 Rainfall manifestation

The average annual rainfall in the study area is 1372.8 mm. The amount of rain that falls has a significant impact on the slope's stability [25, 29, 43]. The instability of slopes increased as the amount and intensity of rainfall increased, which is obvious and reasonable. Due to the nature of the materials exposed in the slopes and the slopes' drainage characteristics, this is not always the case. As a result, rainfall-induced features are useful indicators for assessing the impact of rainfall on slope instabilities. Rainfall-induced slope manifestations (such as gully formation, toe erosion, and stream bank erosion) were taken into account when determining the rainfall rating (**Figure 7a**).

3.2.8 Human activity

Man-made activities, in addition to natural triggering parameters, increase the slopes' potential for instability [44]. Developmental activities such as road construction and cultivation activities are examples of man-made activities that affect slope stability conditions. Such anthropogenic activities increased the moisture content of soil or rock masses, as well as a reduction in slope stability (**Figure 7b**).

3.2.9 Landslide hazard zonation

For landslide hazard zonation, the slopes in the study area were divided into individual land facets. For this purpose, topographical map on 1:50,000 scales was utilized to delineate the land facets. A total of 106 slope land facets were delineated (**Figure 4**). Slope geometry includes relative relief and slope morphometry. From a total 110 km^2, 60.5 km^2 (55%) fall in to very high relative relief (>300 m), 24.2 km^2 (22%) fall in to high relative relief (201–300 m), 14.3 km^2 (13%) fall in to medium relative relief (101–200 m), and 11 km^2 (10%) fall in to moderate relative relief (51–100 m) (**Figure 6d**). Slope morphometry defines the steepness of the slopes. In the study area, about 18.87% of slopes fall under the category of escarpment/cliff (>45°) and 29.25% of slopes are a step (36–45°). The remaining slope falls under moderately steep slope (26–35°), gentle slope (16–25°), and very gentle slope (<15°) which account for 30.19, 15.09, and 6.60%, respectively (**Figure 6c**).

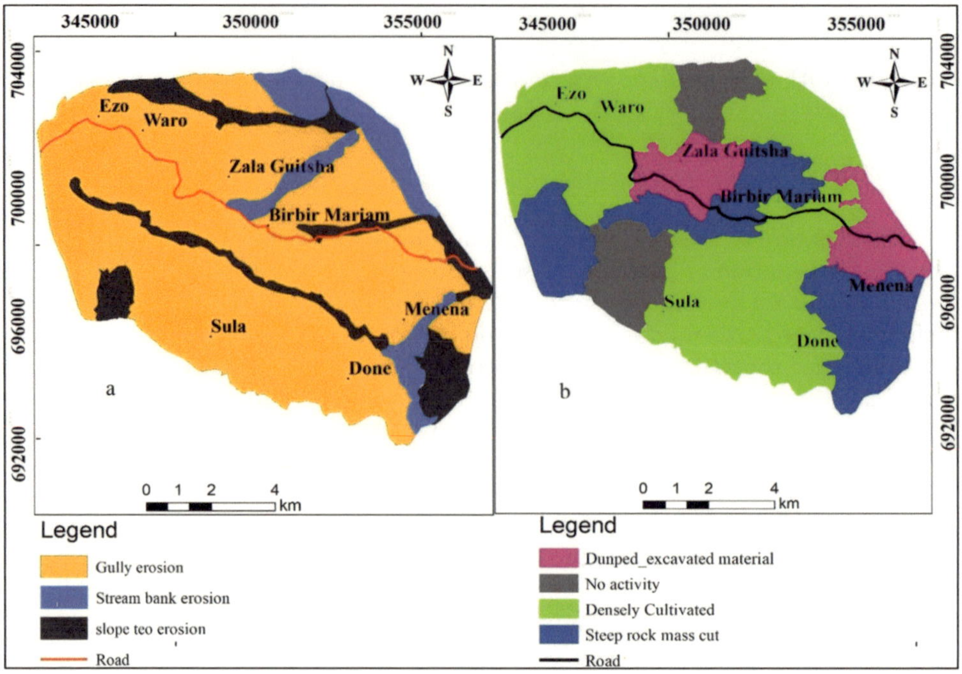

Figure 7.
a. Rainfall-induced surface manifestation map and b. anthropogenic developmental activities affecting slope stability.

The most dominant lithological units of the study area are highly disintegrated basalt. Spheroidal weathering was dominantly observed in the northwest and central parts of the study area (**Figure 2**). Faults, joints, and fractures are the dominant geological structures affecting the slope material. Residual soils and colluvial deposits are the most common slope materials (**Figure 5b**). Based on the map 58.11% of the total area is covered by soil mass, 28.3% is covered by disintegrated rock mass, and 18.87% is covered by medium-strong rock mass (**Figure 5b**).

Based on the land use-land cover, the major portion of the slopes is covered by agricultural land (43.4%). Moreover, areas covered by moderately vegetated, sparsely vegetated, and barren land account for 27.34, 18.87, and 10.39%, respectively. Surface indications such as damp, wet, dripping, and flowing water were considered for each facet to assess and evaluate the groundwater conditions. Watermarks, algal growth, and other features were also noted. As a result, each land facet was given a score (**Figure 6a**). A record of the rainfall in the study area shows that the months from April to June and from July to October received more rain. Rain-induced slope manifestations such as gully erosion, toe erosion, and stream bank erosion were also taken into account. Slope toe erosion, stream bank erosion, and gully erosion all accounted for 24.53, 20.75, and 54.72% of the total erosion (**Figure 7a**).

Manmade activities which affect the slope stability in the study area include cultivation activity, road construction, and unsafe dumping of materials. Based on field data-intensive cultivation activity covered about 43.4%, steep rock-cut for road construction constitute 26.4%, and unsafe dumped materials covered about 22.6%. About 7.6% of the total area falls under no human activities (**Figure 7b**). The study

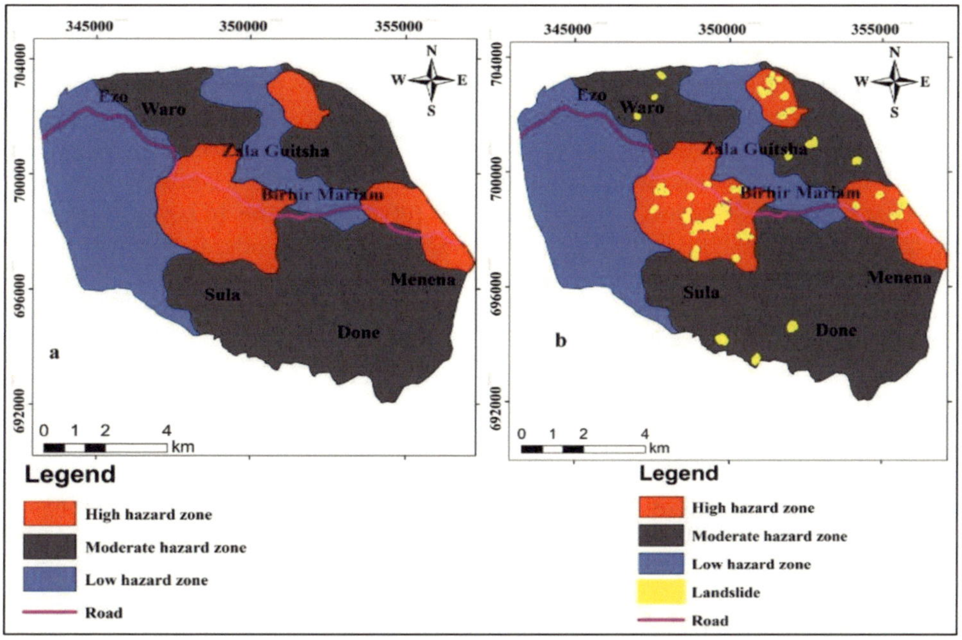

Figure 8.
a. Landslide hazard zonation of Birbir Mariam and b. validation of present landslide hazard zonation map.

area's landslide hazard was determined facet-by-facet using an evaluated landslide hazard, which indicates the cumulative likelihood of instability. The area was divided into three zones based on the values of the assessed landslide hazard: high, moderate, and low hazard zones (**Figure 8a**).

4. Discussion

The selection and validity of the conditioning factors are of major importance for the correct result of the slope susceptibility evaluation parameter. Intrinsic and external causative parameter rating schemes were therefore considered to be responsible for slope instability. The high-hazard zones are mainly located in the central part and north-eastern side of the study area (**Figure 8a**). These localities are dominantly agricultural lands subjected to many anthropogenic activities. Most of the main road that crosses the area falls within the high-hazard zone areas. Along the main road, it is common to observe slope failures also in the form of earth slide rock falls (**Figure 8a**). Such failures mostly occurred following heavy rains. It was also reported that the main road experienced frequent failures and maintenance in the past. The problem still exists in the areas decease a lack of proper understanding of the causes and failure mechanisms of failures in the area. The southern and northeastern parts of the area fall under the moderate hazard zone, while the West and North West areas fall under the low hazard zone (**Figure 8a**). On the landslide hazard map of the study area, 18.87% (20.76 km^2) is classified as high hazard, 54.72% (60.19 km^2) is categorized as a moderate hazard, and 26.41% (29.05 km^2) is divided into a low hazard (**Figure 8a**).

5. Validation of slope susceptibility evaluation rating scheme

By comparing the prepared landslide hazard zonation map of the current study area to existing landslide inventory data, the prepared landslide hazard zonation map was validated. The majority of the landslides occurred along the road and river banks. The landslide inventories were superimposed on the area's landslide hazard map. Based on the validation (by overlying), 37 (80.43%) of the 46 landslide inventory data fall into the high hazard zone, 9 (19.57%) into the moderate hazard zone, and none in the low hazard zone. As a result, the slope susceptibility rating scheme's landslide hazard zonation map is found to be reliable and comparable to real-world ground conditions (**Figure 8b**).

6. Conclusion

In highland and mountain terrain, landslide hazards and susceptibility zonation are critical for land use planning and development. A landslide hazard zonation map was produced using a slope susceptibility evaluation parameter rating scheme. The method takes into account both internal and external causative factors. The slope susceptibility evaluation parameter rating system is a feasible method of assigning numerical ratings to each site-specific intrinsic and external causative factor based on their role and contribution to slope instability. The Birbir Mariam area was subjected to the slope susceptibility evaluation parameter rating method. According to the results, there is a low hazard zone of 26.41% (29.05 km^2), a moderate hazard zone of 54.72% (60.19 km^2), and a high hazard zone of 18.87% (20.76 km^2) in the study area. The generated landslide susceptibility zonation map was compared to actual past landslide activity data to ensure accuracy. According to the comparison, 37 (80.43%) of the 46 landslide inventory data fall into the high hazard zone, 9 (19.57%) into the moderate hazard zone, and none fall into the low hazard zone. As a result, the method used in this study, as well as the resulting landslide susceptibility zonation map, were found to be reliable and can be used in other areas with similar geologic and topographic conditions [45].

Acknowledgements

The authors are grateful to Arba Minch University for providing partial funds to conduct fieldwork and for allowing the first author to pursue his postgraduate studies. The Department of Geology of Arba Minch University is dually acknowledged for various administrative and technical supports during the research.

Conflict of interest

The authors report no potential conflicts of interest.

Author details

Yonas Oyda* and Hailu Regasa
Arba Minch University, Arba Minch, Ethiopia

*Address all correspondence to: yonasoyda777@gmail.com

References

[1] Chi KH, Park NW, Lee K. Identification of Landslide area using remote sensing data and quantitative assessment of landslide hazard BT– Proceedings under IEEE International Symposium on Geoscience and Remote Sensing.

[2] Ghosh S, Carranza EJM, Westen CJ, Jetten VG, Bhattacharya DN. Selecting and weighting of spatial predictors for empirical modeling of landslide susceptibility in Darjeeling Himalaya (India). Geomorphology. 2011;**131**. DOI: 10.1016/j.geomorph.2011.04.019

[3] van Westen CJ, Castellanos E, Kuriakose SL. Spatial data for landslide susceptibility, hazard, and vulnerability assessment: An overview. Engineering Geology. 2008;**102**(3-4):112-131. DOI: 10.1016/j.enggeo.2008.03.010

[4] Anbalagan R. Landslide hazard evaluation and zonation mapping in mountainous terrain. Engineering Geology. 1992;**32**(4):269-277. DOI: 10.1016/0013-7952(92)90053-2

[5] Bernat Gazibara S, Krkač M, Mihalić Arbanas S. Landslide inventory mapping using LiDAR data in the City of Zagreb (Croatia). Journal of Maps. 2019;**15**(2):773-779. DOI: 10.1080/17445647.2019.1671906

[6] Pham BT, Bui DT, Indra P, Dholakia MB. Landslide susceptibility assessment at a part of Uttarakhand Himalaya, India using GIS – Based statistical approach of frequency ratio method. International Journal of Engineering Research And Technical Research. 2015;**V4**(11):338-344. DOI: 10.17577/ijertv4is110285

[7] Calligaris C, Poretti G, Tariq S, Melis MT. First Steps towards a Landslide Inventory Map of the Central Karakoram National Park. 72542017. DOI: 10.5721/EuJRS20134615

[8] Erener A, Düzgün HSB. Landslide susceptibility assessment: What are the effects of mapping unit and mapping method? Environmental Earth Sciences. 2012;**66**(3):859-877. DOI: 10.1007/s12665-011-1297-0

[9] Song Y, Niu R, Xu S, Ye R, Peng L, Guo T, et al. Landslide susceptibility mapping based on weighted gradient boosting decision tree in Wanzhou section of the three gorges reservoir area (China). ISPRS International Journal of Geo-Information. 2019;**8**(1). DOI: 10.3390/ijgi8010004

[10] Lee S, Talib JA. Probabilistic landslide susceptibility and factor effect analysis. Environmental Geology. 2005;**47**. DOI: 10.1007/s00254-005-1228-z

[11] Raghuvanshi TK, Ibrahim J, Ayalew D. Slope stability susceptibility evaluation parameter (SSEP) rating scheme - an approach for landslide hazard zonation. Journal of African Earth Sciences. 2014;**99**(PA2):595-612. DOI: 10.1016/j.jafrearsci.2014.05.004

[12] Mulatu E, Raghuvanshi TK, Abebe B. Landslide hazard zonation around Gilgel gibe-II hydroelectric project, Southwestern Ethiopia. SINET: Ethiopian Journal of Science. 2011;**32**(1):9-20. DOI: 10.4314/sinet.v32i1.68733

[13] Raghuvanshi TK, Negassa L, Kala PM. GIS based grid overlay method versus modeling approach - a comparative study for landslide hazard zonation (LHZ) in meta Robi District of west Showa zone in Ethiopia. Egyptian

Journal of Remote Sensing and Space Science. 2015;**18**(2):235-250. DOI: 10.1016/j.ejrs.2015.08.001

[14] Seid JI, Raghuvanshi TK, Ayalew D. Landslide hazard zonation and slope instability assessment using optical and InSAR data: A case study from gidole town and its surrounding areas, Southern Ethiopia. Hydrology & Meteorology. 2014;**5**(4):7587

[15] Ayalew L, Yamagishi H. Slope failures in the Blue Nile basin, as seen from landscape evolution perspective. Geomorphology. 2004;**57**(1-2):95-116. DOI: 10.1016/S0169-555X(03)00085-0

[16] Kanungo DP, Arora MK, Sarkar S, Gupta RP. Landslide susceptibility zonation (LSZ) mapping – A review. Journal of South Asian Studies. 2009;**2**

[17] Lee S, Pradhan B. Landslide hazard mapping at Selangor, Malaysia using frequency ratio and logistic regression models. Landslides. 2007;**4**. DOI: 10.1007/s10346-006-0047-y

[18] Dahal RK, Hasegawa S, Bhandary NP, Poudel PP, Nonomurac A, Yatabe Y. A replication of landslide hazard mapping at catchment scale. Geomatics, Natural Hazards and Risk. 2012

[19] Canoğlu MC. Deterministic landslide susceptibility assessment with the use of a new index (factor of safety index) under dynamic soil saturation: An example from demirciköy watershed (Sinop/Turkey). Carpathian Journal of Earth and Environmental Sciences. 2017;**12**(2):423-436

[20] Du G l, Zhang Y s, Iqbal J, Yang ZH, Yao X. Landslide susceptibility mapping using an integrated model of information value method and logistic regression in the Bailongjiang watershed, Gansu Province, China. Journal of Mountain Science. 2017;**14**(2):249-268. DOI: 10.1007/s11629-016-4126-9

[21] Silalahi FES, Pamela A, Y., & Hidayat, F. Landslide susceptibility assessment using frequency ratio model in Bogor, West Java, Indonesia. Geoscience Letters. 2019;**6**(1). DOI: 10.1186/s40562-019-0140-4

[22] Ayalew L, Yamagishi H. The application of GIS-based logistic regression for landslide susceptibility mapping in the Kakuda–Yahiko Mountains, Central Japan. Geomorphology. 2005;**65**. DOI: 10.1016/j.geomorph.2004.06.010

[23] Guzzetti F, Reichenbach P, Ardizzone M, Cardinali M, Galli M. Estimating the quality of landslides susceptibility models. Geomorphology. 2006;**81**. DOI: 10.1016/j.geomorph.2006.04.007

[24] Meten M, Bhandary NP, Yatabe R. GIS-based frequency ratio and logistic regression modelling for landslide susceptibility mapping of Debre Sina area in Central Ethiopia. Journal of Mountain Science. 2015;**12**(6):1355-1372. DOI: 10.1007/s11629-015-3464-3

[25] Woldearegay K. Review of the occurrences and influencing factors of landslides in the highlands of Ethiopia: With implications for infrastructural development. Momona Ethiopian Journal of Science. 2013;**5**(1):3. DOI: 10.4314/mejs.v5i1.85329

[26] Azeze AW. Modeling of Landslide Susceptibility in a Part of Abay Basin, Northwestern Ethiopia. March. 2020. DOI: 10.1515/geo-2020-0206

[27] Chauhan S, Sharma M, Arora MK, Gupta NK. Landslide susceptibility zonation through ratings derived

from artificial neural network. The International Journal of Applied Earth Observation and Geoinformation. 2010;**12**. DOI: 10.1016/j.jag.2010.04.006

[28] Guzzetti F, Mondini AC, Cardinali M, Fiorucci F, Santangelo M, Chang KT. Landslide inventory maps: New tools for an old problem. Earth-Science Reviews. 2012;**112**(1-2):42-66. DOI: 10.1016/j.earscirev.2012.02.001

[29] Jeong S, Kassim A, Hong M, Saadatkhah N. Susceptibility assessments of landslides in Hulu Kelang area using a geographic information system-based prediction model. Sustainability (Switzerland). 2018;**10**(8). DOI: 10.3390/su10082941

[30] Santangelo M, Gioia D, Cardinali M, Guzzetti F, Schiattarella M. Landslide inventory map of the upper Sinni River valley, southern Italy. Journal of Maps. 2015;**11**(3):444-453. DOI: 10.1080/17445647.2014.949313

[31] Kannan M, Saranathan E, Anbalagan R. Comparative analysis in GIS-based landslide hazard zonation—A case study in Bodi-Bodimettu Ghat section, Theni District, Tamil Nadu, India. Arabian Journal of Geosciences. 2015;**8**(2):691-699. DOI: 10.1007/s12517-013-1259-9

[32] Kundu S, Saha AK, Sharma DC, Pant CC. Remote sensing and GIS based landslide susceptibility assessment using binary logistic regression model: A case study in the Ganesh ganga watershed, Himalayas. The Journal of the Indian Society of Remote Sensing. 2013;**41**. DOI: 10.1007/s12524-012-0255-y

[33] Legorreta Paulín G, Bursik M, Hubp JL, Mejía LMP, Aceves Quesada F. A GIS method for landslide inventory and susceptibility mapping in the Río El Estado watershed, Pico de Orizaba volcano, México. Natural Hazards. 2014;**71**(1):229-241. DOI: 10.1007/s11069-013-0911-8

[34] Mezughi TH, Akhir JM, Rafek AG, Abdullah I. Landslide susceptibility mapping using the statistical index method and factor effect analysis along the E- W highway (Gerik - Jeli), Malaysia. Australian Journal of Basic and Applied Sciences. 2011;**5**(6):847-857

[35] Sarkar S, Roy AK, Martha TR. Landslide susceptibility assessment using information value method in parts of the Darjeeling Himalayas. Journal of the Geological Society of India. 2013;**82**(4):351-362. DOI: 10.1007/s12594-013-0162-z

[36] Shahabi H, Hashim M. Landslide susceptibility mapping using GIS-based statistical models and remote sensing data in tropical environment. 2015. 1-15. DOI: 10.1038/srep09899

[37] Abbate E, Bruni P, Sagri M. Geology of Ethiopia: A Review and Geomorphological Perspectives. 2015. DOI: 10.1007/978-94-017-8026-1

[38] Bonini M, Corti G, Innocenti F, Manetti P, Mazzarini F, Abebe T, Pecskay Z. Evolution of the main Ethiopian rift in the frame of Afar and Kenya rifts propagation. February. 2005. DOI: 10.1029/2004TC001680

[39] Ebinger C, Yemane T, Woldegabriel G, Aronson J. Late Eocene-Recent volcanism and faulting in the Southern Main Ethiopian Rift. February. 1993. DOI: 10.1144/gsjgs.150.1.0099

[40] Philippon M, Corti G, National I, Sani F. Evolution, distribution and characteristics of rifting in Southern Ethiopia. April. 2014. DOI: 10.1002/2013TC003430

[41] Ermias B, Raghuvanshi TK, Abebe B. Landslide Hazard zonation (LHZ) around Alemketema town, north Showa zone, Central Ethiopia - a GIS based expert evaluation approach. International Journal of Earth Sciences and Engineering. 2017;**10**(01):33-44. DOI: 10.21276/ijee.2017.10.0106

[42] Lee S, Choi J, Min K. Landslide susceptibility analysis and verification using the Bayesian probability model. Environmental Geology. 2002;**43**. DOI: 10.1007/s00254-002-0616-x

[43] Ayalew L. The effect of seasonal rainfall on landslides in the highlands of Ethiopia. Bulletin of Engineering Geology and the Environment. 1999;**58**(1):9-19. DOI: 10.1007/s100640050065

[44] Kanungo DP, Arora MK, Sarkar S, Gupta RP. A fuzzy set based approach for integration of thematic maps for landslide susceptibility zonation. Georisk. 2009;**3**

[45] Das I, Stein A, Kerle N, Dadhwal. Landslide susceptibility mapping along road corridors in the Indian Himalayas using Bayesian logistic regression models. Geomorphology. 2012;**179**. DOI: 10.1016/j.geomorph.2012.08.004

Chapter 21

Underground Coal Mining Methods and Their Impact on Safety

Lingampally Sai Vinay, Ram Madhab Bhattacharjee and Nilabjendu Ghosh

Abstract

In underground coal mining, the working environment is completely enclosed by the geologic medium, which consists of the coal seam and overlying and underlying strata. Depillaring is one of the most dangerous phases of underground coal mining. Even after centuries of experience in underground coal mining, there is no well-established depillaring method w.r.t existing geo-mining method. As far as fatal accidents are concerned, most of the fatal accidents occur mostly during depillaring operations. Method of extraction is one of the keys and underrated parameters, which plays an eminent role in successful depillaring in underground coal mining. This chapter highlights the underground coal mining methods and their impact on safety.

Keywords: underground coal mining, mining methods, coal, room, pillar mining

1. Introduction

Coal plays a critical role in the global economy and day-to-day development by contributing to the electricity sector and the steel and cement industries. Globally, around 37% of electricity and nearly 70% of steel are produced only using coal [1]. Coal is the backbone of both the steel and cement industries. As coal is not only the affordable option for energy in many markets, it will also remain the only feasible option for many critical industries. Coal also has a significant role in the construction of renewable energy infrastructure. It is significant to societal development, particularly in emerging economies. And also, coal is the single largest source of global carbon emissions. Advanced clean coal technologies have been developed to clean the coal by upgrading existing coal plants to the best available technology, which can decrease 2 gigatons of CO_2 emissions. The coal industry has seen consistent improvements against environmental and economic challenges by adopting current and emerging technologies [2].

It can also be observed that coal consumption, primarily in Asian countries, has increased many folds over the past decades due to the growing population and economy. In order to meet the increasing demand and consumption of coal requires a high production of coal through mining activities. The main point to be noted along with the production of coal is safety. Globally, coal mining is a highly regulated industry due to many inherent, operational, and occupational hazards. The work environment in mines is highly dynamic in nature and may change abruptly. The risks to the safety and health of employees due to the unknown and unforeseen hazards pose challenges for occupational safety and health professionals. Further, it requires comprehensive surveillance at workplaces and newer strategies to prevent accidents in mines.

At the same time, the mining sector is critical to a country, such as India, for its economic and social well-being. The mining and quarrying sector contributes around 1.63% of India's gross domestic product (GDP). Indian economy is on the aspirational path of becoming a $5 trillion GDP economy by 2024–25 [3]. The mining industry will contribute to India's envisaged GDP and wealth creation. As a developing economy, India depends on the energy sector's growth to achieve the desired economic growth. Though there is significant global pressure on the immediate reduction of fossil fuels as an energy source, coal is and will remain one of India's primary energy sources in the years to come. The past and present economic reforms have propelled the Indian mining industry into a new and competitive environment. The sector-wise GDP in India is shown in **Figure 1**.

1.1 Role of coal in India

Through a sustained program of investment and greater thrust on the application of modern technologies, it has been possible to raise the all-India production of coal to 716.08 million tonnes in 2020–21 (provisional). The all-India production of coal during 2021–22 was 778.19 MT (provisional), with a positive growth of 8.67%.

Coal India Limited (CIL) and its subsidiaries accounted for 596.221 million tonnes during 2020–21 as compared to the production of 602.129 million tonnes in 2019–20, showing a negative growth of 0.98%. Coal production of CIL, during 2021–22, was 622.634 MT (provisional), with a positive growth of 4.43%.

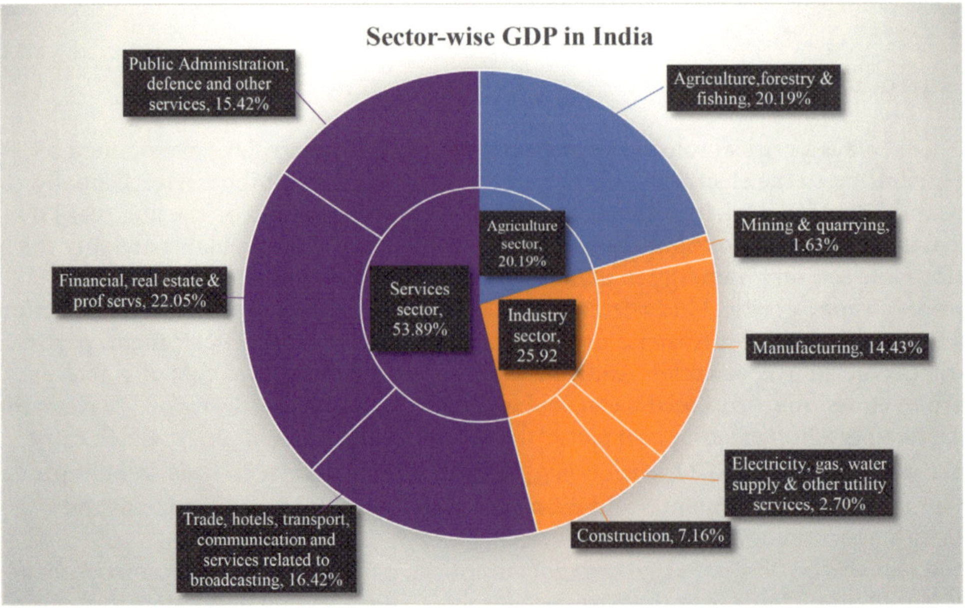

Figure 1.
Sector-wise GDP in India.

Singareni Collieries Company Limited (SCCL) is the main source of the supply of coal to the southern region of India. The company produced 50.580 million tonnes of coal during 2020–21 as against 64.044 million tonnes during the corresponding period last year. SCCL production of coal during 2021–22 was 65.022 MT(Provisional), with a positive growth of 28.55%. Small quantities of coal are also produced by TISCO, IISCO, DVC, and others [4]. In India, the total geological reserves of coal are around 3,44,021 MT, as per reports of the Geological Survey of India (GSI) as of 01.04.2020. The coal reserves of coking and non-coking coal are around 35,004 MT and 3,09,017 MT, respectively. From 1950 to 2020–21, the total coal extracted was around 1,72,96,897 TT [5, 6].

In India, open cast and underground mining are the two major coal mining methods. Opencast mining contributes about 95.74% of the total production. In contrast, underground mining contributes to the rest of the production by 4.26% during 2021–22 (provisional). The major contributors of coal in India are CIL, SCCL, SAIL, etc., in the public sector, and Reliance Power Limited, TISCO, etc., in the private sector. The overall coal production vs. open cast coal production vs. underground coal production is plotted in **Figure 2**.

1.2 Need to increase underground coal production

In India, coal mining is primarily carried out by opencast mining and underground mining. According to the Indian Bureau of Mines, there were 455 operating coal mines in India, out of which 219 were opencast, and 213 were underground. The remaining 23 were mixed collieries [7]. Even though the number of underground coal mines is proportionately equal to opencast coal mines, their production rates are nowhere on a comparable scale. The share of opencast and underground coal

Figure 2.
Opencast and underground coal production from 1996 to 2021–22.

production in overall coal production is nearly 95.64% and 4.36%, respectively [8]. However, the thrust on increasing underground coal production is being felt by the coal companies because of many critical issues, such as difficulty in land acquisition for opencast mining under the current sociopolitical environment, the severe threat of irreparable environmental damage due to opencast mining vis-à-vis the growing concern for mitigating the environmental impact due to opencast mining and its associated cost, depletion of shallower deposits amenable to opencast mining, and at the same time, increasing demand for quality of coal production to maintain economic sustainability of the country [9–12]. In order to increase coal production to meet the country's coal demand, there is no other future option but to go for the exploitation of deeper deposits by underground mining methods.

1.3 Conventional mining

The majority of Indian underground coal mines are still working with conventional mining. This process involves drilling and blasting technology for the extraction of coal and LHD/SDL for conveyance/transport of blasted coal, followed by the installation of supports, which slows the rate of extraction and also becomes a source for adverse strata conditions in case of delay in the extraction of exposed pillar during the depillaring [13, 14]. A rib-and-slice method is commonly practiced in conventional depillaring [15, 16]. A rib/snook is left against the goaf along with breaker-line support in openings at the goaf edge [12, 17]. Left-out rib/snook is judiciously reduced at the time of retreat to facilitate the caving of the roof strata. This depillaring technique used to dominate in the past due to different techno-economic reasons, but the industry found it incapable of further incrementing coal production. The chronic problem of significantly less productivity leads to a considerable loss per ton of coal production and the lack of capability of the mass output by such technologies. Conventional mining methods and adopted low-level technologies result in a low production rate from underground coal mining [18].

It is high time to go for mass production from underground coal mining by adopting suitable technologies, such as longwall mining, shortwall/short longwall mining for developed coal pillars, room and pillar mining, and wongawilli or rib pillar extraction by continuous miner. Earlier attempts to develop mechanized underground mines were not very successful due to improper planning, lack of advanced geotechnical studies and R&D facilities, issues related to maintenance of the mechanized machinery implemented, technical lapses, high cost due to absence of indigenous equipment supplier, lack of understanding the nature of overlying strata concerning the depillaring method adopted and machinery implemented.

2. Mechanized underground coal mining

2.1 Longwall system

In India, the first mechanized longwall mining was implemented in Moonidh in 1978. Longwall mining is a new era in underground coal mining across the globe. This method can be adopted even in weak roof strata conditions with the help of an armoured face conveyor (AFC). Gassy seams are also easier to work on as there are few roadways, and ventilation presents a little problem. Longwall mining can be operated in two ways: longwall advancing and longwall retreating.

Proper scientific exploration is the critical key to the success of longwall mining. It should preferably be implemented in coal seams that are free from geological disturbances and dirt bands. The unforeseen encounter of such dirt band or geological disturbance hampers the underground production and, results in the idle time of machinery and, increases the chances of spontaneous combustion, may even result in substantial human and economic losses.

In a longwall advancing system, a rigid cycle of operations is to be followed. Faults or other geological disturbances cannot be proved in advance. For some distances behind the working face, the roadways are on moving ground. In such conditions, the uninterrupted working of roadway belt conveyors is more challenging to ensure.

When the boundary is reached, the salvage of all support materials and the removal of all machinery must be expedited if serious losses are to be avoided. The gate roads are maintained in solids and, therefore, they are well supported. In advancing the gate roads, as the boundary is reached, the longest length of conveyors, haulage tracks, etc., are required, but in retreating, these lengths go on decreasing, thus reducing the transport cost accordingly.

Packing goaf may be avoided completely, But if there are contiguous seams, packing may be required. The labor required for unproductive work of packing can be reduced. Hence, the number of productive shifts can be increased. There is no leakage of air, so ventilation is less affected [19–22]. The schematic diagram of longwall mining is shown in **Figure 3**.

2.2 Longwall top coal caving

The longwall top coal caving (LTCC) method was first developed in the 1950s and 1960s in the former Soviet Union and France to increase coal resource extraction and

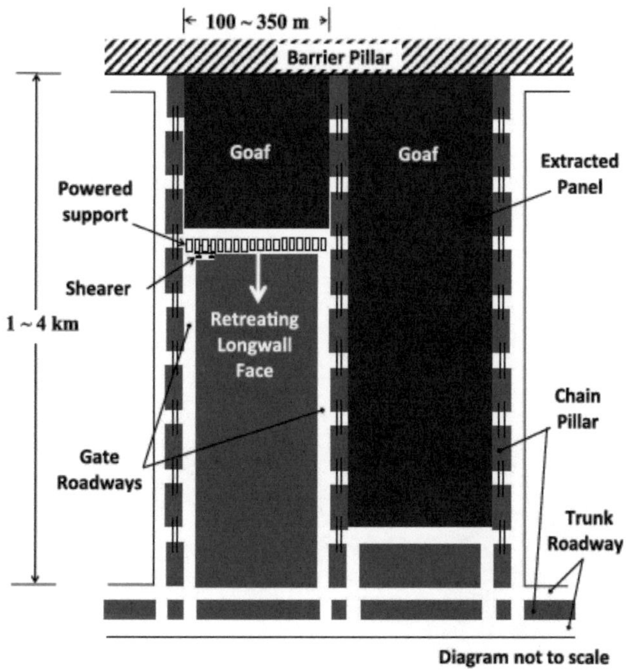

Figure 3.
Schematic diagram of longwall mining.

improve productivity. During the 1970s, it was also adopted in Yugoslavia, Hungary, Romania, and former Czechoslovakia (now the Czech Republic) but generally did not receive much success at the time. The technology was formally introduced in China under extensive research and development. It has become progressively popular in China and rapidly became a significant means of extracting thick coal seams with great success in productivity, cost, and safety [23].

A thick seam is commonly considered to be one that cannot be mined underground safely and economically in single-pass cutting using the current technology, even though the exact thickness defined varies in different countries. Underground thick seam mining technologies have been developed and practiced worldwide for decades. Some methods used for extracting thick seams, including hydraulic mining and room and pillar methods, are less popular.

The multi-slice longwall mining practiced mainly in China is gradually being replaced by the LTCC method because of its lower development and operating costs, given thick seams have favorable caveability. The high-reach single-pass longwall used in Australia has been successful, but the cutting height is limited to less than 4.8 m with current equipment. To increase the recovery of Australian thick seam reserves, the LTCC mining technology is providing an alternative solution to economic underground thick seam mining in Australia [24–27]. The conceptual model of longwall top coal caving (LTCC) is shown in **Figure 4**.

2.3 Wongawilli mining

Australia has a history of successful pillar extraction over 90 years. In 1952, the first continuous miner was introduced at Wongawilli colliery and Huntley colliery. In 1955, they were also introduced in South Clifton colliery, Old Bulli colliery, Nebo colliery, and Kemira colliery, and later on, many continuous miners came into existence. The Wongawilli method was first adopted in Wongawilli colliery in 1958. Since then, this method has been gradually modified to suit both the mining conditions and the development of mining machinery [28].

This method provides a single working place and extraction of coal in stress relieving area. This system can be easily understood and followed by employees. With the Wongawilli system, an overall percentage of extraction of 80–90% can be expected,

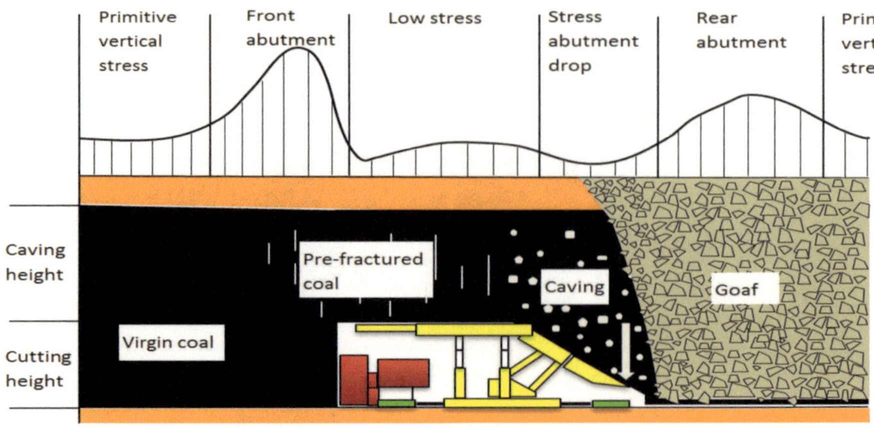

Figure 4.
Conceptual model of longwall top coal caving (after [23]).

with improved worker safety, minimal development workings required, and high production rates achieved. The machinery used for panel development consists of two shuttle cars and a continuous miner. This system is straightforward, repetitive, and easily understood by the employees. The difficulties usually occur in lifting the final stooks at the end of each fender. This method maintains a straight line of extraction, which is helpful in mechanized depillaring in caving. The success of this method was working within distressed areas and can be adopted successfully to a depth of up to 600 meters. This method has been successfully practiced in Australia for over 90 years [29, 30]. The basic layout of the Wongawilli system of extraction is shown in **Figure 5**.

2.4 Room and pillar

In India, underground mining is highly dependent on bord and pillar/room and pillar mining due to their adaptability, flexibility, and accessibility. "room and pillar" mining involves a sequence of activities that are performed to first enter and develop the mine and then progressively extract the coal. The continuous miner extracts the coal as it moves forward, loading it onto an attached loader/shuttle car', which, in turn, transfers the coal to the conveyor system. Bord and pillar method contributes 3–5% of total coal production and 65–75% of total underground production, whereas room & pillar mining, using continuous miners, contributes less than 1–2% of the total output and 20–25% of underground coal production [31].

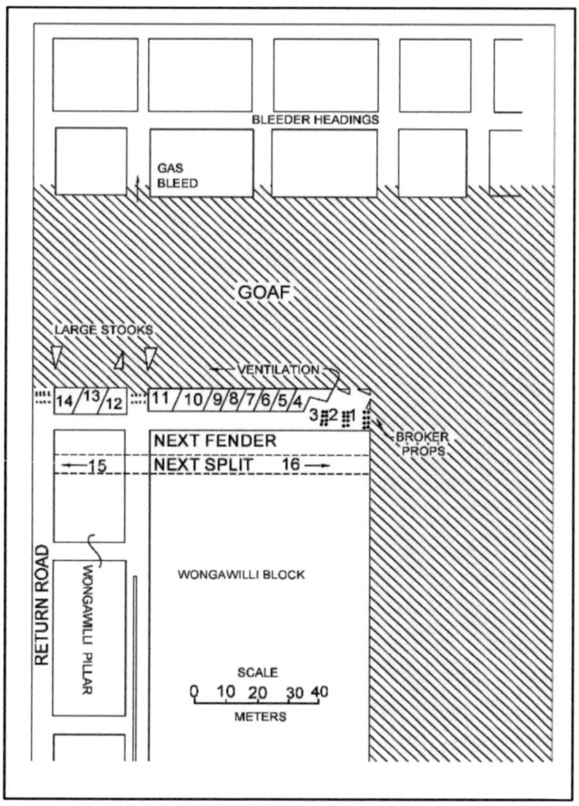

Figure 5.
Basic layout of Wongawilli system of extraction.

The depth at which the pressure of the superincumbent strata reaches nearly the crushing strength of coal may be considered the limiting depth for room and pillar mining. Generally, at depths greater than 2000 ft, longwall mining is usually adopted.

Longwall is not usually suitable in faulted areas, and a room and pillar system is adopted. As it is difficult to dispose of the dirt bands in the room and pillar mining, clean seams free from dirt bands are suitable for the adoption of the room and pillar method. In room and pillar systems, roadways are supported by solid pillars. Thus, the roads must be maintained in safe conditions without· using many artificial supports. The relatively smaller size of the group of workers encourages team spirit. Supervision and maintenance of discipline are facilitated, and work progress can be readily checked shift by shift [31–34]. In room and pillar work, the area is proved in advance by driving headings. In longwall, on meeting a fault, the whole output is lost. In room and pillar, on knowing the direction of the fault, the rest of the headings may be driven accordingly.

The deeper the seams, the greater the difficulty of working by room and pillar due to higher strata-pressure. In this method, the advance rate is sometimes slow because development takes more significant time. There is a danger of crushing during depillaring due to higher stresses of the superincumbent strata acting on the pillars.

This makes supervision difficult. If the mine is extensive, hauling lengths are greatly increased, involving increased haulage costs. In this system, the need for constant flitting of the machinery from· place to place is excellent. So, the number of idle hours of machinery in room and pillar will be greater than longwall. There is difficulty in the extraction of contiguous seams, especially during depillaring operations (without packing). However, this difficulty is reduced by working the seams in descending order. The roof control is comparatively difficult during the extraction of ·pillars when no packing is adopted. If pillars are weak, additional stresses may be thrown on the pillars in the depillaring area, which may cause the overriding of pillars or premature collapse.

If some pillars are left in the goaf, the goaf will not settle quickly, and this will produce an undulating subsided surface. In highly mechanized mines, a high standard of planning and organization and a larger staff of skilled technicians are needed to achieve maximum efficiency in mining operations.

2.4.1 Split and fender

Split and fender include pocket and wing, split and lift, and pocket and fender. Split and fender is the most practiced method in India because it can be implemented in shallow and deeper mines. In this method, initially, roadways are prepared with appropriate size for smooth movement of machinery employed, that is, continuous miner and shuttle car.

In this method, two pillars work simultaneously, one pillar under splitting and another under fender extraction. Unless the first pillar is completely extracted, CM is not allowed to start working on third pillar. A pillar is divided into two or three fenders by driving a split gallery in dip–rising direction into the pillar, depending on the pillar size and local strata conditions. Now the driven split gallery is supported by roof bolting with resin-grouted rock bolts and the installation of breaker lines; meanwhile, CM will be splitting the second pillar. The fender extraction is started by leaving a small portion of coal from the corner of the pillar; thus, slices are driven at an angle of 60 degrees with a suitable cut width depending on the fender size [12, 13, 35, 36].

A rib of 3 m coal is left against the pillar after driving the slices for stability purposes. The rib size may be 2 m or 3 m, and it is completely dependent on the pillar size and local strata conditions. After slices are extracted, a small portion of coal will be left against the goaf edge known as snook. Similarly, fender-B is also extracted.

If the caving in the goaf was delayed, induced blasting of the level junction was done. Before going for induced blasting over the junction of splits and level galleries after extraction of each fender, the snooks/ribs were further reduced or knocked down to facilitate the roof caving in the extracted area. The schematic manner of extraction in the split and fender depillaring method is shown in **Figure 6**.

2.4.2 Christmas tree method

Christmas tree is also known as the left-right method, fish and tail method, or twinning method. This method has been practiced in the Pinoura mine in the Johilla area (Umaria coalfields), South Eastern Coal Fields Ltd. In this method, coal is extracted with the help of continuous miner by cutting slices into the pillar from both level galleries and then extracting pushouts [37]. A straight line of extraction is being maintained and the sequence so followed is to facilitate the caving of the roof in a dip direction. The number of slices driven from both level galleries completely depends on the pillar size and the local strata conditions. The first slice is made most likely with a width of 3.50 m and a suitable length depending on pillar size by leaving a small portion of coal toward the goaf edge.

Similarly, two more slices are extracted in the same manner. The fourth slice will be extracted with a width less than the first. A similar manner of extraction is followed to extract other slices on the other side of the pillar, simultaneously driving the slices into the second pillar. Now, pushouts will be driven from the dip gallery, and the number of pushouts to be lifted is completed dependent on the pillar size and the local strata conditions [36].

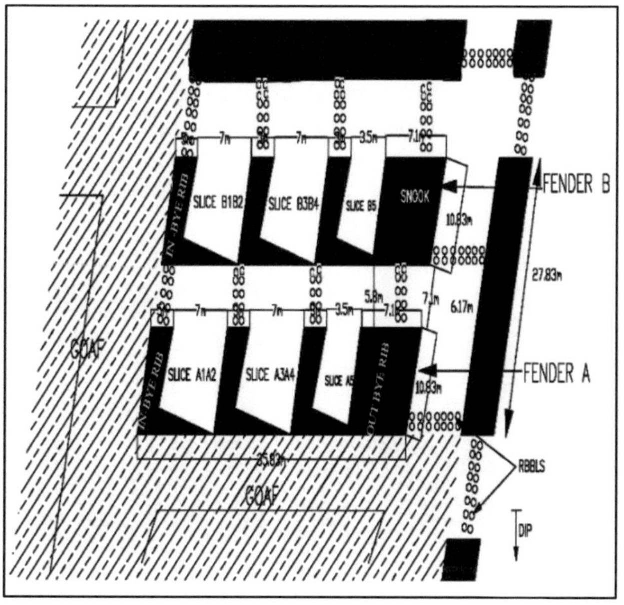

Figure 6.
Split and fender method.

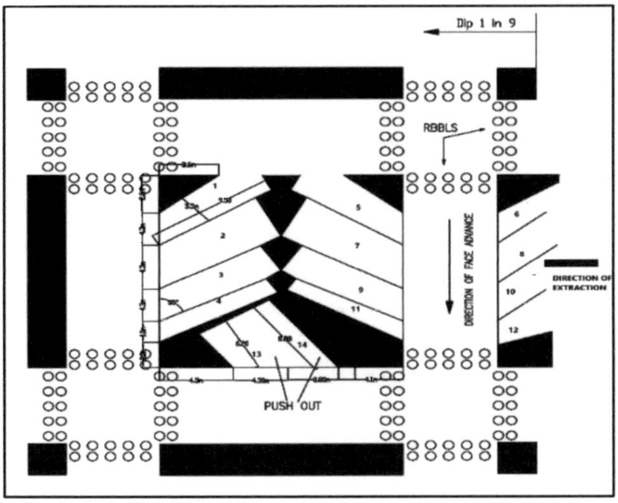

Figure 7.
Christmas tree method.

In India, it is being practiced mainly at shallower depths due to its design, that is, extraction sequence, but it can be successfully implemented even at a depth of around 600 m. This method was successfully adopted at a depth of more than 600 m in the United States. It cannot be practiced for large pillars since the process involves driving slices where we provide supports only at rib pillars, which results in a considerable loss in extraction in case of large pillars and stability reasons. The schematic manner of extraction in the christmas tree depillaring method is shown in **Figure 7**.

2.4.3 Nevid method

Neels Joubert and David Posthma developed this extraction method, so it is named NEVID after them. In this method, the snooks are predesigned in such a way as to ensure that the roof to let down in a controlled and safe manner, that is, adequate strength (width to height ratio) is maintained. For every three rows, a stooper pillar was left in addition to snooks formed. It was found that sometimes the snooks created failed unexpectedly due to inadequate strength. The primary purpose of the stopper

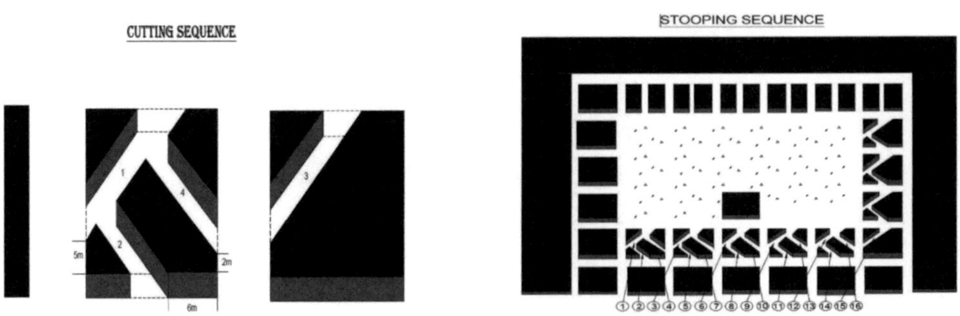

(a) (b)

Figure 8.
Nevid method (a) pillar extraction sequence, and (b) pillar extraction sequence layout for a complete panel [38].

pillar is to take an additional load off the pillars, which are extracted while they are spaced far enough apart to crush as required, mainly when the goaf hangs [38–40].

The main reason for the immediate successful results of this method is that compared to the previous mining methods in South Africa, the cut's length was reduced and maintained appropriately to pillar size. All the lifts were driven into the pillar at an angle of 45 degrees. Another aspect of the method's success was all the lifts in the extraction sequence were marked with the help of survey pegs and direction lines. In 2001, this method adopted panel zoning, which is a process of identifying the hazards based on geological information with the help of geological mapping to zone a panel during the development phase into low-to-high-risk zones [38]. The schematic manner of extraction to be followed in the Nevid method is shown in **Figure 8**.

2.4.4 Yield pillar method

The yield pillar method is one of the oldest extraction methods employed in underground coal mining. The main principle of the yield pillar method is to design the size of remnant pillars in such a way that they will fail in a slow and controlled manner. Thus, the method involves leaving pillars in the back area in a systematic way so that they yield over a period of time. The remnant pillars, which are allowed to yield, are designed in such a way that their width-to-height ratios are greater than five. Slices are made into pillars in a systematic and sequenced way. Along with a remnant pillar for yielding, this method allows leaving snook. The extraction of snook is highly dependent on local strata behavior, and experiences gained [41–43]. The schematic manner of extraction in the yield pillar technique is shown in **Figure 9**.

Although yield pillar behavior is conceptually simple, in-mine implementation is often difficult. These difficulties arise, in part, from-

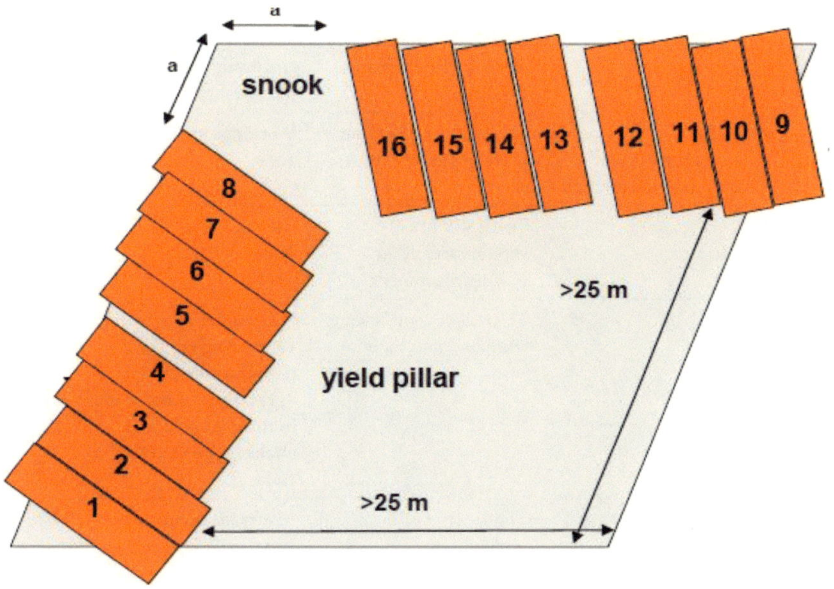

Figure 9.
Yield pillar method.

- A lack of one universally accepted yield pillar technique;
- Implementation of a successful design at operation(s) having different geo-mining conditions;
- Implementation of the yield pillar design under possibly inappropriate conditions, for example, under weak roof conditions or at insufficient depth to induce pillar yielding.

Advantages, disadvantages, and present status of various mining methods are shown in **Table 1**.

3. Advantages, disadvantages, and present status of various mining methods

Parameters	Longwall	Longwall top coal caving	Wongawilli	Room and Pillar
Advantages	Low operating cost	Face relocation frequency is less	Development work is minimum to maximize the work in retreating.	Moderate operating cost
	High production rate	High production rate	High production rate	Rapid development rate/ Moderate production rate
	Low dilution	Less equipment is required	The number of active working faces will be minimum, reducing the interface tramming.	Excellent ventilation
	Applicable to thick seam mining	Adopted at various thicknesses, moderately steep thick coal seams (4–20 m)	Extraction sequences were simple, repetitive, and easily understood by the employees	Not suitable for thick seam mining
	High mechanization/ continuous method	Successful application in steep dipping thick seams	Workings in destressed zone	Continuous production
Disadvantages	High capital investment	Compatibility of support and other mining equipment	Difficulties in removing stooks	Moderate capital costs
	High development	The risk of spontaneous combustion is high	Difficulties in ventilating rib pillar panels when the roof caves, thus filling voids in the goaf area and choking the ventilation flow	The risk of spontaneous combustion is low/limitation on the depth
	Low selectivity and flexibility	The amount of dust is high	Moderate selectivity and flexibility	Moderate selectivity and flexibility
	High subsidence	High subsidence	Moderate subsidence	Moderate subsidence

Parameters	Longwall	Longwall top coal caving	Wongawilli	Room and Pillar
Present status	Highly successful in China, the US, UK, Australia, and semi-successful in India due to a lack of machinery	Successful in countries, such as China, the UK, France, the US, and Russia, and yet to be implemented in India.	Most successful in Australia and New Zealand and yet to be implemented in India	The most common and flexible method in all countries, but liable to less production rate

Table 1.
Advantages, disadvantages, and present status of various mining methods.

4. Role of continuous miner in Indian coal mechanization

The new era of mechanized underground coal mining in India started with the implementation of continuous miner in 2002 [44]. Since then, there has been a steady

1870	English channel machine
1877	Harrison compressed air puncher
1877	Jeffrey (Lechmer) breast machines
1897	Sullivan air operated shortwall cutter
1898	Sullivan electrically operated shortwall cutter
1907	Goodman shortwall machine
1911	Jeffrey track cutter
1920	Jeffrey entry driver
1945	Joy 10-RU cutter
1948	First jeffrey colmol
1948	Joy 3-JCM continuous miner
1950	Lee Norse koal master
1950	Caterpillar introduces its first self-propelled wheel tractor-scraper
1951	Crawler mounted Lee Norse miner
1968	Lee Norse CM-26H miner
1954	Joy 8-CM oscillating continuous miner
1967	Jeffrey heli miner
1968	Joy ripper chain miner
1970	National mine service marietta 3060 miner
1970	Lee Norse 265 continuous miner
1972	Lee Norse 245 continuous miner
1972	Joy high-seam (12CM series) low-seam (14CM series) continuous miners
1974	Remote controlled continuous miner
1974	Joy 14-CM miner with remote control & dust collector (Jewell Ridge Coal Co.)
2000	Mid-seam high-voltage continuous miner (model 14CM27)
2012	Cat CM235 continuous miner

2020s Currently operating CMs	Joy 12 HM series (12HM21-AAA, 12HM31-B, 12HM37-A, 12HM37-B, 12HM37-C)
	Joy 14 HM series (14HM15, 14HM27)
	Joy 12 series (12CM30, 12CM27, 12CM15, 12CM12)
	Joy 14 series (14CM27, 14CM15, 14CM10, 14CM9)
	Caterpillar CM200 series (CM210, CM220, CM230, CM235, CM240)
	Caterpillar CM300 series (CM330, CM340, CM345)
	Caterpillar CM400 series (CM440, CM445)
	Caterpillar miner bolter (CM845)
	Sandvik MC series (MC470, MC430, MC390, MC350, MC250)
	Wildcat JAE42 continuous miner

Table 2.
Evolution of continuous miner.

increase in continuous miners implemented in Indian coal mines with increasing mechanization and safety in underground mines. But most of the 213 underground coal mines are still highly dependent on conventional mining. This is high time to adopt complete mechanized mining in all underground coal mines with continuous miner technology.

Even though continuous miner is a major change in mechanization and safety in Indian coal mines, but still are not able to increase underground coal production due to adverse strata conditions, lack of implementing the suitable mining method prior to existing geo-mining conditions, lack of understanding of the caving nature of overlying strata before starting depillaring, which eventually results in a loss in production by leaving the number of pillars untouched in safety point of view. Undoubtedly, the future of Indian underground coal mining lies in successful pillar extraction with continuous miners. This is high time to study the stability parameters influencing the continuous miner panel workings in underground coal mines. Evolution of continuous Miner [45–48] is shown in **Table 2**.

Sl.no	Mine	Sl.no	Mine
1	Jhanjra mine (ECL)	12	Churcha R.O mine (SECL)
2	Shyamsundarpur UG (ECL)	13	Vijaya west (SECL)
3	Kumardih-B CM UG (ECL)	14	Churi UG mine (CCL)
4	Anjan Hill (SECL)	15	VK-7(SCCL)
5	Vindhya (SECL)	16	GDK-11 (SCCL)
6	Haldhibadi (SECL)	17	Kundapuram (SCCL)
7	Khairaha (SECL)	18	PVK-5 (SCCL)
8	Bangwar (SECL)	19	Shantikhani (SCCL)
9	Pinoura mine (SECL)	20	Vakilpalli (SCCL)
10	NCPH Old mine (SECL)	21	Khottadih CM UG (ECL)
11	Rani Atari (SECL)	22	Sheetaldhara-Kurja mine (SECL)

Table 3.
List of mines operating with continuous miners.

DOI: http://dx.doi.org/10.5772/intechopen.109083

Since the implementation of continuous miner, there has been a steady increase in the deployment of continuous miners in India. Here, some of the operating mines using continuous miners in India is listed below [49, 50].

Some of the lists of mines in Eastern Coalfields Limited (ECL) are Jhanjra mine, Shyamsundarpur UG, Kumardih-B CM UG, and Khottadih CM UG. List of mines in South Eastern Coalfields Limited (SECL) is Anjan Hill, Vindhya, Haldhibadi, Khairaha, Bangwar, Pinoura mine, NCPH Old mine, Rani Atari, Sheetaldhara-Kurja mine, Churcha R.O mine, and Vijaya west mine. The list of Mines in Central Coalfields Limited (CCL) and Western Coalfields Limited (WCL) is Churi UG mine and Tandsi mine, respectively. The list of mines in Singareni Collieries Company Limited (SCCL) are VK-7, GDK-11, Kundapuram, PVK-5, Shantikhani, and Vakilpalli mine. The list of mines operating with Continuous miners in India is shown in **Table 3**.

5. Summary

It is necessary to point out that the many concrete recommendations made, for example, values of factor of safety (FOS), are intended only as a general guide. It is always dangerous to generalize in mining, where geological and other conditions vary widely. The recommendations should always, therefore, be reassessed in terms of local experience. Although the safety factor of the pillars is important to ensure that the pillar strength is adequate to support the overlying strata, it is only the start of the design process. Other factors that play equally important roles are the mining method to be practiced, the equipment to be used, and the behavior of the local strata. In the past, the safety factor was too often the only parameter used to determine the size of the pillars, and other critical factors were ignored. The behavior of the overburden strata is possibly the most critical Factor as it also affects the mining method, the type of mining method that can be used, and the support required during the development and pillar extraction process.

The mining method should be predetermined, and the pillar should be designed to accommodate this method. The type and size of the mining equipment also has a major influence on the type of mining that can be practiced, and the pillar dimensions required for safe and effective extraction. The critical factors that influence the manner of pillar extraction are depth of cover, physicomechanical properties of overlying strata and seam, incubation period, dip and thickness of the seam, and technology available for extraction. In the future, continuous miner is going to be the future of coal mining. It is highly recommended to understand the importance of mining methods and their impact on safety.

Funding acknowledgement

This research has no funding from any source.

Acknowledgements

The authors express their gratitude to IIT(ISM), Dhanbad, for providing all the necessary facilities for carrying out research work at IIT (ISM).

Declaration of competing interest

The authors declare that they have no known competing financial interests or personal relationships that could have appeared to influence the work reported in this paper.

Author details

Lingampally Sai Vinay[1*], Ram Madhab Bhattacharjee[1] and Nilabjendu Ghosh[2]

1 Department of Mining Engineering, IIT (ISM), Dhanbad, India

2 Central Institute of Mining and Fuel Research (CSIR-CIMFR), Dhanbad, India

*Address all correspondence to: dr.vinay439@gmail.com

References

[1] Coal facts 2022, in London: World Coal Association

[2] Arens M, Åhman M, Vogl V. Which countries are prepared to green their coal-based steel industry with electricity?- Reviewing climate and energy policy as well as the implementation of renewable electricity. Renewable and Sustainable Energy Reviews. 2021;**2020**:143

[3] Provisional Estimates of Annual National Income, 2020-21 and Quarterly Estimates (Q4) Of Gross Domestic Product, 2020-21. Gross Domestic Product, 2020-21, Ministry of Statistics and Programme Implementation

[4] Ministry of Coal, Government of India (GOI). Ministry of Coal. 2022

[5] Geological Survey of India (GSI). Geological Survey of India, 2022

[6] Indian Bureau of Mines (IBM). Indian Bureau of Mines. 2022. Available from: https://ibm.gov.in/

[7] Indian Minerals Yearbook 2019. Indian Bureau of Mines. Indian Minerals Yearbook 2019. 2019

[8] Vinay LS et al. Numerical study of stability of coal pillars under the influence of line of extraction. Geomatics, Natural Hazards and Risk. 2022;**13**(1):1556-1570

[9] Das AJ et al. Evaluation of stability of underground workings for exploitation of an inclined coal seam by the ubiquitous joint model. International Journal of Rock Mechanics and Mining Sciences. 2017;**93**:101-114

[10] Mandal PK et al. Underpinning-based simultaneous extraction of contiguous sections of a thick coal seam under weak and laminated parting. International Journal of Rock Mechanics and Mining Sciences. 2008;**45**(1):11-28

[11] Singh R et al. Optimal underground extraction of coal at shallow cover beneath surface/subsurface objects: Indian practices. Rock Mechanics and Rock Engineering. 2007;**41**(3):421-444

[12] Singh R et al. Coal pillar extraction at deep cover: With special reference to Indian coalfields. International Journal of Coal Geology. 2011;**86**(2-3):276-288

[13] Singh R et al. Upshot of strata movement during underground mining of a thick coal seam below hilly terrain. International Journal of Rock Mechanics and Mining Sciences. 2008;**45**(1):29-46

[14] Singh AK et al. Coal pillar extraction under weak roof. Mining, Metallurgy & Exploration. 2020;**37**(5):1451-1459

[15] Sheorey P et al. Pillar strength approaches based on a new failure criterion for coal seams. International Journal of Mining and Geological Engineering. 1986;**4**(4):273-290

[16] Sheorey P et al. Coal pillar strength estimation from failed and stable cases. International Journal of Rock Mechanics and Mining Sciences & Geomechanics Abstracts. 1987:**24**(6):347-355. ISSN: 0148-9062. DOI: 10.1016/0148-9062(87)92256-X. Available from: https://www.sciencedirect.com/science/article/pii/014890628792256X

[17] Ram S et al. Design of rock bolt-based goaf edge support for conventional depillaring with stowing. Arabian Journal of Geosciences. 2021;**14**(21):1-13. DOI: 10.1007/s12517-021-08379-3

[18] Ivaz JS et al. A retrospective comparative study of Serbian underground coalmining injuries. Saf Health Work. 2021;**12**(4):479-489

[19] Peng SS. Longwall Mining. London: CRC Press; 2019. DOI: 10.1201/9780429260049

[20] Pappas DM, Mark C. Behavior of Simulated Longwall Gob Material. United States of America: Department of the Interior; 1993:1-45

[21] Mitchell G. Longwall Mining: Australia: Australasian Coal Mining Practice. 2009. pp. 340-373

[22] Alehossein H, Poulsen BA. Stress analysis of longwall top coal caving. International Journal of Rock Mechanics and Mining Sciences. 2010;**47**(1):30-41

[23] Xu B. Application of the Longwall Top Coal Caving System in Australian Thick Seam Coal Mines. Australia: University of New South Wales; 2004

[24] Ediz IG et al. application of retreating and caving longwall (top coal caving) method for coal production at GLE Turkey. Mining Technology. 2013;**115**(2):41-48

[25] Liu H et al. Comprehensive study of Strata movement behavior in mining a longwall top coal caving panel of a composite coal seam with partings. Applied Sciences. 2020;**10**(15)

[26] Wang J, Wang Z, Li Y. Longwall top coal caving mechanisms in the fractured thick coal seam. International Journal of Geomechanics. 2020;**20**(8)

[27] Yang W et al. Study on the characteristics of top-coal caving and optimization of recovery ratio in steeply inclined residual high sectional coal pillar. Geofluids. 2020;**2020**:1-10

[28] Martin CH. Australia: Australasian Coal Mining Practice. 1986

[29] Lind G. Coal pillar extraction experiences in New South Wales. South African Institute of Mining and Metallurgy. 2002;**102**(4):199-206

[30] Lind G. Key success elements of coal pillar extraction in New South Wales. South African Institute of Mining and Metallurgy. 2002;**102**(4):207-216

[31] Lien L. Advances in coal mining technology. In: The Coal Handbook: Towards Cleaner Production. Cambridge, UK: Woodhead Publishing Limited; 2013. pp. 193-225

[32] Hustrulid WA. Underground Mining Methods Handbook. United States of America. 1982

[33] Zipf Jr RK. Toward Pillar Design to Prevent Collapse of Room-and-Pillar Mines. United States of America. 1900

[34] Jiang B et al. Ground pressure and overlying strata structure for a repeated mining face of residual coal after room and pillar mining. International Journal of Mining Science and Technology. Cambridge, UK: Woodhead Publishing Limited; 2016;**26**(4):645-652

[35] Peng S. Advances in Coal Mine Ground Control. Woodhead Publishing; 2017

[36] Singh R et al. Rib/snook design in mechanised depillaring of rectangular/square pillars. International Journal of Rock Mechanics and Mining Sciences. 2016;**84**:119-129

[37] Galvin JM. Ground Engineering - Principles and Practices for Underground Coal Mining. Springer International Publishing; 2016

[38] Lind G. An integrated risk management approach for underground

coal pillar extraction in South Africa. Journal of the Southern African. Southern Africa: Institute of Mining and Metallurgy. 2005;**105**(2):137-147

[39] Lind GH. Risk management of underground coal pillar extraction in South Africa. International Journal of Surface Mining, Reclamation and Environment. 2005;**19**(3):218-233

[40] Moolman C, Canbulat I. United States of America: COALTECH 2020. 2003

[41] Sheorey P et al. Application of the yield pillar technique for successful depillaring under stiff strata. International Journal of Rock Mechanics and Mining Sciences & Geomechanics Abstracts. 1995:699-708

[42] Li W et al. Numerical modeling for yield pillar design: A case study. Turkey: Rock Mechanics and Rock Engineering. 2015;**48**(1):305-318

[43] Yavuz H. Yielding pillar concept and its design. In: 17th International Mining Congress and Exhibition of Turkey-MCET. 2001

[44] Leeming JJ. Introduction of continuous miner technology into Indian coal mines. Journal of Mines, Metals and Fuels. 2003;**2003**:327-335

[45] Caterpillar Inc. Caterpillar Inc. 2022. Available from: https://www.cat.com/en_US/by-industry/mining.html

[46] Komatsu Limited. 2022. Available from: https://www.komatsu.jp/en/industries-we-support/mining

[47] J A Engineering Australia Pvt. Ltd. 2022. Available from: https://www.jaeaust.com.au/manufacturing/continuous-miners/

[48] Sandvik. 2022. Available from: https://www.rocktechnology.sandvik/

[49] Singareni Collieries Company Limited (SCCL). Singareni Collieries Company Limited (SCCL). 2022. Available from: https://scclmines.com/scclnew/index.asp

[50] Coal India Limited (CIL). Coal India Limited (CIL). 2022. Available from: https://www.coalindia.in/

Section 6

When Disasters Strike: Navigating the Social Challenges of Natural Hazards

Chapter 22

Psychological Reactions after Disasters

Hadis Amiri and Azra Jahanitabesh

Abstract

As the world's population increases and resources are limited, societies become increasingly vulnerable to disasters. Regardless of the objective destructive effects, the psychosocial effects and consequences of natural disasters are quite clear in humans. Natural disasters exert different psychological effects on the exposed people, including but not limited to Post-Traumatic Stress Disorder (PTSD), depression, anxiety, and suicide. Yet, disasters can cause positive reactions. For example, through post-traumatic growth, one takes new meaning from his or her trauma and resumes living in a way completely different from their life prior to the trauma. Additionally, many people display remarkable resilience in the wake of the disasters they struggled with. Many factors such as disaster type, level of destruction, duration of disasters, timing (time of day, day of week, season) also individual indicators such as age, gender, marital status, education, pre-disaster mental health, social and economic status, and resilience are affecting the consequences of disasters. Given the many studies that focus on post-disaster psychological outcomes, in this chapter, we not only describe outcomes but also discuss psychosocial support in disasters.

Keywords: disaster, mental health, psychosocial support, post-disaster psychological outcomes, resilience

1. Introduction

About 200 million people are annually exposed to disasters worldwide, according to the Centre for Research on the Epidemiology of Disasters [1]. Definitions of disasters have varied in the literature. Some authors have defined disaster as a serious disruption in the functioning of society due to an event or natural disaster such as an earthquake, flood, tsunami, etc., leading to environmental, human, economic, and biological damage [2]. On the other hand, Davidson & Baum defined disaster as the subjective psychological response to any event [3]. Regardless of the definition, many studies indicated that disasters have both objective destructive effects and psychosocial effects and consequences [4].

Today, there is extensive information based on studies on psychological and behavioral issues caused by natural disasters. A systematic review showed that the burden of PTSD among persons exposed to disasters is substantial [5]. Other adverse psychological outcomes such as depression, anxiety, stress, and suicide were also mentioned in some studies as disasters' outcomes [6]. For instance, the prevalence

IntechOpen

rates of depression among youth post-disaster reported in a review ranged from 2 to 69%. This rate depended on potential risk factors identified, including female gender, exposure stressors, and post-traumatic stress symptoms [7]. Although mental disorders usually decrease in the second year after the disaster, in some cases, complications remain chronic [8]. Another critical issue is the vulnerability of the people themselves and the possibility of previous disorders in these people, which will undoubtedly lead to more severe reactions [9]. The simultaneous presence of several mental illnesses increases the intensity of mental damage. The most common disorder we deal with is post-traumatic stress disorder (PTSD), although in many cases, we find this disorder together with other disorders [10]. The most common co-occurring disorders are major depression, panic, and phobia. In fact, after a disaster, mental disorders attack a person, such as multiple injuries [11].

In addition to the negative psychological outcomes of disasters, there are some positive consequences, such as post-traumatic growth (PTG). For some people, exposure to disasters or any traumatic events, which may contain great suffering and loss, can lead to very positive changes in the individual [12]. About 30–90% of people with traumatic experiences at some point in their life have reported at least one form of PTG [13]. Different traumatic events have different effects on PTG, with people with severe trauma reporting greater benefits, and chronic events can have very different psychological effects than acute events [14]. In general, PTG refers to positive psychological changes and greater growth than the pre-crisis level of performance, which through cognitive reconstruction makes the person adapt to the new reality [14].

Another object that is related to psychological outcomes after disasters is resilience. There are many resources of stress in life; also, there is so much variability in how people respond to and manage life's stressors. Resilience refers to the process of a person's return to normal functioning after a stressful event or uncomfortable experience, but PTG refers to growth compared to the pre-disaster situation. We will explain this in a separate section [15].

Furthermore, the research results related to mental reactions after disasters give us a view of the common disorder and the needs of the survivors. We tried to explain psychosocial support in disasters according to our experiences after defining resilience.

2. Psychological reactions after disaster

2.1 Common reactions and symptoms following a disaster

Following the occurrence of severe mental stress caused by a disaster, a group of symptoms and disorders appear in people, which can have an adverse effect on their performance. In some people, we are faced with only one symptom, and in others, we are faced with a combination of symptoms that, if not addressed, can lead to chronic mental disorders. The degree of influence of people from the surrounding events is essentially a function of the previous vulnerability, the extent of the disaster, and the amount of destruction and loss, as well as the status of support and timely attention to the psychosocial issues of the people [16]. Symptoms that are commonly seen in disaster survivors include:

1. Disturbing thoughts: Among the most unpleasant phenomena that children and adults experience after traumatic events are disturbing memories, thoughts,

and feelings. These memories may come unread and unwanted at any time of the day due to depression or in response to environmental reminders. They may appear at night in the form of nightmares or terrible dreams [17]. Since these memories are very bright and scary, they impose a lot of psychological pressure on the person. Many people are afraid of going crazy or losing control after such experiences. Therefore, one of the main goals of group counseling sessions is to make people understand that such reactions are normal and do not cause insanity, and they should also be taught skills to control these recurring memories so that they can deal with these memories and have control over them.

2. Over-arousal: Children and adults may show increased psychological arousal after traumatic events. Therefore, people may feel nervous, restless, anxious, and panic very quickly. They may become agitated, suffer from tinnitus, and may experience excessive japer dine or concentration problems [18]. A person's performance in terms of social relationships and academic as well as occupational performance is strongly influenced by over-arousal, which in turn exacerbates the experience of distressing memories. The negative and direct effect of excessive arousal on a person's life and its possible role in strengthening other annoying, stressful symptoms after a disaster means that this area is one of the important parts of intervention and prevention in this field. The training of relaxation skills plays a fundamental role in preparing a person for confrontation work and gradual exposure to factors that cause over-arousal [19].

3. Avoidance: Avoidance may be cognitive and conscious (i.e., trying not to think and remember disturbing events); or it may be in the form of behavior (that is, trying to avoid factors that remind of painful and disturbing events, such as some places and people; and avoiding talking about the incident). Among other avoidance symptoms, the following can be mentioned *Manifestation*: lack of interest or a significant decrease in interest in dealing with important matters, feelings of heartbreak or strangeness among others, a feeling of not knowing everything (for example, he does not expect to have a job or get married and have children) [20]. Avoidance symptoms should be considered targets of interventions for two main reasons. First, avoidance may cause limitations in a person's performance, which in this case has a direct effect on a person's life. Second, although avoidance may cause temporary and short-term relaxation, it will keep problems in the long run. Annoying symptoms can cause avoidance behavior, and following avoidance, the processing of information about the incident is disturbed, and finally, a defective cycle is formed, the result of which is the aggravation of symptoms. Also, we should remember that although avoidance will bring peace in the short term, it will cause more anxiety and tension in the long term [21].

4. Sleep disorder: It can be said that most of the survivors of unexpected events suffer from sleep disorders at some point in time. All the changes that have occurred around people, including loss of loved ones, fear of possible future events, loss of assets, and change of residence, can lead to sleep problems [22]. Sleep problems can exist alone or together with other symptoms, especially symptoms of hyperarousal [23]. The disorder is usually in the form of difficulty falling asleep, lack of continuity of sleep, and nightmares. In addition to being annoying for a person in itself, sleep problems cause a delay in mental balance

and are affected by other symptoms, such as the repetition of disturbing thoughts [24]. In fact, the various symptoms seen in people following disasters can lead to the formation of a vicious cycle and ultimately intensify each other. On the other hand, daily fatigue and lack of energy renewal during the night hours can cause damage to the cardiovascular system and the immune system and make a person susceptible to the occurrence of diseases.

5. Grief reaction: Grief is a reaction that occurs naturally in response to the loss and lack of what a person has depended on. Loss can be due to the loss of relatives and friends, assets and belongings, or a person's future career [25]. It should be remembered and emphasized that grief is completely normal, and we naturally expect such a reaction after unexpected events [25]. Grief reaction is a set of emotions that are different depending on the culture that governs the society. Naturally, this reaction is expected to resolve within two months. The severity of the symptoms and the time of their occurrence also play a role in determining whether this reaction is normal or abnormal. In general, the intensity of the mourning reaction should be proportional to the loss that the person is facing, to be resolved within 2 months, and not started late. Its symptoms should gradually decrease, and the bereaved person should be able to return to their previous level of performance within 2 months. In fact, mourning is a process that ultimately leads to the acceptance of abandonment by humans [26]. In some cases, people are wrongly prohibited from expressing their feelings, crying and mourning, and refraining from expressing feelings is interpreted as "resistance." In certain conditions, the probability of morbid grief increases, including in cases where the loss was very extensive, accompanied by great panic, or happened very quickly, in cases where the person already had a high vulnerability, was isolated, and lacked a social support network, or suffered from disorders such as depression [26]. Also, people who somehow consider themselves involved in the occurrence of the accident or the extent of its effects are more prone to abnormal reactions during mourning. The role of specialists who work in the field of psychosocial interventions is to facilitate the mourning process. In order to achieve this goal, we can also seek help from religious leaders.

2.2 Common mental disorders following disasters

The executive policy of mental health officials should be focused on normalizing reactions caused by disasters. In fact, it should be acknowledged that most of the reactions that occur after disasters are natural reactions to the very unusual incident. One of the general goals of psychosocial support during disasters and unexpected events is to empower the victims and improve their adaptive mechanisms. A disease-oriented approach in community-based psychiatry in disasters will be a serious obstacle to reaching these goals. Therefore, the use of the words disease and disorder should be done very carefully. Of course, there is no doubt that depriving the victims of proper treatment when a definite diagnosis of a psychiatric disorder is involved is also an unethical act, and in necessary cases, there may be a need for more serious treatments, even hospitalization.

The most common disorders that are observed after unexpected events are: acute stress disorder, post-traumatic stress disorder, depression, abnormal grief, somatic disorders, and various anxiety disorders [27].

1. Acute stress disorder: This disorder occurs in situations where a susceptible person has experienced traumatic events with the threat of death or serious harm to his or others' health and, following this experience, a feeling of intense fear and panic accompanied by all or some of the following symptoms:

    - Feeling numb and emotional indifference
    - Decreased awareness of the surrounding environment
    - The feeling of unreality of oneself or the feeling of unreality of the world around us
    - Psychogenic forgetfulness

The traumatic event may be repeated by itself or following thoughts, dreams, sensory errors, repetition of memories by the people around, or the feeling of the incident happening again for the survivor [28]. Avoidance is one of the common and obvious symptoms of this disorder because the person tries to avoid recalling painful memories and re-experiencing the arousal caused by them. Different types of sleep disorders, excitability, problems in concentration, restlessness, excessive ringing, and severe startle reflex are some of the other symptoms that we expect in this disorder [28]. The main difference between this disorder and post-traumatic stress disorder is the duration of the illness, this disorder lasts between 2 days and 4 weeks, and if it is not resolved in this period of time, it turns into post-traumatic stress disorder. The affected person clearly suffers a drop in personal and social functioning [29].

2. Post-traumatic stress disorder: As explained earlier, the symptoms of this disorder are similar to an acute stress disorder, and the main difference between the two is the time of illness. The survivor feels helpless after the accident, and the traumatic incident is repeated for him or her repeatedly. The occurrence of this disorder is possible at any time after the accident; it can be immediately after the accident or even delayed 30 years after the accident. But what is certain is that quick and timely intervention in the case of single symptoms seen in people or intervention in the case of acute stress disorder can be effective in reducing the prevalence of this disorder in the Asian community [29]. The symptoms of this disorder are mainly in the areas of repetition of the accident (the feeling of the accident happening again, in the form of mental images, dreams, repetitive games about the accident in children, etc.), avoidance (avoiding the place of the accident, people who were in that time being close to the person, avoiding talking and discussing, problems in recalling traumatic memories), feeling numb and helpless along with feelings of strangeness and alienation, as well as symptoms of hyperarousal (ears ringing excessively, problems falling asleep or continuity of sleep, problems in concentration and intensification of jumping reflex) are observed [29].

3. Complicated grief: In cases where the mourning reaction does not disappear within 2 months, the severity of the symptoms during the mourning period is more than expected, or there is a delay in the onset of bereavement symptoms, we are facing complicated bereavement that needs immediate intervention. Sometimes the symptoms of mourning do not appear at all [30]. Women are more prone to such complications than men. If there is no timely intervention, there is a possibility of other psychiatric disorders, such as major depression [31].

4. Major depression: Although all the survivors have a natural grief reaction, and after it is resolved, they feel sad by recalling past memories, it should be remembered that this natural psychological reaction to a great loss is completely different from depression [26]. Depression is not a natural reaction to a crisis, and if it occurs, it needs serious and immediate intervention. Its prominent feature is a depressed mood with hopelessness and helplessness, anxiety, decreased energy, sleep disorder, loss of appetite, memory disorder, sense of emptiness and worthlessness, and sometimes suicidal thoughts. Depression is one of the most common disorders [32].

5. Types of anxiety disorders: Panic attacks and anxiety can be seen alone or together with other disorders in people who survived the accident, although more commonly, we find these disorders together with other disorders. The presence of such disorders causes a decrease in performance and a slow return to normal life [33].

6. Somatic disorders: Paying attention to the higher prevalence of pseudo-somatic disorders following disasters is one of the points that should always be considered by the doctors treating the survivors. Because if the psychogenic origin of these disorders is not diagnosed, they are treated as physical symptoms [34]. Usually, in communities that survived an unexpected accident, the number of physical complaints is high. Although the efforts of the medical personnel to find the organic origin of these pains and symptoms are not successful, the patient believes that she or he has an undiagnosed disease and worries too much about her or his body's health. Of course, there may be physical problems and diseases in some cases, but in this case, the amount of complaints from the patient is disproportionate to the severity of the physical problems. The symptoms of this difference cover a wide spectrum, including digestive symptoms, sensory movement, and even paralysis and diplopia. In normal conditions, 20–30% of adult patients referred to health centers have pseudo-physical problems. This amount increases significantly after unexpected events. A point that should be noted is avoiding unnecessary diagnostic measures in these patients. Due to the referral of these people to general practitioners and non-psychiatrist specialists, training this category of healthcare workers is of particular importance.

2.3 Resilience

As mentioned earlier, the effects of disasters are influenced by a variety of variables, including disaster type, degree of destruction, duration, timing (time of day, day of week, season), and individual indicators, including age, gender, marital status, education, pre-disaster mental health, social and economic status, and resilience. Many of studies explained these variables [35, 36]. Nevertheless, the role of resilience in the effects of disasters has not been addressed much. Therefore, in this section, we describe the role of resilience in disasters.

First, we want to ask you to check **Figure 1**. This figure shows a building that was not destroyed in an earthquake in Iran because it was strong against the earthquake, although other buildings have been completely destroyed. We have shown this image from another angle in **Figure 2**. The building captured in the shot is a real-life example of resilience: the ability to become strong, happy, or successful again after a difficult situation or event [37]. This means you will be happy, strong, successful,

Figure 1.
The building that was not destroyed in an earthquake.

Figure 2.
The building that was not destroyed in an earthquake.

and outstanding under high-risk conditions. We imagine that resilience is displaying outstanding strength against problems, and it is the kind of post-traumatic growth (PTG). On the contrary, PTG is positive psychological alterations, and performance improvements over pre-crisis levels enable the individual to adapt to the new environment through cognitive restructuring. It means you have grown compared to before in five aspects including: "Recognizing Personal Strength" occurs when a person feels more confident. "Finding Unknown Possibilities and Opportunities" is experienced when people find a new way of life that would not be available if they did not experience a traumatic event. The domain of "Experiencing Positive Changes in Relationships" that shows a sense of kindness or closeness to others. "Appreciation of life" seems to have more appreciation for each new day of life. Finally, the scope of "Spiritual and Existential Change" understands personal growth that has a much better understanding of spiritual issues [14]. In PTG, you will grow more than before, but in resilience, you come back to the before situations or not negative changes after disasters. Several major studies have documented resilience as the "ability to go on with life," "Bend, but not break," or rebound from adversities; learn to live with ongoing fears and uncertainty [38, 39].

Now that we know the importance of resilience, we should know that psychosocial support is one of the factors that influence the resilience and PTG.

2.4 Psychosocial support

One of the main topics in psychosocial support is familiarity with the basic principles in this field. In this section, we try to explain these main topics. It is necessary to observe the following principles before and during the implementation of the program:

1. Taking a community-based approach: In this approach to build a local resource, including education and services, attention is paid to the whole community. In this way, the people who provide services offer their knowledge and skills to similar groups in the community, and these trained local resources become the cause of the relief and peace of their community. Focusing on groups and social networks instead of individuals can help more people. In addition, paying attention to the ruling values of the society will provide an appropriate answer to the culture in that society.

2. Trained volunteers: The main principle of the support program is to use volunteers. The purpose of training volunteers is to learn basic skills in the field of crisis. Due to the fact that these people are members of the affected society, they can react immediately in times of crisis, and therefore they are considered a valuable resource. In addition, since these people are members of the affected society, the injured have more trust in them.

3. Empowerment: In emergency aid programs that are always there, the danger is to humiliate, passivate, disempower, and make the survivors dependent on the relief forces. In general, relief organizations should be aware of that the quality of relief should be based on the principles of self-respect and independence of people so that it leads to the empowerment of people and emphasizes their abilities and strengths. For this purpose, survivors should actively participate in all programs.

4. Social participation of survivors in relief and reconstruction: Having a sense of control over the environment causes empowerment of people, a sense of belonging and ownership, and more effort to solve problems. Therefore, in programs that use the participation of local people, the results are more stable and prosperous. Because people feel that they have control over their lives and community through participation.

5. Accuracy in the use of specialized terms: Care must be taken in using specialized terms, because, for example, after a disaster such as an earthquake, many people may have symptoms such as re-occurrence or avoidance or feelings of sadness and depression, but all of them do not go beyond one sign, and we cannot make diagnoses such as post-traumatic stress disorder or major depression. Conversely, using words with positive meaning can evoke a feeling of empowerment and participation, such as the term active survivors.

6. Providing services from the first moment: Providing these services from the first moment can be an important factor in helping people to deal with the crisis. Neglecting emotional reactions causes the creation of passive victims. And as a result, the process of recovery and reconstruction of people progresses more slowly than usual.

7. Practical and ongoing interventions: Due to the short-term and long-term effects that disasters have on people, Psychological problems caused by disasters may not appear immediately, but they will take some time. As a result, it is necessary to continue the programs of support and participation of local forces by training local resources.

8. Providing service according to the support pyramid: Interventions and psychosocial support during disasters and accidents should be provided at different levels. This means that different affected people need different support. **Figure 3** shows that different people need interventions in different layers and levels [40].

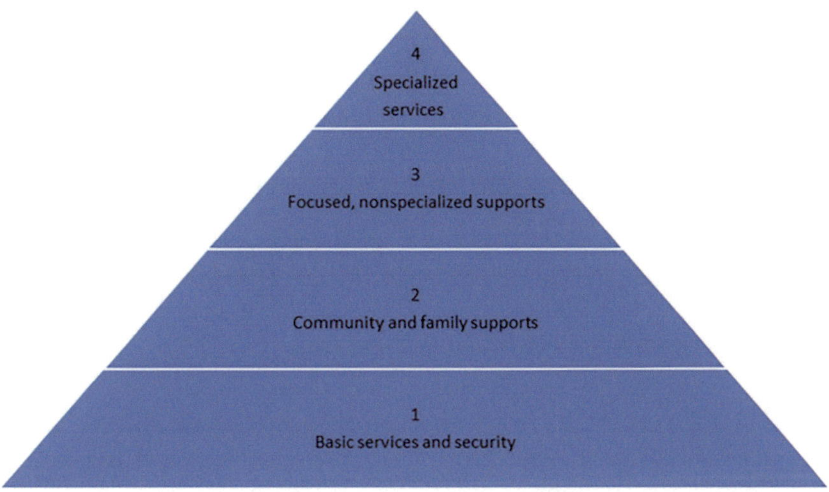

Figure 3.
Intervention pyramid for mental health and psychosocial support in emergencies [40].

3. Conclusions

Following disasters, most people experience unpleasant emotions and experiences. Reactions may be a combination of confusion, fear, despair, helplessness, insomnia, physical pain, anxiety and anger, grief, shock, violence and mistrust, guilt and shame, and loss of trust and self-confidence. Psychological first aid provides a safe, comfortable, connected, self-efficacious, empowered, and hopeful environment.

Acknowledgements

Open-access availability of this work supported by UC Davis.

Conflict of interest

"The authors declare no conflict of interest."

Author details

Hadis Amiri[1*] and Azra Jahanitabesh[2]

1 Health in Disasters and Emergencies Research Center, Institute for Futures Studies in Health, Kerman University of Medical Sciences, Kerman, Iran

2 Department of Psychology, University of California, Davis, United States

*Address all correspondence to: amirih80@gmail.com

References

[1] Amiri H et al. The Long-Term Impact of the Earthquake on Substance Use. International Journal of Emergency Medicine. 2022;**15**:44. DOI: 10.1186/s12245-022-00449-x

[2] Chmutina K, von Meding J. A dilemma of language: "Natural disasters" in academic literature. International Journal of Disaster Risk Science. 2019;**10**(3):283-292

[3] Davidson LM, Baum A. Chronic stress and posttraumatic stress disorders. Journal of Consulting and Clinical Psychology. 1986;**54**(3):303

[4] Sandifer PA, Walker AH. Enhancing disaster resilience by reducing stress-associated health impacts. Frontiers in Public Health. 2018;**6**:373

[5] Neria Y, Nandi A, Galea S. Post-traumatic stress disorder following disasters: A systematic review. Psychological Medicine. 2008;**38**(4):467-480

[6] Pfefferbaum B, Nitiéma P, Newman E. A meta-analysis of intervention effects on depression and/or anxiety in youth exposed to political violence or natural disasters. Child & Youth Care Forum. 2019;**48**(4):449-477

[7] Lai BS et al. Disasters and depressive symptoms in children: A review. Child & Youth Care Forum. 2014;**43**(4):489-504

[8] Copeland WE et al. Impact of COVID-19 pandemic on college student mental health and wellness. Journal of the American Academy of Child & Adolescent Psychiatry. 2021;**60**(1):134-141 e2

[9] Tavaragi MS, Sushma C. Disaster its impact and management. International Journal of Psychology and Psychiatry. 2015;**3**(2):106-116

[10] Nielsen MB et al. Post-traumatic stress disorder as a consequence of bullying at work and at school. A literature review and meta-analysis. Aggression and Violent Behavior. 2015;**21**:17-24

[11] Berenz EC et al. Time course of panic disorder and posttraumatic stress disorder onsets. Social Psychiatry and Psychiatric Epidemiology. 2019;**54**(5):639-647

[12] Tedeschi RG, Calhoun LG. Trauma and Transformation: Growing in the aftermath of suffering. Thousand Oaks, CA: SAGE Publications; 1995

[13] Van Slyke J. Post-traumatic growth. Naval Center for Combat & Operational Stress Control. 2013:1-5

[14] Tedeschi RG, Calhoun LG. The posttraumatic growth inventory: Measuring the positive legacy of trauma. Journal of Traumatic Stress. 1996;**9**(3):455-471

[15] Bonanno GA, Gupta S. Resilience after disaster. 2012

[16] Miller KE, Rasmussen A. War exposure, daily stressors, and mental health in conflict and post-conflict settings: Bridging the divide between trauma-focused and psychosocial frameworks. Social Science & Medicine. 2010;**70**(1):7-16

[17] Norwood AE, Ursano RJ, Fullerton CS. Disaster psychiatry: Principles and practice. Psychiatric Quarterly. 2000;**71**(3):207-226

[18] Yayak A. Terrorism and its effects on human psychology. In: Academic

Research and Reviews in Social Sciences. Duvar Publishing; 2021. pp. 7-19

[19] Stasiak K et al. Delivering solid treatments on shaky ground: Feasibility study of an online therapy for child anxiety in the aftermath of a natural disaster. Psychotherapy Research. 2018;**28**(4):643-653

[20] Stratta P et al. Resilience and coping in trauma spectrum symptoms prediction: A structural equation modeling approach. Personality and Individual Differences. 2015;77:55-61

[21] Golman R, Hagmann D, Loewenstein G. Information avoidance. Journal of Economic Literature. 2017;**55**(1):96-135

[22] Orui M et al. The relationship between starting to drink and psychological distress, sleep disturbance after the great East Japan earthquake and nuclear disaster: The Fukushima health management survey. International Journal of Environmental Research and Public Health. 2017;**14**(10):1281

[23] Wang S et al. Psychological distress and sleep problems when people are under interpersonal isolation during an epidemic: A nationwide multicenter cross-sectional study. European Psychiatry. 2020;**63**(1):1-8

[24] Li X et al. Predictors of persistent sleep problems among older disaster survivors: A natural experiment from the 2011 great East Japan earthquake and tsunami. Sleep. 2018;**41**(7):zsy084

[25] Comtesse H et al. Ecological grief as a response to environmental change: A mental health risk or functional response? International Journal of Environmental Research and Public Health. 2021;**18**(2):734

[26] Math SB et al. Disaster management: Mental health perspective. Indian Journal of Psychological Medicine. 2015;**37**(3):261-271

[27] Beaglehole B et al. Psychological distress and psychiatric disorder after natural disasters: Systematic review and meta-analysis. The British Journal of Psychiatry. 2018;**213**(6):716-722

[28] Bryant RA. Acute stress disorder. Current Opinion in Psychology. 2017;**14**:127-131

[29] Andreasen NC. What Is Post-Traumatic Stress Disorder? Dialogues in Clinical Neuroscience. 2011;**13**(3): 240-243

[30] Hu X-L et al. Factors related to complicated grief among bereaved individuals after the Wenchuan earthquake in China. Chinese Medical Journal. 2015;**128**(11):1438-1443

[31] Kersting A, Wagner B. Complicated grief after perinatal loss. Dialogues in Clinical Neuroscience. 2022;**14**(2): 187-194

[32] Bryant RA et al. Mental health and social networks after disaster. American Journal of Psychiatry. 2017;**174**(3):277-285

[33] Agyapong VI et al. Prevalence rates and predictors of generalized anxiety disorder symptoms in residents of Fort McMurray six months after a wildfire. Frontiers in Psychiatry. 2018;**9**:345

[34] Yang HJ et al. Community mental health status six months after the Sewol ferry disaster in Ansan, Korea. Epidemiology and health. 2015;**37**:e2015046

[35] European C et al. Risk, Hazard and people's Vulnerability to Natural Hazards: A Review of Definitions, Concepts an Data. European Commission Joint Research Centre. EUR. 2004;**21410**:40

[36] Cardona OD, Carreño ML. Updating the Indicators of Disaster Risk and Risk Management for the Americas. Journal of Integrated Disaster Risk Management. 2011;**1**(1):27-47

[37] Lindström B. The meaning of resilience. International Journal of Adolescent Medicine and Health. 2001;**13**(1):7-12

[38] Tedeschi RG, Park CL, Calhoun LG, editors. Posttraumatic Growth: Positive Changes in the Aftermath of Crisis. 1st ed. Mahwah, NJ: Lawrence Erlbaum; 1998. p. 179-213

[39] Meichenbaum D. Resilience and posttraumatic growth: A constructive narrative perspective. In: Handbook of Posttraumatic Growth: Research & Practice. Mahwah, NJ, US: Lawrence Erlbaum Associates Publishers; 2006. pp. 355-367

[40] Bragin M. Clinical social work with survivors of disaster and terrorism: A social ecological approach. In: Essentials of Clinical Social Work. 2nd Edition. One Thousand Oaks, California: Sage Publishing; 2014. p. 366-401

Chapter 23

Spontaneous Volunteers in Emergencies and Disasters

Mustafa Yükseler and Jale Yazgan

Abstract

Spontaneous volunteers are often involved in response and relief efforts during and after emergencies and disasters. These volunteers perform various tasks: search and rescue, first aid, response to fires, distributing food and water, cleaning debris, and caring for animals. It is clear that spontaneous volunteers have played a vital role in preventing the loss of life and property following emergencies and disasters in the past. However, the lack of organization and knowledge about emergencies and disasters among these volunteers creates challenges and risks like liability and security issues. These factors make professional responders reluctant to use spontaneous volunteers in emergencies and disasters. By completing a literature review on the benefits, challenges, and risks of spontaneous volunteers in emergency and disaster response and relief activities, we aim that this study will speed up the fundamental studies on the topic. Creating strategies for these volunteers' efficient usage and incorporating them into the current emergency and disaster management systems can significantly reduce disaster risks.

Keywords: emergency, disaster, volunteer, spontaneous volunteer, resilience

1. Introduction

The frequency and effects of disasters are increasing globally. This increase parallels the increase in human–environment interaction, the number of persons exposed to hazards, and the increase in asset values. As of the 21st century, natural and man-made disasters such as September 1, 2001 attacks, Hurricane Katrina, the 2010 Haiti earthquake, the 2011 East Japan earthquake, the Syria Crisis, and the COVID-19 pandemic and climate migration, climate refugee, flood, drought, and desertification caused by climate change have had a global impact [1]. According to a report published in 2020 by the United Nations Office for Disaster Risk Reduction, there have been 7348 natural disasters in the past 20 years (2000–2019), and as a result of these disasters, approximately 1.23 million people have died, 4.2 billion people have been affected, and 2.97 trillion US dollars have been lost to global economic losses. According to the same report, disasters have become far more frequent and severe than they were in the past 20 years (1980–1999) (while 4212 natural disasters occurred, 1.19 million people died, 3.25 billion people were affected, and 1.63 trillion economic losses were caused globally) [2]. Emergency and disaster authorities worldwide and scientists in the field of emergency and disaster place a strong emphasis on the disaster risk management approach, in contrast to crisis management-oriented disaster management approaches [3–6]. Making societies disaster-resistant is a crucial component of disaster risk management, in addition to reducing structural risks. There are many instances where emergencies and disasters outstrip a nation's response capacity. Volunteering is thus one of the key initiatives for enhancing societal resilience to disasters [3, 4, 7, 8].

Volunteering plays a significant role in many countries' response and recovery efforts during emergencies and disasters. When a disaster strikes, volunteers help in various fields, such as technical rescue, fire response, medical response, debris cleaning, logistical support, and transportation [8]. According to the World's Volunteerism Report, millions of volunteers participate in activities across a range of sectors, and their economic impact is equivalent to almost 61 million full-time employees per month based on the number of full-time employees [9]. These statistics show that volunteering activities are important for achieving sustainable development goals.

Volunteering activities in terms of emergencies and disasters are carried out in two ways as formal and informal [10]. The activities of the Red Cross or the Red Crescent organizations can be examples of formal volunteering. On the other hand, informal volunteering is when people go to the affected area voluntarily and engage in response and relief activities after an emergency or disaster event, called *"convergence reaction"* [11] in the literature. This has been happening instinctively since the existence of human beings. These volunteers who participate in emergency and disaster response and relief activities are referred to by several terms in the literature, including spontaneous volunteers, unorganized volunteers, unaffiliated volunteers, or informal volunteers [12, 13]. These volunteers are referred to in this study as "spontaneous volunteers". Various definitions of spontaneous volunteers can be found in the literature. Generally, they are defined as: *"spontaneous volunteers are individuals or groups of people who seek or are invited to contribute their assistance during and/or after an event, and who are unaffiliated with any part of the existing official emergency management response and recovery system and may or may not have relevant training, skills or experience"* [14]. During past emergencies and disasters, it can be argued that these volunteers performed crucial roles. Below are some of the disasters that spontaneous volunteers participated in:

- The majority of rescue efforts following the 1976 Tangshan earthquake, which killed 242,000 people, was carried out by survivors [15, 16], and approximately five months after the earthquake, 420,000 temporary shelters were completed, with an estimated 100,000 voluntary citizens participating [17].

- Most rescue activities after the 1980 Southern Italy earthquake were carried out by local citizens [18, 19].

- After the 1985 Mexico City earthquake, almost ten percent of the city's two million residents participated in recovery efforts [20].

- After the 1989 Loma Prieta earthquake, seventy percent of the population of Santa Cruz and sixty percent of the population of San Francisco participated in response efforts [21].

- After the 1995 Kobe earthquake, approximately 630,000 to 1.4 million people are estimated to have engaged in various voluntary activities [22, 23].

- After the 1999 Turkey Marmara earthquake, most of the initial responses, especially search and rescue, were done by spontaneous volunteers [24].

- After the September 11 attacks, approximately 30,000 to 40,000 volunteers went to Ground Zero in New York for relief work, while the American Red Cross recruited 22,000 spontaneous volunteers in the two-and-a-half weeks following the attacks [25].

- After Hurricane Katrina in 2005, Harris County received an average of 3500 spontaneous volunteers' registrations per day during emergency operations [26], while American Red Cross used 50,000 spontaneous volunteers [27].

- After the 2013 Khartoum flood, with the initiative of a Sudanese group, more than 7000 volunteers and hundreds of thousands of dollars' worth of aid were collected quickly [28].

- After the 2013 Canadian Alberta flood, approximately 15,000 spontaneous volunteers were used [29].

Although spontaneous volunteers participate in various activities and make financial contributions during emergencies and disasters, their lack of training and skills in response and recovery activities might provide obstacles rather than benefits. Alexander [8] defined the reasons for this situation as illustrated in **Figure 1**.

According to the historical experiences above, spontaneous volunteers have significantly contributed to response and recovery studies, both qualitatively and quantitatively. For the reasons outlined in **Figure 1**, these volunteers nevertheless posed threats to not only the affected individuals but also the professional responders in the same disaster events. For instance, it was determined that 2.2 rescuers per person pulled alive from the debris in the 1985 Mexico City earthquake died because they lacked the necessary skills, knowledge, and tools for urban search and rescue [30]. Similar obstacles prevented emergency response teams from reaching the scene following the 1999 Turkey Marmara earthquake because spontaneous volunteers caused a 32-kilometer traffic jam [24, 30].

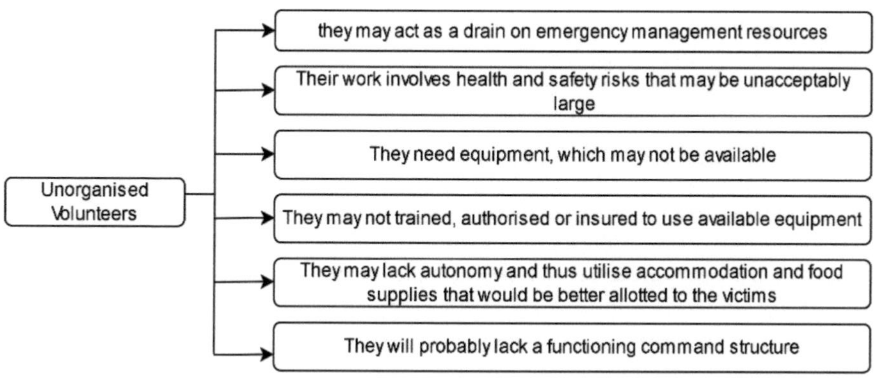

Figure 1.
Barriers that unorganized volunteers may create in emergencies and disasters [8].

The function of spontaneous volunteers in emergencies and disasters is discussed in this paper. The study aims to advance fundamental research on spontaneous volunteerism in emergencies and disasters. The study is divided into the following sections: volunteers as human resources in emergencies and disasters, spontaneous volunteers, benefits of spontaneous volunteers in emergencies and disasters, challenges and risks of spontaneous volunteers in emergencies and disasters, and conclusion.

2. Volunteers as human resources in emergencies and disasters

Disasters are becoming more frequent and more severe, which has an increasingly detrimental impact on human life and the environment. Additionally, it is becoming more challenging to predict and manage the effects of disasters due to various factors, including the growth in human population, technological advancements, greenhouse gas emissions, excessive consumption habits, destruction of forest areas, and inappropriate use of land. Therefore, studies on disaster risk management are the main emphasis of the integrated disaster management system that is employed in the majority of countries. However, it is clear from past disasters that governmental resources alone were insufficient to manage disasters in the nations where the disaster happened. For instance, the cascading disasters that occurred in 2011 in Japan, which has one of the greatest models for disaster management, resulted in almost 22,000 fatalities, and the country went through its worst crisis since the Second World War [31]. At this point, emergency and disaster management organizations require both human resources and modern technologies in addition to a strong institutional structure. In summary, human resources are one of the crucial components of the integrated disaster management system.

Dynes' classification of these structures in his "Organized Behavior in Disaster" study [32] is the most well-known when it comes to the organizational structures associated with emergencies and disasters. In this classification, which provides a valuable framework for comprehending the various categories of emergency and disaster volunteers [12], there are four types: Type I *(established organizations)*, Type II *(expanding organizations)*, Type III *(extending organizations)*, and Type IV *(emergent organizations)*. Type I organizations perform regular tasks within existing structures, such as police forces routing traffic in the affected area when the disaster occurs in a region or the firefighting personnel responding if a fire has occurred. Type II organizations are often formed due to a community or organizational planning and perform regular tasks. Examples of these organizations are the Red Cross and the Red Crescent. These organizations are defined as organizations that carry out humanitarian aid activities other than disaster events and undertake various tasks in disaster situations. Type III is defined as organizations that undertake non-regular work. These organizations are not expected to participate in emergency and disaster responses but can operate in the crises in their area of expertise. For example, in an emergency and disaster, a sports club or a faith group may mobilize its members to distribute clothing and food to people in the affected area, or a logging company may send the machines and their staff for post-disaster debris removal work [12]. Although Type III organizations can work in coordination with Type I and II organizations, they do not operate under the effective control of these organizations. Finally, Type IV organizations perform non-regular tasks as Type III and generally assist affected people before Type I and II organizations reach the affected area. This group is the group that is not included in the official emergency and disaster management systems, has a lack of information about disaster responses, and makes the first response in matters such as search and rescue and first aid.

In the most general sense, volunteering is defined as "any activity in which time is given freely to benefit another person, group, or organization" [33]. Types of volunteering in the literature are broadly considered as formal and informal. Formal volunteering is considered as the activities carried out actively through a formal organization or government program, while informal volunteering is defined as the activities carried out without the umbrella of a formal organization or government [10]. Regarding emergencies and disasters, it is possible to see that volunteering is classified differently [34, 35]. Shaskolsky [34] discussed volunteering in disaster situations in four forms. According to these:

1. *Anticipated individual volunteers*—those who individually meet broad expectations of society—for example, doctors using their expertise to respond to disaster-affected people medically

2. *Anticipated organization volunteers*—those who are regularly associated with an organization, such as volunteer Red Cross or Red Crescent, whose participation in these organizations is expected and planned

3. *Spontaneous individual volunteers*—those who provide individual assistance in the emergency and early stages of a disaster—for example, individuals who conduct search and rescue activities or first aid

4. *Spontaneous organization volunteers*—those who join an organization themselves following an emergency or disaster. These spontaneous volunteers can be connected with four separate organizations:

    a. Those who help a regular emergency and disaster organization;

    b. Those who are in an ad hoc organization formally created to deal with specific disasters;

    c. Before the disaster, those who use an organization not related to disasters for disaster assistance;

    d. Those who respond to disasters as part of an informal group.

Volunteers play a crucial role because of the uncertainty of the effects of emergencies and disasters that occur today and the limited capacity of nations to respond. For instance, the International Federation of Red Cross and Red Crescent Societies reported that their organization's *"disaster response and early recovery program"* had a reach of 650.2 million people in 2020 and that they had roughly 14.9 million volunteers [36]. In 2018, more than 12.7 million Canadians (aged 15 and over) officially spent 1.6 million hours for charities, nonprofits, and community organizations. This number was said to be equivalent to about 858,000 full-time jobs [37]. This shows that volunteering increases social cohesion and is a significant labor and economic resource for societies.

Volunteers typically perform a variety of tasks in the wake of emergencies and disasters, including search and rescue operations, technical assistance, and psychological support [8, 13, 38, 39], and they can offer support to those affected outside the local area, mainly through social media, with materials that the society may require and fundraising campaigns. As previously mentioned, these volunteer efforts for people affected by emergencies and disasters is carried out by expanding organizations or spontaneous volunteers.

3. Spontaneous volunteers

The term spontaneous volunteer [26, 40–45] used in this study is also widely used in the literature with different terms such as helpers [46, 47], unaffiliated volunteer [48], informal volunteer [12, 49], walk-in disaster volunteer [50], and emergent groups [25, 51]. It is possible to say that the use of these various terminologies typically developed as a reflection of people's actions following emergencies and disasters. People frequently go to the affected area in a converging reaction to take part in response and relief operations when an emergency or disaster event happens. Given that the topic is how people behave after a disaster, it shows that sociological research constitutes the earliest study on the behaviors that literature refers to as "convergence" and "emergent" [11, 14, 20, 32, 46]. While Dynes et al. [20] claimed that the majority of people's activities following the disaster were a significant and positive impact on civic responsibility, Fischer [52] classified these actions as "altruism" in his study on people's behavior following the September 11 attacks. Disaster volunteers may participate in crisis response and relief efforts for various reasons, while altruism is often a major driving force for those who have experienced a crisis [53]. There are three ways to observe convergence behaviors: *personal convergence*, *informational convergence*, and *material convergence* [46]. The direct physical human movement to the disaster-affected area is represented by the personal convergence form mentioned here. According to Fritz and Mathewson [46], there are five different types of motivation for this personnel convergence reaction following a disaster: *"the returness"* (victims or survivors at the time of crisis), *"the anxious"* (those who want to be empowered through mobility, such as by learning about family or friends), *"the helpers"* (people who are motivated by altruism), *"the curious"* (disaster tourists), and *"the exploiters"* (opportunistic individuals looking to gain recognition or, at worst, access to vulnerable individuals to exert power in any number of ways). Kendra and Wachtendorf [54] added to these types of personnel convergence the motivations of *"the supporters"* (encouraging and expressing appreciation to first responders) and *"the mourners or memorializers"*. Although the reasons people move to disaster-affected areas can vary due to the unpredictable effects and consequences of modern disasters as well as the sociocultural differences of the affected communities, the types of personnel convergence discussed here act as an essential framework for comprehending these motivations.

Different definitions of spontaneous volunteering exist, similar to the terms that are used for these volunteers. One of the most accepted definitions of spontaneous volunteering in the literature is as follows: *"spontaneous volunteers are those who seek to contribute on impulse- people who offer assistance following a disaster and who are not previously affiliated with recognized volunteer agencies and may or may not have relevant training, skills or experience"* [14]. In their study, Whittaker et al. [12] defined spontaneous volunteering in a broader sense as follows: *"people who work outside of formal emergency and disaster management arrangements to help others who are at risk or are affected by emergencies and disasters. Such volunteerism may take place before, during, or after an event. Informal volunteers may participate as individuals or as part of a group, on short or long-term basis, regularly or irregularly, and in situ or ex situ. Their participation may be spontaneous and unplanned, or deliberate and carefully planned".* According to the criteria provided, it can be concluded that following a crisis, spontaneous volunteers appear as a natural phenomenon with various motivations and can contribute significantly to the labor force and the economy through their activities. As an illustration, professional cleanup efforts following the Florida tornadoes of February 1998 were anticipated to cost $8 million and require 90 days to complete in Osceola County, but spontaneous volunteer efforts cost roughly $1.4 million and took 55 days to complete [55]. Modern mass media, such as today's technologies and social media, play a significant part in the development of this reaction in addition to human instincts. For instance, 2.3 million people participated in the 2014 Malaysian Airlines search for the missing flight MH370 by scanning 24,000 square kilometers of satellite imagery published on the Tommod website [56]. After the crisis, it is also possible to argue that contemporary mass media has drawbacks. Even if the disaster is relatively modest, the media's dramatization and exaggeration can turn it from a local emergency to a national or international event [11]. While the efforts of spontaneous volunteers during the times of crisis cannot be discounted, their lack of organization, planning, knowledge, and skills regarding emergencies and disasters and uncertainty regarding the legal responsibility of professional organizations can result in these volunteers doing more damage than good. The types of activities that these volunteers have participated in as a result of previous emergencies and disasters, as well as the benefits, challenges, and risks associated with these efforts, will be covered in the next sections of the study.

3.1 The benefits of spontaneous volunteers in emergencies and disasters

When countries' disaster management systems are reviewed, it can be concluded that they are getting more specialized and that the response studies are being carried out more effectively. However, first response tasks are typically carried out by the locals who are already there in the period between the disaster's occurrence and the professional disaster managers' arrival at the affected area [12, 13, 55]. According to a Canadian research, 37% of people who experienced a major emergency or disaster resorted to family members, 24% to neighbors, and 15% to friends for assistance, while just 15% turned to the local government, 9% to first responders, 9% to the police, and 5% to the state government [57]. Undoubtedly, established organizations are essential in the times of crisis, but this study demonstrates how vital spontaneous volunteers are to response and relief operations.

Spontaneous volunteers participate in various tasks throughout emergency and disaster response and recovery phases. These activities may change depending on the kind of emergency or disaster and the location where it occurs. Spontaneous volunteers can be a valuable human resource for identifying vulnerable groups in the social structure, cultural content, and demographic structure and for more access to the necessary resources and geographic information that may be required by response teams and accelerate response efforts [13, 29]. Twigg and Mosel [25] presented the activities of spontaneous volunteers in emergency and post-disaster events in 9 main categories. **Table 1** illustrates that these activities show that spontaneous volunteers are generally involved in basic or special skills. In particular, they play a significant role in carrying out many tasks that call for only basic skills, such as gathering, transporting, and distributing relief supplies and clearing debris, in addition to

Disaster response activities of emergency groups and spontaneous volunteers	
Medical • search and rescue • first aid and emergency medical care • donating blood	*Shelter* • shelter provision • hosting displaced people
Information and communications • registration of victims, displaced persons, and evacuees • making lists and searching for missing people • translating • issuing and sharing messages and information	*Security and Coordination* • informal coordination with other organizations and activities • preserving the safety of property • controlling traffic and crowds
Psychological and bereavement • psychological guidance • handling the deceased • ensuring proper rites are observed at burials	*Preparedness* • issuing warnings and helping with evacuation • emergency safety (e.g., protective measures against flooding such as sandbags)
Buildings and services • removing debris and clearing streets • damage assessment • building inspection • restoring services (e.g., communications) and equipment • cleaning up after disasters	*Advocacy* • challenging actions and practices of official response organizations • presentation of survivors' complaints and compensation advocacy • influencing policymakers to minimize disasters in the future
Provisions and Supplies • collecting, transporting, unloading, storing, and distributing relief supplies, clothing, etc. • providing food and drinks to people and emergency responders (e.g., setting up kitchens and canteens)	*Other* • collecting funds for victims • caring for animals

Table 1.
Activities of emergent groups and spontaneous volunteers [25].

performing critically essential responses such as search and rescue, first aid, and fire-fighting until professional teams arrive at the affected area. According to emergency managers in the study by Daddoust et al. [13] on the use of spontaneous volunteers, they can be used for tasks such as *"collecting and organizing donations"*, *"assisting with evacuations"*, *"food distribution"*, *"meet and greet"*, *"providing food and drink"*, and *"pet services"*. As another illustration, following the 2009 "Black Saturday" bushfires in Victoria, Australia, more than 22,000 volunteers provided online support [27], and in response to the call of farmers whose farms were damaged, citizens helped to rebuild farms [12]. A community group called "BlazeAid," which aims to carry out recovery operations and assist rural areas devastated by natural disasters, was founded due to these actions [12].

Given that both the procedural restrictions of official organizations and the affected society may not be aware of these official procedures and rules, spontaneous volunteers can be tools that can lead to beneficial results for both groups after emergencies and disasters [13]. Effectively utilizing spontaneous volunteers can reduce the number of deaths and minimize labor costs [13, 55] while also bringing about intangible advantages for the affected community's mental health, social cohesion, trust, and harmony [58]. Orloff [22] proposed that the efficient use of spontaneous volunteers will aid professional teams' response activities. In view of this, spontaneous volunteers, who typically originate from the affected community, might be a valuable tool in response activities due to their familiarity with its linguistic, cultural, and structural aspects. Additionally, it is possible to make efficient use of community-based and religious leaders and organizations that can assist not only in the response effort but also in the procurement and evacuation for the organizations established pre-disaster and expanding organizations. According to Orloff [22], because spontaneous volunteers can provide advantages such as "*local experts*", "*cultural competency*", "*language skills*", "*specialized skills and resources from work or other experience*", "*official responders capable of doing more advanced work*", "*official responders capable of assisting community in healing and emotional recovery process*", and "*savings money and time*", relevant institutions can perform more effective response and resource management if they are integrated into emergency and disaster management systems. There are guidelines and tool kits published by the US, New Zealand, and Australia [27, 59–61] as well as the "*Security and resilience-Community resilience-Guidelines for planning the involvement of spontaneous volunteers*" [62] published by the International Organization for Standardization (ISO) for the effective use and management of spontaneous volunteers in addition to academic studies suggesting models [7, 13, 43, 44, 63–65]. It should be noted, however, that the process of integrating spontaneous volunteers into management systems is challenging and complex due to variety of factors, including emergency managers' resistance to using them, the various personality traits and skill sets of these volunteers, the variety of disasters, and social and cultural differences.

In summary, spontaneous volunteers are human resources that can help with a variety of problems during emergencies and disasters. At this point, it can be stated that societal resilience will improve and disaster damage will decrease if this unavoidable human resource can be integrated into the current emergency and disaster management systems and appropriately utilized.

3.2 The challenges and risks of spontaneous volunteers in emergencies and disasters

Without being a part of an established or expanding organization and going to the disaster area with various motivations, spontaneous volunteers cause risks and challenges to themselves, those affected by the disaster, and professional organizations. According to the research, these volunteers could present the affected area with a wide range of risks and challenges. They generally run the potential of being damaging due to several issues, such as a lack of coordination, knowledge, and skills; security issues; and legal liabilities. These factors make professional responders reluctant to work with these volunteers, and they refer to them as "disaster with disaster" [66].

Fernandez et al. [53] identified two main categories of risks: emergency and disaster response volunteers are not used effectively and they are unorganized and do not have adequate training. The first of these risks, ineffective use of spontaneous volunteers by emergency and disaster managers, may create a poor public perspective of emergency and disaster response and increase the probability of deaths, serious injuries, and financial loss [12, 53]. The second risk is that unorganized and untrained spontaneous volunteers can hinder the efficient use of resources and endanger both professional first responders and disaster victims as well as themselves. The rescue of 800 individuals from the debris following the earthquake in Mexico City in 1985 by untrained, spontaneous volunteers—100 of whom died in the process—is one of the most commonly mentioned instances of this situation in the literature [20, 30]. Another illustration is that after the Hebei Spirit oil spill in South Korea in 2007, physical harm and infections were brought on by volunteers taking part in the response operations without having the proper personal protective equipment and not knowing the toxic effects of the oil [67]. Integrating spontaneous volunteers into the current emergency and disaster management systems can reduce the abovementioned risks. As a result, professional organizations do not have the time during a crisis to identify who has the necessary abilities and to train them for field response activities [25]. On the other hand, if they can be integrated, both these human resources and other resources will be used efficiently, easing the burden on professional organizations that work in the disaster area.

Particularly for the people affected by a disaster, the first hours after it strikes are crucial. After a disaster, there is less chance of people being rescued as time goes on. Therefore, given the importance of professional rescue teams reaching and responding to the area as soon as possible, the presence of numerous individuals, equipment, support materials, and vehicles in the area can cause obstructions and hinder professional responders [25]. With more than 22,000 deaths, the 1999 Marmara earthquake (also called Gölcük) in Turkey is one of the deadliest disasters in history. International recovery was necessary because this disaster struck the most populated region of the country, which at the time had a crisis management-focused disaster management system. Spontaneous volunteers who arrived from the disaster's area and other parts of the country were crucial to many operations, including search and rescue. However, professional disaster responders were unable to access the area due to a 32 km-long traffic jam caused by spontaneous volunteers who arrived in the area after the earthquake [24]. Nearly 20 years after this earthquake, on October 30, 2020, an earthquake hit in Izmir, one of the largest metropolitan cities. Since then, the country has adopted an integrated disaster management system and improved its organization and coordination of disaster response. However, as in the Marmara earthquake, the access of the emergency response teams was delayed as the volunteers came to the disaster-affected area with their vehicles and caused a traffic jam. These experiences demonstrate the requirement of incorporating spontaneous volunteers into the current disaster management system in order to provide the quick and efficient response strategy that is essentially intended.

According to Orloff [22], there are three drawbacks to using spontaneous volunteers during emergencies and disasters: *"liability," "physical and emotional concerns,"* and *"the potential lack of internal readiness within agencies"*. First, liability is viewed as a drawback in many studies in the literature because of the legal uncertainties surrounding the use of spontaneous volunteering in emergencies and disasters. According to Twigg and Mosel [25], this circumstance is as follows. "There may be uncertainty about legal liability of volunteer responders (or official organizations they assist) for deaths, injuries or damages suffered by volunteers, or by disaster-affected people as a result of their actions. A related issue is lack of insurance cover for volunteers." Liability is one of the reasons professional emergency and disaster managers are reluctant to engage spontaneous volunteers. In the study conducted by Daddoust et al. [13], emergency managers indicated *"liability issues"* as one of the most problematic and challenges, while *"sued for spontaneous volunteer actions"* and *"sued by spontaneous volunteer"* were the subjects they saw as the most potential risks. Additionally, the literature also discusses the problem of security [12, 13, 25, 39].

While spontaneous volunteers create a security problem for themselves and those affected by the disaster, they can also be seen as abusing the disaster situation. In addition to looting materials of economic value, they can also abuse the current situation differently. For instance, after the 2010 Haiti earthquake, a group of Baptist missionaries from the US who had gone to the disaster area was apprehended at the Haiti-Dominican Republic border with 33 children [12, 68]. Later, this group claimed in their statements that they had taken orphaned and abandoned children to give them the opportunity to rescue, care for, and adopt. Eventually, it was confirmed that the children were not abandoned nor orphaned [68]. The dimension of physical concern is the situations that can hinder the work of professional organizations because spontaneous volunteers are typically unorganized and uneducated and can pose a risk of death and injury to both themselves and those affected by the disaster. On the other hand, the emotional component, the involvement of people who have already experienced emotional harm before an emergency and disaster or as a result of this event, and the work to overcome these feelings can both worsen one's situation and cause emotional harm to those who are affected by the disaster. Emergency and disaster managers will need to be more careful about using spontaneous volunteers at this stage because their appearance on the scene is almost inevitable. Following the September 11 attacks, response and recovery operations involving 30,000–40,000 volunteers posed security risks, and a large number of the first volunteers who came on the spot to offer help were emotionally exhausted during the search and rescue operations [12]. A study on the management of spontaneous volunteers after the attacks found that emergency managers rarely take spontaneous volunteers into account when planning, and because they do not self-care, spontaneous volunteers may be traumatized, become victims of disaster, and need services [69]. Finally, it was stated that the potential lack of internal readiness within agencies, which is seen as a disadvantage, will be ensured by the inclusion of spontaneous volunteers in the studies and their use, if they are compatible with the current institutional structure. At this point, within the institutional structure, there may be preparation issues such as limited formal staff to manage spontaneous volunteers, limited affiliated volunteers to work with and mentor these volunteers, limited time to create and implement a management plan, language barriers between employees and the affected community, and existing organizational culture [22].

4. Conclusion

Volunteerism has a significant impact during emergencies and disasters. The significance of volunteering activities is emphasized in both scientific studies and initiatives to reduce disaster risks. Community resilience is crucial in reducing disaster risks considering the limited resources of countries and the increasingly complex effects of disasters. Therefore, spontaneous volunteers can be a valuable human resource in reducing risks. It is clear from the literature that this subject is gaining more and more attention [43]. In this study, which the spontaneous volunteers are in emergencies and disasters, what kind of activities they take part in, and the benefits, challenges, and risks they provide as a result of past emergencies and disasters are given as a result of the literature review. This study is aimed to contribute to the studies to be done on this subject. Spontaneous volunteers, who come to the affected region with various motivations after emergencies and disasters and operate in many sectors, provide tangible and intangible benefits. However, professional organizations are reluctant to use spontaneous volunteers because they have challenges and risks such as insufficient knowledge and skills, lack of organization, security and liability problems, and limited resources, equipment, and personnel in existing management systems. Because spontaneous volunteers cannot be ignored in the affected area and the response resources are limited, it is clearly seen that it is a critical requirement to integrate these volunteers into disaster management systems. It can be said that this integration is a challenging due to the complex nature of emergencies and disasters, different types of disasters, different personal characteristics and skills, and the social and cultural differences of spontaneous volunteers.

In summary, integrating spontaneous volunteers, who are essential human resources, into emergency and disaster management systems and establishing a legal framework will significantly reduce disaster risks.

Conflict of interest

The authors declare no conflict of interest.

Author details

Mustafa Yükseler* and Jale Yazgan
Mugla Sıtkı Kocman University, Mugla, Turkey

*Address all correspondence to: mustafayukseler@mu.edu.tr

References

[1] Usta E, Yükseler M. Afetlerde Sosyal Medya Kullanımı ve Etik İkilemler: İzmir Seferihisar Depremi Örneği. Journal of Disaster and Risk. 2021;**4**(2):249-269

[2] UNDRR. Human Cost of Disasters: An Overview of the Last 20 Years (2000-2019). UN Office for Disaster Risk Reduction; 2020. Available from: https://www.preventionweb.net/files/74124_humancostofdisasters20002019reportu.pdf?_gl=1*1objzgk*_ga*NTY4NDEwMzA2LjE2NjcxMTE0MzQ.*_ga_D8G5WXP6YM*MTY3MjA2NDQwMi42LjAuMTY3MjA2NDQwMi4wLjAuMA

[3] United Nations Office for Disaster Risk Reduction. Hyogo Framework for Action 2005-2015: Building the Resilience of Nations and Communities to Disasters. United Nations; 2007. Available from: https://www.unisdr.org/files/18197_midterm.pdf

[4] United Nations Office for Disaster Risk Reduction. Sendai Framework for Disaster Risk Reduction 2015-2030. United Nations; 2015. Available from: https://www.preventionweb.net/files/43291_sendaiframeworkfordrren.pdf

[5] Cardona OD, van Aalst MK, Birkmann J, Fordham M, McGregor G, Perez R, et al. Determinants of risk: Exposure and vulnerability. In: Field CB, Barros V, Stocker TF, Dahe Q, editors. Managing the Risks of Extreme Events and Disasters to Advance Climate Change Adaptation. Cambridge: Cambridge University Press; 2012. pp. 65-108

[6] Cutter SL, Burton CG, Emrich CT. Disaster resilience indicators for benchmarking baseline conditions. Journal of Homeland Security and Emergency Management. 2010;7(1):1-22

[7] Fernandez LS. Strategies for managing volunteers during incident response a systems approach. Homeland Security Affairs. 2006;**2**(3):9

[8] Alexander D. The voluntary sector in emergency response and civil protection: Review and recommendations. IJEM. 2010;7(2):151

[9] United Nations Volunteers (UNV) programme. 2022 State of the World's Volunteerism Report. Building equal and inclusive societies. Bonn; 2021. Available from: https://swvr2022.unv.org/

[10] Carson ED. Comment: On defining and measuring volunteering in the United States and abroad. Law and Contemporary Problems. 1999;**62**(4):67

[11] der Heide EA. Convergence behavior in disasters. Annals of Emergency Medicine. 2003;**41**(4):463-466

[12] Whittaker J, McLennan B, Handmer J. A review of informal volunteerism in emergencies and disasters: Definition, opportunities and challenges. International Journal of Disaster Risk Reduction. 2015;**13**:358-368

[13] Daddoust L, Asgary A, McBey KJ, Elliott S, Normand A. Spontaneous volunteer coordination during disasters and emergencies: Opportunities, challenges, and risks. International Journal of Disaster Risk Reduction. 2021:**65**. Available from: https://www.sciencedirect.com/science/article/pii/S2212420921005070

[14] Drabek TE, McEntire DA. Emergent phenomena and the sociology of disaster: Lessons, trends and opportunities from the research

literature. Disaster Prevention and Management: An International Journal. 2003;**12**(2):97-112

[15] Chen Y, editor. The Great Tangshan Earthquake of 1976: An Anatomy of Disaster. 1st ed. New York: Oxford; Pergamon Press; 1988 153 p.

[16] Noji EK. The public health consequences of disasters. Prehospital and Disaster Medicine. 2000;**15**(4):147-157

[17] Zhang Y, Zhang C, Drake W, Olshansky R. Planning and recovery following the great 1976 Tangshan earthquake. Journal of Planning History. 2015;**14**(3):224-243

[18] de Bruycker M, Greco D, Annino I, Stazi MA, de Ruggiero N, Triassi M, et al. The 1980 earthquake in southern Italy: Rescue of trapped victims and mortality. Bulletin of the World Health Organization. 1983;**61**(6):1021-1025

[19] Bruycker MD, Greco D, Lechat MF. The 1980 earthquake in southern Italy— Morbidity and mortality. International Journal of Epidemiology. 1985;**14**(1):113-117

[20] Dynes RR, Quarantelli EL, Wenger D. Individual and Organizational Response to the 1985 Earthquake in Mexico City, Mexico. Disaster Research Center; 1990. Available from: https://udspace.udel.edu/items/62493ff1-9bf1-4182-8e06-f227d6d5e0f8

[21] O'Brien P, Mileti DS. Citizen participation in emergency response following the Loma Prieta earthquake. International Journal of Mass Emergencies and Disasters. 1992;**10**(1):71-89

[22] Orloff L. Managing Spontaneous Community Volunteers in Disasters: A Field Manual. Boca Raton, FL: CRC Press; 2011 289 p

[23] Shaw R, Goda K. From disaster to sustainable civil society: The Kobe experience. Disasters. 2004;**28**(1):16-40

[24] Jalali R. Civil society and the state: Turkey after the earthquake. Disasters. 2002;**26**(2):120-139

[25] Twigg J, Mosel I. Emergent groups and spontaneous volunteers in urban disaster response. Environment and Urbanization. 2017;**29**(2):443-458

[26] U.S. Department of Homeland Security. Good Story: "Harris County, Texas Citizen Group' Response to Hurricane Katrina." Lessons Learned Information Sharing. U.S. Department of Homeland Security; 2005. Available from: https://docplayer.net/11542652-Harris-county-texas-citizen-corps-response-to-hurricane-katrina.html

[27] Australian Government. Spontaneous Volunteer Management Resource Kit. 2010; Available from: https://www.dss.gov.au/our-responsibilities/communities-and-vulnerable-people/publications-articles/spontaneous-volunteer-management-resource-kit. [Accessed: November 25, 2022]

[28] Hazeldine S, Smith MB. Global Revire on Volunteering Report. Geneva: International Federation of Red Cross and Red Crescent Societies; 2015. p. 100

[29] Mackwani Z. Volunteers and the Alberta Flood Response: Insight into Voluntary Sector's Readiness and Surge Capacity. 2015. Available from: https://www.academia.edu/22347512/Volunteers_and_the_Alberta_Flood_Response_Insight_into_Voluntary_Sectors_Readiness_and_Surge_Capacity. [Accessed: November 25, 2022]

[30] Helsloot I, Ruitenberg A. Citizen response to disasters: A survey of literature and some practical implications. J Contingencies & Crisis Man. 2004;**12**(3):98-111

[31] Kingston J, editor. Natural Disaster and Nuclear Crisis in Japan: Response and Recovery after Japan's 3/11. London; New York: Routledge; 2012. p. 302 (The Nissan Institute/Routledge Japanese studies series)

[32] Dynes RR. Organized Behavior in Disaster: Analysis and Conceptualization. Disaster Research Center, Ohio State University; 1969. p. 264. Available from: https://books.google.com.tr/books?id=IeT48KFQkVMC

[33] Wilson J. Volunteering. Annual Review of Sociology. 2000;**26**(1):215-240

[34] Shaskolsky L. Volunteerism in Disaster Situations. 1965. Available from: http://udspace.udel.edu/handle/19716/1165 [Accessed: November 25, 2022].

[35] Wolensky RP. Toward a broader conceptualization of volunteerism in disaster. Journal of Voluntary Action Research. 1979;**8**(3-4):33-42

[36] The International Federation of Red Cross and Red Crescent Societies (IFRC). Everyone Counts Report 2022. Geneva: IFRC; 2022

[37] Hahmann T, Plessis V, Fournier-Savard P. Volunteering in Canada: Challenges and Opportunities during the COVID-19 Pandemic. Statistics Canada; 2020, p. 8. Report No.: 45280001. Available from: https://www150.statcan.gc.ca/n1/en/pub/45-28-0001/2020001/article/00037-eng.pdf?st=woH_CB5s

[38] Crane M. The invisibles: The role of volunteer emergency service members in human health emergency response. Global Biosecurity. 2019;**1**(1):116

[39] Sauer LM, Catlett C, Tosatto R, Kirsch TD. The utility of and risks associated with the use of spontaneous volunteers in disaster response: A survey. Disaster Medicine and Public Health Preparedness. 2014;**8**(1):65-69

[40] Harris M, Shaw D, Scully J, Smith CM, Hieke G. The involvement/exclusion paradox of spontaneous volunteering: New lessons and theory from winter flood episodes in England. Nonprofit and Voluntary Sector Quarterly. 2017;**46**(2):352-371

[41] Ludwig T, Kotthaus C, Reuter C, Dongen S, van Pipek V. Situated crowdsourcing during disasters: Managing the tasks of spontaneous volunteers through public displays. International Journal of Human-Computer Studies. 2017;**102**:103-121

[42] Mayorga ME, Lodree EJ, Wolczynski J. The optimal assignment of spontaneous volunteers. Journal of the Operational Research Society. 2017;**68**(9):1106-1116

[43] Paciarotti C, Cesaroni A, Bevilacqua M. The management of spontaneous volunteers: A successful model from a flood emergency in Italy. International Journal of Disaster Risk Reduction. 2018;**31**:260-274

[44] Paciarotti C, Cesaroni A. Spontaneous volunteerism in disasters, managerial inputs and policy implications from Italian case studies. Safety Science. 2020;**122**:104521

[45] Rivera JD, Wood ZD. Disaster relief volunteerism: Evaluating cities' planning for the usage and management of spontaneous volunteers. JEM. 2016;**14**(2):127

[46] Fritz CE, Mathewson JH. Convergence Behavior in Disasters. Washington, D.C.: National Academy of Sciences-National Research Council; 1957. p. 102 (Disaster Study)

[47] Barsky LE, Trainor JE, Torres MR, BenignoE A. Managing volunteers: Fema's urban search and rescue Programme and interactions with unaffiliated responders in disaster response. Disasters. 2007;**31**(4):495-507

[48] White HP. Understanding the Role of Spontaneous Volunteers in Disaster: The Case Study of the World Trade Centre on 9/11. Auckland, New Zealand: Auckland University of Technology; 2016

[49] Dean J. Informal volunteering, inequality, and illegitimacy. Nonprofit and Voluntary Sector Quarterly. 2022;**51**(3):527-544

[50] Emergency Management Division Justice Institute of British Colobia. Managing Walk-In Disaster Volunteers Participant Guide. Justice Institute of British Colombia; 2022. Available from: https://www.mapleridge.ca/DocumentCenter/View/6940/Managing-Walk-in-Disaster-Volunteers--Participant-Manual?bidId=

[51] Strandh V, Eklund N. Emergent groups in disaster research: Varieties of scientific observation over time and across studies of nine natural disasters. J Contingencies and Crisis Management. 2018;**26**(3):329-337

[52] Fischer PDHW III. Behavioral and organizational responses to terrorism: A model based on 9/11. JEM. 2003;**1**(3):42

[53] Fernandez LS, Barbera JA, van Dorp JR. Spontaneous volunteer response to disasters: The benefits and consequences of good intentions. Journal of Emergency Management. 2006;**4**(5):57-68

[54] Kendra JM, Wachtendorf T. Elements of resilience after the world trade center disaster: Reconstituting new York City's emergency operations Centre: Elements of resilience after the world trade center disaster. Disasters. 2003;**27**(1):37-53

[55] Volunteer Florida. Unaffiliated Volunteers in Response and Recovery. Volunteer Florida; 2000. Available from: https://www.volunteerflorida.org/wp-content/uploads/2013/03/UnaffiliatedVolunteers.pdf

[56] Fishwick C. Tomnod-the Online Search Party Looking for Malaysian Airlines Flight MH370. The Guardian; 2014. Available from: https://www.theguardian.com/world/2014/mar/14/tomnod-online-search-malaysian-airlines-flight-mh370

[57] İbrahim D. Canadian's Experience with Emergencies and Disaster. Canadian Centre for Justice Statistics; 2004 Report No.: 85-002–X. Available from: https://www150.statcan.gc.ca/n1/en/pub/85-002-x/2016001/article/14469-eng.pdf?st=QOkBUI7o

[58] Bachner G, Seebauer S, Pfurtscheller C, Brucker A. Assessing the benefits of organized voluntary emergency services: Concepts and evidence from flood protection in Austria. Disaster Prevention and Management. 2016;**25**(3):298-313

[59] Australian Government Initiative. Spontaneous Volunteer Strategy. 2015. Available from: https://knowledge.aidr.org.au/media/2140/national-spontaneous-volunteer-strategy.pdf [Accessed: June 10, 2022]

[60] FEMA. Developing and Managing Volunteers. 2006. Available from:

https://training.fema.gov/emiweb/downloads/is244.pdf

[61] Ministry of Civil Defence & Emergency Management (MCDEM). Volunteer Coordination in CDEM Director's Guideline for Civil Defence Emergency Management Groups [DGL15/3]. Wellington: Ministry of Civil Defence & Emergency Management; 2013. p. 86

[62] International Organization for Standardization. ISO 22319: 2017 Security and Resilience-Community Resilience-Guidelines for Planning the Involvement of Spontaneous Volunteers. Geneva: International Organization for Standardization; 2017. p. 16

[63] Lorenz DF, Schulze K, Voss M. Emerging citizen responses to disasters in Germany. Disaster myths as an impediment for a collaboration of unaffiliated responders and professional rescue forces. J Contingencies and Crisis Management. 2018;**26**(3):358-367

[64] Lodree EJ, Davis LB. Empirical analysis of volunteer convergence following the 2011 tornado disaster in Tuscaloosa, Alabama. Natural Hazards. 2016;**84**(2):1109-1135

[65] Marana P, Labaka L, Sarriegi JM. A framework for public-private-people partnerships in the city resilience-building process. Safety Science. 2018;**110**:39-50

[66] Points of Light Foundation. Volunteer Center National Network. Preventing a Disaster within the Disaster: The Effective Use and Management of Unaffiliated Volunteers. Points of Light Foundation; 2002. Available from: https://www.ojp.gov/pdffiles1/Archive/202852NCJRS.pdf

[67] Hur J. Disaster management from the perspective of governance: Case study of the *Hebei Spirit* oil spill. Disaster Prevention and Management: An International Journal. 2012;**21**(3):288-298

[68] Atzet I. Post-crisis actions to avoid international child trafficking. Journal of Law & Family Studies. 2010;**12**(2):499-510

[69] Sharon L. Averting a disaster within a disaster: The Management of Spontaneous Volunteers Following the 11 September 2001 attacks on the world trade Center in new York. Voluntary Action: The Journal of the Institute for Volunteering. 2004;**6**(2):11-29

Chapter 24

Trends in Urban Planning and Measures of Disaster Risks in Japan – Roles of the Activities for Disaster Mitigation by Using Planning Support Systems

Kazuki Karashima and Akira Ohgai

Abstract

Based on the history of urban form and the history of disasters, and the current issues in Japanese society such as depopulation and expansion of the cities in the areas with high disaster risk, the formation of a compact city with safety to natural disasters is required. To realize the sustainable cities mentioned above, the necessity for plans and activities to (1) induce residents from areas with high disaster risk to safer areas with long-standing perspectives, (2) to improve the safety of areas that promote induce residence and increase population density, and the areas with the difficulty of induce residents are explained. As one of the means to support such plans and activities, the importance of the development and utilization of planning support tools was shown. Especially, through some examples, it showed the usefulness for substantiating discussions and considering safe and sustainable urban structures.

Keywords: urban planning, depopulation, natural disaster risk, compact city, planning support system, Japan

1. Introduction

The country where the aging society is progressing due to depopulation, such as Japan, many researchers have suggested that urban planning should consider an intensive urban structure, such as the "compact city," as the future vision of an aging society. In addition, Japan is one of the countries with high risk of natural disasters, such as the 2011 Great East Japan Earthquake and frequent huge flood damage. This chapter first explains the recent trends of urban planning in Japan and the policy of mitigation of natural disaster risks in this context. After that, we will explain the measures to reduce the risk of natural disasters in urban areas, and the role of planning support system (PSS) for their examination, including some examples.

2. History of urban planning in Japan

2.1 History of urban formation and urban planning

During the Edo era (1603–1868), Edo, the city, with one of the largest populations in the world, was formed in Japan. Nationwide, many cities such as castle towns and post towns were created. According to the formulation of the cities, the population increased in whole country and concentrated into the cities.

During the Meiji era (1868–1912), the modernization of urban infrastructures such as railways, roads, buildings, and industry progressed, and urbanization and population growth accelerated.

World War I (1914–1918) brought rapid growth to the Japanese economy. As a result, the urban environment has also changed significantly. Buildings made of bricks began to be built on the wooden urban area constructed in Edo era. In addition, various functions of the city were spatially differentiated, and the formation of a complex city began.

Against the background of such urban growth, the Urban Planning Low and Urban Building Law was promulgated in 1919. For this, application of building layout and structural standards, classification by zoning into residences, factories, commercial areas, fire zones, esthetic zones, scenic zones, etc., restrictions on construction in each zone, creation of urban facilities such as roads and parks have been possible [1].

Around this time, the concentration of the population in the big cities became noticeable. In addition, the problems such as the mixture of housing and factories, overcrowding of population and buildings, and the occurrence of slums in inner urban areas began.

After World War II (1939–1945), 115 cities were designated as war-damaged cities because major cities had extensive damage.

The basic policy for the reconstruction of war-damaged areas was officially announced as follows: (1) to curb the growth of oversized cities and (2) to promote the development of small and medium-sized local cities.

In addition, following were officially announced:

(1) Industrial location and population distribution, (2) guidelines for land use planning, and (3) planning standards for major facilities such as streets, plazas, and green areas. According to these policies, reconstruction plans were drafted in each city.

Against the background of the confusion in land use such as height control (high-level), advanced use of existing urban areas, and the emergence of urban sprawl due to rapid urbanization accompanying rapid economic growth, the enactment of the New Urban Planning Law (1968) and revision of the Building Standards Law (1970) were conducted. As a result, the division of urbanization promotion areas and urbanization control areas, and the development permit system linked to the division of areas were implemented.

After that, making the master plan of urban planning as the basic policy on urban planning in each municipality (1992) was implemented.

Then, due to the appearance of the signs of low growth and maturity such as the decline of central urban areas, the policy for restructuring existing urban areas, compact cities, and intensive urban structures had begun to emphasize.

In 2014, the system of the Location Normalization Plan was established. This aim is to reorganize the urban structure into a "compact city plus network of public transportation" model by gently guiding the development of private facilities.

Since then, many local governments have drawn up a Location Normalization Plan to designate urban function induction areas and residential promotion areas within urbanization promotion areas [2].

2.2 History of urban disaster prevention

In Japan, during the Edo era, the main issue of urban disaster prevention is fire spread. The towns in this era consisted of one-story wooden buildings although having one of the most populations in the world. During the 250 years of Edo, about 50 large fire spreads occurred.

Through the Meiji Era, the modernization of the urban structure and buildings was emphasized as one of the goals of modernization.

In the Great Kanto Earthquake in 1919, huge damage was caused by the collapse of buildings. In addition, the damage caused by fire spread with unfortunately strong wind speeds of 10 m/s or more was remarkable. After that, as the policy to improve the urban structure, the strong urban structure for fire spread (noncombustible and fireproof cities) was promoted.

After WW2, Japan experienced a lot of typhoon damage. These damages were expanded due to the damage to urban infrastructures by WW2. In particular, the damage caused by the Isewan Typhoon in 1959 was enormous. This became an opportunity, and the Disaster Countermeasures Basic Law to prevent damage by huge natural disasters was enacted in 1960. This Law promotes each municipality preparing comprehensive disaster prevention including making disaster prevention plans.

Around the 1980s, the areas with high risk of fire spread by huge earthquakes such as densely built-up areas by wooden apartment buildings and the areas not designated as fire prevention areas were aimed to promote safety. Especially, improvement of the evacuation routes and the areas around the designated evacuation areas was emphasized. Therefore, construction of fire prevention zones such as surrounded by wide-width roads and fire-proof buildings around designated evacuation sites was promoted [1].

In addition, the concept of "disaster prevention living area" was proposed. This aims to promote the activities for disaster mitigation such as the planning and development of disaster prevention roads and parks and support the activities for disaster mitigation through community-based activities. This area was based on the daily living area, the same as an elementary school area surrounded by fire prevention roads.

During the period to promote the activities mentioned above, in 1995, the Great Hanshin-Awaji Earthquake occurred. Due to the weak wind speed, the damage caused by the collapse of the building was more serious than the damage caused by the fire. The inner area surrounded by wide roads, which is a fire barrier, had serious damage.

For this reason, after the Great Hanshin-Awaji Earthquake, urban disaster prevention measures for the inner areas such as improvement of densely built-up areas and making houses earthquake-resistant were actively discussed.

In addition, "mutual assistance," the community-based activities to mitigate the damage caused by disasters such as the support of evacuation and rescue to the person buried by the collapsed building, has begun to be emphasized. Most of the communities such as disaster prevention organizations formed by neighborhood associations carry out disaster mitigation activities such as firefighting drills and evacuation drills [3].

In 2011, the Great East Japan Earthquakes occurred. The damage by the tsunami was enormous.

In recent years, many flood disasters and landslides due to heavy rain such as localized downpours of short duration occurred.

3. Measures to reduce natural disaster risk in residential areas

Already explained in Section 2.1, the system of the Location Normalization Plan was established in 2014. To achieve the urban structure into a "compact city plus networks of public transportation", designating (1) urban function induction areas and (2) residential promotion areas and considering the accessibility to public transportation are required.

The policy to designate these areas that the national government has explained, the areas with high risk of natural disasters should be excluded. However, it is difficult for many municipalities to fulfill the policy (**Figure 1**).

For example, the results of the survey on the presence or absence of disaster risk in residential promotion areas have been published [5]. The survey targeted 275 cities that announced Location Normalization Plan by December 2019. There were 13 cities with red zones and 254 cities with yellow zones.

Based on the result, the national government published the policy that municipalities should make the "guideline for disaster mitigation" as the countermeasure to the risk of the designated induction areas (**Figure 2**).

Specifically, to accelerate community-based activities and the improvement of urban structure that incorporates the perspective of disaster prevention, the "disaster prevention guideline," the countermeasure for disaster mitigation in the designated residential promotion areas is described as one of the items in the Location Normalization Plan. This is expected that promote the mainstreaming of disaster prevention in formulating compact cities.

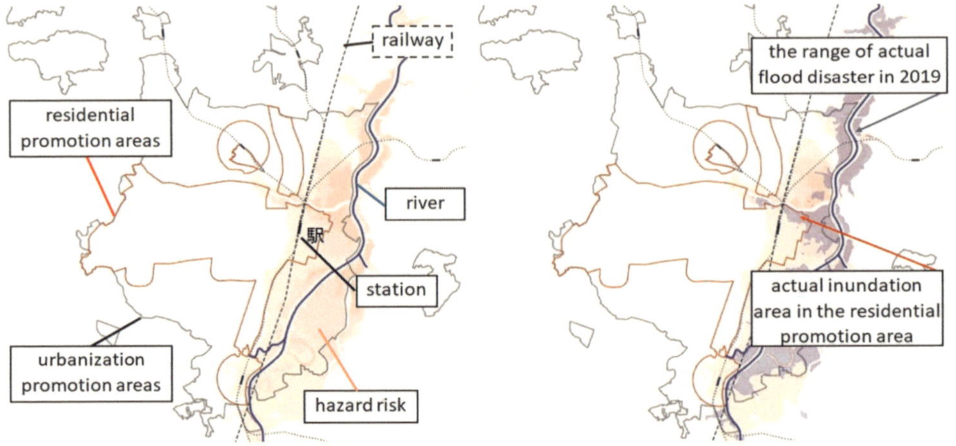

Figure 1.
Example Designation of residential promotion area [4]. Residential promotion areas are designated considering to avoid flood inundation areas. However, the central area around the railway station is overlapped with the flood risk.

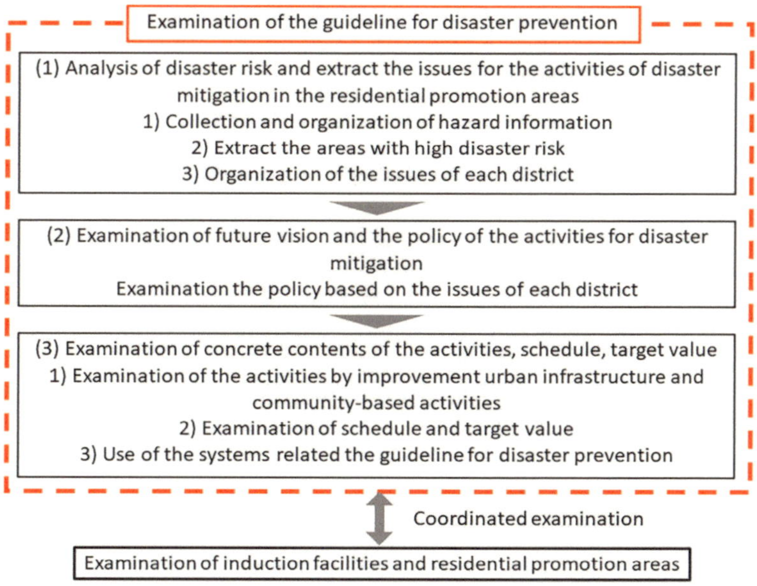

Figure 2.
Summary of examination of guideline for disaster prevention [6].

The following criteria for making the guideline for disaster mitigation were published.

The future vision and aims of disaster mitigation activities should be clearly explained. The countermeasures for disaster mitigation and both urban infrastructure improvement and activities to respond to disasters are necessary (**Figure 3**).

In this way, the designation of the induction areas excluding high disaster risk areas from the long-standing perspective, the difficulty of excluding high disaster risk

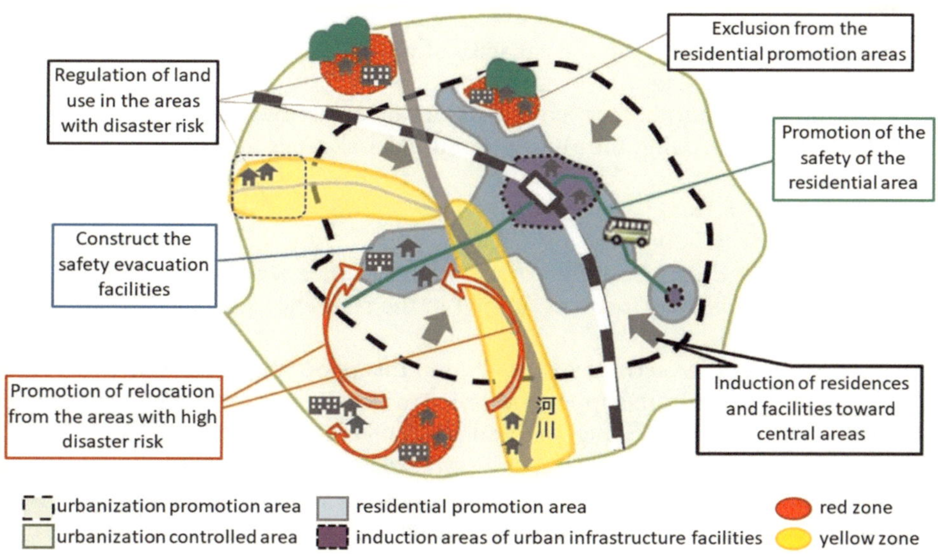

Figure 3.
Image of countermeasures for disaster mitigation [6].

areas within the induction areas, mitigating the disaster risk by the countermeasures for disaster mitigation and both urban infrastructure improvement and activities to respond to disasters are important.

4. Role of planning support tools for urban planning

As mentioned in sections 2 and 3, in order to induce residents to the areas with a low risk of natural disasters and to conduct measures for disaster mitigation in the areas where it is difficult to induce residents, it is necessary to proceed with planning through discussions involving not only the local government but also residents and the relevant parties.

One of the support methods for that purpose is the development and utilization of planning support systems (PSSs). This section introduces examples of PSS and indicates its role.

4.1 Induction of residents considering improvement of mutual assistance ability

4.1.1 Background and objective

When a large-scale earthquake occurs, many residents need rescue and evacuation support at the same time. In this situation, only assistance activities by organizations of the national government and local government (public help) are not able to support all persons who need someone's help. This generates an increase in human damage. In order to reduce such human damage, mutual assistance activities among residents are important. Therefore, enhancing the mutual assistance capacity of the district is important. For this reason, it is important to secure people who can carry out mutual assistance activities, such as leading young households to the areas with low capability of mutual assistance.

Based on that mentioned above, in this section, through the practice in the model area, the examination method of improving the capacity of mutual cooperation introduces the concept of improving the capacity of mutual assistance and the role of planning support technology.

In addition, the importance of a compact urban structure will be mentioned from the perspective of improving the ability of mutual assistance activities.

4.1.2 Outline of the PSS and the concept of the mutual assistance

The PSS for promoting the ability of mutual assistance was developed by the authors. The details of the PSS are described in the reference [7]. This section shows the outline of the PSS.

The mutual assistance activities described in this paper mainly consisted of these two activities: (1) evacuation support behavior to nearby residents who were unable to evacuate alone due to declining physical functions due to aging or disabilities. (2) When a resident discovered a victim such as a buried person by rubble in need of help within the evacuation route during the evacuation, they take part in the rescue activities within a reasonable range while ensuring safety.

In order to extract the areas where the ability of mutual assistance activities is low, the authors' laboratory has developed a PSS to evaluate the ability of mutual assistance using GIS that enables quantitative evaluation by building units (**Figure 4**). In this

Figure 4.
Image of the PSS.

evaluation method, based on the number of local residents, gender, and age, physical strength, implementation rate, and activity rate (1) are multiplied according to **Table 1**). Then the expected value of mutual assistance activities under the situation of a large-scale earthquake disaster is calculated. After that, the expected value is weighted by distance based on the assumption that residents take some time to discover or recognize those persons who cannot evacuate without some assistance, in accordance with the distance.

It is indicating that if one person is caught in the rubble generated by the collapse of the building with less than one ability of mutual assistance capacity, the person does not receive sufficient mutual assistance from the surroundings.

Age	Men's strength	Women's strength	Executing rate	Men's activity rate	Women's activity rate	Men's expected value	Women's expected value
10	1	0.85	0.228	0.76	0.24	0.1733	0.0465
20	1	0.76	0.228	0.76	0.24	0.1733	0.0416
30	0.96	0.76	0.229	0.72	0.28	0.1583	0.0487
40	0.93	0.73	0.298	0.72	0.28	0.1995	0.0609
50	0.9	0.72	0.228	0.63	0.37	0.1293	0.0607
60	0.84	0.7	0.191	0.74	0.26	0.1187	0.0348
70-	0.78	0.65	0.129	0.75	0.25	0.0755	0.021

Table 1.
The expected value in accordance with age and gender.

4.1.3 Evaluation of the ability of mutual assistance in model district

The model district is located in the Ushikubo district of Toyokawa City, Aichi prefecture in Japan. The Ushikubo district was selected according to the result of the evaluation and extraction of the districts with a high risk of the earthquake disaster in Toyokawa city, which was carried out in cooperation with Toyokawa city and the authors' laboratory in 2014 [8].

Based on the results of the survey, from 2015, the district started specific community-based activities for disaster mitigation with a view to improving densely built-up areas. The model area is one of the residents' associations included in the Ushikubo district. It has a high awareness of disaster prevention activities and has been proactively continuing disaster prevention efforts for more than 10 years.

The location of the model area is within walking distance from the railway station (JR Ushikubo Station). It is included in the residential promotion area announced in Toyokawa City's location optimization plan (**Figure 5**).

Regarding the data collection, first, a questionnaire was done for all households in the district in order to obtain the necessary information for evaluating mutual assistance. Distribution and collection of the questionnaire were carried out through the cooperation of the neighborhood association.

For those households that did not join the neighborhood association, questionnaires were distributed directly to their mailboxes and collected by mail. The recovery rates were 97% (128/132) and 14% (7/50), respectively. The items of the questionnaire were the number of household members, gender, age group, and the number of people who required special care.

The evaluation results of the ability of mutual assistance and the number of people who need assistance are shown in **Table 2**. The followings are the details of each items in **Table 2**. The maps of the evaluation are shown in **Figure 6**.

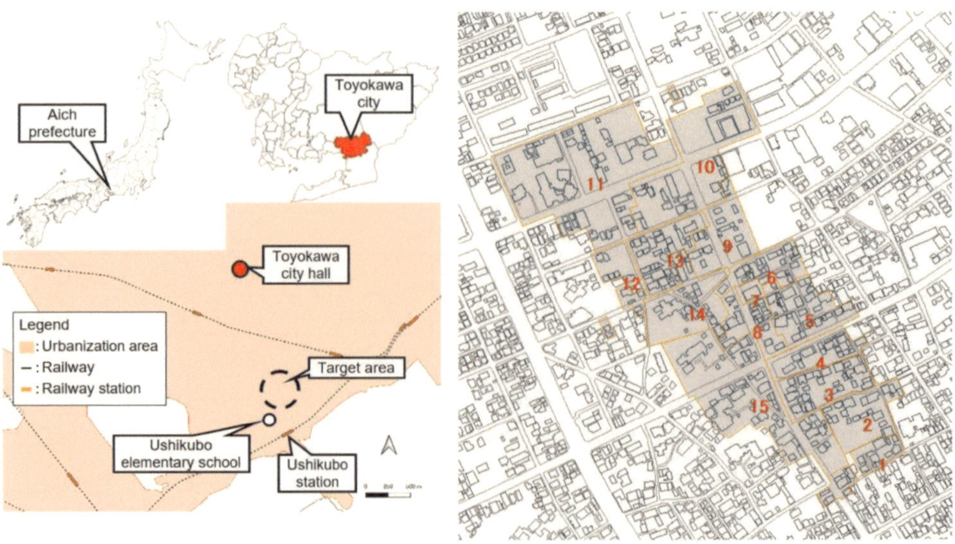

Figure 5.
Location of the target area.

DOI: http://dx.doi.org/10.5772/intechopen.109461

Items	The number of small groups															Total
	1	2	3	4	5	6	7	8	9	10	11	12	13	14	15	
1): People who need assistance	6	2	6	1	1	3	2	0	3	5	6	5	2	3	2	47
2): 1) - people under 10 years old	6	2	4	1	1	2	0	0	3	4	3	3	2	2	0	33
3): Ability of mutual assistance	1.3327	0.9839	3.0331	1.6542	2.8272	1.3194	0.4674	1.3514	2.5286	2.2181	2.4041	3.451	1.2214	2.3447	3.0414	30.1786
4): Surplus or shortage by 3) - 1)	−4.6673	−1.0161	−2.9669	0.6542	1.8272	−1.6806	−1.5326	1.3514	−0.4714	−2.7819	−3.5959	−1.549	−0.7786	−0.6553	1.0414	−16.821
5): Surplus or shortage by 3) - 2)	4.6673	−1.0161	−0.9669	0.6542	1.8272	−0.6806	0.4674	1.3514	−0.4714	−1.7819	−0.5959	0.451	−0.7786	0.3447	3.0414	−28,214

Table 2.
Evaluation results of the ability of mutual assistance.

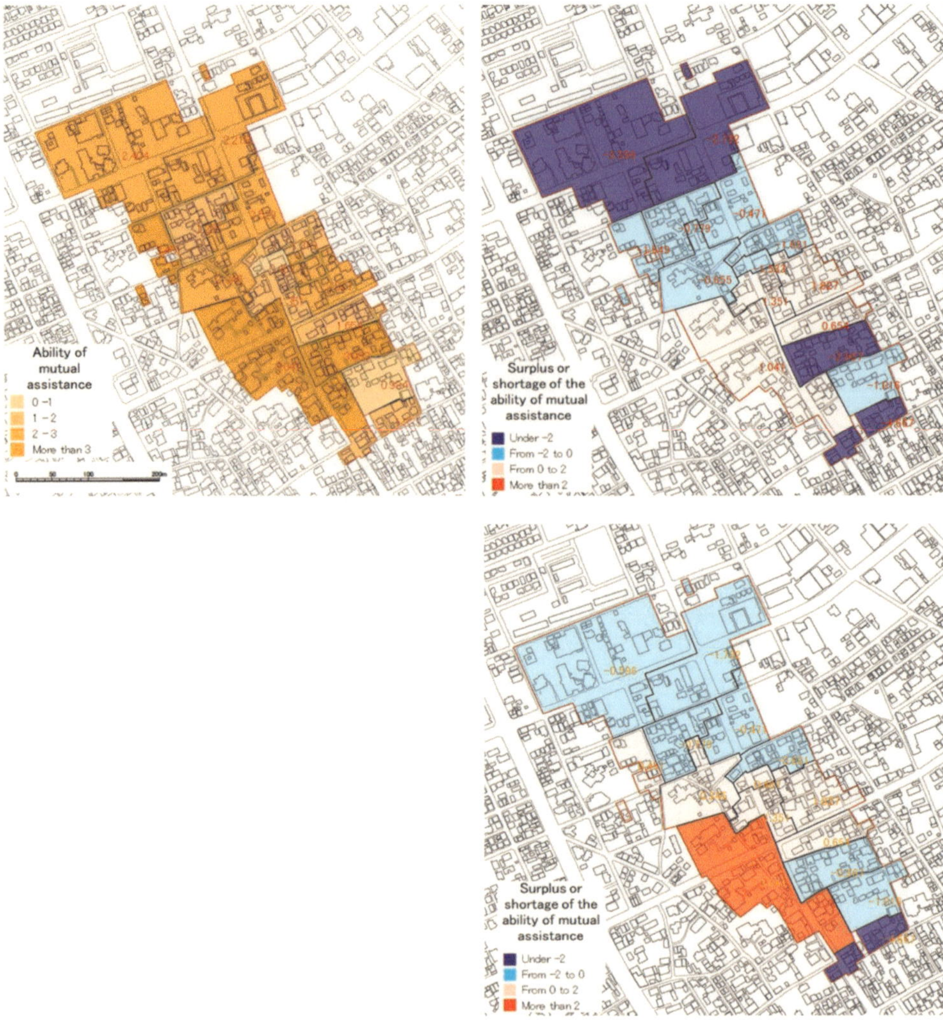

Figure 6.
Map of evaluation results of the ability of mutual assistance.

1. Person who needs assistance: Person who is difficult to evacuate when a huge earthquake occurs, such as the elderly, disabled people, infants, pregnant women, and foreigners without Japanese language knowledge.

2. The value subtracting the number of people under 10 years old from the number of people who need assistance.

In the questionnaire, a person under 10 years old was regarded as a person who needs assistance. However, if it is an infant, it can be carried by one adult. If it is older than an infant who can walk, it is thought that one adult can hold hands and guide him/her to evacuate.

Therefore, the number of people under 10 years old was grasped in the questionnaire.

1. The ability of mutual assistance: If the value is more than 1, a person can receive enough assistance from neighbors. If the value is more than 2, two people can receive enough assistance.

2. The value subtracting the number of people who need assistance from the rescue expectation value. The blue letter means shortage. The red letter means surplus.

3. The value subtracting the number of people who need assistance from the rescue expectation value according to the consideration of (2).

4.1.4 Possibility of improving mutual assistance through residential induction

From the perspective of improving the ability of mutual assistance, if residential induction of young households, such as those of the child-rearing generation, is encouraged in the areas where the ability of mutual assistance is low, it is thought that the ability of the area will increase.

As a method of residential induction, for example, in Toyokawa City, the model area is located, new residents who move into the urban function induction area from other cities or from the designated disaster-predicted areas in Toyokawa city have been given Machinaka subsidy.

The subsidy is such as equivalent to the property tax of housing and land for up to 3 years (with an area limit) and a child-rearing incentive (100,000 yen per person for junior high school students and younger, one-time only).

By applying such a subsidy system to the areas where improvement of mutual assistance is necessary within residential promotion areas where improvement of mutual assistance, it is conceivable to strategically induce young households.

Of course, it seems not easy to guide the selection of residential areas into the residential promotion areas. Furthermore, induction into the narrowed areas is more difficult.

However, the areas like the model district of this study, which are close to railway stations, have high population density even within the city and are close to facilities for child-rearing such as elementary schools, have a high potential for inducing residents. Therefore, it seems to be worth considering induction the residence.

As already mentioned in Section 3, in June 2020, the preparation of "disaster prevention guidelines" was added to the Location Normalization Plan.

The areas with high disaster risk should be excluded from residential promotion areas in order to suppress new location in principle. In the areas with high risk in the residential promotion area, the plan to mitigate the disaster risk should be incorporated in the Location Normalization Plan and conducted to achieve the contents in the plan.

Considering the policy, although it is not a designated disaster risk area, a district like a model district with a high density of old wooden buildings in a densely built-up area and a high disaster risk against earthquakes might be examined to exclude from the residential promotion area.

On the other hand, the countermeasures for disaster mitigation as the disaster prevention guideline, the model area has enough activities for disaster mitigation such as grasping the people who cannot evacuate alone, examining evacuation support for those who need support, and activities for improving the ability of mutual assistance.

The areas with active activities for disaster mitigation such as the model area might be examined to include in the residential promotion area as an incentive. The promotion of activities for disaster mitigation might be expected if the discussion of this incentive system is deepening.

4.2 Improvement in densely built-up areas of the residential promotion area

4.2.1 Background and Objective

In Japan, there still remain many densely built-up areas, and the promotion of safety in these areas is a pressing issue for urban planning. However, improvement in these areas has been very slow. Exploring draft plans through collaborations between local governments and local residents is essential in order to promote improvements aimed at the improvement of safety, living environment, and townscape.

Therefore, citizen participatory workshops are used nationwide to reach a consensus among stakeholders. However, it is difficult to reach a consensus. This is a key factor to promote improvements in these areas [9].

This study aimed to try to develop a PSS for discussion of workshops to examine draft plans and to verify the usability of using PSS to reach a consensus.

Authors focus on the following two issues as inhibiting consensus building: (1) The lack of a method that can quantitatively and objectively evaluate the earthquake disaster vulnerability of the subject area in order for participating residents to easily understand the effects of improvement plans; (2) The lack of a method for participants to share their understanding of spatial townscape images in combination with disaster mitigation performance evaluation (DMPE) during the discussion exploring the draft plans. The authors explain the outline of a PSS for exploring improvement draft plans of unified DMPE1) in real time with a function providing spatial townscape images by coupling Web-GIS and Virtual Reality in order to solve the above issues.

4.2.2 Outline of the developed PSS

A. Constitution

The PSS we propose in this paper primarily consists of a Web-GIS system incorporating the DMPE method (Web-GIS support system) and a system for exploring spatial townscape images using a VR viewer (VR support system) (**Figure 7**). The details of the PSS are described in the reference [10]. This section shows the outline of the tool.

B. Function of the Web-GIS support system

The Web-GIS system we developed has two functions. The first function computes DMPE and compares the results. By using this function, users can evaluate current disaster risk conditions or improvements in safety after the contents of draft plans have been implemented in the form of visual information. The other function is the support for the exploration of draft improvement plans. This system can be used in various types of community-based activities such as those geared toward increasing disaster mitigation awareness and exploring draft plans for disaster risk improvements. We

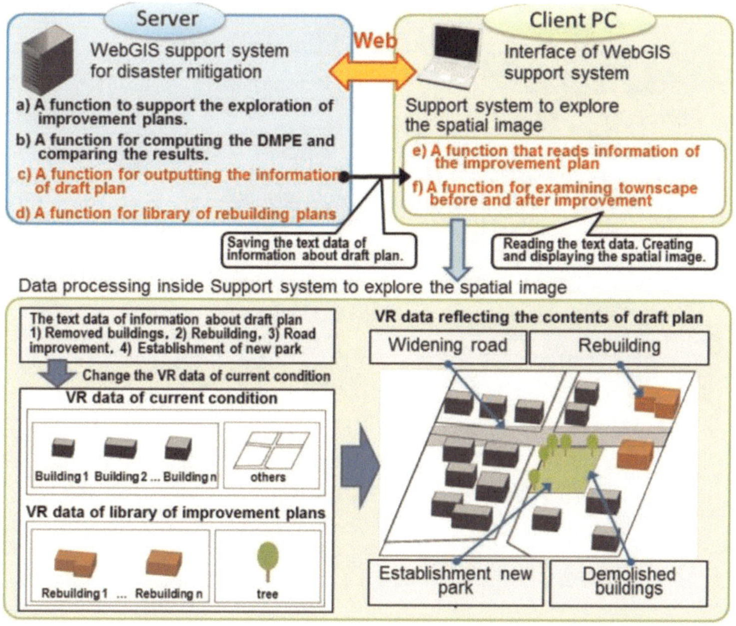

Figure 7.
System configuration of the PSS.

assume that this system is operated by local governments and helps community-based activities geared toward disaster mitigation.

In order to tightly couple the system that uses a VR viewer to explore spatial images to the Web-GIS support system, we added a function that outputs the information contained in draft plans to the system (Function c). Using this function, users create "the text data to transfer the information in the draft plan" and save it on the client PC. After that, users read it and display spatial images reflecting the contents of the draft plan using the VR support system. In the draft plans, the proposed improvements can include road improvement such as the widening of narrow roads and the establishment of new roads in densely built-up areas; building improvements such as the reconstruction of those with high fire hazard risk and risk of collapse; establishment of new parks to create evacuation sites, fire prevention zones, green space, community space, etc. Thus, users can visualize the effect of building reconstruction and removal, the establishment and widening of roads, and the removal and establishment of parks as specified in the VR data applied to the current condition by using the support tool. When reconstructing a building, ideally the support tool needs to have a function that allows users to freely change the elements of the building such as figure, size, texture of wall and roof, etc. To achieve this, the support tool needs vast amounts of VR data. However, if the method adopted directly allows users to select the texture of a building from an overwhelming amount of data, users may feel that the tool is too difficult to use. Thus, a method that allows users to select the data from some options carefully selected in advance is more useful than the way mentioned above. Moreover, our aim is to contribute to community-based planning by facilitating collaboration between local governments and local residents using a method that

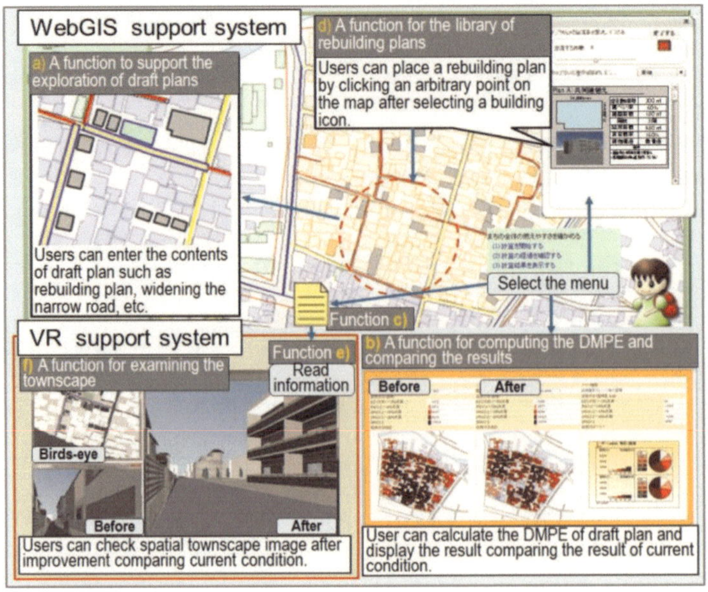

Figure 8.
Using image of the support tool.

enables gradual improvements over time until consensus among participants is achieved. We cannot assume that the improvement method will achieve consensus all at once. Therefore, the range of design for reconstruction needs options that blend in with the around present townscape so as to keep the district townscape harmonized after improvement. Consequently, we added a function for a library of rebuilding plans with a tool that users can use to select building data from the library when exploring a draft plan (Function d). This library consists of several rebuilding explored by expert in advance and register them in the database of the server.

We assume that these rebuilding data are changed and updated based on opinions participants said in the WS exploring draft plans.

4.2.3 Using method of the PSS

Before exploring the draft plans, users access the Web-GIS support system from the client PC and download the VR support system. Users start by exploring the draft plans and create a draft plan. They enter the contents of the draft plan on the interface of the tool using a function that facilitates exploration of the improvement plans (function (a)). Using function (d), a function for the library of rebuilding plans users can enact a rebuilding plan by clicking an arbitrary point on the map of the interface after selecting a building icon.

After entering the contents of the draft plan, users can calculate the DMPE of the draft plan and display the result of comparison with the current condition using (b) a function for computing and comparing the DMPE. When creating the spatial townscape image, first, users save information about the improvement plan as text data using function (c) a function for outputting the information about the improvement plan. After that, users can display the VR image reflecting the draft plan read in as text data using function (e) a function for reading information about the improvement plan (**Figure 8**).

5. Conclusions

Grasping natural disaster risks is very important for planning urban sustainability.

In the cities like Japanese cities, where the population is declining, it is necessary to induce residents to withdraw from the areas with high disaster risk, and the induction of the relocation to the areas without disaster risk from a long-term perspective was explained.

In order to promote the formation of such cities, the development of planning support technology introduced in this chapter is one of the useful means. Especially, it is expected that the PSS is useful to provide quantitative information such as the ability of mutual assistance, to support the examination of initial response activities, and to examine the plans that minimize damage from natural disasters.

In the future, with the spread of technologies such as AI and big data analysis technology, the role of PSS is expected to become more common. It is preferable to include a Conclusion(s) section, which will summarize the content of the book chapter.

Acknowledgements

This research was funded by Japan Society for the Promotion of Science (JSPS) KAKENHI, grant number 15 K18177".

Conflict of interest

The authors declare no conflict of interest.

Author details

Kazuki Karashima* and Akira Ohgai

1 Faculty of Engineering, Department of Environment and Design, Maebashi, Japan

2 Toyohashi University of Technology, Toyohashi, Japan

*Address all correspondence to: k-karashima@maebashi-it.ac.jp

References

[1] Naoto Nakajima N et al. Urban Planning. Kyoto: Gakugei; 2018. p. 34 (in Japanese)

[2] Ministry of Land, Infrastructure, Transport and Tourism. Location Normalization Plan [Internet]. 2014. Available from: https://www.mlit.go.jp/en/toshi/city_plan/compactcity_network.html [Accessed: 2022-11-30] (in Japanese)

[3] The Japan Society of Community Disaster Management Plan [Internet]. 2014. Available from: https://gakkai.chiku-bousai.jp/english.html [Accessed: 2022-11-30]

[4] Japan Institute of Country-ology and Engineering. Proposal of urban improvement of pre-disaster prevention utilizing the disaster prevention guideline in the location optimization plan [Internet]. 2019. Available from: https://www.jice.or.jp/cms/kokudo/pdf/reports/recital/2021/gj2021_05.pdf [Accessed: 2022-11-30] (in Japanese)

[5] Ministry of Land, Infrastructure, Transport and Tourism. Survey of disaster risk in residential promotion areas [Internet]. 2019. Available from: https://www.cas.go.jp/jp/seisaku/resilience/dai56/siryou3.pdf [Accessed: 2022-11-30] (in Japanese)

[6] Ministry of Land, Infrastructure, Transport and Tourism. Disaster prevention guideline of location normalization plan [Internet]. 2014. Available from: https://www.mlit.go.jp/toshi/city_plan/content/001379331.pdf [Accessed: 2022-11-30] (in Japanese)

[7] Karashima K, Ohgai A. A Methodology of Workshops to Explore Mutual Assistance Activities for Earthquake Disaster Mitigation. International Journal of Environmental Research and Public Health MDPI. 2022; **18**:26. DOI: 10.3390/ijerph18073814

[8] Karashima K, Ohgai A. AIP Conference Proceedings: An attempt to implement tools to support examination of community-based activities for disaster mitigation: A case study in Toyokawa city, Japan. API Publishing; 2018. DOI: 10.1063/1.5005773

[9] Sato S. Machizukuri Design Games. Gakugei Shuppansha. Kyoto: Gakugei; 2005

[10] Karashima K, Ohgai A, Tadamura K. A support tool coupling WEB-GIS and virtual reality to explore improvement draft plans in densely built-up areas. Journal of Architecture and Planning Architectural Institute of Japan. 2014;**79**: 745-754. DOI: 10.3130/aija.79.745 (in Japanese)

Chapter 25

Analysis of Social Vulnerability to Earthquake Hazards in Dhaka South City, Bangladesh Using a Modified PAR Model

Mohammad Mahbubur Rahman
and Mohammad Harun-Or-Rashid

Abstract

The study aims at investigating the root causes, dynamic pressures, and unsafe conditions to the progression of the vulnerability of earthquake hazards among the people residing in the 22nd ward of *Hazaribagh* Thana of Dhaka South City (DSC) in Bangladesh. Following a purposive sampling procedure, the study selected the 22nd Ward of Dhaka South city. As part of qualitative methods, a total of 20 Key Informants Interviews (KII) were carried out and six focus group discussions (FGDs) were conducted. By adopting a modified Pressure and Release (PAR) model, the study revealed that the low-income status of the family, lack of proper health services, single-headed family, gender, differently able person, and the interpersonal good relation members, the availability of urban emergency center has found root causes for the progression of the vulnerability of earthquake hazards. Lack of appropriate skills, unplanned residences, and industry contribute to the production of dynamic pressures of the vulnerability of earthquake hazards. The unsafe conditions involve poor housing materials, dangerous locations for residents, and a lack of first aid documents associated with the production of the vulnerability of earthquake hazards.

Keywords: earthquake, risks and threats, vulnerability, perceptions, the modified PAR model, Bangladesh

1. Introduction

Bangladesh is located close to two active tectonic plates: the Indian plate and the Eurasian plate [1]. The country suffers from the seismic or earthquake zone of which two-thirds exists under major and moderate fault [2–4]. These fault zones were instrumental in causing some of the world's severest earthquakes in the past. According to the earthquake zoning map of 1993, it is observed that 26% of areas of Bangladesh lie in high risk, 38% moderate, and 36% in low-risk zone regarding earthquake vulnerability [5].

As one of the most densely populated mega-cities of the world, Dhaka becomes a high-risk earthquake zone [6, 7]. Earthquake Vulnerability Index (EVI) revealed that Dhaka city is 2nd among the 20 most vulnerable cities on the earth [8]. Over the past decades, it is estimated that the factors of susceptibility and vulnerability of earthquakes were the population density, haphazard migration, unplanned rapid urbanization, quicker made of new buildings in every available space, and most garment factory buildings built in congested areas without open spaces, non-compliance of building codes and proper guidelines, fire incidences from gas and electricity line, narrow spaces of road construction, insufficient of preparedness from the responding agencies and lack of awareness among city dwellers and decision makers [9, 10]. Around 13 million larger inhabitants, enormous poorly constructed, and dilapidated structures in unplanned Dhaka city signify extremely destructive and vulnerable situations for the massive loss of lives and property during a moderately large earthquake event [5, 11].

The United Nations IDNDR-RADIUS (International Decade for Natural Disaster Reduction) identified that Dhaka city is located in the high-risk zone of earthquake hazards (Integrated Regional Information Networks News 2008) considering its massive population density along with unplanned building apartments, narrow streets, poorly constructed infrastructure, exposed and risky electrical lines, lack of coordination between institutions concerned, the inadequacy of recovery tools and lack of awareness among the people [9]. An unprecedented human disaster like moderate to heavy tremor may happen in the city at any time [7]. An overwhelming number of multi-story buildings have been created without following the standard building codes set by the *RAJUK (Rajdhani Unnayan Kortripokkho*, the English meaning Capital Development Authority) for meeting the millions of people's housing demands [9].

A few studies have been conducted so far for exploring the social causations of earthquake disasters. Therefore, the study aimed to explore the linkage between social causations of vulnerability and earthquake disaster perceptions in Dhaka South City. The research shed light on exploring the root causes of the progression of the vulnerability of earthquake hazards. In addition, we focus on finding out the dynamic pressure and unsafe conditions of the progression of vulnerability to observe how these factors lead to aggravating the vulnerability level to an earthquake disaster in Dhaka South City.

2. The theoretical framework: the modified pressure and release (PAR) model

Since an earthquake hazard is a natural event that hits nature, people do not have much control over it. However, we have adequate control over the production of vulnerability because our normal activities create our vulnerability consciously and/or sometimes unconsciously. We have adopted a few factors from the PAR model [10]; which seem to be relevant to the creation of human vulnerability in the study area. Some factors have been taken out from the modified model as they are found difficult to show the linkages with the production of vulnerability. The research area, Dhaka South City, considerably differs from some factors of the macro level of 'Dynamic pressures' of the model to the creation of vulnerability. The macro level factors i.e. factors of arms expenditure, debt repayment, and deforestation may be found a significant relationship with the creation of vulnerability in other contexts, however not appropriate in our study area (**Figure 1**).

The political ideology determines certain people's access to take part in major decision-making regarding disaster planning and mitigation strategies (Ibid). It limits certain people and groups' access to political participation at local and national levels.

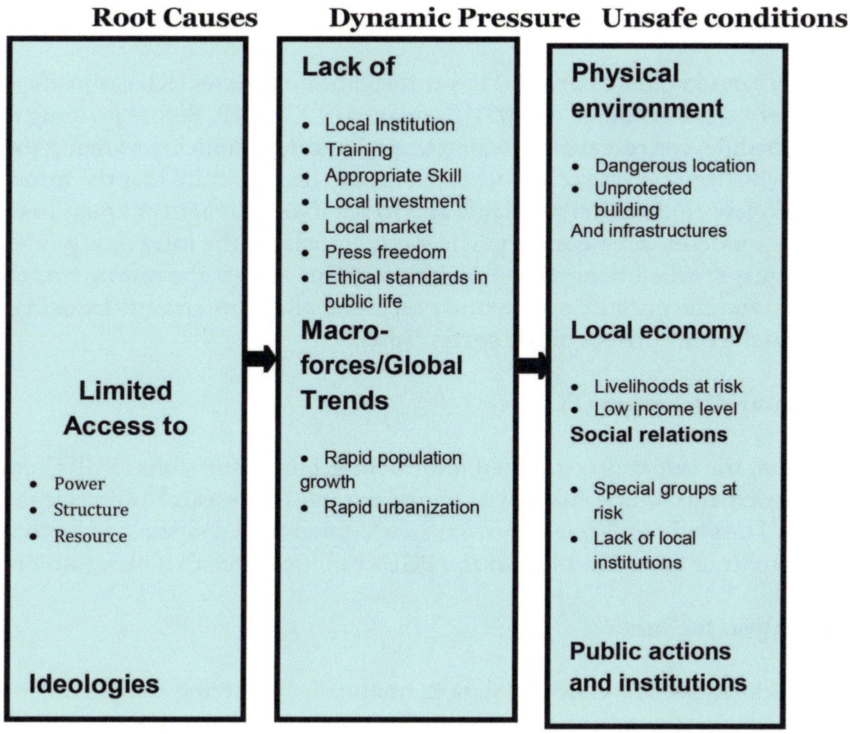

Figure 1.
The progression of vulnerability: root causes dynamic pressure unsafe conditions (source: reprinted from the book of [10]).

Some people receive less attention from local NGOs and local government personnel in the event of natural hazards [5]. The economic pattern of society originates and maintains unequal access to resources and opportunities. Some people do not have equal access to land or farm, adequate access to safe water, and basic health services [10].

In terms of unsafe conditions, some segments of people live and work in dangerous locations, as they have no other choices left (Ibid). It also includes the initial level of well-being of the people which involves access to tangible (e.g. cash, shelter, food stock, etc.) and intangible resources like; networks of support, knowledge regarding survival and sources of assistance, ability to function in a crisis [10].

3. Methods and procedures

3.1 Selection of the study area

The study selected the 22nd ward of the *Hazaribagh* area of Dhaka South City followed by purposive sampling, which comprises about 1.07 square kilometers with 19,599 households [5]. The density of the population in the area is 160,316 among the total population of Dhaka is 6,539,704 with 3,741,220 population in Dhaka North City Corporation (DNCC) and 2,798,484 in DSCC (Population Census—2014). The research site is in the former tannery industrial area of *Hazaribagh* thana, just adjacent to the *Hazaribagh* Police Station and *Hazaribagh* bazaar, houses- accommodation, schools and Rickshaw stations, etc. in Dhaka South City of Bangladesh.

3.2 In-depth interviews

The study carried out a total of 20 key informant interviews (KII) as in-depth interviews between December 16, 2021, and January 22, 2022. Before running the interview schedule, the researchers tested three interview guidelines among the informants who lived outside of the research area to demonstrate that the informants find the interview guide understandable and to see if any changes or amendments to the interview guideline are necessary. The study translated the interview guide into the local language, called Bengali, before interviewing so that the informants could easily understand the questions. The study recorded all the interviews by using a tape recorder with the full consent of the participants.

3.3 Focus group discussion (FGD)

In addition, the research conducted four focus group discussions (FGDs) consisting of 30 participants which enabled us to understand the research objectives at a deeper level. The study took support from a note-taker with a tape recorder that facilitated us to carry out the FGD. Most of the FGDs ran for about an hour or a bit more.

3.4 Data analysis techniques

The study analyzed the collected data thematically by sorting out the salient themes and patterns from the narratives of the interviews. Tracing the thematic analysis, the study showed patterns of similarities and differences in data. In the next step, the study segmented, coded and presented the data. The study coded the data followed by the ideas of interpretive and narrative analysis. After transcribing the collected data, the study returned the interview information to the research participants for 'member checks' to establish credibility and reliability.

3.5 Ethical considerations

The study received oral consent before conducting in-depth interviewing, Focus Group Discussion (FGD), and tape-recording from our participants. The study informants were assured that their responses will be held with strict confidence and shall remain anonymous. In addition, the research confirmed that they can stop answering our questions whenever they feel embarrassed.

4. Root causes and earthquake disaster

The findings revealed that socio-economic factors along with dynamic challenges and unsafe conditions interact to pose challenges to successfully facing the potential earthquake hazards in the area. Followed by the modified PAR model, the present research categorized the causes of the progression of vulnerability to disasters into three themes: root causes, dynamic factors, and unsafe conditions. The factors have been clearly explained by Wisner et al. [10] in their research model called the Pressure and Release model (PAR).

The economically marginal people in the environmentally vulnerable areas are isolated, and disaster-prone urban locations [10]. This leads to the three often mutually reinforcing sources of vulnerability. First, Low access to resources triggers an insecure and unrewarded livelihood for the people. The vulnerability produced by their activities. Secondly, these people get less priority to reduce hazard intensity from government interventions. Finally, People lost confidence and trust for self-protection who are economically and politically marginal. This area is commonly known as a dirty and polluted (*'pocha o nongra'* in Bengali term) area due to the absence of proper waste management in these industries. Low-income poor people live here as the house rent, and education expenses (tuition fees) of a child are comparatively cheaper than that of other places of the city. In this connection, an in-depth interviewee stated that.

> *My parents have been suffering from many chronic diseases like heart disease, hearing problems, etc. for the last couple of years. They need to be treated properly but we do not have the required amount of money to do so. I can never think of getting treatment from a private clinic as this is quite expensive. Hence, how can I expect to avail of medical services for me and my family? Considering my poor socio-economic conditions, I am not even able to preserve a small amount of food for my children for the next day. So, how can I stock dry foods thinking of hitting potential earthquake hazards? (A 38-year-old female, own translation).*

One FGD informant stated that the Health system is critical in this area. If they become ill, they must go to the Dhaka Medical College which is not near to them. However, there are a lot of private clinics which are so expensive. Finding no other alternatives, they must take treatment from non-governmental hospitals. As they are not so rich people, they must pay higher for health services that are not good for them. Getting proper health treatment from public hospital people requires strong political affiliation. It is well-documented that poor people in the study area have a lack of good connection with political parties which deprive them of health services [12].

Differently able people do not receive any funds from the Government of Bangladesh (GoB), and they become vulnerable seriously during their emergencies concerning health concerns. By highlighting a few socio-demographic conditions, like, single income generated family, gender, and disabled family members, an informant remarked that.

> *"When the earthquake struck in late at night on January 4, 2019, almost everyone was asleep and thrown out of their beds. I ran out with my four children onto the street, and I found the area had a blackout. Unfortunately, I forgot to help my elderly parents come out of the room and since they are disabled; they were unable to come out of their room without my help. Thanks to Allah that he saved us from any major damage caused by the earthquake. Now, I go to bed every night thinking of the earthquake and feel helpless, thinking of how I will evacuate my family in the event of an earthquake. (A 33-year-old female, own translation).*

Ethnic identity played a vital role to maintain affiliation with mainstream society. The ethnic background obstructs people to receive minimum help and assistance during the time of their need. If they need money for emergencies, ethnic people can hardly get it from colleagues and neighbors. They create a lack of faith among people from different socio-cultural backgrounds in their place. Besides, the people get in trouble for speaking due to a lack of their native (Bengali proficiency) languages. Although they can speak minimal *Bengali*, unfortunately, they do not know the ethnic language that well. In this connection, a santal (an ethnic people) stated being a member of an ethnic community faced arrange of discrimination to get their basic needs tackling potential earthquake disasters. In this regard, a male informant said that.

> *"If I need money for an emergency, I can hardly get it from my colleagues and neighbors. I get in trouble due to my lack of Bengali proficiency. I think due to cultural differences (food habits, beliefs, religion, ethnicity, etc.), I cannot interact with them well. I feel uncomfortable about sharing my perceptions, regarding the precautions to take during an earthquake. (A 42-year-old Male, own translation).*

The people living in this area have been neglected by the GoB officials from receiving any information regarding earthquake hazards. The people do not get any awareness message for avoiding the disastrous situation of the hazards to have occurred in the future. They do not get proper attention from either GoB personnel or NGO people. In this connection, a participant from an in-depth interviewee stated that,

> *"I strongly believe that people do not have any evacuation plan during an earthquake. Just a few days ago, a 5.9 earthquake struck Dhaka city. I just heard that people are screaming and that is all about! No initiatives by the government and NGOs have been taken after that! This indicates that people living in this area huge negligence from the respective authorities". (A 40-year-old male, own translation).*

The lower-income people lack the required knowledge for tackling potential earthquake disasters. They do not find time for learning lessons and preparing tools for successfully coping with probable earthquake hazards. In this connection, a rickshaw puller voiced out the following statement from an in-depth interview.

> *"Just a few days ago, I found some people were reading leaflets and discussing what should be done during an earthquake, whether they are inside or outside. I felt hopeless because I could neither buy helmets nor provide necessary instructions to my family, due to my illiteracy. (A 48-year-old Male, own translation).*

One of our FGD participants stated that they can assert their minimum knowledge of reserving dry food, and safe water and should keep their cattle safe from the disaster. They learned it from watching television and getting information from social media. In terms of maintaining interpersonal good relations among the neighbors in the local area are vital [13]. Neighbors help people in trouble during earthquake hazards. A college teacher stated the following statement in an in-depth interview,

> *We need to make our relationship much stronger than before by helping each other in the event of an earthquake. If an earthquake occurs, there will be a need for cooperation and assistance from each other, to evacuate safely; provide food and shelter. (Female, 48 years).*

Given the consequences, one focus group (Male guardians) pointed out that they have to work for their family members when they are in danger. Again, they live together in a building as a community. So, they have great social networking influences. All members of this building help each other at their utmost need. Social networking and kinship can work here perfectly to help each other in the face of earthquake hazards. Each of them understands to be more tolerant at the time of making buildings and during a disaster. One FGD participant opined that they have

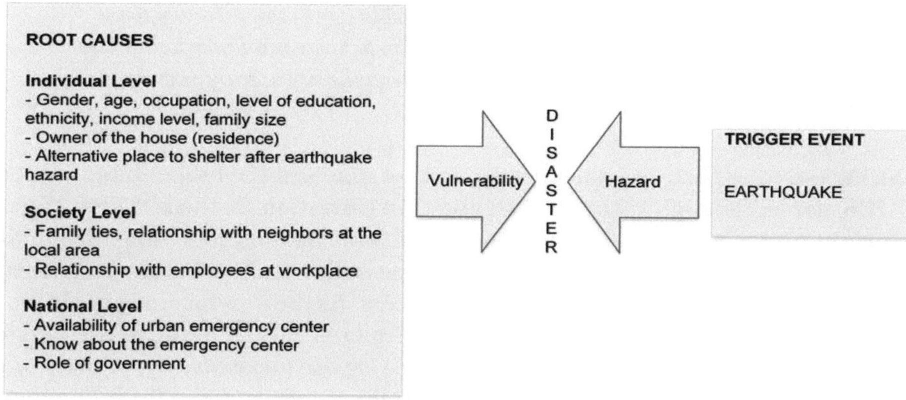

Figure 2.
Root causes of vulnerability (Source: Authors' framework, 2022).

great social networking, family bondage, and kinship ties that influence them to mitigate the earthquake disaster. They think that they have to help and work for their family members, living in the same building, neighbors, and community members when they are in danger and in their utmost need.

The following diagram has been developed based on the informants' thoughts and statements regarding their perceptions of the root causes of earthquake hazards (**Figure 2**).

5. Dynamic pressures and earthquake disaster

The dynamic pressures include rapid urbanization, foreign debt, and certain structural adjustment programs [10]. Dynamic pressures channel the root causes into forms of unsafe conditions that then must be considered concerning the different types of hazards facing people.

The dynamic factors are the appropriate skills and knowledge such as "dos and don'ts" while staying at home (e.g. under chair or table, corner to the wall, narrow space, besides hard and tall wardrobe, close the window, behind the door, on the balcony, on the bed, jump off the building), [5], while staying outside the home (i.e.- besides the building, the center of the road, beside the tree, run here and there), elderly people at home, unplanned tannery industries, unplanned commercial institutions, lack of govt. and non-govt. emergency center, and the rapid growth of unplanned residences [10].

The role of government or nongovernment organizations to make conscious people about dos and don'ts is far beyond their thinking [14]. People observe that no organizations have trained local people on how to find out escaping routes during earthquake hazards. Ironically, many unplanned high-rise masonry buildings have been built very close to the narrow roads; therefore, people have much doubt about how and where they can find out any escaping routes. In this connection, one informant stated that:

> *"Neither any governmental or nongovernmental organizations have made us conscious of this nor are training far beyond our thinking. I have much doubt about how and where we can find out any escaping routes. I heard that Dhaka city is*

very vulnerable to earthquake hazards; thus, I would like to call attention to the government that they should take the responsibility to arrange training for the local people and build some emergency centers throughout the city as early as they can. (A 50-year-old female, own translation).

During the construction, many people neither followed the proper building code of RAJUK nor left enough space for road and transportation. In the last 12 to 15 years, hundreds of multi-storied apartments have been built without following the building code and they have also left no gap in between the buildings. It is hardly possible to see the daylight from any room in these apartments. As the low-income people mostly live in this area, the house owners build the apartments only thinking of their accommodation. *An interview with a house owner describing the alternative escape routes due to the occurrence of a potential earthquake.* A household owner stated the following statement from the in-depth interview that.

I observed that in the last 15 years, hundreds of buildings have been built only for profit and not taken into consideration, what would happen if an earthquake struck. These compact apartments have been illegally built by bribing the authorities; otherwise, these buildings would not have been there. I think if a strong earthquake hits the area, it will damage the buildings and endanger thousands of lives. (A 68-year-old Male, own translation).

One of our informants from FGD told us that all the buildings are adjacent to one another. Besides, they think the tannery industry buildings were not built according to the rules and regulations of RAJUK and City Corporation. All building owners tried their best to capture all their room and space to make the best use of their expenditure. All the participants of the 3rd FGD think that all building owners of this area had the intention to make a profit from their building. All the buildings are too near to all others. Sometimes it is difficult to see the sky as they are so adjacent to one another. So, they have the potential to cause great damage or destruction if an earthquake happens.

Female guardians in the 4th FGD discussion echoed the present critical situation. She demonstrated that "road networks are not well constructed. The high density of the population creates a traffic jam in the area. At the time of the school closing,

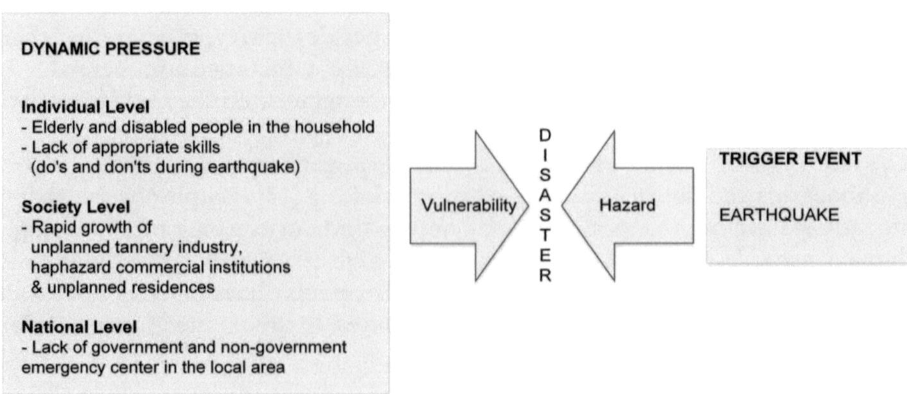

Figure 3.
Dynamic pressures of vulnerability (Source: Authors' framework, 2022).

the road faces severe traffic jams. Government has little or no control over it". She added that roads are full of rubbish, and they face a lot of problems even while walking in the road. The following diagram associated with dynamic pressures and earthquake disaster perceptions has been made based on the participants' thoughts and comments on earthquake hazards (**Figure 3**).

6. Unsafe conditions and earthquake disaster

Unsafe conditions are the specific forms in which the vulnerability of a population is expressed in time and space in conjunction with a hazard. Examples include people having to live in hazardous locations, the weak pattern/structure of the house, aged and dilapidated buildings, narrow staircases, house building materials, building without fire protection, hazardous chemical industries, and residential use in others [10]. The unsafe conditions also involve congested electric wares with buildings, lack of required open spaces, overcrowded high density of population, and lack of enforcement of the law [15]). Rahman et al. [5] found that having to engage in dangerous livelihoods such as prostitution, gambling with its attendant health risks, or having minimal food entitlements poor sanitation, lack of preservation of boiled drinking water, and keeping telephone number of govt. or non-govt. emergency center, the pattern of the weak drainage system, and insufficient first aid equipment e.g., helmet, torch, and radio have a strong connection with the creation of unsafe conditions of vulnerability [16]. It is important to consider the pattern of access to tangible resources (e.g., cash, shelter, food stocks, agricultural equipment) and intangible resources (networks of support, knowledge regarding survival and sources of assistance, morale, and the ability to function in a crisis) [10]. In this sense, an in-depth interviewee illustrated the following story.

> *"Most dangerous thing is that maximum tanneries do not have the proper drainage system of their toxic wastages. Moreover, most of them are kept uncovered here. I am afraid thinking of that if these poisons leak out on the ground due to the earthquake tremor, it will cause extremely hazardous living conditions for the people as the whole area will be enormously contaminated. (A 45-year-old-male, own translation).*

In line with the previous interview, another in-depth interview informant added the following issues:

> *"As the people in this area are comparatively poor, they build their housing in such a way so that they can collect rent from the people. The house owners do not intend to let City Corporation to have enough place for road construction; as a result, the road pattern remains very narrow here". (A 45-Year-old male, own translation).*

Few FGD participants echoed the statement of the previous in-depth interviewee. They said that the authority of *RAJUK* had never come to them. They have neither any idea about *RAJUK* policy nor about following the proper guidelines of *RAJUK*/Government. One participant said that even though there is any regulation from RAJUK/Government authority they do not abide by these regulations specifically in the tannery industry. Tannery industries are, undoubtedly, posing an

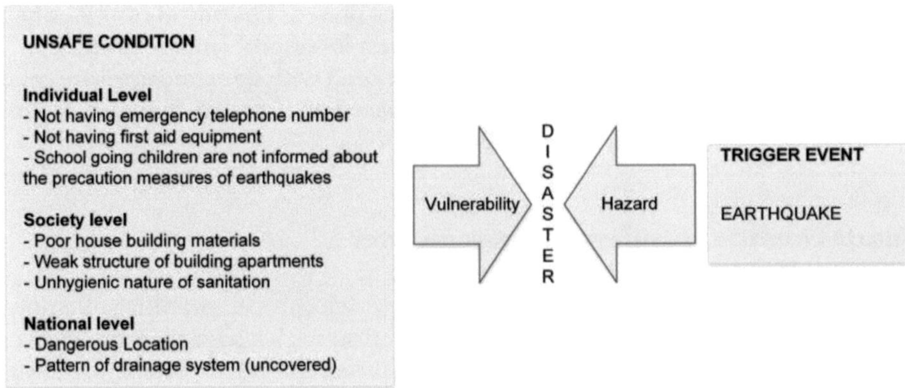

Figure 4.
Unsafe conditions of vulnerability. (Source: Authors' framework, 2022).

environmental threat in the area as well as, causing health hazards among the people. In this connection, a tannery owner stated the following case.

> *"During the initial stages of this industry, we did not follow the Central Effluent Treatment Plant (CETP) process. As a result, huge amounts of toxic liquid wastage have been discharged into the rivers, canals, and ponds that run alongside the road, followed by uncovered drains. Recently, the Government of Bangladesh decided to relocate the tannery industry from Hazaribagh to the city outskirts called Savar. Hence, we have to buy most of the machines for Savar anew. A huge amount of money is required to build new buildings which we lack at the moment. Moreover, mortgaging the tanneries to banks for the loan has prompted several complexities to relocate the industry (A 67-year-old male, own translation).*

Focusing on the previous statement, one FGD participant stated that on a rainy day the road is drowned with the dirty color water from the tannery, and water logging arises this day. The participants of the 2nd FGD told that most of the drains are interrupted for wastage of tanning industries before following into the canal. The association between unsafe conditions and the progression of earthquake hazards has been developed relying on the current perceptions of the people involved with the research. The diagram revealed that factors of unsafe conditions are closely connected with the earthquake hazards in the studied area (**Figure 4**).

7. Discussion

Studies pointed out that the socio-economic conditions of the people significantly impact people to understand and perception exposure of to earthquake hazards. People's economic class, age group, gender, ethnicity, disability, and migration status have vital roles in the progression of the vulnerability of earthquake hazards [3, 10]. Social vulnerability is a multidimensional concept that is historically produced by excluding some groups of people from accessing the production and allocation systems of economic resources [5, 10]. The socio-demographic features enable people to successfully respond to earthquake hazards [5]. Recent studies showed that the geographic location, unplanned establishment of buildings, and high density of development contribute to the production of the vulnerability of some areas [3, 15]. The present study revealed that the dangerous location, unplanned buildings, and narrow road patterns contributed to the progression of the vulnerability of the people living in the area.

The city has experienced massive population growth, high density of population, unplanned urbanization, high level of corruption, political instability and violation, gender discrimination, and lack of social services. Certain individual characteristics and social factors contribute to the production of vulnerability in the event of earthquake hazards [5]. People who have adequate access to a decent home, safe water, and other economic, as well as political resources, are less susceptible to any natural hazard events. The study pointed out that individual characteristics and social factors have a very significant role in determining where people will live and work, what kind of buildings, their level of hazard protection, information, preparedness, wealth, and health which place some segments of people most at risk in the process of earthquake disasters.

The gendered people, such as women, differently able and aged people living in poverty [17], cannot stay safely puts them vulnerable in the event of earthquake hazards [5]. The study showed that women and aged people suffer from a range of chronic diseases and become susceptible in the face of earthquake hazards. Rahman et al. [3] argued that these people are more at risk to the impacts of earthquake hazards as they are more dependent on others and are less capable of evacuation. Earthquake vulnerability largely depends on its geology, population density, building density and quality, and finally the coping strategy of its people and it shows clear spatial variations [18, 19].

Therefore, the study pinpointed that earthquake hazards become a disaster only for the people who have a certain pattern of social, physical, and other facilities. The study revealed that the rich have more money and have a high status which can be used to solve the critical condition better than the poor people as they are rich and know well. Social participation in natural hazards plays a crucial role to mitigate hazardous events [20]. The study showed that relatives can be more tolerant and sacrifice their shared room after shaking the earthquake and help them to evacuate from the earthquake and after the earthquake. During earthquake hazards, safe people can work for saving people as a part of a family and social networking system, etc.

People's prior knowledge of potential earthquake hazards would help the people to take necessary steps before, during, and after earthquake hazards [3, 15]. Educated people possessed a comparatively clear understanding of 'do's and don'ts' in the event of an earthquake disaster and explained more details about earthquakes. Rahman et al. [3] revealed that people can save themselves hiding under the wooden furniture during the time of earthquake. They added that by watching television, and social media, people acquired the relevant knowledge like reserving the dry food, safe water and should keep our cattle safe from the disaster. The absence of basic social services, polluted air, and water resources; and the absence of educational opportunities make people more vulnerable in the face of earthquake hazards [10]. The study found that all the roads are narrow and congested and full of rubbish. Thus, these all factors contribute to the deteriorating physical environment in the study area.

8. Conclusion

The study is an effort to explore the social vulnerability of earthquake disasters from the insights of the people living at the 22nd Ward of Dhaka South City, Bangladesh. The socio-demographic features of the people have significant relations with the production of the vulnerability of earthquake hazards in the studied area. The root causes of the production of vulnerability found that socio-economically poor people have less capacity to face the potential earthquake hazard. Moreover, ethnic identity, differently able people, and social networking play a vital role to produce vulnerability once they are hit by the hazard. The dynamic pressure includes the increasing number of people living in unplanned congested buildings exposed to the earthquake hazard. The study showed that the risk is high in Dhaka South City, as the area is located in the high natural seismicity along with unplanned urbanization, poorly designed and constructed structures, extremely dense population, and poor preparation for the eventually necessary response to an earthquake have produced social vulnerability among the respondents.

There has been an increasing concern about the earthquake vulnerability of the present built environment of Dhaka South city. There have been many campaigns run by the government to make people aware of earthquakes and its devastating effects. However, people living in the study area are less concerned about the dos and don'ts of the potential threat of earthquake hazards. The poor housing materials, unsafe drainage systems, and lack of emergency equipment in the building make the people more vulnerable in the face of an eminent earthquake hazard. The Government of Bangladesh has been trying to provide necessary knowledge and information about the earthquake hazards among its people. However, people living in dangerous areas having less socioeconomic strength and poor political connections get less attention from the higher authority of the government.

Author details

Mohammad Mahbubur Rahman[1*] and Mohammad Harun-Or-Rashid[2]

1 International Standard University (ISU), Dhaka, Bangladesh

2 Additional District and Sessions Judge, Law and Justice Division, Ministry of Law, Justice and Parliamentary Affairs, Bangladesh

*Address all correspondence to: mahbubur@isu.ac.bd

References

[1] Ali MH, Choudhury JR. Assessment of seismic hazard in Bangladesh. In: Nizamuddin K, editor. Disaster in Bangladesh: Selected Readings. Dhaka: University of Dhaka; 2001. pp. 109-126

[2] Ministry of Disaster Management and Relief. Seismic Risk Assessment in Bangladesh for Bogra, Dinajpur, Mymensingh, Rajshahi. Bangladesh: Rangpur and Tangail City Corporation; 2015

[3] Rahman MM, Rahman S, Jerin T. Social vulnerability to earthquake disaster: Insights from the people of 48th ward of Dhaka South City. Bangladesh, Environmental Hazards. 2022;**21**:1-21

[4] Zaman MAA, Sifty S, Rakhine NJ, et al. Earthquake risks in Bangladesh and evaluation of awareness among the university students. Journal of Earth Science Climate Change. 2017;**9**:482

[5] Rahman MM, Abdullah ABM, Murad MW. Community perceptions of and vulnerability to earthquake disaster: Insights from the City of Dhaka, Bangladesh. Journal of Environmental Assessment Policy and Management. 2018;**20**(4):1-27

[6] Islam MS, Hossain MT, Ameen SF, Hoque E, Ahamed S. Earthquake induced liquefaction vulnerability of reclaimed areas of Dhaka. Journal of Civil Engineering (IEB). 2010;**38**(**1**):65-80

[7] The Daily Star. Bangladesh at Extreme Risk: Disaster Preparedness Needs Beefing Up. Dhaka: The Daily Star; 2010

[8] Rahman MM. Natural hazard and human induced hazard: A discourse of disaster and social vulnerability analysis. Journal of BanglaVision. 2012;**2012**:1-19

[9] Jahan A, Ansary M, Ara S, Islam I. Assessing social vulnerability to earthquake hazard in Old Dhaka, Bangladesh. Asian Journal of Environment and Disaster Management. 2011;**3**(3):285-300

[10] Wisner B, Blaikie P, Cannon T, Davis I. At Risk: Natural Hazard, People's Vulnerabilities and Disaster. London: Routledge; 2004

[11] Khan AA, Hussain MS. Recurrence of 1885 Bengal earthquake and hazard vulnerability status of Dhaka Metropolitan City, Bangladesh. Oriental Geography. 2005;**49**(**2**):205-216

[12] Islam MS, Sultana N, Bushra N, Banna LN, Tusherand TR, Ansary MA. Effects of earthquake on urbanization in Dhaka City. Journal of Environmental Science & Natural Resources. 2013;**6**(1): 107-112

[13] Islam M, Lim S. When nature strikes: Sociology of climate change and disaster vulnerabilities in Asia. Nature and Culture. 2015;**10**(1):57-80

[14] Rashed T, Weeks J. Assessing vulnerability to earthquake hazards through spatial multicriteria analysis of urban areas. International Journal of Geographical Information Science. 2003;**17**(6):547-576

[15] Barua U, Mannan S, Islam I, et al. People's awareness, knowledge and perception influencing earthquake vulnerability of a community: A study on Ward no. 14, Mymensingh municipality, Bangladesh. Natural Hazards. 2020;**103**:1121-1181

[16] Zaman MDA, Monira NJ. A study of earthquakes in Bangladesh and the

data analysis of the earthquakes that were generated in Bangladesh and its' very close regions for the last forty years (1976-2016). Journal of Geological and Geophysics. 2017;**6**:300

[17] Islam ZR, M. M. Human-environment relations in disaster analysis: An anthropological review. Journal of Sociology. 2013;**1**(2):1-16

[18] Rahman MM. Occupational health hazards of tanners at Hazaribagh tannery industry of Dhaka city in Bangladesh. Journal of Social Science Review. 2014;**2**(2):41-62

[19] Sarraz A, Khorshed Ali M, Das DC. Seismic vulnerability assessment of existing building stocks at Chandgaon in Chittagong city, Bangladesh. American. Journal of Civil Engineering. 2015;**3**(1):1-8

[20] Witvorapong N, Muttarak R, Pothisiri W. Social participation and disaster risk reduction behaviors in tsunami prone areas. Journal of Introducing Plos Climate. 2015;**10**:1-28